CIRIA CONSTRUCTION INDUSTRY RESEARCH AND INFORMATION ASSOCIATION

CENTRE FOR CIVIL ENGINEERING RESEARCH AND CODES CUR

KV-025-454

Manual on the use of rock in coastal and shoreline engineering

CIRIA SPECIAL PUBLICATION 83

CUR REPORT 154

CIRIA/CUR
Manual on the use of rock in coastal and shoreline engineering
Construction Industry Research and Information Association
Special Publication 83, 1991
Centre for Civil Engineering Research and Codes
Report 154

Reader Interest

Consultants, contractors, coastal authorities, hydraulic engineers, NRA, MAFF

© CIRIA/CUR, 1991

ISBN: 0-86017-326-7

Classification

AVAILABILITY — Unrestricted
CONTENTS — Advice for non-specialist
STATUS — Committee-guided
USERS — Design & site staff

All rights reserved. No part of this publication may be reproduced or transmitted in any form or by any means, including photocopying and recording, without the permission of the copyright holder, application for which should be addressed to the publisher. Such written permission must also be obtained before any part of this publication is stored in a retrieval system of any nature.

CIRIA and CUR and those associated with this publication have exercised all possible care in compiling and presenting the information contained in it. This information reflects the state of the art at the time of publication. Nevertheless the possibility that inaccuracies may occur in this publication cannot be ruled out. Any one wishing to use the information in it will be deemed to do so at his or her own risk. CIRIA and CUR decline—also on behalf of all persons associated with this publication—any liability whatsoever in respect of loss or damage that may arise in consequence of such use.

Published by CIRIA, 6 Storey's Gate, London SW1P 3AU, UK
   Tel: 071-222-8891, Fax: 071-222-1708

   CUR, PO Box 420, 2800 AK Gouda, The Netherlands
   Tel: 01820-39600, Fax: 01820-30046

Printed in Great Britain by Bell & Bain Ltd., Glasgow

C627.58/Man

AC 0038975 7

or before the last date

SAC Aberdeen
AC 0038975 7

Preface

This manual was prepared under a collaborative project carried out by CUR (the Netherlands Centre for Civil Engineering Research and Codes) and CIRIA (the UK Construction Industry Research and Information Association) during the period April 1988 to August 1991. The joint project was supported by the Netherlands and the UK governments.

Funding was provided by the Department of Public Works (Rijkswaterstaat), Ministry of Economic Affairs, Association of Manufacturers and Importers of civil engineering materials (VPI) and private companies in the Netherlands, and the Department of the Environment (Construction Industry Directorate) and the Ministry of Agriculture, Fisheries and Food (with responsibility for coastal defence) in the UK. The work was carried out with the active support of organisations involved in the Netherlands and UK construction industries, and the British Aggregate Construction Materials Industries (BACMI).

The manual was produced under the guidance of a joint Steering Committee comprising:

M. E. Bramley (to September 1989)	CIRIA
G. Stephenson (from October 1989)	CIRIA
W. Leeuwestein	CUR; Secretary CUR-C67
K. W. Pilarczyk	Rijkswaterstaat; Chairman CUR-C67
J. D. Simm	Robert West & Partners; Research Supervisor CIRIA RP402

Authorship of the manual was divided between CIRIA and CUR, who also set up national Steering Groups to advise on and review its content. The following were involved as authors:

N. W. H. Allsop	HR Wallingford
F. B. J. Barends	Delft Geotechnics
A. J. Bliek	Svasek Consult Ltd
M. E. Bramley	CIRIA
H. F. Burcharth	University of Aalborg, Denmark
C. Davies	Bullen and Partners
L. de Quelerij	Fugro McLelland
C. P. Orbell-Durrant	May Gurney & Co. (formerly Shephard Hill & Co. Ltd.)
G. J. Laan	Rijkswaterstaat
J.-P. Latham	Queen Mary & Westfield College
W. Leeuwestein	CUR
H. Ligteringen	Frederic Harris Consult Ltd
M. D. McKemey	Environmental Assessment Services
K. W. Pilarczyk	Rijkswaterstaat
A. B. Poole	Queen Mary & Westfield College
H. Postma	Boskalis Westminster
K. Powell	HR Wallingford
J. D. Simm	Robert West & Partners
J. E. Smith	EPD Consultants (Balfour Beatty)
C. J. Stam	ACZ Marine Contractors
J. W. van der Meer	Delft Hydraulics
J. H. van Oorschot	Aveco Infrastructure Consultants/BMK
D. Yelland	ARC Ltd.

CIRIA and CUR acknowledge the assistance (technical information, review of draft text) given by the following:

J. F. Agema	Delft University of Technology
P. C. Barber	Shoreline Management Partnership
M. Barnes	Martin Barnes Project Management
J. G. Berry	Bertlin & Partners
L. de Bruyn	Ballast Nedam Engineering Ltd.
F. O'Hara	Costain International
T. S. Hedges	University of Liverpool, UK
K. R. Hall	University of Ontario, Canada
D. A. Harlow	Bournemouth Borough Council
M. H. Lindo	Van Oord ACZ
M. W. Owen	HR Wallingford
G. Sexton	Ministry of Agriculture, Fisheries and Food
J. V. Smallman	HR Wallingford
R. P. Thorogood	Department of the Environment
F. Tyehurst	Christchurch District Council, UK
E. A. Wilson	Mersey Barrage Company (Costain)
H. Velsink	Nedeco/Delft University of Technology
J. K. Vrijling	Delft University of Technology/ Rijswaterstaat

Principal contributions to the various chapters and appendices in the manual were as follows. Chapter 1 was written jointly by M. E. Bramley, J. D. Simm and W. Leeuwestein. Chapter 2 was largely written by W. Leeuwestein and J. D. Simm with technical information and review by J. K. Vrijling. Section 2.4 on environmental assessment was prepared by M. D. McKemey. Careful review work by H. Velsink was appreciated. Chapter 3 was written and co-ordinated by J.-P. Latham with contributions from A. Poole (Sections 3.1 and 3.5) on rock types and other artificial rock materials, from G. J. Laan (Section 3.5) on grouts, geotextiles and composite materials and from D. Yelland (Section 3.4) on production and handling. G. J. Laan was the principal Netherlands reviewer. Chapter 4 (with the exception of Section 4.3) was prepared from material largely assembled by W. Leeuwestein. J. D. Simm provided additional material and the final version of the chapter. Much useful thought and technical contributions were received from H. F. Burcharth, H. Ligteringen, J. W. van der Meer, M. W. Owen, and K. W. Pilarczyk. Section 4.3 was prepared by L. de Quelerij and reviewed by H. Velsink and F. B. J. Barends. Chapter 5 was written by J. W. van der Meer and N. W. H. Allsop (Introduction and Section 5.1) and F. B. J. Barends (Section 5.2). Additional contributions to Section 5.1 were provided by K. W. Pilarczyk, C.-J. Stam, K. Powell, H. Burcharth. A detailed review of Section 5.2 was provided by T. S. Hedges. Chapter 6 was prepared by a different contributor for each main section. Section 6.1 was written jointly by H. Ligteringen and J. H. van Oorschot with a contribution on cost optimisation by H. F. Burcharth and reviewed by F. O'Hara. Section 6.2 was written by J. D. Simm with additional technical information provided by K. W. Pilarczyk, P. C. Barber, C. Davies, C. P. Orbell-Durrant and F. Tyehurst. Section 6.3 was written by J. E. Smith and J. D. Simm with a technical contribution on diffraction by J. V. Smallman. Section 6.4 was written by A. J. Bliek and H. Postma and reviewed by D. A. Harlow. Section 6.5 was written by C.-J. Stam and reviewed by J. D. Simm. Chapter 7 was prepared by J. D. Simm and C. Davies with technical input by N. W. H. Allsop. Appendix 1 was prepared by G. J. Laan, J.-P. Latham and J. D. Simm and reviewed by a number of the principal authors. Appendix 2 was prepared by J.-P. Latham with review by J. D. Simm using, in many cases, draft NEN standard as a starting point. Appendix 3 was prepared by C. P. Orbell-Durrant and reviewed by M. Barnes and J. D. Simm. Appendix 4 was prepared by

W. Leeuwestein and reviewed by J. D. Simm. Appendix 5 was prepared by
L. de Quelerij and reviewed by H. Velsink. Appendix 6 was prepared by
N. W. H. Allsop and Appendix 7 by M. D. McKemey.

General reviews of text in its final stages were provided by W. Leeuwestein,
G. Sexton, G. Stephenson, K. W. Pilarczyk and by the editor, J. D. Simm.

The manual was prepared for publication by:

J. D. Simm	General Reporter and Editor
W. Leeuwestein	CUR Reporter and Secretary CUR-C67
G. J. H. Vergeer	CUR Coordinator

The assistance of Mr C. van Beesten in the translation of Dutch-language
references into English is gratefully acknowledged. CIRIA and CUR also wish to
thank the following individuals and organisations for providing photographs and
other illustrative material:

ACZ; Aveco Infrastructure Consultants; Central Regional Council, Scotland, UK;
Delft Geotechnics; Director General—Ports and Coasts, Spain; EPD Consultants;
HR Wallingford; G. J. Laan; J.-P. Latham; M. D. McKemey; Nicolon b.v.,
Netherlands; C. P. Orbell-Durrant; J. D. Simm; Shephard Hill & Co. Ltd;
F. Tatsuoka, University of Tokyo, Institute of Industrial Science; Queen Mary &
Westfield College, London, UK; Robert West & Partners; Zanen Dredging.

A useful information exchange visit to the Waterways Experiment Station,
Vicksburg, USA, by J. D. Simm and G. Stephenson—funded by the US Army
Corps of Engineers—took place in February 1990.

The hours of patient effort by Mrs D. Manning in preparing a technically correct
typescript is gratefully acknowledged.

November 1991	Construction Industry Research and Information Association
	Centre for Civil Engineering Research and Codes

Contents

List of Figures	xi
List of Tables	xxi
List of Boxes	xxiii
Glossary	xxv
Notation	xxix
1 INTRODUCTION	**1–6**
1.1 Background, use and approach	1
1.2 Use of the manual	2–6
1.2.1 Design process	2
1.2.2 Structure and contents of this manual	4
1.3 Structure types covered by this manual	6
2 PLANNING AND DESIGNING	**7–66**
2.1 Design process	7–31
2.1.1 Problem identification	7
2.1.2 Boundary conditions	7
2.1.3 Functional Analysis	12
2.1.4 Generation of alternative solutions	14
2.1.5 Comparison and selection	18
2.1.6 Final design and detailing	21
2.1.7 Cost Assessment	22
2.1.8 Quality Assurance	26
2.2 Structure types and failure modes	31–45
2.2.1 Structure types	31
2.2.2 Principal failure modes	33
2.2.3 Fault tree analysis	38
2.2.4 Risk-level assessment	44
2.3 Design approach	45–56
2.3.1 Deterministic approach	47
2.3.2 Probabilistic approach	48
2.4 Environmental assessment	56–66
2.4.1 Reasons for carrying out EA	56
2.4.2 EA as an integral part of the design process	56
2.4.3 Method of assessment	59
2.4.4 Additional sources of information	59
2.4.5 Assessing the impacts of the proposed works	61
2.4.6 Impacts from the choice of materials and construction	64
2.4.7 The Environmental Statement (ES)	66
3 MATERIALS	**67–182**
3.1 Rock types	68–76
3.1.1 Generation considerations	68

3.1.2 Evaluation of rock types at source	68
3.1.3 Alluvial, glacial and marine sources of rock materials	75

3.2 Properties and functions — 76–113

3.2.1 Intrinsic properties	78
3.2.2 Production-affected properties	86
3.2.3 Execution-induced properties	101
3.2.4 Durability and armourstone degradation rates	106

3.3 Testing and evaluation — 114–134

3.3.1 Size and weight gradations	115
3.3.2 Shape	117
3.3.3 Petrographic examination	118
3.3.4 Density and water absorption	118
3.3.5 Intact strength	122
3.3.6 Resistance to weathering	128
3.3.7 Block integrity	133

3.4 Production and handling — 134–155

3.4.1 Resource appraisal	135
3.4.2 Production	136
3.4.3 Transportation	148

3.5 Other materials — 155–169

3.5.1 Concrete	155
3.5.2 Industrial by-products	156
3.5.3 Grouts	157
3.5.4 Geotextiles	160
3.5.5 Composite materials	167

3.6 Quality control — 169–182

3.6.1 Objective	169
3.6.2 Quality control in the quarry and during transportation	169
3.6.3 Quality control during construction	179

4 PHYSICAL SITE CONDITIONS AND DATA COLLECTION — 183–233

4.1 Bathymetry and morphology — 186

4.2 Hydraulic boundary conditions and data collection — 186–227

4.2.1 Water levels	186
4.2.2 Deep-water wave conditions	194
4.2.3 Transformation of waves in shallow water	207
4.2.4 Seabottom orbital velocities	220
4.2.5 Currents	220
4.2.6 Ship-induced water movements	221
4.2.7 Joint probabilities	222
4.2.8 Modelling	226

4.3 Geotechnical boundary conditions and data collection — 227–233

4.3.1 General aspects of planning a soil-investigation programme	228
4.3.2 Desk studies	229
4.3.3 Preliminary soil investigations	230
4.3.4 Detailed soil investigations	230

5 PHYSICAL PROCESSES AND DESIGN TOOLS — 235–350

5.1 Hydraulic interactions — 237–306

5.1.1 Governing parameters	237
5.1.2 Hydraulic response	246
5.1.3 Structural response	261
5.1.4 Physical and numerical modelling	304

5.2 Geotechnical interactions	307–350
5.2.1 Approach	307
5.2.2 Parameters	308
5.2.3 Scope of geotechnical aspects	313
5.2.4 Physical background	314
5.2.5 Modelling and simulation	319
5.2.6 Computer model types	325
5.2.7 Computer model applications	325
5.2.8 Practical formulae and engineering experience	327
5.2.9 Application of probabilistic analysis	348

6 STRUCTURES — 351–478

6.1 Rubble-mound breakwaters	351–395
6.1.1 Definitions	352
6.1.2 Layout	354
6.1.3 General design considerations for breakwater cross-section	358
6.1.4 Structure-specific design aspects	361
6.1.5 Construction aspects	373
6.1.6 Cost aspects and project optimisation	383

6.2 Seawalls and shoreline protection structures	395–443
6.2.1 Definitions	396
6.2.2 Plan layout and overall concept selection	397
6.2.3 General considerations for cross-section design of rock coastal structures	413
6.2.4 Structure-specific design aspects	420
6.2.5 Construction aspects	440

6.3 Dam-face protection	443–453
6.3.1 Protection concept selection	443
6.3.2 General design considerations	445
6.3.3 Special design considerations	449
6.3.4 Special construction aspects	451
6.3.5 Measurement and cost aspects	452
6.3.6 Maintenance aspects	452

6.4 Gravel beaches	453–463
6.4.1 Design process	454
6.4.2 Coastal processes	455
6.4.3 Gravel beach profile	456
6.4.4 Construction and cost aspects	457
6.4.5 Monitoring and renourishment	460
6.4.6 Cost optimisation	461

6.5 Rockfill in offshore engineering	463
6.5.1 Principal design considerations	465
6.5.2 Construction aspects	476
6.5.3 Cost aspects	478

7 MAINTENANCE — 479–487

7.1 Monitoring	479–484
7.1.1 Types of monitoring	480
7.1.2 Frequency of monitoring	481

7.2 Appraisal of structure performance	484–485
7.3 Repair/replacement construction methods	486–487

Appendices

1 Model specification for quarried rock applications in coastal and shoreline engineering — 489–507
2 Standards for quarried rock materials applications in coastal and shoreline engineering — 509–530
3 Measurement of quarried rock in coastal and shoreline engineering — 531–542
4 Hydraulic data measurement and instrumentation — 543–550
5 Instrumentation for geotechnic data collection — 551–570
6 Structure-monitoring techniques — 571–583
7 European, British and Dutch legislation/authorities/designated sites relevant to environmental assessment of projects involving the use of rock in coastal and shoreline engineering — 585–587

References — 589–602

Index — 603

List of Figures

Chapter 1

Figure 1	Design process	2
Figure 2	Manual on use of rock in coastal and shoreline engineering—logic diagram	3
Figure 3	Rock breakwater	4
Figure 4	Rock coastal revetment	4
Figure 5	Typical rock dam-face protection	5
Figure 6	Typical rock groynes and gravel beach	5

Chapter 2

Figure 7	Detailed description of design process with reference to parts of this manual	8
Figure 8	Exposure zones on coastal structures	9
Figure 9	Risk level in Western countries	12
Figure 10	Interrelationship of construction considerations in design of rock structures	17
Figure 11	Distribution of rubble mound breakwater construction costs	25
Figure 12	Cost optimisation	26
Figure 13	Quality loop for rock structures in coastal and shoreline engineering	27
Figure 14	Principle of a quality control system	29
Figure 15	Some basic breakwater concepts	32
Figure 16	Some basic seawall concepts	33
Figure 17	Applications of rock scour or bottom protection	34
Figure 18	Response and failure of a rock system	34
Figure 19	A rock system and its responses	35
Figure 20	Failure modes of rock structures	36
Figure 21	Relation between loading and failure	39
Figure 22	Typical damage curve for rock armour	40
Figure 23	Evolution of damage as a function of time	41
Figure 24	Example of a fault tree	41
Figure 25	Example calculation of probability of failure	43
Figure 26	P/p_{fi} as a function of n and ρ_c for a series system	44
Figure 27	P/p_{fi} as a function of n and ρ_c for a parallel system	44
Figure 28	Failure systems	45
Figure 29	Example of risk-level assessment for a seawall front rock slope	46
Figure 30	Example of a deterministic approach	47
Figure 31	Load and strength distribution	48
Figure 32	Joint distribution of loading and strength	49
Figure 33	Contours of constant $f_{x1,x2}$, partial distribution functions for x_1 and x_2 and limit state of failure	50
Figure 34	Determination of joint probability density function	51
Figure 35	Normalised Gaussian pdf	53
Figure 36	Distribution of Z based on Gaussian distributed parameters	55
Figure 37	Typical coastal environmental impacts	57
Figure 38	EC Directive 85/337 aspect interaction scoping chart	58
Figure 39	Flow diagram for EA as part of project planning and design	60
Figure 40	Example of an assessment matrix	65

Chapter 3

Figure 41	Logic diagram relating chapter sections to the materials evaluation and design processes	67
Figure 42	Idealised sketches of common natural outcrop forms	69
Figure 43	Three idealized quarries showing different types of weathering	73
Figure 44	Flow diagram for rock properties and functions	77
Figure 45	The influence of rock density on design parameters	80

Figure 46	Scan line mapping in progress at a quarry	82
Figure 47	Measurement of the angles of θ, ϕ and α between the mean orientations of rock mass discontinuities	83
Figure 48	Type 1 breakage of gabbro block with many joints	84
Figure 49	(a) Mode 1 crack geometry; (b) Synoptic diagram of the variation of future toughness of rocks with various mineralogical compositions	85
Figure 50	Conceptual scheme for weight reduction of armour blocks during handling phases	88
Figure 51	Illustration of blocks with equal volume and $l/d = 3$, compared with a cube with $l/d = 1.73$	90
Figure 52	Visual comparison of block shapes	92
Figure 53	Harmonic contributions to shape	93
Figure 54	Comparison of log-linear, Rosin Rammler and Schumann equations for stone gradations and fragmentation	95
Figure 55	Explanation of the grading class limits of a standard grading	98
Figure 56	Weight gradings and size relationships for the standard light and heavy grading classes	100
Figure 57	Size gradings and relationships for the standard fine and light grading classes	100
Figure 58	Factors affecting degradation of rock materials	107
Figure 59	Conceptual scheme for structural performance of armour blocks in service	108
Figure 60	Prototype production model of QMW abrasion mill apparatus	108
Figure 61	QMW abrasion test fractional weight loss plot zonations for different rock types tested	110
Figure 62	Degradation model principles	111
Figure 63	Application of abrasion mill test results to the degradation model	114
Figure 64	Flow diagram for in-service rock durability evaluation	116
Figure 65	Sieve test for fine gradings	118
Figure 66	Weighing devices being employed in grading tests for light/heavy gradings	119
Figure 67	Fracture toughness test	124
Figure 68	Point load test machine	126
Figure 69	Schmidt hammer	129
Figure 70	Rocks affected by the 'Sonnenbrand' effect	131
Figure 71	Blasting	138
Figure 72	Drilling equipment used in drilling blast holes	139
Figure 73	Schematic illustration of definition of the bench blasting terms	139
Figure 74	Splitting oversize blocks with a hydraulic 'pecker'	142
Figure 75	Selection by machine operator of 60–300 kg and 300–1000 kg basalt	142
Figure 76	Production screen for 10–60 kg and 60–300 kg (falling through) and greater than 300 kg (sliding over)	143
Figure 77	Handling rock with chains	144
Figure 78	Epoxy grouted transport bolt	145
Figure 79	Stockpiles of standard gradings and basalt	146
Figure 80	Loading rock using a front-end loader	147
Figure 81	Road truck with steel body	149
Figure 82	Rail transport of rock at buffer stockpile, Europoort	150
Figure 83	Loading 10–15 tonne rock into a 2000-tonne barge using epoxy-grouted hooks	150
Figure 84	A split box for loading a ship	151
Figure 85	Loading rock into barges	152
Figure 86	A 20 000 tonne pontoon	153
Figure 87	Unloading 1–3 tonne rock from a 20 000-tonne pontoon	154
Figure 88	Unloading a 10 000 tonne pontoon directly onto a beach at Fairlight Cove, UK	154
Figure 89	Grouting methods	157
Figure 90	Comparative properties of general polymer families	160
Figure 91	Geotextile classification groups	161
Figure 92	Tape fabric	162

Figure 93	Fibrillated split-film fabric	163
Figure 94	Gabions	167
Figure 95	Rejected blocks with flaws and cracks	171
Figure 96	Blocks of poor integrity revealed by rough handling in stockpile	173
Figure 97	Measurement of thickness and length	176
Figure 98	Dynamometer weighing device	180

Chapter 4

Figure 99	Diagram of data flow	184
Figure 100	Hydraulic boundary conditions	187
Figure 101	Variation in water level due to storm surge and astronomic tide	188
Figure 102	Wind set-up	189
Figure 103	Wave set-up	190
Figure 104	Heights and periods of tsunamis related to earthquake magnitude	192
Figure 105	Example of storm surge levels for a location on the Dutch coast	193
Figure 106	Definition sketch for regular wave parameters	195
Figure 107	Applicability range for wave theories	195
Figure 108	Typical record of irregular sea	196
Figure 109	Rayleigh distribution of wave heights compared with a shallow-water distribution	197
Figure 110	Scatter diagram indicating joint distribution of H_s and T_m	198
Figure 111	Weibull fit to long-term wave height distribution	198
Figure 112	Pierson–Moskowitz and JONSWAP spectra	200
Figure 113	Deep water wave forecasting diagram for a standard wind field	203
Figure 114	Shallow-water wave forecasting diagram in a standard wind field	203
Figure 115	Prediction of significant wave height from wind speed for JONSWAP spectrum	204
Figure 116	Prediction of peak wave period from wind speed for JONSWAP spectrum	204
Figure 117	Example calculation of effective fetch length by Saville's method	206
Figure 118	Transformation of waves in shallow water (refraction, diffraction, shoaling and breaking)	208
Figure 119	Examples of wave refraction	209
Figure 120	K_R for an irregular directional wave field on a coast with straight, parallel depth contours	210
Figure 121	Shallow-water significant wave heights for uniform foreshore slopes	211
Figure 122	Effect of angle of wave incidence on shallow water significant wave height for uniform foreshore slopes	212
Figure 123	Shallow-water significant wave heights for a two-slope foreshore profile	213
Figure 124	Diffraction diagrams for a semi-infinite breakwater for random waves of normal incidence	215
Figure 125	Diffraction diagrams for a breakwater opening with $B/L=1.0$ for random waves of normal incidence	215
Figure 126	Diffraction diagrams for a breakwater opening with $B/L=4.0$ for random waves of normal incidence	216
Figure 127	γ_{br} as a function of ξ	217
Figure 128	Spilling breakers on Welsh coast, UK	218
Figure 129	Wave reflection from a breakwater	219
Figure 130	Diffraction approach to determine the height of waves reflecting from a limited-length structure	219
Figure 131	Bottom orbital velocity for monochromatic waves and a spectrum of waves	220
Figure 132	Definition sketch of waterway characteristics and ship-induced water movements	222

Figure 133	Schijf's chart for estimating return current and water level depression	223
Figure 134	Conditional probability for densities $p(H_s, z)$ for entrance to eastern Scheldt	224
Figure 135	Design process using joint probability distribution of z and H_s	225
Figure 136	Example of conditional distributions of H_s obtained from joint Gaussian distributions of correlated parameters H_s and z	225
Figure 137	Relational scheme for geotechnical structural performance	228

Chapter 5

Figure 138	Basic scheme for assessment of coastal structure response	235
Figure 139	Governing hydraulic parameters	237
Figure 140	Breaker types as a function of ξ	238
Figure 141	Type of structure as a function of $H/\Delta D$	240
Figure 142	Governing parameters related to the structure cross-section	243
Figure 143	Notional permeability factor P for various structures	244
Figure 144	Damage S_d based on erosion area A_e	244
Figure 145	Schematised shingle beach/rock slope profile on a 1:5 initial slope	245
Figure 146	Comparison of relative 2% run-up for smooth and rubble slopes	247
Figure 147	Comparison of relative significant run-up for smooth and rubble slopes	248
Figure 148	Relative 2% run-up on rock slopes	250
Figure 149	Relative significant run-up on rock slopes	250
Figure 150	Relative 2% run-down on impermeable and permeable rock slopes	251
Figure 151	Influence of oblique and long- and short-crested seas on run-up on a 1:4 smooth slope	252
Figure 152	Critical overtopping discharges	253
Figure 153	Generalised profile for smooth, bermed seawalls	255
Figure 154	Overtopping discharges for smooth, straight slopes	256
Figure 155	Overtopping of a smooth 1:4 slope described using T_p	256
Figure 156	Overtopped rock structures with low crown walls	257
Figure 157	Studies of tested cross-sections	257
Figure 158	Wave transmission over and through low-crested structures	258
Figure 159	Comparison of data on rock slopes	260
Figure 160	Wave run-up data	261
Figure 161	Effective relative berm length to local wave length on reflection	262
Figure 162	Cross sections and reflection coefficients for rubble protection placed against a vertical wall	263
Figure 163	Various parts of a structure whose hydraulic design is described in this chapter	263
Figure 164	Examples of stability number versus slope angle plots	266
Figure 165	Probability of exceedence of the damage level S_d in the lifetime of the structure	271
Figure 166	Definition sketch of splash area	272
Figure 167	Cross-sections of low-crested structures	273
Figure 168	Forces and pressures on a crown wall showing typical assumed pressure distributions	277
Figure 169	Cross-section giving empirical coefficients a and b for horizontal wave forces on crown walls	278
Figure 170	Wave force data plots for crown walls in cross-sections shown in Figure 169	279
Figure 171	Toe stability as a function of h_t/h	280
Figure 172	Stability of a breakwater head armoured with tetrapods	282
Figure 173	Example of erosion of a berm breakwater head	282
Figure 174	Examples of dynamically stable profiles for different initial slopes	283
Figure 175	Example of a computer profile for a berm breakwater	284

Figure 176	Example of influence of wave climate on a berm breakwater profile	284
Figure 177	Simple schematised profile of rock and gravel beaches	285
Figure 178	Schematic shingle beach profile	286
Figure 179	(a) and (b) Berm breakwater profiles	288
Figure 180	Stability increase factors, f_i, for stepped or bermed slopes	289
Figure 181	Stability increase factors, f_i, for a composite slope with gradients: upper slope 1:3, lower slope 1:6	289
Figure 182	Stability increase factors, f_i, for smooth upper slopes and rock lower slopes	290
Figure 183	Longshore transport relation for gravel beaches	294
Figure 184	Dimensionless critical bed shear stress versus shear Reynolds number	297
Figure 185	Modified Shield's curve for unsteady flow	298
Figure 186	Additional (local) scour immediately in front of a seawall due to storms	301
Figure 187	Example of predicted scour depths for a vertical wall	303
Figure 188	Model wave basin after oblique wave test on offshore breakwater	306
Figure 189	Measured flow rules	310
Figure 190	Permeability versus grain size	310
Figure 191	Conglomerate rock particle (polished surface)	311
Figure 192	Equivalent roughness R'	312
Figure 193	Equivalent strength S'	312
Figure 194	Liquifaction potential (lab tests)	313
Figure 195	Effective stress principle	315
Figure 196	Simulated transient turbulent porous flow field	316
Figure 197	Centrifuge modelling of failure	316
Figure 198	Dynamic effects	318
Figure 199	Simulation of earthquake response	319
Figure 200	A sinkhole	320
Figure 201	Mechanism by sub-surface cavities	320
Figure 202	Pore-pressure and displacement responses to loading	323
Figure 203	Measuring hydraulic boundary conditions	326
Figure 204	Two-dimensional consolidation	326
Figure 205	Phreatic surfaces under wave action	327
Figure 206	Principle of Bishop's method	328
Figure 207	Parameters used in description of surface wave penetration	329
Figure 208	Diagram for internal set-up	330
Figure 209	Illustration of internal set-up under wave action	331
Figure 210	Example of field storage and dissipation of excess pore pressure	331
Figure 211	Rock slope failure	332
Figure 212	Stability factors during rapid drawdown	333
Figure 213	Porous flow during overtopping	333
Figure 214	Coverlayer failure	334
Figure 215	Mechanics of a coverlayer	334
Figure 216	Principle of coverlayer hydraulic loading	335
Figure 217	Upgrading diagram	336
Figure 218	Consolidation ratio versus time	337
Figure 219	Earthquake-induced residual settlements	338
Figure 220	Densification by wave impacts	339
Figure 221	Excess cyclic pore pressures	340
Figure 222	Dilatancy cyclic pore pressure generation in undrained cyclic triaxial tests	341
Figure 223	Slope stability under waver action	342
Figure 224	Stability against squeezing	342
Figure 225	Squeezing with a Tresca approach	342
Figure 226	Filter behaviour under hydraulic loading	343
Figure 227	Mechanisms related to N_f	344
Figure 228	Nomogram for filtered design	344
Figure 229	A wide-graded material	345
Figure 230	Internal stability curve	345

| Figure 231 | Grainsize distribution curves | 346 |
| Figure 232 | Cross-section of breakwater | 349 |

Chapter 6

Figure 233	Breakwater design logic diagram	352
Figure 234	Cross-section of a typical rubble-mound breakwater	352
Figure 235	Examples of various types of breakwaters	353
Figure 236	Location and layout of Port d'Arzew El Djedid	353
Figure 237	Schematic relation of annual wave climate to total downtime costs	355
Figure 238	Example of use of ship movement simulation model in design of breakwater extension at Taichung Port, Taiwan	357
Figure 239	Definition sketch for a rubble-mound breakwater	362
Figure 240	Failure mechanisms of a rubble-mound breakwater	362
Figure 241	Alternative arrangements of a toe berm in rubble-mound breakwater	363
Figure 242	Shallow-water breakwater cross-section	364
Figure 243	Concept of a rubble mound with crown wall	365
Figure 244	Superstructure—west breakwater, Sines	365
Figure 245	Crown wall configurations	366
Figure 246	Berm breakwater with two stone categories	367
Figure 247	Reef breakwater	368
Figure 248	Low-crested/submerged breakwater	369
Figure 249	Typical caisson-type breakwaters	369
Figure 250	Failure mechanisms of a caisson breakwater	370
Figure 251	Wave pressures on a caisson	371
Figure 252	Failure mechanisms of a horizontally composite breakwater	373
Figure 253	End-tipping rock to repair a breakwater breach	374
Figure 254	Typical equipment capacities for slope trimming	375
Figure 255	Space requirements at a breakwater crest	375
Figure 256	Lifting capacity of common types of cranes	376
Figure 257	Relative weight of grab versus payload	377
Figure 258	Orange-peel grab placing rock at breakwater	377
Figure 259	Lifting capacity of two typical heavy cranes	378
Figure 260	Placing rock from floating plant using a rock tray	379
Figure 261	Types of self-unloading core barges	380
Figure 262	Example of output from STORTISM rock-dumping simulation model	382
Figure 263	Tolerances at transition berms	383
Figure 264	Produced and required quarry output	384
Figure 265	Example breakwater and quarry yield curve	386
Figure 266	Simulation diagram for breakwater and quarry operations	387
Figure 267	The variation of expected total expenses of a breakwater armour layer during a lifetime of 100 years, depending on the armour unit weight, W, and rate of return, r	388
Figure 268	Typical cross-section of outer breakwaters of Zeebrugge, Belgium	389
Figure 269	Influence of choice of armour block on breakwater costs	391
Figure 270	Construction road, Zeebrugge breakwater	392
Figure 271	Examples of accessibility evaluations for rubble-mound breakwaters at Zeebrugge, Arzew and Ras Lanuf	394
Figure 272	Conventional rock revetments	399
Figure 273	Scour protection	399
Figure 274	Bastion rock groynes retaining shingle beach	400
Figure 275	Beach classification	401
Figure 276	Rock groynes on shingle beach, Barton-on-Sea, UK	402
Figure 277	Offshore breakwaters retaining sand beach	403
Figure 278	Conditions for tombolo formation with decreasing relative distance from breakwater to shore	404
Figure 279	Rihuete beach, Murcia, Spain, after creation of pocket beaches	405
Figure 280	Fishtailed breakwaters retaining sand beach	406
Figure 281	Basic geometry of the fishtailed breakwater	406
Figure 282	Clacton Sea Defence Scheme	408

Figure 283	Sand beach retained by T- or L-shaped breakwaters and offshore sill	407
Figure 284	Use of rock groyne for fishing, Melford-on-Sea, UK	411
Figure 285	Offshore breakwater, Rhos-on-Sea, UK	412
Figure 286	Seawall rehabilitation in progress. Morecambe, UK	414
Figure 287	Rock and timber groyne cost comparison by cross-section, Christchurch, UK	415
Figure 288	Quarry locations evaluated for Morecambe Bay Coast Protection Scheme, UK	417
Figure 289	Collapse of a rip-rap revetment on Australian Gold Coast	420
Figure 290	Boulder wall design (Gold Coast, Australia)	421
Figure 291	(a) Gabion box wall structure; (b) combined box wall/mattress structure	422
Figure 292	Coast-protection revetments	423
Figure 293	Sea-defence revetment	423
Figure 294	Land-reclamation revetment	423
Figure 295	Typical revetments for rehabilitation of existing vertical seawalls	424
Figure 296	Fault tree: events leading to flow under seawall	426
Figure 297	Fault tree: events leading to flow over seawall	426
Figure 298	Fault tree: events leading to damage to slope protection	427
Figure 299	Fault tree: events leading to geotechnical instability	427
Figure 300	Rock revetment at Buckhaven, Fife, UK	430
Figure 301	(a) Determination of dike height; (b) Dike settlement as a function of time	431
Figure 302	Toe detail: revetment on rock beach	432
Figure 303	Revetment toe protection—erodible beach	432
Figure 304	Erosion at seawall termination	433
Figure 305	Example of groyne cross-section from Atlantic Coast, North Carolina, USA	433
Figure 306	Rock groyne on shingle beach, Barton-on-Sea, UK	434
Figure 307	Alternative longitudinal profiles of rock groynes	434
Figure 308	Offshore breakwater, Leasowe Bay, UK	435
Figure 309	Transition details, Leasowe Bay offshore breakwater, UK	436
Figure 310	Eastness breakwater, Clacton-on-Sea, UK	437
Figure 311	Pedregalejo Beach, Malaga, Spain	439
Figure 312	Sill and perched beach design, Lido di Ostia, Rome, Italy	440
Figure 313	'Orange-peel' grab placing rock	441
Figure 314	Grapple placing rock	442
Figure 315	Hydralic bucket placing rock	442
Figure 316	Wave action at Carron reservoir and dam	446
Figure 317	Typical protection to dam face	448
Figure 318	Diffraction patterns	450
Figure 319	Placing armourstone on dam faces	452
Figure 320	Ranges of simplified gravel beach profiles (vertically exaggerated scale)	457
Figure 321	Self-healing process of toe gravel beach (vertically exaggerated scale)	457
Figure 322	Maximum discharge length for gravel pumping	458
Figure 323	Artist's impression of gravel beach construction at Seaford, UK, in 1987	459
Figure 324	Isolines of overall ratio, R, versus phi mean difference and phi sorting ratio	462
Figure 325	Examples for a pipeline rock cover protection/stabilisation	464
Figure 326	Simplified logic diagram of the design process	465
Figure 327	Distribution of kinetic energy of falling objects just before impact (kNm)	468
Figure 328	Anchor on rock protection lying on seabed	469
Figure 329	Axial compressive force along a typical pipeline	471
Figure 330	Temperature profile along a typical pipeline	472
Figure 331	Pullout mechanism for a shallow pipe	473
Figure 332	Shear displacement pattern for a horizontal pipe pulled longitudinally	473

| Figure 333 | Common methods of dumping stone offshore | 474 |
| Figure 334 | Dynamically positioned, flexible fall-pipe vessel *Trollnes* | 477 |

Chapter 7

Figure 335	Maintenance programme flow chart	480
Figure 336	Preventative condition-based maintenance	482
Figure 337	Measuring watch for a selected damage pattern	485
Figure 338	Safety control panel for seawall management	485

Appendix 1

Figure A1	Explanation of the grading class limits for a standard grading	492
Figure A2	Sampling locations in the load on floating equipment	500
Figure A3	Sampling locations in a non-segregated load on floating equipment	501
Figure A4	Sampling locations in a spread-dumped load	501
Figure A5	Halving a sample by means of a separation plane	502
Figure A6	Dividing a sample with two separation planes	502

Appendix 2

Figure A7	Illustration and notation used in bulk weighing method	513
Figure A8	Aggregate impact test machine	518
Figure A9	The QMW abrasion mill (schematic)	525
Figure A10	Determination of abrasion resistance index, k_s	527

Appendix 3

| Figure A11 | Sample page of bill of quantities prepared according to CESMM (2nd edition) and Annexes A and B of this appendix | 532 |

Appendix 4

Figure A12	Principle of echo sounding	543
Figure A13	The principle and typical product of side-scan sonar	544
Figure A14	Principle of a float water level measuring system	545
Figure A15	Wave measurement using a step gauge	548

Appendix 5

Figure A16	Principle of seismic reflection (SCB) and refraction (SDEF)	552
Figure A17	Seismic refraction profile	552
Figure A18	Principle of electrical resistivity surveying	553
Figure A19	Principle of electromagnetic surveying	553
Figure A20	Comparison of electrical resistivity and electromagnetic measurements	554
Figure A21	Cross-section of SPT sampler	554
Figure A22	SPT results in coarse sand and fine sand	555
Figure A23	Principle of the electrical CPT on land	556
Figure A24	Presentation of CPT results of electrical cone with local friction measurement	557
Figure A25	Soil identification by cylindrical electrical cone resistance versus local friction	558
Figure A26	New developments in cone penetrometers by Fugro McClelland	558
Figure A27	Results obtained using the piezocone penetrometer	559
Figure A28	Field vane test performed in a borehole on land	560
Figure A29	Field vane test in seabed mode as developed by Fugro McClelland	561
Figure A30	Schematic diagram of a pressuremeter test set-up	562
Figure A31	Dilatometer (Marchatti)	562
Figure A32	Percussion boring	563
Figure A33	Hydraulic rotary drilling in unconsolidated soil	564
Figure A34	Good-quality borehole record	565
Figure A35	Drill string with Weissen cone penetrometer	566

Figure A36	Remote control boring and penetration systems for seabed investigation	567
Figure A37	Principle of triaxial test apparatus	570
Figure A38	Stress/strain diagram and critical Mohr circle obtained in triaxial tests	570
Figure A39	Determination of internal angle of friction, ϕ, and cohesion, c, from triaxial tests	570

Appendix 6

Figure A40	Comparison of armour slope profiles	573
Figure A41	Contours of equal settlement on a breakwater	574
Figure A42	Taking comparative photographs from a boat	577
Figure A43	Armour survey scheme	578
Figure A44	Effect of armour placement on side-scan sonar return signal	582

List of Tables

Chapter 2

Table 1	Categories of losses	11
Table 2	Functions of rock structures	13
Table 3	Functions of typical component parts of a rock breakwater	13
Table 4	Fixed and variable costs related to rock construction	25
Table 5	Process involvements by parties to a rock structure project	29
Table 6	Quality control system for rock structure projects	30
Table 7	Failure mechanisms and characteristic parameters	37
Table 8	Example of a FOMVA calculation	54
Table 9	Example of an AFDA calculation	55
Table 10	Possible social and socio-economic considerations in the environmental appraisal of rock structures	62

Chapter 3

Table 11	Igneous rocks: strong rocks with interlocking crystals	70
Table 12	Sedimentary rocks: bedded rocks with grains cemented by interstitial material	70
Table 13	Metamorphic rocks: crystals usually interlocking but grain alignment common	71
Table 14	Some generalised engineering characteristics of unweathered common rocks	72
Table 15	Weathering and alteration grades	74
Table 16	Idealised typical rock quality parameter ranges for marine structures	74
Table 17	Generalised evaluation of the use of fresh rock in marine structures	75
Table 18	Shape classification for armourstone	91
Table 19	Requirements and supplementary information for standard gradings	97
Table 20	Thickness and porosity in narrow gradation armour layers	105
Table 21	Guide to rock durability from test results	109
Table 22	Typical QMW mill abrasion results for rocks of differing durability zonings	110
Table 23	Ratings estimate for factors in degradation model	111
Table 24	MBA values (g/100 g) of some clay minerals and rocks	133
Table 25	Industrial by-products and wastes, and their potential for use in marine structures	156
Table 26	Frequency of tests at quarry supplying large total quantities ($>25\,000$ t)	174

Chapter 4

Table 27	Wind speed (U_w) conversion factors related to duration of wind speed	200
Table 28	Estimates and relations of wave parameters in a stationary (Gaussian) sea state	201
Table 29	Required soil data for evaluation of the geotechnical limit states	231
Table 30	*In-situ* test methods and their perceived applicability	232

Chapter 5

Table 31	Design values of S_d for a $2D_{n50}$ thick armour layer	245
Table 32	Values of the coefficients a and b in equation (5.26) for straight, smooth slopes	254
Table 33	Values of the coefficients a and b in equation (5.26) for bermed smooth slopes	255

Table 34	Coefficients a and b in equation (5.29) for overtopping discharges over cross-sections in Figure 157	258
Table 35	Parameters used in Level II probabilistic calculations	267
Table 36	Indicative stability comparison for various rock-protection systems	291

Chapter 6

Table 37	Relative importance of various considerations for maintenance and major repair	384
Table 38	Relative construction costs for Zeebrugge breakwaters	389
Table 39	Example of influence of grading of core material on costs	390
Table 40	Cost comparison of use of 25- and 30-tonne cubes as armour units	392
Table 41	Comparison of land-based and floating cranes	395
Table 42	Assessment of some commonly expressed concerns relating to coastal armouring	410
Table 43	Critical modes of failure of rock-based revetment systems	419
Table 44	Frequency of dropped objects with regard to weight (accidents per 100 crane-years)	467

Chapter 7

Table 45	Measures of the state of a rock structure	481
Table 46	Measures of environmental conditions or loadings	482
Table 47	Measures of hydraulic/geotechnic response of rock structure to wave loading	482
Table 48	Frequency of planned monitoring above low-tide level	483
Table 49	Outputs from comparison of measures of the state of rock structures over a period of time	484
Table 50	Construction equipment for repair of rock armour layers	486

Appendix 1

Table A1	Fine-grading classes	491
Table A2	Light-grading class requirements	492
Table A3	Heavy-grading class requirements	493
Table A4	Vertical placing tolerances for rock materials	506

Appendix 2

Table A5	Weight intervals for the cumulative weight plot	511

Appendix 6

Table A6	Summary of survey data at three UK sites	576

List of Boxes

Chapter 2

Box 1	Exposure zones for coastal structures	9
Box 2	Personal risk	12
Box 3	Multi-criteria analysis (MCA)	19
Box 4	Benefit-cost analysis (BCA)	20
Box 5	Rock materials specifications and bills of quantities	23
Box 6	Calculation of repair costs	24
Box 7	Relative costs in a typical rubble-mound breakwater	25
Box 8	Key definitions (ISO 8402) related to quality assurance	27
Box 9	Main aspects of a typical client/owner quality plan	27
Box 10	Main aspects of a typical designer's quality plan	28
Box 11	Main aspects of a typical contractor's quality plan	28
Box 12	Main aspects of a typical quarry operator's quality plan	28
Box 13	European Community (EC) environmental assessment regulations	58–59

Chapter 3

Box 14	Influence of climate on weathering of rock in quarries	73
Box 15	Effect of water absorption on relative buoyant density	79
Box 16	Effects of rock density and water absorption on required armour block weight and total armour weight	80
Box 17	Prediction of *in-situ* block size distribution	83
Box 18	Weathering in geological and engineering time	84
Box 19	Fracture toughness and other intact strength measures	85–86
Box 20	Brittle influence on Type 1 breakage of large blocks	87
Box 21	Block aspect ratio data	89
Box 22	The Fourier asperity roughness parameter, P_R	93
Box 23	Useful theoretical equations for stone gradations and fragmentation curves	95
Box 24	Explanation of class limit system of standard gradings	98
Box 25	Median and effective mean weight	99
Box 26	Derivation of non-standard specification for narrow heavy gradings	101
Box 27	Derivation of non-standard specification for wide, light and light/heavy gradings	102–103
Box 28	Interlock	103
Box 29	Consequences of using different rock density	106
Box 30	Calibration of degradation model	112
Box 31	Application of degradation model	113
Box 32	Rock durability indicators	117
Box 33	Bond's third theory equation	140
Box 34	Rosin–Rammler equation	141
Box 35	Shape control	176
Box 36	Grading plan: 3–6 tonne	177
Box 37	Example of grading control for a large barge using grading plan of Box 36	178
Box 38	Weighing devices	180
Box 39	Verification test panel calculations	182

Chapter 4

Box 40	Supporting equipment for physical site data collection	185
Box 41	Theoretical extreme value probability distributions	193
Box 42	Simple approach to correlating extreme water levels	194
Box 43	Formulae due to Goda (1985) for wave height estimation in the surf zone	212

Chapter 5

Box 44	Wave height–period parameters	242
Box 45	Examples of gradings	242
Box 46	Critical overtopping discharges	253
Box 47	Example of reflection from rubble placed against a vertical wall	263
Box 48	Thickness of layers and number of units	264
Box 49	Comparison of Hudson and van der Meer formulae	266
Box 50	Example of plot of H_s versus ξ_m showing influence of damage levels	268
Box 51	Example of plot of H_s versus ξ_m showing influence of permeability	268
Box 52	Example of wave height–damage graph	269
Box 53	Stability of armour layers with very wide gradings	271
Box 54	Protection against overtopping	272
Box 55	Example of stability relation for reef-type breakwater	275
Box 56	Design curves for low-crested breakwaters ($R_c > 0$)	275
Box 57	Design curves for submerged breakwaters ($R_c < 0$)	276
Box 58	Powell's parametric model for shingle beach profiles	286
Box 59	Berm breakwater profile model due to Kao and Hall (1990)	288
Box 60	Longshore transport formulae and criteria for rock and gravel	296
Box 61	Hydraulic stability formulae for seabed rockfill material	300
Box 62	Dynamic soil–water–structure interaction	323
Box 63	Wave-induced internal water motion in a rock structure	327
Box 64	Bishop's stability analysis	328
Box 65	Surface wave penetration in a porous rock structure	329
Box 66	Internal set-up under time-variant conditions	331
Box 67	Dynamic excess pore pressures: a simple approach	331
Box 68	Liquefaction of seabed sand under waves; a simple approach	341
Box 69	Example of probabilistic analysis of a breakwater (CIAD, 1985)	349

Chapter 6

Box 70	Breakwater layout development	357
Box 71	Most significant characteristics of land-based and marine execution	360
Box 72	Non-breaking wave pressures and forces on caissons	371
Box 73	Simulation of rock dumping	381
Box 74	Matching demand for stone to quarry fragmentation curves	386
Box 75	Example of breakwater construction logistic simulation	387
Box 76	Design of pocket beaches	405
Box 77	Clacton Sea Defence Scheme, UK	408
Box 78	Design of beach using rock sill concept	409
Box 79	Cost comparison of timber versus rock for bastion groynes	415
Box 80	Quarry selection, Morecambe Bay Coast Protection Scheme, UK	417
Box 81	Use of gabions and mattresses	422
Box 82	Case study in sizing armourstone for a seawall rehabilitation	425
Box 83	Dutch practice for determination of dike height	431
Box 84	Maintenance of UK water-retaining dams	453
Box 85	Construction details for Seaford gravel beach-nourishment project, 1987	460
Box 86	Estimation of beach-nourishment losses	462

Appendix 3

Box A1	Example of use of principles in this appendix within the context of common UK methods of measurement	531
Box A2	Measurement of armourstone by weight rather than volume	541

Glossary

Accretion Build-up of material solely by the action of the forces of nature by the deposition of water- or airborne material.
Alongshore see **longshore**.
Anisotropic Property of the mineral fabric of a rock block or piece of rock such that its strength and visual appearance is not the same in all directions.
Apron Layer of stone, concrete or other material to protect the toe of the seawall against scour.
Armour layer Protective layer on rubble-mound breakwater composed of armour units.
Armour unit Large quarry stone or special concrete shape used as primary wave protection.
Artificial nourishment, beach replenishment/recharge, beach feeding Supplementing the natural supply of beach material to a **beach**, using imported material.
Back-rush The seaward return of the water following **run-up**.
Bastion A massive **groyne**, or projecting section of seawall normally constructed with its crest above water level.
Bathymetry Topography of sea/estuary/lake bed.
Beach By common usage, the zone of **beach material** that extends landward from the lowest water line to the place beyond the high water line, where there is a marked change in material or physiographic form, or to the line of permanent vegetation.
Beach material Granular sediments, usually sand or shingle moved by the sea.
Berm Relative small mound to support or key-in armour layer.
Berm breakwater Rubble mound with horizontal berm of armour stones at about sea-side water level, which is allowed to be (re)shaped by the waves.
Breastwork Timber structure, generally parallel to coast.
Bull nose Substantial lip or protuberance at the top of the seaward face of a wall, to deflect waves seaward.
Bypassing Moving of **beach material** from the accumulating updrift side to the eroding downdrift side of an obstruction to longshore transport (e.g. in inlet or harbour).
Caisson Concrete box-type structure.
Coastal defences, coastal works Collective terms covering protection provided to the coastline. These include **coast protection** and **sea defences**.
Coastal processes Collective term covering the action of natural forces on the coastline and adjoining seabed.
Coastal regime The overall system resulting from the interaction on the coast and seabed of the various **coastal processes**.
Coast protection Works to protect land against **erosion** or encroachment by the sea.
Crenulate An indented or wavy shoreline beach form, with the regular seaward-pointing parts rounded rather than sharp, as in the **cuspate** type.
Crest Highest part of a breakwater.
Crown wall Concrete superstructure on a rubble mound.
Cuspate Form of beach shoreline involving sharp seaward-pointing cusps (normally at regular intervals) between which the shoreline follows a smooth arc.
Deep water Water so deep that waves are little affected by the bottom. Generally, water deeper than one half the surface **wave length** is considered to be **deep water**.
Design storm Seawalls will often be designed to withstand wave attack by the extreme design storm. The severity of the storm (i.e. **return period**) is chosen in view of the acceptable level of risk of damage or failure.
Diffraction Process by which energy is transmitted laterally along a wave crest. Propagation of waves into the sheltered region behind a barrier such as a breakwater.
Discontinuity Any actual or incipient fracture plane in a rock mass, including bedding planes, laminations, foliation planes, joints and fault planes.
Downdrift The direction of predominant movement of **littoral drift** along the shore.

Durability The ability of a rock to retain its physical and mechanical properties (i.e. resist degradation) in engineering service.
Duricrust A hard layer formed at a present or past desert surface where salts carried in solution by capillary action have precipitated and cemented the surface layer sediments.
ELCL, LCL, UCL, EUCL Extreme lower, lower, upper and extreme upper class limits used to define standard grading classes. Each class limit is a particular weight for which the cumulative percentage passing by weight must fall within a specified range (e.g. for ELCL, between 0% and 2%).
Erosion The wearing away of material by the action of natural forces.
Fetch Relative to a particular point (on the sea), the area of sea over which the wind can blow to generate waves at the point.
Filter Intermediate layer, preventing fine materials of an underlayer from being washed through the voids of an upper layer.
Flaws Discontinuities and voids within a piece of rock.
Flood wall, splash wall Wall, retired from the seaward edge of the seawall crest, to prevent water from flowing onto the land behind.
Foreshore The part of the shore lying between high water mark and low water mark (see **tides**).
Fracture toughness The characteristic level of stress intensity ahead of a crack tip that is required to propagate the new crack catastrophically through the mineral fabric of the rock.
Freeboard The height of a structure above **still-water level**.
Geotextile A synthetic fabric which may be woven or non-woven used as a filter.
Gradings Heavy gradings
        Light gradings   see Appendix 1
        Fine gradings
Groyne A structure generally perpendicular to the shoreline built to control the movement of **beach material**.
Hard defences In common usage, normally taken to describe concrete, timber, steel, asphalt or rubble shoreline structures. Rubble or rock structures are often considered **soft defences** because of their ability to absorb wave energy.
Head End of breakwater or dam.
Hydraulics Science of water motion/flow/mass behaviour.
Hydrology Science of the hydrological cycle (including precipitation, runoff, fluvial flooding).
Igneous rocks Formed by the crystallisation and solidification of a molten silicate magma.
***In-situ* block** A piece of rock bounded by discontinuities located within the rock mass prior to excavation.
Intact fabric strength Strength of rock as a consequence of strength and fabric of the rock's minerals.
Integrity The degree of wholeness of a rock block as reflected by the degree to which its strength to resist impacts is reduced by the presence of flaws.
Littoral drift, littoral transport The movement of **beach material** in the **littoral zone** by waves and currents. Includes movement parallel (longshore transport) and perpendicular (onshore–offshore transport) to the shore.
Littoral zone Beach and **surf zone**.
Longshore Along the shore.
Mach-stem wave Higher-than-normal wave generated when waves strike a structure at an oblique angle.
Maintenance Repair or replacement of components of a structure whose life is less than that of the overall structure, or of a localised area which has failed.
Metamorphic rocks Formed by the effect of heat and pressure on igneous or sedimentary rocks for geological periods of time with the consequent development of new minerals and textures within the pre-existing rock.
Monochromatic waves A series of waves, each of which has the same **wave period**.
Morphology River/estuary/lake/seabed form and its change with time.
Offshore breakwater A breakwater built towards the seaward limit of the **littoral zone**, parallel (or near-parallel) to the shore.
Orthogonal (wave ray) In a wave refraction/diffraction diagram, a line drawn perpendicular to the wave crest.
Overtopping Water passing over the top of the seawall.

Parapet Solid wall at crest of seawall projecting above deck level.
Parapet-wall See **crown-wall**.
Porosity Laboratory-measured property of the rock indicating its ability to retain fluids or gases.
Porous In terms of **revetments** and **armour**, cladding that allows rapid movement of water through it such as during wave action (many geotextiles and sand asphalt can be non-porous under the action of waves but porous in terms of soil mechanics).
Prototype The actual structure or condition being simulated in a model.
Quarry Site where natural rock stone is mined.
Quarry run Waste of generally small material, in a quarry, left after selection of larger gradings.
Random waves The laboratory simulation of irregular sea states that occur in nature.
RDI_d, RDI_s Rock-durability indicators of Fookes et al (1988).
Reef breakwater Rubble mound of single-sized stones with a crest at or below sea level which is allowed to be (re)shaped by the waves.
Reflected wave That part of an incident wave that is returned seaward when a wave impinges on a **beach**, seawall or other reflecting surface.
Refraction (of water waves) The process by which the direction of a wave moving in **shallow water** at an angle to the contours is changed so that the wave crests tend to become more aligned with those contours.
Refurbishment, renovation Restoring the seawall to its original function and level of protection.
Regular or monochromatic waves (1) *Fetch length*: A length used to characterise the distance over which wind may act to generate waves. The fetch length will depend upon the shape and dimensions of fetch area, and upon the relative wind direction. (2) *Mean and peak wave periods*: The mean period of the wave defined by zero-crossing. The peak period is given by the inverse of the frequency at which the energy spectrum reaches a maximum.
Rehabilitation Renovation or upgrading.
Replacement Process of demolition and reconstruction.
Return period In statistical analysis an event with a **return period** of N years is likely, on average, to be exceeded only once every N years.
Revetment A cladding of stone, concrete or other material used to protect the sloping surface of an embankment, natural coast or shoreline against erosion.
Rip-rap Wide-graded quarry stone normally used as a protective layer to prevent **erosion**.
Rock-degradation model (for armour stone) A model under research and development at Queen Mary and Westfield College, London University, which attempts to predict yearly weight losses from the armour, taking account of rock properties and site conditions.
Rock weathering Physical and mineralogical decay processes in rock brought about by exposure to climatic conditions either at the present time or in the geological past.
Rubble-mound structure A mound of random-shaped and random-placed stones.
Run-up The rush of water up a structure or **beach** as a result of wave action.
Run-up, run-down The upper and lower levels reached by a wave on a structure, expressed relative to still-water level.
S-slope breakwater Rubble mound with gentle slope around still-water level and steeper slopes above and below.
Scour protection Protection against erosion of the seabed in front of the toe.
Sea defences Works to prevent or alleviate flooding by the sea.
Secular changes Long-term changes in sea level.
Sedimentary rocks Formed by the sedimentation and subsequent lithification of mineral grains, either under water or, more rarely, on an ancient land surface.
Shallow water Commonly, water of such depth that surface waves are noticeably affected by bottom topography. It is customary to consider water of depths less than half the surface **wave length** as shallow water.
Shoulder Horizontal transition to layer of larger stones which is placed at a higher elevation.
Significant wave height The average height of the highest of one third of the waves in a given sea state.

Significant wave period An arbitrary period generally taken as the period of one of the highest waves within a given sea state.
Soft defences Usually refers to **beaches** (natural or designed) but may also relate to energy-absorbing structures, including those constructed of rock, considered as **hard defences** because of their stability.
Stationary process A process in which the mean statistical properties do not vary with time.
Still-water level Water level which would exist in the absence of waves.
Stochastic Having random variation in statistics.
Storm surge A rise in water level in the open coast due to the action of wind stress as well as atmospheric pressure on the sea surface.
Surf zone The area between the outermost breaker and the limit of the wave **run-up**.
Suspended load The material moving in suspension in a fluid, kept up by the upward components of the turbulent currents or by the colloidal suspension.
Swell (waves) Wind-generated waves that have travelled out of their generating area. Swell characteristically exhibits a more regular and longer period and has flatter crests than waves within their **fetch**.
Tides (1) *Highest astronomical tide (HAT), lowest astronomical tide (LAT)*: The highest and lowest levels, respectively, which can be predicted to occur under average meteorological conditions and under any combination of astronomical conditions. These levels will not be reached every year. HAT and LAT are not the extreme levels which can be reached, as storm surges may cause considerably higher and lower levels to occur. (2) *Mean high water springs (MHWS), mean low water springs (MLWS)*: The height of mean high water springs is the average, throughout a year when the average maximum declination of the moon is $23\frac{1}{2}°$ of the heights of two successive high waters during those periods of 24 hours (approximately once a fortnight) when the range of the tide is greatest. The height of mean low water springs is the average height obtained by the two successive low waters during the same periods. (3) *Mean high water (MHW), mean low water (MLW)*: For the purpose of this manual, mean high water, as shown on Ordnance Survey Maps, is defined as the arithmetic mean of the published values of mean low water springs and mean low water neaps. This ruling applies to England and Wales. In Scotland the tidal levels shown on Ordance Survey Maps are those of mean high (or low) water springs (MH (or L) WS).
Toe Lowest part of sea- and portside breakwater slope, generally forming the transition to the seabed.
Type 1 breakage Breakage of rock blocks into major pieces along flaws.
Type 2 breakage Breakage of rock blocks along new fractures.
Upgrading Improved performance against certain criteria.
Up-rush, down-rush The flow of water up or down the face of a structure.
Vesicular Term used to describe basalt and other volcanic rocks containing many spherical or ellipsoidal cavities produced by bubbles of gas trapped during solidification.
Wave return face The face of a **crown wall** designed to throw back the waves.

Notation

The notation used in this manual differs from that in many previous design manuals for coastal structures. This is intentional. To do otherwise would lead to considerable duplication and confusion.

The starting points for the notation adopted have been recent reports prepared by IAHR/PIANC (1986) on sea state terminology and PIANC (1990) on rubble-mound breakwaters. Where practical, notation used in the *SPM* (1984) and Dutch and British standards have been incorporated with little change. The user of the manual is cautioned that some familiar terminology may have been changed, modified or redefined.

A_c	Armour crest freeboard, relative to still-water level
A_e	Erosion area on profile
a_0	Amplitude of horizontal wave motion at bed
B	Structure width, in horizontal direction normal to face
B_{wl}	Structure width at still-water level
C	Chezy coefficient
C_i	Compression index
C_{pi}, C_{si}	Primary (or secondary) compression index
C_r	Coefficient of wave reflection
C_t, C_{to}, C_{tt}	Coefficient of total transmission, by overtopping or transmission through
c'	Soil cohesion
c_v	Consolidation coefficient (units of m²/s)
D	Particle size, or typical dimension
D_e	Effective particle diameter
D_f	Degree of fissurisation (French)
D_n	Nominal block diameter $=(M/\rho_a)^{1/3}$
D_{n50}	Nominal diameter $(M_{50}/\rho_a)^{1/3}$
D_p	Diameter of profiler or survey head
D_s	Size of the equivalent volume sphere
D_z	Sieve diameter
D_{50}	Sieve diameter, diameter of stone which exceeds the 50% value of sieve curve
D_{85}	85% value of sieve curve
D_{15}	15% value of sieve curve
D_{85}/D_{15}	Armour grading parameter
d	Thickness or minimum axial breadth (given by the minimun distance between two parallel straight lines between which an armour block can just pass); thickness of a layer of rock or material in geotechnical calculations
E	Young's Modulus
E_i	Incident wave energy
E_r	Reflected wave energy
E_t	Transmitted wave energy
E_d	Energy absorbed or dissipated
$E_{\eta\eta}$	Energy density of a wave spectrum
e, e_0	Void ratio $=n_v/(1-n_v)$, initial void ratio
F	Fetch length; or geotechnical factor of safety, defined as ultimate resistance/required resistance
$\tilde{F}$	Dimensionless fetch length $=gF/U_w^2$ or fetch length in wind wave-generation formulae
F_c	Difference of level between crown wall and armour crest $=R_c-A_c$
F_s	Shape factor $=M/\rho_a D_z^3$
F_*	Dimensionless freeboard parameter $=(R_c/H_s)^2(s/2\pi)^{1/2}$
f	Frequency of waves $=1/T$
f_p	'Peak' frequency of waves at which maximum wave energy occurs

f_w	Wave friction factor
G	Shear strength granular skeleton
G_c	Width of armour berm at crest
g	Gravitational acceleration
H	Wave height, from trough to crest
H_{max}	Maximum wave height in a record
H_{mo}	Significant wave height calculated from the spectrum $=4m_o^{1/2}$
H_o	Offshore wave height, unaffected by shallow-water processes
H_{os}	Offshore significant wave height, unaffected by shallow-water processes
H_s	Significant wave height, average of highest one-third of wave heights
$\tilde{H}_s$	Dimensionless significant wave height $=gH_s/U_w^2$
$H_{2\%}$	Wave height exceeded by 2% of waves
$H_{1/10}$	Mean height of highest one-tenth of waves
h	Water depth
h_c, h_c'	Armour crest level relative to seabed, after and before exposure to waves
h_f	Height of crown wall over which wave pressure acts
h_s	Water depth over half wave length, or five times the maximum wave height, seaward of structure toe
I	Potential gradient: $I = \nabla p/\gamma$
I_c	Continuity index (French)
I_d	Drop test breakage index
$I_{s(50)}$	Point load strength index
i_w	Wind-induced gradient of still-water surface
K	Hydraulic conductivity or permeability
K_b	Compressive strength granular skeleton
K_D	Stability coefficient in Hudson formula
K_{IC}	Fracture toughness
k	Wave number, $2\pi/L$
k_s	Bed roughness length
k_t	Layer thickness coefficient
L	Wave length, in the direction of propagation
L_o	Deep water or offshore wave length, $gT^2/2\pi$
L_{om}, L_{op}	Offshore wave length of mean, T_m, and peak, T_p, periods, respectively
L_s	Wave length in (shallow) water at structure toe
L_{ms}, L_{ps}	Wave length of mean or peak period at structure toe
l	Maximum axial length (given by the maximum distance between two points on the block)
M	Mass of an armour unit
$MAIV$	Modified aggregate impact value from Hosking and Tubey (1969) method
m_s	Exponent in the Schumann equation for weight distribution
M_{50}, M_i	Mass of unit given by 50%, i%, on mass distribution curve
m	Seabed slope
m_0	Zeroth moment of wave spectrum
m_n	nth moment of spectrum
N	Number of waves in a storm, record or test, $=T_r/T_m$
N_a	Total number of armour units in area considered
N_d	Number of armour units displaced in area considered
N_{od}, N_{or}	Number of displaced, or rocking, units per width D_n across armour face
N_r	Number of armour units rocking in area considered
N_s	Stability number $=H_s/(\Delta D_{n50})$
N_s^*	Spectral stability number $=(H_{mo}^2 L_{ps})^{1/3}/(\Delta D_{n50})$
n	Volumetric porosity
n_a	Area porosity, void area as proportion of total projected area
n_c	Critical porosity (density of skeleton)
n_r	Rock porosity
n_{rr}	Exponent in Rosin–Rammler equation for weight distribution
n_v	Volumetric porosity, volume of voids as proportion of total volume

O, O_i	Opening size in geotextile, $i\%$ opening size
P	Notional permeability factor, defined by van der Meer
P_s	Fourier shape factor based on the first to tenth harmonic amplitudes
P_R	Fourier asperity roughness based on the eleventh to twentieth harmonic amplitudes
P_x	Probability that x will not exceed a certain value; often known as cumulative probability density of x
p	Pore water pressure, or wave pressure
p_a	Atmospheric pressure at sea level
p_x	Probability density of x
Q	Overtopping discharge, per unit length of seawall
Q^*	Dimensionless overtopping discharge $= Q/(T_m g H_s)$
q_o	Volume of overtopping, per wave, per unit length of structure
q_s	Superficial velocity, or specific discharge, discharge per unit area, usually through a porous matrix
R	Strength descriptor in probabilistic calculations
R'	Equivalent rock roughness
R^*	Dimensionless freeboard $= R_c/T_m(gH_s)^{1/2}$
R_a	Average roughness of a surface from profile data
R_c	Crest freeboard, level of crest relative to still-water level
R_e	Reynolds number $= U_d/u$
$R_{d2\%}$	Run-down level, below which only 2% pass
Rf	Particle surface roughness
R_u	Run-up level, relative to still-water level
R_{us}	Run-up level of significant wave
$R_{u2\%}$	Run-level exceeded by only 2% of run-up crests
r	Roughness value, usually relative to smooth slopes
S	Loading descriptor in probabilistic design
S'	Equivalent rock strength
S_c	Settlement or compression distance
S_d	Dimensionless damage, A_e/D_{n50}^2; may be calculated from mean profiles or separately for each profile line, then averaged
Sh	Particle shape factor
SST	Magnesium sulphate soundness value from Hosking and Tubey's (1969) method
s	Wave steepness, H/L_o
s_m	Wave steepness for mean period, $2\pi H_s/gT_m^2$
s_{op}	Offshore wave steepness for peak period, $H_{os}/L_{op} = 2\pi H_{os}/gT_p^2$
s_p	Wave steepness for peak periods, $2\pi H_s/gT_p^2$
T	Wave period
T_m	Mean wave period
$\tilde{T}_m$	Dimensionless mean wave period $= gT_m/U_w$
T_p	Spectral peak period, inverse of peak frequency
$\tilde{T}_p$	Dimensionless peak wave period $= gT_p/U_w$
T_R	Duration of wave record, test or sea state
t	Time, variable
t_a, t_f, t_x	Thickness of armour, underlayer or other layer in direction normal to face
U	Horizontal depth mean current velocity
U_*	Shear velocity $= \tau_b/\rho$
U_c	Coefficient of uniformity, D_{60}/D_{10}
U_o	Maximum seabed wave orbital velocity
U_z	Wind speed at z metres above sea surface
U_{10}	Wind speed 10 metres above sea surface
u, v, w	Local velocities, usually defined in x, y, z directions; u sometimes refers to porous matrix velocity
W	Armour unit weight, $= Mg$
W_{em}	Effective mean weight (of a standard grading) i.e. the arithmetic average of all blocks excluding those which fall below the ELCL weight for the grading class
W_0, W_{15}, W_{50}, W_y	Weight for which a fraction or percentage y is lighter on the cumulative weight distribution curve

$W_{63.2}$	Location parameter in the Rosin–Rammler equation for weight distribution
w_{ab}	Water absorption
X	Equivalent wear factor in the QMW rock-degradation model equal to the number of years in service divided by the equivalent number of revolutions in the QMW mill abrasion test
$X_1, X_2, \ldots, X_9$	Parameters which are given rating values in the QMW rock-degradation model
X, Y, Z	Block dimensions of enclosing cuboid system
x, y, z	Distances along orthogonal axes
Z	Reliability function in probabilistic design; $Z = R - S$
z	Sieve size (i.e. the smallest square hole that a block can pass through with optimum orientation)
z_a	Static rise in water level due to storm surge
z_s	Internal set-up in a mount above still-water level
α	Structure front face angle
β	Angle of wave attack with respect to the structure
γ	Weight density $= \rho g$
γ_a	Weight density of armour
γ_b	Bulk weight density of an armour layer as laid, i.e. where the enclosing volume is calculated by surveying the average armour profiles
γ_{br}	Breaker index $= [H/h]$ max
γ_{b1}	Actual tonnage per cubic metre in a test panel as measured from weighings and using agreed profile survey method
γ_{b2}	Calculated tonneage per lin. metre in a test panel from block count, D_{n50} specified and using agreed profile survey method
γ_w	Weight density of water
Δ	Relative buoyant density of material considered, e.g. for rock $= (\rho_r/\rho_w) - 1$
ε	Strain, relative displacement
η	Instantaneous surface elevation
θ	Mean direction of waves, usually to grid north
λ	Geometric scale ratio in physical models; or penetration length in set-up calculations
μ	Coefficient of friction
$\mu(x)$	Mean of x
v	Coefficient of kinematic viscosity; or Poisson's ratio
ξ	Surf similarity parameter, or Iribarren number, $= \tan \alpha / s_m^{1/2}$
ξ_p	Modified surf parameter $= \tan \alpha / s_p^{1/2}$
ρ	Mass density, usually of fresh water
ρ_r	Mass density, oven-dried density
ρ_{ssd}	Saturated surface dry density
ρ_w	Mass density of sea water
ρ_r, ρ_c, ρ_a	Mass density of rock, concrete, armour
ρ_b	Bulk density of material as laid
$\rho_{x,y}$	Correlation coefficient between x and y
σ	Stress
σ'	Effective stress in soil or rubble $= \sigma - p$
σ_c	Uniaxial compressive strength
σ_x	Standard deviation of x
σ_x^2	Variance of x
τ	Shear strength of rubble or soil
τ_c	Bed shear stress exerted by a steady current
τ_w	Bed shear stress due to wave orbital water motion
τ_{cw}	Bed shear stress due to both currents and wavers
Υ	Dimensionless damage $= N_d/N_a$, may be expressed as a percentage
ϕ	Angle of internal friction of rubble or soil or angle of wind direction in wind wave-generation calculations
ψ	Liquefaction potential
ψ_w	Dimensionless shear parameter or Shields number
ω	Angular frequency of waves ($= 2\pi/T$)

1. Introduction

1.1 Background, use and approach

This manual has been produced jointly by the Netherlands and UK construction industries to provide practical guidance on the use of rock in coastal and shoreline engineering. It reflects British and Dutch national and international experience in applications where protection against the action of wind-generated waves is one of the dominant design considerations. Typical applications are described in Section 1.3. The term 'rock' refers principally to irregularly shaped quarried rock material, but includes other materials such as industrial by-products and naturally occurring shingle whose hydraulic performance under wave action is similar to that of quarried rock.

The manual sets out an integrated approach to the planning and design process by considering a range of related parameters (e.g. availability and durability of materials; environmental implications; method of construction; future management strategy; economic factors) as well as the basic engineering requirements. This enables optimum use to be made of locally available materials and helps to avoid design decisions being made which inadvertently constrain the evolution of the best overall solution to the problem.

Specific objectives of the Anglo-Dutch initiative in producing the manual were as follows:

- To collate available research data and technical information, together with practical experience gained by practitioners from the Netherlands and the UK in order to facilitate better dissemination and use of existing knowledge;
- To describe the best of present design practice (i.e. the state of the art). In doing this, care has been taken to indicate the limitations in present understanding of the processes involved, and the extent to which empirical methods and engineering judgement are related to present design procedures;
- To set out a procedural framework for planning and design activities which guides the practitioner in an integrated approach to the use of rock yet allows an appropriate site-specific solution to be adopted. It is emphasised that standard solutions do not generally exist in this field of engineering.

The manual is written for the practising engineer with a first degree in civil engineering and some experience in construction procedure. It will also be useful to practitioners of other disciplines involved in the production and use of rock in coastal and shoreline engineering. The manual is for the non-specialist in that it aims to provide the reader with an understanding of the principles and procedures involved. It is, however, emphasised that the manual itself cannot convert the non-specialist into an expert, and is not a substitute for experience and judgement.

Depending on the nature of the problem and the experience of the reader, the manual might be used in one of the following ways:

1. As a source of information or reference on specific points. In this respect, the reader should refer to the Contents and/or the index at the end of the manual;
2. As a procedural guide to carrying out the planning and design process which is introduced in Section 1.2. Note that design includes the specification of materials and workmanship, and the approach to future construction and management;
3. As an aid to interfacing with specialists and practitioners of other disciplines who need to be involved in solving the problem.

In all circumstances the reader is advised to study this introductory chapter before using the manual.

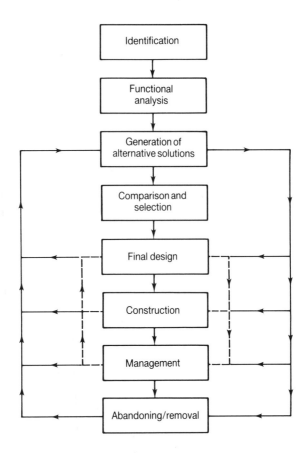

Figure 1 *Design process*

The structure of the manual can be seen from the Contents and the design process logic diagram in Section 1.2, which indicates how the different sections are interrelated. Particular attention is drawn to the use of 'Boxes' to present information—such as empirical design methods, worked examples, detailed information and case histories—which are outside the general run of the main text.

Standards quoted in this manual are, in general, those of the ISO (International Standards Organisation). Where these do not exist, the equivalent Netherlands (NEN) or British (BS) standards are indicated. The reader must ascertain whether any other local or national standards and similar regulations pertain to the problem in question, and follow these as appropriate.

1.2 Use of the manual

1.2.1 DESIGN PROCESS

In general, a rock structure is planned as a practical measure to solve an identified problem. Examples are seawalls, planned to reduce the occurrence of inundation due to storm surges, or a shore protection to reduce erosion. Starting with identification of the problem (e.g. inundation or shoreline erosion), a number of stages can be distinguished in the design process for (and life cycle of) a structure. After problem identification, the subsequent stages are determined by a series of decisions and actions culminating in the creation of a structure or structures to resolve the problem. Post-design stages (to be considered *during* design!) are the construction and maintenance (monitoring and repair) of the structure and, finally, its removal or replacement.

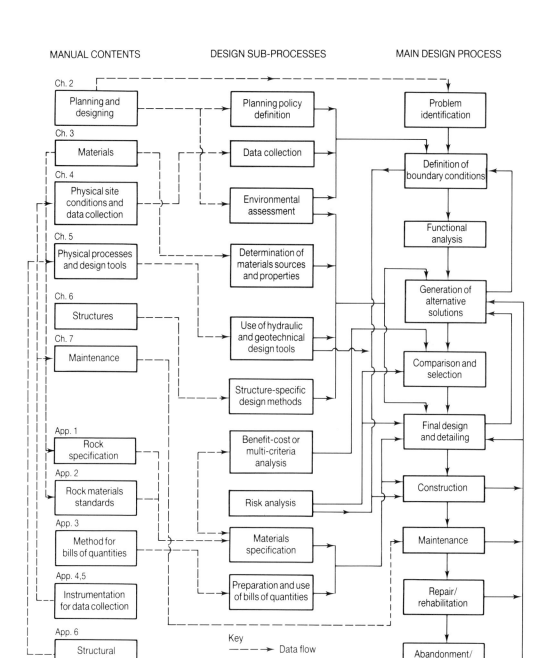

Figure 2 *Manual on use of rock in coastal and shoreline engineering—logic diagram*

A general overview of a modern design process is given in Figure 1 as a cyclic process covering eight stages. The design problem is solved iteratively following this process. During each cycle of the design process the description of the design and construction and maintenance methods becomes less abstract and is defined with an increasing degree of detail.

From the design process/life cycle of a structure the manual user should be aware that the design of a structure may easily develop into a multidisciplinary process, extending beyond the description of the mere technological boundary conditions and the resulting structural design. Additional non-technical aspects that may affect the eventual outcome of the design process include:

Social conditions;
Economics;
Environmental impact;
Safety requirements.

Figure 3 *Rock breakwater (courtesy HR Wallingford)*

Figure 4 *Rock coastal revetment (courtesy HR Wallingford)*

Most of these matters are discussed briefly in this manual, but the reader should always be aware of the need to take a wider view in developing a design concept.

1.2.2 STRUCTURE AND CONTENTS OF THIS MANUAL

The manual is structured to follow the design process shown in Figure 1 and elaborated in more detail in the logic diagram in Figure 2. This shows how the chapter contents interrelate with the main design process and the design sub-

Figure 5 *Typical rock dam-face protection (courtesy EPD Consultants)*

Figure 6 *Typical rock groynes and gravel beach (courtesy J. Simm, Robert West & Partners)*

processes which support it. A full discussion of the design process is given in Chapter 2, guiding the reader to the relevant parts of the manual as appropriate.

Chapter 2 gives the reader an overview of the design process, including an outline of the main considerations the designer should be aware of in an application of rock in coastal and shoreline engineering. In addition, it guides readers to the relevant part of this manual where they will find more information on any particular subject. Chapter 2 also provides more information on the basic structure types and failure modes that should be considered and the approach of fault tree analysis is introduced. Deterministic and probabilistic methods are compared.

Chapter 3 contains all the basic materials information, including types and sources of rock, rock properties and function, rock testing, rock production, handling and transport, other materials information and an approach to quality control. The chapter is closely tied to Appendices 1–3, which provide a model specification, suite of tests and method of preparing bills of quantities.

Chapter 4 provides information on how to determine the physical site conditions (bathymetry, hydraulic and geotechnic) both offshore and up to the site of the rock structure of interest, but on the basis that the structure itself, in terms of interaction with the hydraulics and geotechnical conditions, is not present. Instrumentation for data collection is dealt with in Appendices 4 and 5.

Chapter 5 then describes the physical processes whereby the structure itself interacts with the hydraulic and geotechnical boundary conditions producing both hydraulic, geotechnic and structural responses. It also provides tools for designing the structure, given these interactions. Thus practical guidance is given on such matters as:

- Wave run-up, run-down, overtopping, transmission
- Structural stability of rock in armour layers
- Underlayer and filter design
- Slope stability.

Chapter 6 describes the specific structure types mentioned in Section 1.2 above and how design tools are applied together with practical considerations, of which construction plays a significant part, to provide an overall design approach.

Chapter 7 provides information on maintenance (monitoring, structure condition appraisal and repair work) and explains how this interacts with the design. Structure state monitoring techniques are described separately in Appendix 6.

1.3 Structure types covered by this manual

This manual covers applications of rock to the following types of structures (see Figures 3–6):

- Breakwaters
- Seawalls, groynes and shoreline protection structures
- Dam face protection
- Gravel beaches
- Rockfill offshore engineering.

2. Planning and designing

This chapter presents an overview of the entire design process (Section 2.1) for rock structures in coastal and shoreline engineering. This ranges from initial problem-identification, boundary condition definition and functional analysis to design concept generation, selection, detailing and costing, and includes an examination of construction and maintenance considerations and Quality Assurance/Quality Control aspects. The overview is supported (Section 2.2) by a more in-depth study of the structure types considered together with their potential failure modes and by a review (Section 2.3) of deterministic and probabilistic design approaches illustrated with examples. A final section (2.4) considers the important area of Environmental (Impact) Assessment in relation to coastal and shoreline rock structures.

2.1 Design process

This section addresses the details of the overall design process described in the design process logic diagram (Figure 2 in Section 1.2) which relates the design process to the contents of this manual. It also indicates the principles and methods which support the design procedure, making reference, as appropriate, to other parts of the manual. It must be recognised that the design process is a complex iterative process and may be described in more than one way. Another overall formulation in flow chart form is given in Figure 7 with cross-references to various parts of this chapter and the manual.

2.1.1 PROBLEM IDENTIFICATION

At the problem-identification stage the presence of an existing or future problem is acknowledged and defined. The acknowledgement of an existing or future problem is generally accompanied by a determination to find an appropriate solution to that problem. In the context of this manual this solution will probably be by means of constructing some coastal or shoreline structure. Future problems may be foreseeable as a result of predictable changes or may be generated by proposed engineering works. A simple example of the latter would be the problem of protecting the upstream face of a planned water-retaining dam.

2.1.2 BOUNDARY CONDITIONS

In conjunction with identification of the problem, all the boundary conditions which influence the problem and its potential solution must also be identified. These boundary conditions are of various types, and include aspects of the following:

- Planning policy (including environmental impact aspects)
- Physical site conditions
- Construction and maintenance considerations
- Cost considerations.

Planning policy aspects involve political, legislative and social conditions and include a definition of acceptable risk of failure/damage/loss of life and acceptable/desirable environmental impacts. Environmental impact aspects are discussed later in Sections 2.1.4.1 and 2.4, as these tend to be predominantly boundary conditions on the solution and, as such, effectively become part of the iterative design

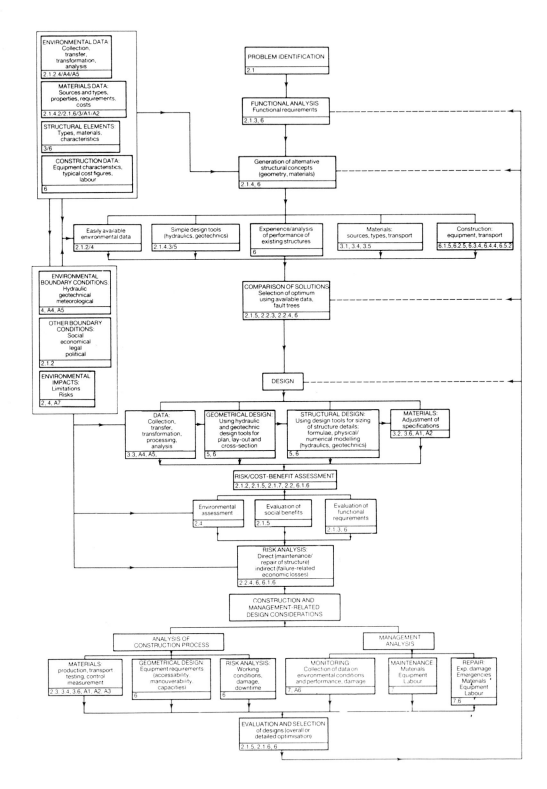

Figure 7 *Detailed description of a design process with reference to parts of this manual*

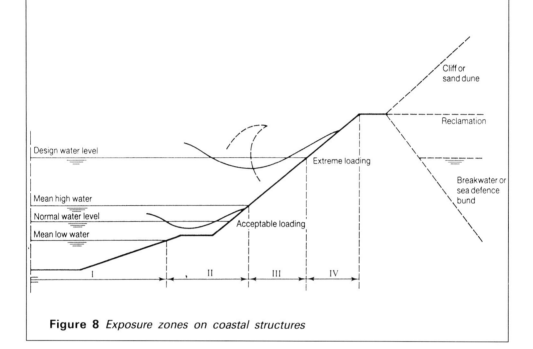

process, evaluating the potential beneficial or adverse effects of each proposed solution. Other planning policy aspects are described in Section 2.1.2.1 below.

Physical site conditions of principal concern will be those, the determination of which is described in some detail in Chapter 4 (with instrumentation details in Appendices 4 and 5). These include bathymetry and morphology, hydraulic conditions (water levels, winds, waves, currents), site conditions and geotechnical conditions (foundation soils characteristics and pore water pressures). They may also, in certain special situations, include other effects such as chemical, biological and ice impact which are outside the scope of this manual.

Figure 99 in Chapter 4 gives an overview of important site conditions and their evaluation in the design of coastal and shoreline rock structures. Chapter 4 also discusses how to determine the appropriate combination of physical site conditions to be used in design, both for normal (serviceability limit state) and extreme (ultimate limit state) conditions. The influence of the probable sequence of these normal and extreme conditions must be recognised in design and the zone of their influence on the (proposed) structure understood (see Box 1).

Construction and maintenance considerations only really act as boundary conditions to the solution rather than the problem and, as such, like many aspects of environmental impact, become part of the iterative design process. They are therefore reviewed in a general way as part of this process later in this chapter (Sections 2.1.4.1 and 2.1.4.5). Detailed descriptions of construction, both land-based and waterborne, and of maintenance (including monitoring and repair), will be found in Chapters 6 and 7, respectively.

Expenditure constraints act as a boundary condition to the cost of the solution and thus to the solution itself. Cost aspects are closely related to most other aspects of the design process. For an introduction see Section 2.1.7.

2.1.2.1 Planning policy

Projects involving rock in coastal and shoreline engineering will normally have to meet the requirements of some governmental or company policy. Governments or local authorities may initiate the design of a structure because of their responsibility for the management of coastal defence, harbour facilities, etc. Companies may do the same as part of the investment in a new plant. In both cases a number of procedures will have to be followed, related to established planning requirements, legislation, decision-making procedures, and benefit-cost analysis for financing of the project. Key policy matters that will have a very significant impact on the design that are often predefined include:

- Acceptable risk of failure/damage/loss of life (see Section 2.1.2.2);
- Minimum benefit-cost ratio or rate of return for the project.

However, many aspects of planning policy are not predefined at the start of a project and in many cases permission ('planning permission') from the relevant governmental organisation to proceed with the project, if given, is subject to constraints imposed once a scheme concept has been presented for approval. Therefore it is necessary to involve at a preliminary stage of the design process the decision makers, authorities, politicians, public and any groups or individuals who may have an interest in the existing problem and/or the way to solve it (including planning, design, construction and management of a structure). Experience has shown that a technical solution to a problem may not be accepted, by any of the parties of interest, if it is presented as an independent and predefined closed solution. The background of the various interested parties should be acknowledged, as these may relate to various social (individual), political and/or economical interests.

The impact of constraints imposed by planning authorities can be considerable. While a project can usually be undertaken using a variety of structures, materials, equipment and labour, the interested parties and/or the planning authority may seek to limit the freedom to choose from some of the available options. This limitation may have a significant influence on the design, construction and maintenance of the future structure.

One method of overcoming the problem of non-acceptance may be that of policy analysis. This, in essence, implies a representative involvement of all interested parties at an early stage of the design. The involvement will typically aim to first discuss and agree the following points:

1. The proper description of the problem;
2. Criteria and weightings to be used in a subsequent Multi-Criteria Analysis (see Section 2.1.7) of solutions.

This being achieved, the designer may later, having reached the stage of needing to compare solutions and select a suitable design (Section 2.1.7), propose a number of policy or design options to the interested parties. They can then judge these using their own previously defined and agreed criteria matrix. Once a decision is made on the preferred option or options, this should then be written up in a formal document of agreement that will also contain the earlier two agreed points listed above. The agreement should also take account of possible future changes in attitude of the interested parties involved. Once in place, such

Table 1 Categories of losses

	Quantifiable	Unquantifiable
Direct	Repair, replacement and rehabilitation of structure Structure-related repair, rehabilitation and replacement Failure-related repair, rehabilitation and replacement of other objects	Loss of human life Injuries Loss of irreplaceable matter Environmental damage
Indirect	Failure-related lack of production at the structure Failure-related lack of production in the vicinity of the structure Lack of production due to failure-induced disruption of economic system	Suffering and disruption of social system Stress, fear and increased susceptibility to disease

an agreement will then allow final design and detailing to proceed, covered by a general political acceptance.

2.1.2.2 Acceptable risk

The acceptable risk of failure/damage/loss of life when the design parameters for a rock structure are exceeded is a central boundary condition to any design (Vrijling, 1990), both for serviceability and ultimate limit state conditions.
According to a common definition, risk is the product of failure probability and the consequences of failure, and thereby the consequences of failure are often expressed in terms of a capital cost. The first factor, the *probability* of failure, can be defined quite objectively as the probability that the functional requirements are not met. An objective quantitative definition of the consequences, however, is not easy. The consequences are multi-dimensional and may be difficult to relate explicitly to the structure in concern. Therefore a generally agreed scale and units to measure consequences may be impossible. Examples of possible consequences (with different dimensions) are:

- Social stress
- Loss of human life
- Human injuries
- Loss of property
- Loss of investments
- Loss of (expected) future income.

These losses can be categorised as shown in Table 1.
Sometimes an acceptable risk level for these various losses is proposed by the owner of the project or by the society. A risk evaluation can be done by comparing with an agreed predefined risk level. However, subjective and ethical problems are likely in predefining this risk level, when so-called unquantifiable losses are involved. One method of assessing acceptable risk level in relation to loss of human life is to identify the *personal* risk, as described in Box 2. If an acceptable risk level is predefined, any structure being considered as an option for the project should then be capable of meeting this risk.

If no acceptable risk level is predefined, then the question of risk may be assessed using economic benefit-cost analysis (see Box 4 in Section 2.1.5), with an attempted subjective financial evaluation of such important intangible or unquantifiable costs as loss of human life. These costs can then be included in the same way as other costs such as maintenance and repair. In this method, the value of human life may be expressed by the present value of the national product per head of the population. However, the ethical problems of assigning a value to such intangibles as suffering and the loss of human life are considerable. Intangibles are therefore often omitted from the economic evaluation and compared in a separate reasoned argument.

> **Box 2** Personal risk
>
> One option is to set the acceptable probability of failure equal to the probability of mortality, accepted by the average individual. Statistical analysis of risks within a society will often give an indication of the average risk to die that is individually accepted. It is assumed that individuals balance risk and expected profit of an activity in both a rational and a subconscious way. Depending on the outcome of this process, the action is undertaken or not. A distinction can thereby be made between activities that are undertaken voluntarily and those that are not. According to this model, the degree of voluntariness can be related to the probability of a fatal accident. Figure 9 gives an indication of personal risk during a year evaluated for Western countries and for a number of activities. The figures are obtained from statistics on fatal accidents.
>
>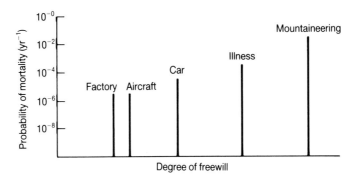
>
> **Figure 9** *Risk level in Western countries*
>
> The statistics of causes of death reveal a probability of approximately 10^{-4} for activities of average freewill. This is used as an indication for the personally accepted risk. For non-Western societies the outcome of such an exercise may differ significantly from the given figure (for example, due to geographical, cultural or economical reasons). The personal risk acceptance (P_{di}) for a particular activity can be found from:
>
> $$P_{di} = \frac{N_{di}}{N_{pi}} = \frac{N_{pi} P_{fi} P_{d/fi}}{N_{pi}}$$
>
> where N_{pi} = number of participants to activity i
> N_{di} = number of deaths with activity i
> P_{fi} = probability of accident with activity i
> $P_{d/fi}$ = probability of a death, given the occurrence of an accident

Other intangibles, such as the value of environmental damage, due to implementation of a structure can be expressed as the cost of a shadow project to mitigate the effect of adverse impacts (see Section 2.1.4.1).

2.1.3 FUNCTIONAL ANALYSIS

An essential stage in the design process is the analysis of the functions the structure has to fulfil in order to remove the stated problem. In conducting the analysis one has to see whether there are any unstated elements in the problem which should be taken care of.

The outcome of the functional analysis is a set of functional requirements for the future structure. The degree to which this structure will perform satisfactorily depends to a considerable extent on the requirements thus defined. Functional requirements can be defined in relation to the structure as a whole and in relation to the component parts of its cross-section.

The main functions which rock structures, as a whole, can potentially solve are listed in Table 2. In addition, rock structures may form the seaward boundary of

Table 2 Functions of rock structures

Function of rock structure	Rock structure providing function						
	Breakwater	Seawall	Rock groyne	Offshore breakwater	Gravel beach	Reservoir dam protection	Offshore seabed structure
(1) Shelter from waves/currents for vessels	*			(*)			
(2) Sediment trap to preserve navigation channels	*		*				
(3) Flood protection in hinterland		*					
(4) Prevention of erosion of coast or shoreline		*	*	*	*	*	
(5) Prevention of undermining of foundation							*
(6) Protection against impact, including ballasting							*

Notes
Item (3) includes reducing presently too frequent inundation of low-lying ground and properties, as well as providing an adequate defence against extreme storm-surge events.
Item (4) may include protection of erodable cliff formations, wind-blown dunes or beaches where the littoral processes are leading to denudation rather than accretion of beach material. In the latter context it is important to note that man-made influences may be leading to erosion, such as wave reflections generated by existing sea walls.

Table 3 Functions of typical component parts of a rock breakwater

Component	Function
(1) Scour protection	Prevention of erosion
(2) Core	Attenuation of wave transmission
	Support to armour
	Geotechnical stability
(3) Berm/toe	Attenuation of wave overtopping
	Provision of additional geotechnical stability
	Provision of stable footing to armour layer
(4) Underlayer	Filtration
	Erosion protection of subsoil/core
	In-plane drainage
	Regulating course
	Separation and reduction of hydraulic gradient into subsoil/core
(5) Armour layer	Prevention of erosion of underlayer by wave action
	Wave energy dissipation
(6) Crest	Attenuation of wave overtopping
	Access for maintenance
(7) Crown wall	Attenuation of wave overtopping
	Access for maintenance
	Support to facilities such as pipelines

a larger structure (e.g. cliff or sand dune) to which a major function can be assigned. Here the function of the rock structure is to prevent the larger structure from being affected by hydraulic loadings from the sea. The functional requirements for rock structures as a whole will largely determine the plan layout (see Chapter 6).

The functions of component parts of a structure are best appreciated by an example. Table 3 lists component parts of a rock breakwater (see e.g. Figure 233) along with the primary functions which they perform. Examination of the functions of these component parts reveals that they fall into two categories:

1. Functions related to the primary function of the structure;
2. Functions related to the necessity of maintaining the structural integrity of the protected structure. Thus the core of a breakwater fulfils a primary function in that it prevents or significantly attenuates wave transmission but also provides support to the armour layer and overall geotechnical stability.

Having considered the functions required for various structures and structural components, the other main aspect of functional analysis is to consider at what times, over what durations and, if appropriate, at what rates these functions need to be fulfilled. Compliance with these requirements is perhaps best expressed in terms of risk of non-performance, which, in turn, will enable them to be expressed as reliability, serviceability, risk of loss or injury and an overall management strategy. Practical examples of such translation of functional requirements include:

- *Serviceability*
  Example: the requirement that ships of a prescribed dead weight should be able to enter a harbour safely and to be serviced at the berth, except for a given average number of days per year;
- *Maintenance*
  Example: the requirement of a maximum expenditure on maintenance during a prescribed period of time.

Functional analysis should also consider possible changes of functions or magnitudes of functional requirements within the projected life of the structure so that appropriate flexibility can be provided in the envisaged structure and its planned maintenance.

Examples of the need to change or modify initial functional requirements or to introduce new functional requirements during the lifetime of a structure may include:

1. Changing hydraulic and morphologic or other boundary conditions, such as varying wave climate, rising water levels, differing scour/sedimentation rates, increasing traffic, changing availability of local materials for maintenance and labour, etc.;
2. End of the primary function of a temporary works structure upon completion of construction.

Acknowledgement of the functional requirements can be regarded as the 'functional design' of a structure.

2.1.4 GENERATION OF ALTERNATIVE SOLUTIONS

The next step in the design process is the generation of alternative design concepts to meet the boundary conditions and functional requirements. As indicated in Figure 2, this process draws on a wide range of technical experience and knowledge, much of which is summarised in this manual, and covers the following principal areas:

- Environmental assessment (Section 2.4);
- Determination of materials sources and properties (Chapter 3);
- Understanding the relevant hydraulic and geotechnical processes (Chapter 5);
- Structure-specific design methods (Chapter 6);
- Construction considerations (Chapter 6);
- Maintenance considerations (Chapter 7).

These areas are described in the following subsections in more detail in relation to the design process. In this discussion, the reader should appreciate that definition of the plan layout of a structure will normally precede definition of its cross-section (see Chapter 6 for practical examples). The process of concept generation will also highlight the need for refinements in predictions of the physical site boundary conditions, and this would normally be the stage to complete the appropriate data collection and analysis (see Chapter 4).

2.1.4.1 Environmental assessment

Environmental Assessment (EA), also known as Environmental Impact Assessment

(EIA), is a procedure whereby the likely effects of both plan layout and cross-section of proposed engineering works on the wider environment are considered as an important and integral part of the planning and design of the project. It is discussed in more detail in Section 2.4.

Environmental impacts from the use of rock in coastal and shoreline engineering will include construction and long-term impacts. Construction impacts, in common with most construction sites, will comprise:

- Quarrying and dredging of rock materials;
- Short-term impacts of materials transport;
- Noise, vibration, dust, odour and (if not controlled) pollution from equipment operations, etc.;
- Effects on local community in regard to employment, commerce, recreation and access to site.

Long-term impacts, which are often of more interest, include:

- Changes in bathymetry and landscape of construction;
- Changes in existing physical process (e.g. littoral transport, cliff erosion);
- Effect on ecology (flora and fauna) (e.g. the favourable impact of coastal rock structures providing a valuable marine habitat);
- Visual impression of the landscape and horizon;
- Social and socio-economic;
- Geological, archeological, historical and cultural;
- Pollution of air, water and soil (mostly project rather than rock structure related).

In many countries, including all those in the European Community, environmental (impact) assessment is a legal obligation for major projects and an environmental (impact) statement is required (see Section 2.4). Even where not called for by law, the designer/proponent of the scheme involving use of rock in coastal and shoreline engineering will find it cost-effective in the long term and certainly, through the liaison and consultation involved, will assist in securing public support and planning consent. Environmental assessment then becomes involved in the policy analysis (Section 2.1.2.1) process and environmental factors can be given appropriate weightings in a multi-criteria analysis (Section 2.1.5).

Environmental aspects can also be included in benefit-cost appraisal (Section 2.1.5), where they appear as the costs of a shadow project (in the case of additional measures to mitigate the effect of adverse impacts) or as benefits (where the environment is improved or protected). Examples of mitigating measures might include the following:

- Purification, for the duration of the existence of the structure, of polluted soil, water and air up to the quality that it would have had if the structure had not been built;
- Provision and maintenance of an equivalent substitute for the environmental space and systems that are lost by building the structure;
- Processing and/or storage of waste materials originating from construction and maintenance of the structure during its lifetime.

2.1.4.2 Materials availability and properties

Determination of materials availability and properties, together with the means of testing and evaluating these and taking account of production, transport and handling, are dealt with extensively in Chapter 3. Materials clearly represent a fundamental consideration in the generation of alternative solutions for a structure's cross-section.

In materials evaluation, available rock sources and types must first be established (Section 3.1) and an assessment made as to whether existing quarries can be used or whether a temporary one may need to be opened (Section 3.4). A choice must be made between local, easily available rock and that imported from a distant source. Also worth consideration in this context (see Section 3.6) are alternatives to rock (concrete, industrial by-products), rock with stability improvement (by asphalt or colloidal cement grouts) and composite rock systems (gabions and mattresses).

In determining rock properties the intrinsic properties of the rock (Section 3.4.2) such as colour, density, porosity, degree of weathering and strength will first need to be determined, density being vital for early design work and colour being an important visual environmental consideration.

Production-affected properties must also be assessed to review the practical range of weight, size, shape and grading that is available. Where rock is being produced from a permanent quarry (Section 3.3.2) it will be important both to take account of existing gradings being produced and to encourage standardisation, to select from one of the standard gradings specified in Appendix 1. Selection of standard grading of fine and light gradings of rip-rap (W_{50} under 300 kg) which are handled and placed in bulk is particularly important. Where rock will be produced from a temporary quarry opened for the purpose (Section 3.5.2), the gradings (and quantities of each) that are selected should ideally be based on blasting trials or, if these are not possible, on predictions based on a geological assessment of the *in-situ* rock. The properties that will be created by construction (execution-induced properties, Section 3.3.3), which have implications for the resulting porosity, density and layer thickness, must also be recognised.

Durability of rocks vary, and this factor may be incorporated into the design (Section 3.2.4) using the QMW rock-degradation model presently under development. This attempts to relate the rate at which rock wears to measurable rock parameters and to the aggressiveness or otherwise of the wave climate and abrasive power of bed/shoreline sediments.

Methods of testing and assessing all the above characteristics are given in Section 3.3 and the need for quality control must be recognised (Section 3.6).

2.1.4.3 Understanding the hydraulic and geotechnical processes

Understanding the physical processes which may be involved in any rock structure solution that may be being generated is clearly essential in designing both the plan layout and structure cross-section to meet the functional requirements. Chapter 5 presents a comprehensive review of all processes and design tools necessary to determine these. The designer must recognise the need to combine the bathymetric, hydraulic and geotechnical boundary conditions (Section 2.1.2 and Chapter 4) with the parameters of the structure (both materials (Section 2.1.4.2 above, and Chapter 3 and Section 5.2) and geometric (Section 5.1.2)) in order to determine the hydraulic and geotechnic response (e.g. wave run-up, pore pressures), the consequent loading on the structure and the potential resistance or otherwise of the structure to damage or failure.

2.1.4.4 Structure-specific design methods

Some of the physical processes and design methods involved may be structure-specific and preparation of a range of alternative solutions for the identified problem will necessarily draw on experience of performance of existing structures, some of which are included in Chapter 6. The designer must be fully aware of the functional possibilities and limitations of the various structures that he considers. In assessing various solutions (e.g. a seawall versus rock groynes versus offshore breakwater for a coastal erosion problem) the designer should be aware

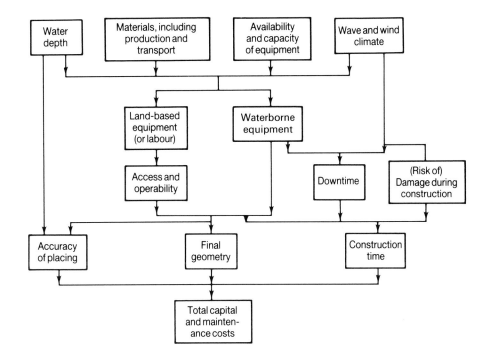

Figure 10 *Interrelationship of construction considerations in design of rock structures*

of the known potential failure modes for these structures and their component structural elements (see Section 2.2 and Chapter 6).

2.1.4.5 Construction considerations

Construction considerations may be viewed, as they are for convenience in this manual, as all those activities (primarily, in this case, placement of rock materials) taking place at the site of the structure and elaborated in some depth in Chapter 6, particularly in Sections 6.1.3.4 and 6.1.5. The temporary stability of the placed rock materials during construction is a vital consideration. Production and transport of rock materials discussed above are, however, clearly important, and interrelated issues.

The interrelationship of various construction aspects affecting the design concept are shown in summary in Figure 10. This figure illustrates the key influence of the decision as to whether to use land-based or waterborne equipment on a number of factors which ultimately influence the resultant duration of construction and project cost. This decision will be influenced by a number of initial constraints such as availability and capacity of plant to handle the selected materials, time for mobilisation/demobilisation, and accessibility and operability of the plant in the given wind, wave and water-depth conditions. It will also be affected by the likely risks of damage to the structure during construction, which are in turn significantly influenced by the selected construction sequence. Where land-based equipment (cranes and transport) are chosen, the design must make space provision for their access and operation during construction and ensure that they will be protected from wave action. The influence of water depth on accuracy of placing (see also Appendix 1) must also be recognised.

The design should also be such as to give contractors as much flexibility as possible in matters which are properly their responsibility (e.g. capacity of means of transport, fluctuations of material supply). The possibility should also be left open, where possible, for contractors to suggest alternatives to construction materials within the requirements of the basic design parameters in order to save cost in reduced basic materials prices or by efficiency of placing.

2.1.4.6 Maintenance considerations

Maintenance considerations, discussed in depth in Chapter 7, should be viewed as equally important as construction considerations in the design process and may indeed have a greater bearing on the design concept, as they are strongly influenced by the engineering and financial resources of the (future) owner of the rock structure.

Maintenance comprises all those periodic activities that are required following construction to ensure that the structure performs to an acceptable standard during its lifetime. These include inspection and monitoring, appraisal of monitoring data and the repair or replacement of components of the structure whose life is estimated to be less than the overall structure or of a localised area which is assessed to have failed.

The essential first step for the designer therefore is to determine from the owner the level of maintenance that he will be able (or wishes to) afford in terms of his own practical engineering and financial resources. While a low capital cost, high maintenance cost solution might prove to be theoretically attractive in a given situation, it cannot be adopted if the maintenance cannot be implemented.

Within this overall constraint, a range of possible solutions may be generated, setting the maximum permitted damage (or risk of damage) level appropriately in relation to the known physical site considerations. Solutions must also take account of problems of obtaining rock materials for maintenance after construction completion (e.g. allowing for stockpiling of armourstone at site) and provide adequate access for equipment to undertake repair work.

When subsequent costings have been carried out at the detailed design stage (see Section 2.1.7) it will then be possible to calculate a total whole-life cost for the structure based on the capital and discounted (net present value) maintenance (monitoring, appraisal, repair) costs. Normally, this total cost will be minimised, but the maintenance option finally adopted may be subject to owner or political choice.

Within the costings it will be important to allow during the financial period of time being assessed for any major expenses such as rehabilitation or major repair. Theoretically, it is also possible to allow for the costs of removing a structure or of the consequences of abandoning it at a certain point in time, but this is rare in practice.

2.1.5 COMPARISON AND SELECTION

The process of concept generation moves without any clearly defined boundary into the process of comparing these concepts and selecting preferred options using techniques such as Multi-Criteria Analysis (MCA) or Benefit-Cost Analysis (BCA) described below. In order to achieve this process, an outline design of each of the identified range of possible structures will be necessary using the tools described in Section 2.1.4 above. The simpler techniques available should be employed and use of more refined methods restricted to the stage of final design and dimensioning. At this stage, therefore, the designer might use rules of thumb, simple analytical methods or empirical equations and graphs, many of which are presented in this manual. Numerical and physical modelling is generally not appropriate for structural design at this stage.

After a series of concepts have been generated (for example, the range of breakwater cross-sections discussed in Section 6.1), unrealistic or uneconomic concepts can be eliminated in a selection process involving both objective and subjective criteria. Impractical concepts will be eliminated on the objective

> **Box 3** Multi-criteria analysis (MCA)
>
> The method basically consists of making a matrix, with the various alternatives listed horizontally (A to E in the tables below) and the selection criteria listed vertically. An appreciation of alternatives is made with respect to each criterion. The appreciations are expressed in a mark using a predefined scale (e.g. 0,1,...,10), as shown in the limited example table below.
>
> Example of an MCA (weight factors are given in the table below)
>
> | CRITERIA:/alternatives: | I | II | III | IV | V | Weight |
> |---|---|---|---|---|---|---|
> | A Rock volume | 2 | 5 | 1 | 1 | 8 | 1 |
> | B Environmental impact | 7 | 7 | 6 | 2 | 2 | 4 |
> | C Construction time | 9 | 0 | 5 | 4 | 1 | 2 |
> | D Maintenance | 8 | 8 | 3 | 1 | 1 | 3 |
> | E Risk level | 6 | 3 | 10 | 7 | 5 | 3 |
> | F Initial cost | 6 | 5 | 3 | 6 | 6 | 3 |
> | Resulting appreciation: | 108 | 81 | 83 | 59 | 54 | |
>
> Introduction of weighting factors improves the method, and these factors can be adjusted by agreement (e.g. within the project management team or with a group of interested parties as part of a policy analysis process). A more objective approach to determining weight factors which can also be carried out by the project management or policy analysis team is to assign priority to one or all possible pairs of criteria. For example, assigning 1 to each dominating criterion, thus leaving the other with 0, weight factors can be found by adding all 1's for each criterion (instead, use of 2, 1 and 0 also anticipates equally important criteria).
>
> Weight factors for MCA table above
>
> | Criteria | A | B | C | D | E | F | Weight |
> |---|---|---|---|---|---|---|---|
> | A | — | 0 | 0 | 1 | 0 | 0 | 1 |
> | B | 1 | — | 1 | 1 | 0 | 1 | 4 |
> | C | 1 | 0 | — | 0 | 0 | 1 | 2 |
> | D | 0 | 0 | 1 | — | 1 | 1 | 3 |
> | E | 1 | 1 | 1 | 0 | — | 0 | 3 |
> | F | 1 | 0 | 1 | 0 | 1 | — | 3 |

engineering criteria of performance (capability to meet the functional requirements assessed using the tools described in Chapters 5 and 6), materials availability, construction time and equipment available, and maintenance, discussed in Section 2.1.4 above.

Uneconomic concepts will be eliminated on the basis of (objective) comparisons of whole-life costs (including capital and discounted maintenance costs) calculated as described in Section 2.1.7. The selection process will also include more subjective criteria such as:

- Risk-level assessment during construction (see Section 2.1.4.5) and operation, compared with acceptable risk levels (Section 2.1.2.2);
- Comparison with political, social and legislative conditions (Section 2.1.2);
- Environmental impact assessment (Section 2.1.4.1);
- Complexity of operation and maintenance relative to local technological experience and resources

Policy analysis criteria established with interested parties as described in Section 2.1.2.1 can be used here, together with the objective criteria, to carry out a screening of the more economic solutions using MCA (see Box 3). The advantage of MCA is that it is the only known method of including both quantitative and qualitative criteria in one assessment.

Having used minimum cost and MCA principles to reduce the number of options

> **Box 4** Benefit-cost analysis (BCA)
>
> Benefit-cost analysis assesses economic 'worthwhileness' by comparing the 'present values' (PVs) of streams of future costs with future scheme benefits calculated using a test discount rate approximately equal to the real interest rate (market interest rate minus inflation rate, generally 4-6% in most European countries in 1991).
>
> The benefits of a scheme are generally taken to be the difference in economic assets with and without the scheme, and in assessing these care must be taken to avoid double-counting. In the case of government-funded projects such as seawalls, the economic assets to be considered are the national assets, and may include quantification of intangibles (see Section 2.1.2.2).
>
> The present value of costs should include:
>
> - All costs associated with initial construction;
> - All maintenance and other operational costs; and
> - The possible cost of major repairs.
>
> Benefit-cost analysis then allows the comparison of all possible different schemes, and the comparison of any one scheme with the 'do nothing' situation, on the basis of those benefits that are calculable.
>
> A number of measures of scheme-worthwhileness may be computed and used for comparison. Where alternative schemes have the same benefits then a direct comparison of the costs, on a present value basis, is adequate. If alternative schemes have different benefits then the difference between the present value of benefits and costs—the net present value—should be used as the basis for comparison. The ratio of benefits to costs may be considered for economic comparison of alternatives. International agencies often use the internal rate of return (the discount rate at which present benefits equals present cost) which is derived from the investment.
>
> *Loss and damage probability*
>
> If an event occurs of a severity greater than that for which the structure is designed, thus involving loss or damage that could not be predicted, then it can be assumed that such an event has an equal chance of occurring in any year in the future. The present value of such loss or damage should be assessed on the basis of appropriate probability relationships and deducted from the benefits of the scheme.
>
> *Sensitivity analysis*
>
> There may be a considerable variation in the reliability and accuracy of input data for economic appraisal. In many instances adequate data are not available and assumptions have to be made based on engineering judgement. Where there is doubt as to the impact of likely variations in data (or a range of assumptions is possible) then the sensitivity of the results to such variations should be examined.
>
> Upper and lower likely data values should be used to evaluate the upper and lower results that could be obtained. Where the results are found to change significantly within the likely range of data, then refinement to the data can be sought while those results showing only tolerable change can be adopted without further effort. Such a sensitivity analysis leads to both greater effectiveness in data collection and a better appreciation of the strengths and weaknesses in the cost and benefit comparisons made.
>
> In making a comparison between initial construction and subsequent costs, the relative accuracy of the estimates should be considered. Operational and replacement costs are usually less well defined than intial construction costs both in timing and extent.
>
> In the case of projects with a particularly long construction period, especially those including a substantial foreign-exchange element, the justification should be sufficiently robust to withstand reasonable exchange-rate fluctuations.

to the two or three most promising, a more detailed design and costing can be carried out of these solutions using the methods described in Sections 2.1.4 and 2.1.7 and the final solution selected using a minimum whole-life cost approach or an MCA. At this stage, a more rigorous financial–economic appraisal can be undertaken and the whole-life costs (capital plus discounted maintenance and operations costs) compared with the benefits that will accrue in order to assess the economic 'worthwhileness' of the project, using a technique known as Benefit-Cost Analysis (BCA). BCA is described in Box 4 and discussed in much more detail in the forthcoming CIRIA/Butterworth–Heinemann publication *Seawall Design* to which the reader is referred for more information.

2.1.6 FINAL DESIGN AND DETAILING

Having selected an appropriate solution to the identified problem, the final design and detailing can proceed, taking into account all the previous design thinking. At this stage, further alternatives may appear, but these will generally be minor variations on the basic option that has been selected arising from the interplay of optimisation of functional efficiency and minimisation of total cost. Variations and adjustments will tend to concentrate on minor details of plan layout and detailing of the cross-section. The final design essentially consists of a series of calculations and model tests to check and adjust as necessary all details of the structure and to produce contract documents and a design report.

Before commencing the final design, a decision must be taken as to whether to proceed on a deterministic or probabilistic basis. Preliminary designs at the stage of generation of alternative solutions will preferably have been carried out using deterministic or simple probabilistic methods, whereas in the final design process a more thorough probabilistic approach may be adopted. Probabilistic methods are described in Section 2.3 and have the advantage of providing the designer with a quantifiable list of probabilities, the interrelation of which is identified in a so-called 'fault tree' or 'failure tree' (see Section 2.2). A knowledge of the significance of an individual failure mechanism in relation to the overall functioning of the structure is particularly important in the structure optimisation process.

The hydraulic and geotechnical tools used to check and adjust the hydraulic and structural performance in the final design will be a combination of established theoretical and empirical approaches along with numerical and physical modelling (see Chapter 5). The calculations and model tests will have the objective of ensuring that the final structural design meets all the functional requirements, given the physical site conditions and other boundary conditions. In this latter regard, it will naturally incorporate all the latest information on boundary conditions, particularly in relation to physical site conditions (results from surveys commissioned earlier in the design process may only become available at this stage).

The process of checking and adjusting the selected design will involve each of the sub-processes involved in the stage of generation of alternative solutions (see Section 2.1.4), but carried out to a greater degree of refinement. The evaluation of the hydraulic and structural performance is carried out for each of two categories of performance, in each case related to a limit state (a limit state being a loading condition), the exceedance of which will lead to a significant decrease in performance. Probabilities of failure hence represent probabilities of exceeding a given limit state. The two categories of performance are:

1. *Performance under extreme conditions:* Here the ability of the structure to survive extreme conditions is checked. This is done by evaluating all failure mechanisms likely to occur under the specified extreme conditions. In this case, the limit state is the ultimate limit state (ULS). This limit state defines collapse or unacceptable serious deformations of the structure (e.g. sliding, breakage or removal of armour elements) for conditions exceeding the ULS.
2. *Performance under normal conditions:* Here the performance is evaluated under the 'normal' or daily conditions to which the structure will be exposed during most of its lifetime. In this case, the limit states are the serviceability limit states (SLS). These limit states define (mostly hydraulic) conditions, the exceedance of which will disable various activities or services provided by the structure (e.g. unacceptable ship movements due to wave penetration into a harbour basin). In addition, the long-term performance of the structure under 'normal' conditions (analogous to fatigue) will be evaluated here (e.g. degradation of a scour protection, deterioration of armour elements).

In ensuring that both the whole structure and its component parts have complied with ULS and SLS requirements, it is suggested that a list of aspects be prepared and a check made to ensure that limit state criteria for each aspect are satisfied. Such a checklist might include for a typical rock structure such aspects as:

- Overall plan geometry (e.g. side slopes, crest level)
- Armour (seaward face, toe, crest, rear face)
- Underlayers and filters
- Core design and foundation drainage
- Arrangements at limits of/transitions in structure

  with limit state criteria including:

- Run-up, overtopping and reflection of waves
- Armour stability
- Filter criteria
- Pore pressures for geotechnical stability, wave transmission, allowance for settlement
- Avoidance of outflanking.

On completion of design and detailing, there will be two main products: a design report and a set of contract documents. The design report will contain a summary of the design process as described above but specific to the structure in question, explaining the reasons for the various choices made. It will generally contain within it the following key components:

- Description of selected structure and selected process
- Materials to be used, reasons for selection and anticipated method of production and transport to site
- Description of how the selected structure meets the functional criteria up to defined limit states
- Probabilities of failure in various hydraulic and geotechnic failure modes, ideally linked by a fault tree
- Construction methods and equipment envisaged
- Description of maintenance strategy agreed with future owner
- Environmental (impact) statement
- Cost estimates
- Economic benefit-cost justification.

The contract documents will be of a standard form and will contain drawings, specifications, bills of quantities and conditions of contract. Both specifications and bills of quantities should, as far as possible, be prepared on a common basis throughout the quarrying and production industries in order to avoid unrealistic or impractical requirements or measurement techniques being imposed, while still retaining proper control over the quality of the construction work. As indicated in Box 5, use of Appendices 1–3 will ensure that such a standard approach is adopted.

2.1.7 COST ASSESSMENT

As for all projects, a fundamental principle in the design of rock structures in coastal and shoreline engineering is usually the minimisation of total cost within the limits of the functional requirements and boundary conditions. Normally, all costs are expressed in terms of their present values using an economic technique called discounting (see Box 4). This permits the calculation of a total whole-life cost based on the capital and discounted mean annual maintenance (monitoring, appraisal, repair) costs together with the discounted costs of any major repairs or rehabilitation works expected during the lifetime of the structure.

Normally, this whole-life cost will be minimised by the design, and this may suggest that a reduced capital cost and more frequent maintenance expenditure is the economic optimum. However, obtaining funding for maintenance is often time consuming and complicated, and owners or governments may prefer greater than optimum initial investments giving more safety as a matter of policy choice (see Sections 2.1.2.1 and 2.1.4). The calculation of repair costs (described in Box 6) is also often a very uncertain matter because of the number of variables involved, and this uncertainty may also influence policy choice.

Box 5 Rock materials specifications and bills of quantities

The specification and measurement for payment of rock armour in structures has historically been one in which there has been no standardisation and few agreed areas of common approach. The quarrying and contracting industries were thus presented with a bewildering array of different and ill-defined specifications for rock materials, particularly in relation to the geometric requirements for rock shape and grading. As a result, quarry producers have not been able to achieve the cost/price savings which might be possible from stockpiling if standard gradings and rock shapes were used. Equally, contractors have been obliged to place rock in some very different ways, some highly impractical.

It was one of the main objectives of this manual to seek to rectify this situation by providing the industry as a whole with a common framework and database with which to work. The background to this framework is given in Chapter 3 (particularly Sections 3.2, 3.3 and 3.6) and the framework itself is expressed in the following appendices to this manual:

1. Model specification for quarried rock applications in coastal and shoreline engineering—the basic framework.
2. Standards for quarried rock materials for use in coastal and shoreline engineering—the supporting materials tests.
3. Measurement of quarried rock in coastal and shoreline engineering—the method of preparing bills of quantities consistent with Appendix 1.

The framework can be viewed at a number of levels:

1. The basic level of providing common definitions and methods of testing, measuring and surveying, rock and rock structures, specifically covering the following points:
    - Intrinsic properties, providing (mainly in Appendix 2) both new tests and those adapted from tests originally developed for aggregates, including tests for: density, freeze–thaw resistance, dynamic crushing resistance, water absorption, 'Sonnenbrand' in basalt, presence of clay minerals;
    - Production-affected properties, providing standard definitions and methods of determining such fundamental geometric properties as: relationships between rock size parameters and weight, rock shape (by a definition of axial ratio L/d), rock grading;
    - Construction-induced properties, such as: layer thickness, porosity, and tonneage per cubic metre placed.

   The last set of properties is based on a standard method of surface profile measurement and providing the basis for the suggested standard method of measurement for payment (i.e. preparation of bills of quantities), making allowance for settlement and the need to categorize work in relation to tidal levels.
2. The next level of providing guidance of recommended or desirable values for the defined properties in list 1, including tolerances on surface profile and later thickness.
3. The final level of recommending standard gradings for adoption in production processes in permanent quarries, which will then be available rapidly and economically from quarries when needed for works in coastal and shoreline engineering. This final level of standard gradings must not be adhered to without thought, and some situations will dictate use of non-standard gradings, particularly:
    - In a temporary quarry opened up for a specific project where the design should be tailored to the expected yield of rock of various sizes;
    - Where local permanent quarries have highly variable seams and/or little capacity for stockpiling and yet can offer an economic and useable source of rock.

The contractor is also given sufficient guidance to enable him to propose alternative rock sources of different density and durability for inclusion in the works, so long as he complies with basic hydraulic stability requirements. A corresponding method of adjusting payment rates is given in Appendix 3.

Examination of all the costs involved in a rock structure project suggest they can normally be broken down into the following approximately chronological categories:

- Initial engineering:
    Feasibility study
    Design and design report
    Contract documents and maintenance Manual

- Construction (materials, equipment and labour):
    Rock production at quarry
    Transport of rock
    Placing of rock

- Maintenance:
    Monitoring
    Appraisal
    Repair
    Rehabilitation or replacement.

> **Box 6** Calculation of repair costs
>
> The contribution of discounted repair costs to the overall whole-life cost of a rock structure can be seen to be a function both of the frequency of the repair activities and to the design strength (R)/loading (S) ratio. In addition, the larger the uncertainties or spread in the probability distributions of loading (S) or the strength (R) expressed respectively by their standard deviations σ_S and σ_R, the greater the overlap in their probability distributions and hence the larger the expected damage.
>
> The cost involved with the necessary repair on each occasion may be approximated as follows:
>
> $$C_r = C_0 + C'_e D\{f_m, S, R\}$$
>
> where C_r = cost of repair
> D = damage or volume of rock for repair/maintenance
> C_0 = initial cost (mobilisation and demobilisation of repair equipment)
> C'_e = repair cost per unit volume of rock
> f_m = frequency of repair maintenance
>
> The cost of mobilisation (C_0) principle is a constant for each unit of equipment. The expected frequency of repair or maintenance (f_m) is not an independent variable and also depends on S and R. It will be clear that, taking this into account, the volume of rock material to be replaced (D) and the cost involved will be largely determined in the design through the choice of S and R.
>
> When costs with low probabilities have to be estimated the uncertainties σ_S and σ_R will be the principal contributions to these (low-probability) costs. The cost (C'_e) will depend on a variety of (related) parameters, the most evident of which are:
>
> - Type of equipment used;
> - Capacity of equipment;
> - Accessibility of structure (geometry);
> - Material characteristics;
> - Spatial distribution of damage.
>
> Repair costs calculated from the above equation can be incorporated into the overall stream of costs at appropriate time intervals given by ($1/f_m$). It may be possible to determine a minimum total repair cost if sufficient information is known about all the cost and other factors involved.

To these costs often have to be added the interest charges involved in financing a project. The relative balance of the cost components are not only project- and site-specific but will also be affected by the economic conditions in the countries in which (or from which) engineering, material production, construction, maintenance and financing resources originate.

Another way of describing the component parts of the total cost is to divide this into those elements which are fixed and those which are variable, the latter either related to time expended or to quantity of material handled. In terms of construction activities involving rock, a detailed breakdown along these lines is given in the tables for the formatting of bills of quantities in Appendix 3 and these categories are summarised in Table 4.

A typical overall breakdown of some fixed and variable costs for a breakwater is given in Box 7. Further detailed information on costs is given in Section 3.5 in relation to production and transport and in Section 6.1 in relation to construction at site.

Generally, cost assessment should concentrate on those items that give the largest contribution to the total cost and to elements that are subject to important uncertainties. For example, time-related charges may be important when there is a risk that the programme will not be met. Again, the economic consequences of higher or lower production rates may have to be considered where planning and cash flow consequences for the owner are critical. In this situation, a reduced construction time at the expense of higher production cost may still be economical when the overall economics of the owner are considered.

In the overall design process, cost optimisation takes place at different levels and

Table 4 Fixed and variable costs relating to rock construction

Fixed costs
Opening/closing of dedicated quarry
Mobilisation/demobilisation of floating/land-based plant
Establishment/removal of accesses
Trials and testing

Variable (time-related) charges
Maintenance of dedicated quarry
Maintenance of floating/land-based plant
Maintenance of accesses
Designer's quarry inspections

Variable (quantity-related) charges
Excavation
Excavation ancillaries (e.g. trimming of slopes)
Filling (e.g. placing rock in bulk or individually)
Filling ancillaries (e.g. trimming slopes, placing geotextiles)

Box 7 Relative costs in a typical rubble-mound breakwater

An example of the distribution of construction cost across the main cost items is shown in Figure 11. The project concerned was a breakwater in the Near East, constructed with $3.6 \times 10^6 \, m^3$ quarry run 0–1 t in the core (C), $1.6 \times 10^6 \, m^3$ quarry rock (1–5 t) in underlayers (U) and $1.4 \times 10^6 \, m^3$ concrete blocks (B). Cost items refer to opening of the quarry (Q), materials production (C, U and B) and transport, including mobilisation (TC, TU and TB). It should be noted that relative figures will differ considerably for other projects.

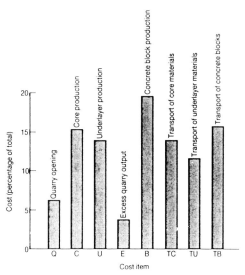

Figure 11 *Distribution of rubble-mound breakwater construction costs*

---

different phases of the design (Burcharth and Rietveld, 1987) which can loosely be divided into three phases. The first phase commences at the stage of generation of alternative solutions (Section 2.1.4) and involves the cost analysis of alternatives using unit prices per type of structure. Selections based on cost estimates made during this phase often have the most significant influence on the final cost of the project. The second phase is carried out when the basic structure type has been selected and involves the optimisation of the structure itself. Design and construction (production, transport, phasing) costs are estimated with greater accuracy than maintenance and repair costs (see comment in Box 4) because of the uncertainties involved with the latter (see Box 7). In this second phase, the cost of design, construction and maintenance can be presented as in Figure 12. In assessing the optimum cost, one should not only consider the minimum of curve A but also the risks due to errors in the estimates at this stage, and discontinuities due to

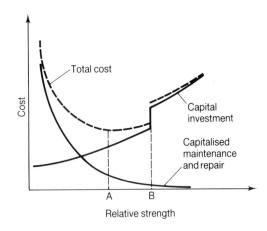

Figure 12 *Cost optimisation*

transition to the more expensive equipment (B). In view of the uncertainties, it is often advisable in case of a flat curve to take a position just left of point B, even if this is not the absolute minimum. This then secures the advantages (see above) of a slightly higher capital cost and lower maintenance cost without incurring the penalty of the additional equipment cost. The third and final phase of the cost optimisation is related to the construction and involvement of the contractor's capabilities and execution experiences to arrive at the optimum procedure. This is normally achieved in a tendering procedure based on contract documents prepared by the designer.

2.1.8 QUALITY ASSURANCE

Quality assurance is a style of management or a management philosophy which, when properly applied, affects every aspect of working life. It was originally developed for the manufacturing industry, but its application within the construction industry is now well established and growing. Quality assurance must therefore be covered in a manual such as this, although space precludes more than a brief treatment. The reader is strongly advised to consult the relevant national and international standards on the subject. The international standards on quality systems are the ISO 9000 series (9000–9004) which is identical to the European EN 29000 series. The internationally accepted quality vocabulary used in these standards is set out in ISO 8402. To assist the reader, some of the key definitions contained within ISO 8402 which are essential to the understanding of the concept of quality assurance are listed in Box 8.

In terms of the use of rock in coastal and shoreline engineering, the *quality loop* shown in Figure 13 is seen to contain all the activities listed in the logic diagram in Figure 2. The *quality system*, set up using the principles laid down in the international standards, will ensure that quality is assured. Central to the application of quality assurance to specific projects involving uses of rock in coastal and shoreline engineering is the drawing up of the relevant quality plan or plans (see Boxes 9–12). These plans should define:

1. The quality objectives to be attained;
2. The specific allocation of responsibilities and authority during the different phases of the project;
3. The specific procedures, methods and work instructions to be applied;
4. Suitable testing, inspection, examination and audit programmes at appropriate stages (e.g. design, development);
5. A method for changes and modifications in a quality plan as projects proceed;
6. Other measures necessary to meet objectives.

For full implementation, quality assurance will need to be adopted by the client/

Box 8 Key definitions (ISO 8402) related to quality assurance

Quality: The totality of features and characteristics of a product or service that bear on its ability to satisfy stated or implied needs.

Quality loop; quality spiral: Conceptual model of interacting activities that influence the quality of a product or service in the various stages ranging from the identification of needs to the assessment of whether these needs have been satisfied.

Quality policy: The overall quality intentions and direction of an organisation as regards quality, as formally expressed by top management.

Quality management: That aspect of the overall management function that determines and implements the quality policy.

Quality assurance: All those planned and systematic actions necessary to provide adequate confidence that a product or service will satisfy given requirements for quality.

Quality control: The operational techniques and activities that are used to fulfil requirements for quality.

Note: Quality control involves operational techniques and activities aimed both at monitoring a process and at eliminating causes of unsatisfactory performance at relevant stages of the quality loop (quality spiral) in order to result in economic effectiveness.

Quality system: The organisational structure, responsibilities, procedures, processes and resources for implementing quality management.

Quality plan: A document setting out the specific quality practices, resources and sequence of activities relevant to a particular product, service, contract or project.

Design review: A formal, documented, comprehensive and systematic examination of a design to evaluate the design requirements and the capability of the design to meet these requirements and to identify problems and propose solutions.

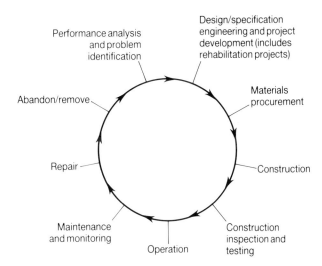

Figure 13 *Quality loop for rock structures in coastal and shoreline engineering*

Box 9 Main aspects of a typical client/owner quality plan

- The quality objectives of the client/owner
- The planning policy and management objectives for the project (Section 3.2)
- The allocation of responsibilities within the client's organisation
- The contractual arrangements to be adopted with the designer and contractor
- Selection of designer/contractor
- Agreements required with designer/contractor on quality assurance

Box 10 Main aspects of a typical designer's quality plan

- The designer's quality objectives
- The allocation of responsibilities within the design organisation
- The procedures to be adopted:
  (a) To determine conditions at the site
  (b) To carry out functional analysis
  (c) To generate alternative solutions
  (d) To compare and select solutions
  (e) To carry out final design and dimensioning
- Definition of appropriate testing and inspection regimes to be included within any contractual arrangements with the contractor
- Method for validation and review of the design
- Method for design change control

In addition, insofar as the designer is procuring on behalf of the client a contractor for the works, the plan will include:

- Requirements for specifications, drawings and contract documents
- Section of qualified contractors
- Agreement on verification methods
- Provisions for settlement on quality disputes
- Receiving inspection plans and controls of the works
- Receiving quality records of the works

Box 11 Main aspects of a typical contractor's quality plan

- The contractor's quality objectives
- Requirements for materials to be purchased, including drawings where appropriate
- Selection of qualified suppliers of rock and other materials
- Agreements with suppliers on quality assurance, verification methods, settlement of quality disputes, receiving inspection plans, controls and quality records
- Planning for controlled construction
- Construction capability, including materials, equipment, procedures and personnel
- Supplies, utilities and working environments
- Materials control and traceability
- Equipment control and maintenance
- Documentation
- Construction change controls
- Verification of incoming materials, construction processes, *in-situ* construction verification
- Control of measuring and test equipment
- Assessment of non-conformity
- Maintenance strategy

Box 12 Main aspects of a typical quarry operator's quality plan

- The quarry operator's quality objectives
- Purchaser's equipment
- Planning for controlled quarrying
- Quarrying capability, including equipment, procedures and personnel
- Supplies, utilities and working environments
- Materials control and traceability
- Equipment control and maintenance
- Documentation
- Quarrying technique/material change controls
- Verification of quarrying methods and produced material
- Assessment of non-conformity
- Corrective action plan in the event of non-conformity

owner, the designer, the contractor and the operator of the quarry supplying the rock. Each of these organisations will require an appropriate quality system.

2.1.8.1 Quality control

Quality control systems are essential as a part of the quality assurance measurement philosophy, in order to be able to guarantee successful completion and lifetime functioning of a rock structure. A typical quality control system will consist of four main elements:

1. A set of specification standards;
2. Measuring systems and procedures;
3. Quality control or comparison of standards and results of measurements;
4. Procedures to correct or change the production process.

These elements are schematised in Figure 14.

The principles of the above system should be applied from the design process through the construction to the management of the revetment system. The processes that can, from the designer's viewpoint, be subjected to quality control systems are listed in Table 6. It should be noted that these processes usually consist of several sub-processes. Often, many of the (sub)processes are, to some extent, interrelated. Each of these may involve a separate party in the project (e.g. contractors, materials producers, transport companies—see Table 5). Such parties may have their own internal quality assurance systems (see Boxes 9 to 12).

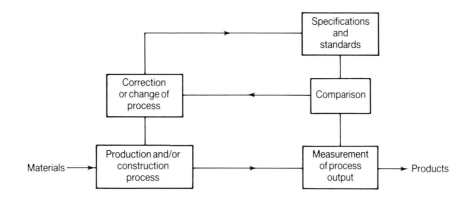

Figure 14 *Principle of a quality control system*

Table 5 Process involvements by parties to a rock structure project

Process	Constituent processes	Parties involved
Design	Data collection	Survey company
	Physical modelling	Laboratory
	Numerical modelling	Consultant
	Making of drawings	Contractor/consultant
	Analysis of construction methods	Contractor/consultant
	Economic analysis	Contractor/consultant
	Legislative implementation	Consultant/authority
		Owner/legal consultant
Material production	Quarrying	Quarry company
	Industrial processes	Industrial company
Material transport	Loading/unloading	Transport company
		Contractor
Construction	Material placement	Contractor
	Measurements	Contractor/
		Survey company
Management	Maintenance	Contractor/owner
	Repair	Contractor/owner
	Monitoring	Survey company/owner

Table 6 Quality control system for rock structure projects

Process	Input	Output	Standards
Design	Concepts Ideas Formulae Models Designers	Design data Drawings Materials Specification Manual for management	Experience Economy Formulae Performance data
Physical modelling	Site data Materials Scale(s) Instruments	Scale model Data for verification of design Design data Scale effects	Experience Scale relations
Numerical modelling	Site data Other data Theories Schematisations Numerical scheme Constants Coefficients	Design data Numerical effects Other data	Measurement data Reference cases Reference models Literature Verification Reports
Material production	Conditions on production site Rough material Equipment Labour Energy	Construction materials Fabrication materials	Materials specifications Standard test procedures Reference projects
Material transport	Materials Equipment Distance Conditions Labour	Materials Capacity	Equipment standards Material-handling procedures
Product fabrication	Materials Machinery Equipment Conditions Working method	Materials Construction Elements	Test specifications Product checklist Storage specifications
Placement	Materials Products Equipment Environment Geometry Energy Labour	Materials arrangement Cross-sections Geometry Alignment (Un)evenness	Drawings Procedures Tolerances
Monitoring	Procedures Equipment Instruments Data system Labour Energy	Displacements Geometry Materials state State of environment Other damage Loading data	Procedures Manuals Design report Data bank Instructions
Maintenance	Materials Equipment Labour Energy	Geometry Materials (Re)arrangement alignment Cross-sections	Drawings Specifications Procedures Tolerances Management Manual
Repair	See 'Maintenance'	See 'Maintenance'	See 'Maintenance' Design report (Emergency) Instructions Repair scenarios

A contractor's internal quality control system can be used by the designer to assess the probability of a proper realisation of the construction process, i.e. that specifications will be answered and that construction will proceed according to the programme. The designer's own quality control system may, for example, be concerned with the reliability of the boundary conditions (obtained by measurements, numerical modelling and statistical analysis) and other data to be used for design and with accurate recording of data, specifications, drawings, procedures, etc. Organisations should have their internal quality assurance manual available for other parties involved with the project.

In order to provide for a proper quality control in a rock structure project, the project quality assurance system and organisation should be capable of integrating the various quality control systems listed in Table 6. It should be noted that contract documents produced at the detailed design stage should provide a sound basis for quality control during the construction contract. As such, the model specifications (Appendices 1 and 2) and method for preparing bills of quantities (Appendix 3) will provide a useful basis. Further information on quality control of rock materials and placements is given in Section 3.7.

2.2 Structure types and failure modes

Generation of design concepts (Section 2.1.4) is based on both the functional requirements and the experience and creative thinking of the designer. An important criterion in selecting alternatives for further development into well-defined structural concepts is the failure risk involved in the various alternatives, and the relation of this risk to their corresponding benefits.

In Section 2.2.1 the main categories of structures are introduced, which are covered by this manual, i.e.:

- Breakwaters
- Seawalls (including scour mats), groynes and shore-protection breakwaters
- Reservoir dam face protection
- Gravel beaches
- Offshore bottom or scour protection.

In Section 2.2.2 the parameters of the principal failure mechanisms of these structures are described and in Section 2.2.3 the principle of a fault tree analysis is given, fault tree analysis enabling the relevance of the various mechanisms and their mutual relations to be assessed.

2.2.1 STRUCTURE TYPES

In a preliminary investigation a number of concepts are generated. These are often very sketchy, which may be selected to be developed in well-defined structural concepts. This manual is confined to concepts which consist entirely or partly of rock materials. It will be clear that only some principal concepts can be treated and that for any specific case variations in these concepts may prove to be feasible. The following categories of structures are treated in Chapter 6.

Breakwaters

Both the alignment and the cross-section of a breakwater affect, to a certain extent, the hydraulic loading of the coverlayer. Moreover, the bulk volume of a breakwater is mainly determined by these geometrical characteristics. In many cases breakwaters are exposed to relatively heavy wave loadings because of their protruding situation. Given the strong dependency of the required armour strength on the wave height, high demands must often be made upon the armour elements, construction techniques and equipment. Depending upon the specific function of the breakwater, overtopping may or may not be allowed, which has important consequences for the design of the structure. The most common breakwater concepts incorporating rock, described in detail in Section 6.1, are (Figure 15):

- Conventional rubble-mound breakwater
- Berm breakwater
- Reef breakwater
- Low-crested/submerged breakwater
- Caisson breakwater on rock foundation
- Composite caisson/rock breakwater.

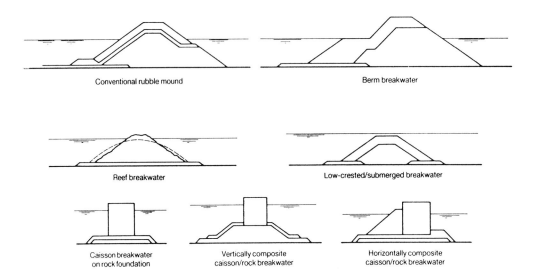

Figure 15 *Some basic breakwater concepts*

In relation to composite breakwaters only those aspects of the design which relate to the rock are treated here.

Coastal/shoreline defence structures (seawalls, etc.)

In this category are included seawalls (dikes, revetments, etc.) and other littoral coastal protection structures such as groynes and forms of breakwaters. Characteristics for all coastal and shoreline defence structures are similar to those of the land, both in relation to functions and for construction. These structures usually border on shallow water with the corresponding hydraulic loadings.

Seawalls have been constructed with a wide variety of materials and cross-sections. The most common types of seawall cross-sections incorporating rock are described in Section 6.2 (Figure 16):

- Slope protection (with or without berm) to erodible coast or sea defence
- Reclamation bund
- Rehabilitation mound to existing vertical wall
- Anti-scour mat to existing vertical wall.

Only the design of the rock elements of seawalls is treated in this manual.

Groynes and various forms of breakwater used for coastal/shoreline protection employ the basic breakwater cross-sections (Figure 15), mostly using the non-composite forms and often with the cross-section simplified due to their more modest scale.

Dam face protection

Water-retaining dams created out of mounds of relatively impermeable but wave-erodible material need face protection, which can be rock used in a similar way to its application to seawalls. The design is, however, governed by the needs to retain often considerable depths of water and to predict wave conditions before the reservoir is filled, and the difficulties of subsequent maintenance. This manual therefore treats design of dam face protection in its own right (Section 6.3).

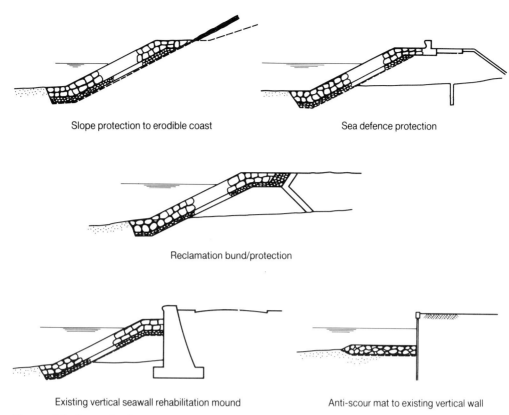

Figure 16 *Some basic seawall concepts*

Gravel (or shingle) beaches

Gravel beaches often act as coastal defence structures and dredged gravel materials may be supplied artificially to maintain a prescribed beach profile (see Section 6.4). A major characteristic of gravel beaches is that movements of individual stones are accepted and consequently there is a dynamic development of the beach profile.

Bottom protection

Bottom or scour protection is often needed to prevent instability of offshore oil platform foundations due to local scour. In addition, protection of the erodible bottom itself can be necessary because it contains and protects, for example, an offshore pipeline. The principle rock-based concepts for bottom protection are:

- Rock-type (single- or multi-layered);
- Container-type mats (e.g. gabions).

2.2.2 PRINCIPAL FAILURE MODES

This section describes some principal initiating failures for the various failure mechanisms or modes of rock structures as far as these are relevant to rock structures. Then in Section 2.2.3 these failures are built into the concept of fault tree analysis.

Failure can be simply defined as the exceedance of a predefined limit state (Section 2.3.1 and 2.1.6), which occurs when the loading exceeds the strength. When this exceedance occurs, a failure response of the structure (or parts of it) can be defined (Figure 18).

Failures can occur during both construction and operation. Typical loadings and responses for rock are wave height and displacement relative to the as-built

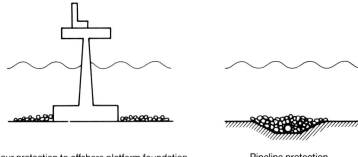

Scour protection to offshore platform foundation   Pipeline protection

Figure 17 *Applications of rock scour or bottom protection*

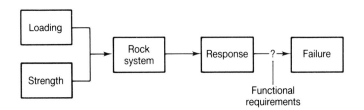

Figure 18 *Response and failure of a rock system*

position. Both loading and response are functions of time. The response is determined by the loadings and by the rock system characteristics such as weight and shape. Also, the loading may, to a certain extent, be affected by the system (for example, through permeability). The loadings are mainly determined by the (hydraulic and geotechnic) boundary conditions. The description of the physical boundary conditions is given in Chapter 4 and their use as loading descriptors in the design formulae for various types of structures (interactions and responses) is treated in Chapter 5.

In Figure 19 a rock system is schematised with its hydraulic and structural responses (Section 5.1). This schematisation or response model applies to elements of the structure (stones, sublayers) or to the structure as a whole. Failure occurs when the response exceeds a value (ULS or SLS, Section 2.1.6) which relates to the functional requirements of the structure. Failure thus corresponds to a defined loading, the failure loading. In general, failure mechanisms are named after their consequent displacements or movements. Failure modes or mechanisms are thus characterised by a relatively large increase in response due to a minor increase in loading. An overview of the principal failure mechanisms for rock structures and corresponding loadings is given in Figure 20.

Unfortunately, only few of the failure modes given in Figure 20 can, at the present state of the art, be properly described in terms of well-defined limit states of loads and responses. The failure modes which can at present be modelled to a certain extent will be described below. For each mode the characteristic loadings, the principal and secondary loading parameters, the governing system characteristics and the resulting responses are given first. This is followed by a brief description of the failure mechanisms. A summary of the failure mechanisms and their characteristic parameters is given in Table 7.

Settlement

Loading(s): weight.
Loading parameters (principal): specific density of materials.
Loading parameters (secondary): pore(water) pressures, time.
System characteristics: soil compressibility and permeability, thickness of compressible layers.
Response(s): crest lowering and horizontal deformations.

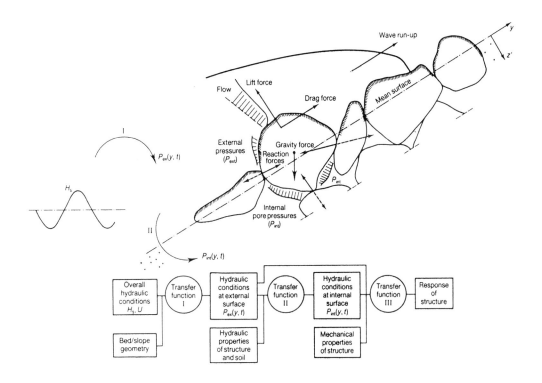

Figure 19 *A rock system and its responses*

The weight of a structure causes an extra load on the subsoil, which may then be compacted or squeezed, either instantaneously or (for low-permeability compressible layers) in a retarded way. In karstic terrain a further consequence may be the collapse of underground cavities. In addition, the structure itself may become densified during construction or in the first stages of its operation.

As a consequence of all the above processes, the crest level settles and the structure's capability to limit overtopping under conditions of high water levels and wave attack is reduced. Differential settlements lead to uneven surfaces, which make some rocks more susceptible to being washed out, and to undermining of support for crest structures. For submerged structures, however, settlement often leads to increases in armour stability.

Movement of coverlayer elements

Loading(s): waves, currents.
Loading parameters (principal): wave height/period, velocity.
Loading parameters (secondary): time, angle of incidence.
System characteristics: stone diameter/density, permeability.
Response(s): rocking, sliding, lifting, rolling.

Waves and currents determine the lift and drag forces acting on the stones of the coverlayer. The inertial forces are also determined by the stone characteristics. The stone weight as well as forces due to friction and interlocking with other stones are stabilising factors.

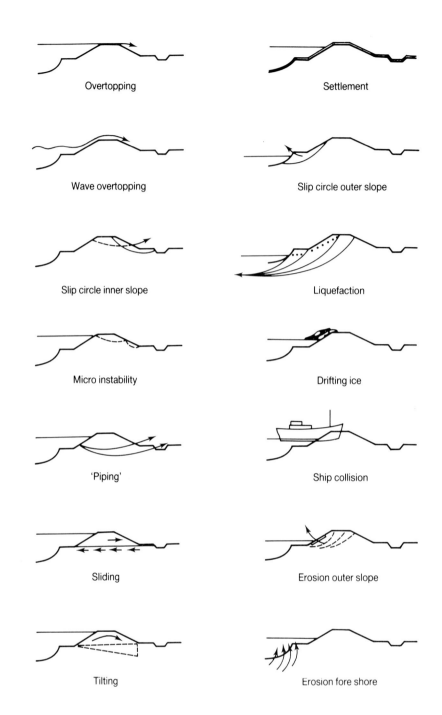

Figure 20 *Failure modes of rock structures*

The dynamic (loss of) balance of all these forces may result in a great variety of the above-mentioned stone movements. These responses may be allowed for in the design, but care is needed to avoid responses large enough to initiate other failure modes such as damage of the filter layer.

Migration of sublayer or core material

Loading(s): hydraulic gradients, internal flow.
Loading parameters (principal): water pressures/velocities.
Loading parameters (secondary):
System characteristics: layer permeabilities/thicknesses, grain size.
Response(s): material transport out of structure.

Due to a difference in water level or to local (generation of) excess pore water

Table 7 Failure mechanisms and characteristic parameters

Mechanism	Loading	(*Principal*) loading parameters	System characteristics	Response characteristics
Settlement	Weight	*Specific density of materials;* saturation degree; pore water pressure; time	Soil compressibility; soil permeability; layer thicknesses	Crest lowering Horizontal deformations
Movement of rock cover	Waves Current Ice	*Wave height; wave period;* angle of incidence; time velocities; turbulence strength; velocities; thickness	Stone diameter and density; permeability	Rocking; sliding; lifting; rolling
Migration of sublayers and/ or filters	(Tidal) waves; ship-induced water movements; other dropping water levels	*Hydraulic gradients;* internal flow velocities	Layer permeabilities and thicknesses; grain sizes	Internal material transport rate
Piping	Hydraulic gradient	Internal channel flow velocities	Pipe (internal channel) length; hydraulic resistance; grain size	Internal material transport rate
Sliding of structure	Weight of structure or structure elements	Weight of construction materials; pore-water pressures (influenced by wave height/period); slope angle	Friction angle, cohesion and permeability of soil/core and cover layer(s)	Sliding of (a significant part of) the structure; collapse
Scour	Waves; currents	Orbital and current velocities; turbulence intensity	Sediment grain size; structure slope; permeability of structure	Degradation of seabed in front of structure
Liquefaction	Waves; earthquakes	Wave height and period; *pore-water pressures;* (relative) shear stress amplitude; acceleration; frequency; number of loading cycles	Permeability; compaction; thickness of layers; friction angles	Serious deformation of structure; collapse

pressures, an internal flow may be established. When a certain critical hydraulic gradient and the corresponding flow velocities occur, the finer grains are transported out from the inner layers through the coarser material of the upper ones. Often these finer grains can also easily pass through the coverlayer, resulting in a loss of material from the filter layer and/or from the core, which may finally lead to (local) settlements.

Piping

Loading(s): hydraulic gradient.
Loading parameters (principal): water velocities/pressures.
Loading parameters (secondary):
System characteristics: grain size, 'pipe'-length.
Response(s): material transport out of structure.

Piping refers to the formation of stable open channels in a granular skeleton created by migration of particles out of the system. These short 'pipes' may connect up and thus allow progressive internal erosion with eventual consequent collapse of the structure. This phenomenon may occur preferentially at structural interfaces, such as at boundaries between permeable and less permeable materials or where loosely packed and densely packed granular materials adjoin one another.

Sliding of (parts of) structure

Loading(s): weight of structure: waves.
Loading parameters (principal): density of building materials.

Loading parameters (secondary): pore water pressures (influenced by wave height/
period); slope angle.
System characteristics: soil friction angle/cohesion.
Response(s): sliding of (a significant part of) structure, collapse.

The stability of a rock slope is determined by slope angle, specific weight, pore
pressures (related to the wave motion) and internal friction and cohesion
(interlocking). Horizontal accelerations are also important, for example, during
earthquakes or wave shock loading. Sliding may occur preferentially along
interfaces between different materials (e.g. armour and underlayer) because the
local friction here is reduced.

The subsoil also has a part in supporting the structure and the generation of
excess pore pressures and liquefaction in any fine layers beneath rock structures
may be important for toe stability and slope support. The crest structure may
also move (slide) under wave loading so adequate friction between the structure
and the underlying rock is critical for stability.

Scour

Loading(s): waves, currents.
Loading parameters (principal): orbital/current velocity, turbulence intensity.
Loading parameters (secondary): wave period, angle of incidence.
System characteristics: sediment grain size, structure slope, stone size.
Response(s): degradation of seabed in front of structure.

Waves and currents cause resulting water movements near the seabed, which may
generate a sediment transport. Interactions with the structure (wave reflection,
generation of turbulence) may affect the natural sediment transport of bed or
beach materials. Relative to the natural sediments, most structures can be
considered to be rigid and non-erodible, although some may be permeable to
sediment, and thus impose a physical boundary condition on the transport
processes.

Liquefaction

Loading(s): pore-water pressures, waves, earthquakes.
Loading parameters (principal): pore water pressure, wave height/period, (relative)
shear amplitude, acceleration, frequency.
Loading parameters (secondary): number of waves/loading cycles.
System characteristics: thickness of compressible layers.
Response(s): serious deformations of structure, collapse.

Cyclic loadings generate excess pore-water pressures when the deformations
resulting from the loading cause compaction at the same time as the drainage
capacity for dissipation of the resulting increases in pore pressure is low.
Liquefaction refers to a situation in (fine) granular materials where excess pore
pressures are generated to such a degree that intergranular contact is lost. The
whole medium then loses all its shear strength and behaves like a thick fluid.
Under these circumstances any shear loading causes sliding/stability failure.

2.2.3 FAULT TREE ANALYSIS

A structure is generally planned to fulfil its functions for a certain prescribed
period of time, normally called the lifetime of the structure. When its principal
functions can no longer be fulfilled the structure can be said to have failed. Such
failures may result from structure degradation (allowed for in the design),
increased or excessive loadings or a design fault.

In order to find all possible relevant mechanisms that might lead to failure of a
structure a fault tree analysis must be carried out. This is a logical diagram
showing all the (partial) failure mechanisms as separate branches (or roots) of the
tree that might either cause or contribute to failure of the structure (the trunk of
the tree).

Failure does not necessarily imply a total collapse or vanishing of the structure. Some reduced level of functioning and/or a certain residual strength may remain after failure.

2.2.3.1 Definition of damage

Damage must be properly defined before the design process can proceed and allow, for instance, for different structural concepts to be compared. In general, damage is defined as a certain change in the state of the structure. The state of a structure is reflected by the following three characteristics:

1. The external boundaries or contours of the structure;
2. Typical cross-sections of the structure and their configuration;
3. The integrity of constituent elements (e.g. rocks, crown wall).

Changes of types (1) and (2) often correspond to a certain physical loss from or displacement of material of the structure, resulting in a loss of functions. Often such damage can easily be observed or measured by setting up an efficient monitoring programme.

In practice, a gradual loss of functions will be observed when damage increases. Failure can be defined as to correspond to an ultimate degree of damage which relates to an unacceptable loss of functions (Section 2.2.2). This point is usually reached after a period of time, depending on the evolution of the damage. At this stage functional requirements (e.g. wave transmission, scour protection) can no longer be met and the structure is said to have failed. For example, degradation of a breakwater crest may lead to exceedance of a critical value of wave disturbance in the harbour basin behind it. Resulting downtime in ship handling may reduce the 'economy' of the harbour to below the design level, so the structure fails. The failure process may therefore be summarised in the simple flow diagram in Figure 21.

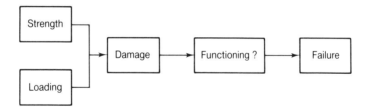

Figure 21 *Relation between loading and failure*

The physical relation between damage and loading links failure to a loading level, which can be expressed, for example, as wave height or wave-induced pressure. Therefore by clear definitions of required functions and damage, failure is related to a certain damage level. The latter can, for example, be practically expressed in terms of displacement of armourstone, deformation or settlement.

Practical dimensionless damage and loading parameters for rock armouring structures will be described in Chapter 5, where the damage volume (S_d) and the loading parameter ($H_s/\Delta D_{n50}$) will be defined. An example of damage as a function of loading is given in Figure 22 showing test results by van der Meer (1988a). In this figure $S_d = 8$ represents moderate damage. S_d values above about 8 in this case would represent failure which is defined to correspond to the point at which the underlayer becomes visible. When a (sudden) progressive increase of damage as a function of the loading level cannot be observed, the point of failure is, for practical reasons, assumed to be reached at one particular degree of damage.

Failure can occur in parts of a structure and in an entire structure (e.g. partial failure of an armour layer and total failure of a breakwater due to liquefaction of

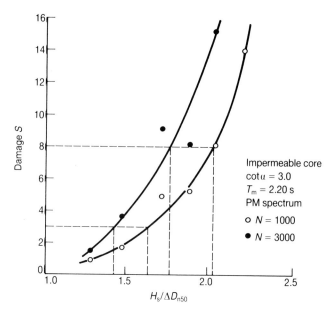

Figure 22 *Typical damage curve for rock armour*

the subsoil). Partial failure as such is generally regarded as less serious than total failure.

Damage as a function of time

Some failure mechanisms can be allowed to occur repeatedly up to a certain limit (e.g. the displacement of an armourstone in a dynamically stable rock slope). For other mechanisms not even a single occurrence can be accepted (e.g. liquefaction of the subsoil under a breakwater).

Repeated occurrences of one mechanism will lead to increasing damage and the frequency of repetition will determine the rate of damage development. Consequently, the damage will increase not only with the loading level but also with time. An important question, therefore, is whether (at a certain constant loading level) the *rate* of damage will decrease or increase with time.

In Figure 23 examples are shown of damage as a function of time, expressed as the cumulative number of waves. Tests on coverlayer stability by Thompson and Shuttler (1975) and van der Meer (1988a) have shown an exponential decrease of the damage rate (curve 1). A similar decline in damage rate has been found in stability tests with bottom protection under random wave attack (curve 2—Redeker, 1985). The damage data have been related to the damage level observed after 5000 waves. Increasing damage rate, as shown by a (fictitious) curve (3), is typical for armour layers, which derive an important part of their strength from interlocking forces and which are exposed to loadings beyond the design value of the individual element (e.g. interlocking breakwater armour units). For such layers, once the loading conditions have exceeded the total resistance provided by both the weight of an individual element and the interlocking between the elements, and once initial damage has occurred, progressive displacements of elements may happen more easily.

In cases such as the first two in Figure 23 a practical (asymptotic) limiting damage level can often be defined, which can be incorporated into the design process. Cases such as the third in Figure 23, in which *progressive damage* is involved, deserve careful attention in design. For most structures progressive damage cannot be accepted and is usually considered to indicate failure.

When a (sudden) progressive increase of damage does not apply, the point of failure can, for practical reasons, be said to be reached at a predefined degree of damage (see Figure 23). Such a damage point may be set to take account of the fact that

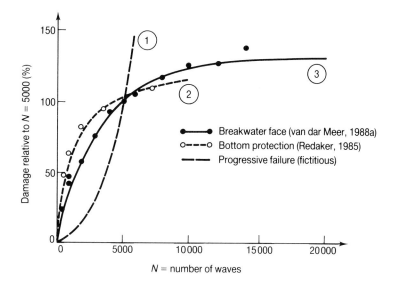

Figure 23 *Evolution of damage as a function of time*

damage due to one partial failure mechanism may eventually lead to the initiation of other such mechanisms or eventually even to total failure. For such cases and where certain damage is accounted for (e.g. dynamic profiles—Section 5.1.3.7), proper assessment of the damage time curve is needed.

2.2.3.2 Fault tree systems and mechanisms

A structure can be schematised as a system, consisting of subsystems and components, which can either fail or function. Failure of any subsystem or component is related to an initial or basic failure mechanism. Fault trees are used to schematise the relations between failure of a system (the trunk of the tree), through failures of subsystems and components (the roots), to the initial failure mechanisms (the ends of the roots). An initial failure occurs when a load exceeds the counteracting strength or resistance. This exceedance of the limit state will have a certain corresponding probability, which is the probability of failure.

Figure 24 shows an example of the principle of a fault tree, this being part of a much more extended tree that can be made for breakwaters and seawalls. In the figure the 'AND' and 'OR' gates describe two different relations between subsystems and are explained below.

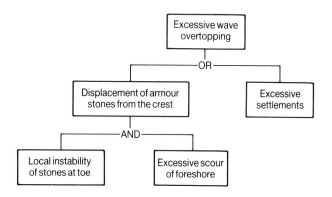

Figure 24 *Example of a fault tree*

Initial mechanisms

Each initial failure mechanism is determined by one or more physical processes and their basic variables. With these variables the limit, or the reliability function $(Z = R - S)$, state functions are defined. These functions, in terms of loading (S) and strength (R) describe the limit states and are used to calculate the failure probability. In fact, this failure probability is the probability of exceeding the safety domain, $P\{S > R\}$ or $P\{Z < 0\}$.

Depending on the state of the knowledge of the physical processes, three options exist to find P:

1. S and R are available as an analytical or empirical formulae. This is the most objective and theoretically the best option. Numerical models that relate loading and response (expressed as damage—e.g. deformations, movements) can also be used for this purpose. In conventional deterministic design practice a design loading (e.g. a design wave height, H_{des}) is chosen and statistics of the loading variable are used to find the corresponding exceedance probability for the design loading, which is set equal to the failure probability.
2. The probability of failure is estimated from performance evaluations (particularly from failure evaluations) of built structures. Here the problem arises of choosing structures that are sufficiently representative (with regard to structural design and environmental conditions).
3. The probability of failure is estimated by experts or from engineering judgement. This option is in fact very similar to (2) except for the absence of documentation of the underlying evaluations.

For many failure mechanisms, research has not resulted yet in analytical, empirical or numerical tools to describe processes and limit states and, subsequently, to determine failure probabilities. A proper use of performance and failure evaluations can be made by setting up data banks containing relevant strength, loading and failure data over a sufficiently long period of time. Such a data bank should contain information on the type and strength of structures and their behaviour (damage, failures) under specified loading conditions.

It should be remembered that failure mechanisms are often dependent on each other. The performance of one mechanism may initiate, either immediately or after a delay, another mechanism. Such a sequence of failure mechanisms is known as *progressive failure* this process is discussed in Section 2.2.3.1 above.

Basic system types

In the structural design, failure probabilities are determined for each initial mechanism and for all components of the fault tree. Finally, the probability of failure for the system of interest (e.g. a breakwater) is calculated following the tree of initial mechanisms and components.

In this process a major distinction is made between parallel or series (sub)systems. Consider a system consisting of n components with failure events that exclude each other. Let p_{fi} be the probabilities of failure of the subsystems. Failure of a *series system* requires failure of only one component. In Figure 24, series components are related to higher levels by using 'OR' gates. The probability of failure of a series system can be found from:

$$P_F = \sum_{i=1}^{n} p_{fi} \qquad (2.1)$$

A *parallel system* consisting of n independent subsystems fails when all n c components fail. In Figure 24, parallel components are related to higher levels by using 'AND' gates. The failure probability of the system is found from:

$$P_F = \prod_{i=1}^{n} p_{fi} \qquad (2.2)$$

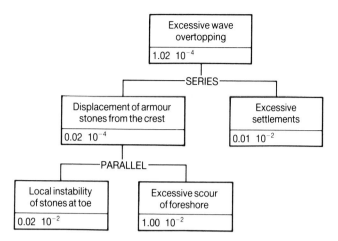

Figure 25 *Example calculation of probability of failure*

An example of the calculation of failure probability of the system with the fault tree of Figure 24 is shown in Figure 25. The subsystems are considered to be uncorrelated.

Coincident and correlated mechanisms

When mechanisms can occur simultaneously or when these are not independent (because common variables are involved) the simple procedures described above cannot be applied. Suppose two components of a *series system* have probabilities of failure p_{f1} and p_{f2}. If coincident failure cannot be excluded, the system fails with a probability P, which is a function of p_{f1}, p_{f2} and of the probability of coincident failures f_1 and f_2:

$$P = p_{f1} + p_{f2} - p\{f_1 \text{ and } f_2\} \tag{2.3}$$

It should be noted that where f_1 and f_2 are fully independent, then $p\{f_1 \text{ and } f_2\} = p_{f1} \cdot p_{f2}$. For the general case, however, an upper and a lower limit of P can be found as follows:

$$\max(p_{f1}, p_{f2}) \leq P \leq p_{f1} + p_{f2} \tag{2.4}$$

The actual value within this range will depend on the correlation coefficient for f_1 and f_2. For small p_{f1} and p_{f2} (practically: $p_{fi} \leq 0.1$) the upper limit provides a good approximation.

This principle can be applied as follows to find limiting values for a n-component series system:

$$\max(p_{f1}, \ldots, p_{fn}) \leq P \leq \sum_{i=1}^{n} p_{fi} \tag{2.5}$$

For n components each having the same probability of failure (p_f) and mutual correlation coefficient (ρ_c), a further simplification can be made, resulting in the value of P reducing to a function of n and of the correlation coefficient ρ_c (see Figure 26). It is then possible to express P in terms of a variation coefficient, defined as the ratio of the resulting and component failure probabilities, P/p_{fi}.

For a treatment of the correlation between components of a *parallel system* the residual strength of a component (when the limit state is exceeded) is an important factor. The simplest situation is where the residual resistance of a component remains constant and equal to the ultimate limit value (ductile parallel system). The failure probability (expressed as P/p_{fi}) of an n-component parallel system, with the components having the same probability of failure (p_f) and mutual correlation (ρ_c), is shown in Figure 27.

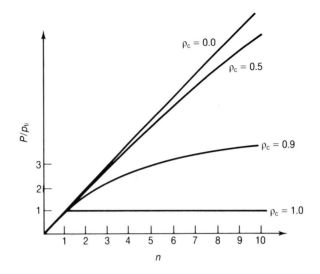

Figure 26 P/p_{fi} as a function of n and ρ_c for a series system

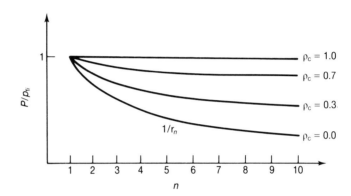

Figure 27 P/p_{fi} as a function of n and ρ_c for a parallel system

2.2.4 RISK-LEVEL ASSESSMENT

In the previous section fault tree analysis has been described. Using the results of a fault tree analysis a risk assessment can be carried out, which subsequently may enable a risk(cost)-benefit evaluation of the design. The combination of probability and consequences is the risk, which can usually be expressed in a certain amount of cost (e.g. cost of replacement). All possible failure mechanisms should be covered by a failure tree. The top of the tree is the event of total failure of the structure, whereas on the lower levels partial and minor failure events are ranged (see Figure 28). At the roots initiating failure events are shown, which occur when a load exceeds the strength $R_k < S_k$. Total failure of a structure may ultimately lead to the consequence of rebuilding the structure and the corresponding cost.

The total risk is defined as the product of the probability of total failure (p_t) and the (economic) consequences of total failure (c_t):

$$r_t = p_t c_t \tag{2.6}$$

Similarly, the partial risk corresponding to the kth partial failure on level i (e.g. partial damage) can be generally expressed as:

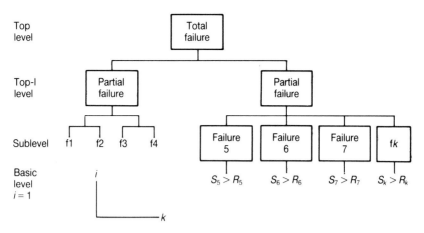

Figure 28 *Failure systems*

$$r_{i,k} = p_{i,k} c_{i,k} \qquad (2.7)$$

In general, consequences will accumulate when the system level (i) increases. Because indirect consequences (e.g. loss of production due to malfunctioning of the structure) may be involved with each failure, the risk due to k subsystems at level i can be expressed as:

for series and parallel subsystems respectively $r_i \geq \sum_{k'=1}^{k} p_{i,k} c_{i,k} \quad r_i \geq \prod_{k'=1}^{k} (p_{i,k}) \sum_{k'=1}^{k} c_{i,k}$ (2.8)

When only risks related to repair and maintenance of the structure are considered (direct risks), equation (2.8) becomes an equality and the minimum risk is obtained.

In general, a risk assessment can be made by working through the failure tree, accounting for the various costs involved with partial failures. It should be mentioned that along each path within the fault tree the number of levels from the base ($i=1$) to the top ($i=t$) can be different. A more elaborate example is shown in Figure 29, assuming uncorrelated mechanisms and systems. The example concerns the failure of the front slope of a seawall (see also fault trees in Section 6.2). Risks are relatively small for the basic events (limited repair) and increase rapidly with higher failure levels. In Figure 29 'OR' gates are indicated by the symbol ∩ and PM branches have not been elaborated. In this figure only direct risk is considered, related to repair or replacement of (parts) of the rock slope protection system.

When the probability is expressed as 1/(return period in years) the total risk of failure for the protection system amounts to $2.2 * 10^{-2} * 100 = 2.2$ ($ yr^{-1}). In this example, partial failures have a risk of 0.006 (piping), 0.400 (sliding), 10^{-8} (flow slide), 0.065 (failure of part of protection) and 0.005 (failure of transitions), respectively (all expressed in ($ yr^{-1})). The contributions of the various initial failures are proportional to the respective probabilities (e.g. instability of filter layer elements contributes 0.03 $ yr^{-1} in total). It should be stressed that the numbers quoted are only an example and must not be applied generally. Individual risks vary widely from site to site and from one project to another.

2.3 Design approach

Strength (R) and loading (S) were introduced in Section 2.2.3.2 as being important parameters to determine the expected damage of a structure and the consequent maintenance costs. Loading and strength are both functions of a (large) number of variables and parameters. In general, basic strength variables are determined by material characteristics (including the subsoil) and by the geometry of the structure.

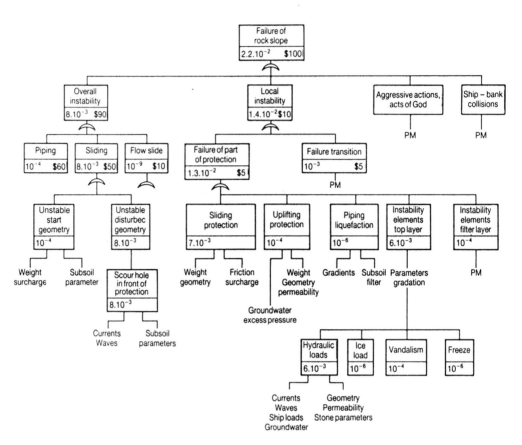

Figure 29 Example of risk-level assessment for a seawall front rock slope

Basic loading variables of marine structures mainly originate from the environmental boundary conditions (wind, waves, currents, ice). The general formulations for the strength and load functions, that are used in this manual are, respectively:

Strength function: $\qquad R = R(x_1, x_2, \ldots, x_m)$ (2.9)

Loading function: $\qquad S = S(x_{m+1}, \ldots, x_n)$ (2.10)

where $x_1, x_2, \ldots, x_m$ are the strength variables and $x_{m+1}, \ldots, x_n$ are the loading variables.

These basic variables of the hydraulic and geotechnical processes usually appear in various stability formulae, describing a limit state (for example, 'no movement', 'no displacement' or 'no deformation'). Often the basic variables of both R and S are themselves stochastic variables. When no movement or displacements are allowed the limit state condition is $R = S$ is applied. An example of a limit state criterion would be van der Meer's armour stability formulae (equations (5.44) and (5.45), in Section 5.1.3.1).

The probability of failure is then equal to the probability that the loading exceeds the strength $(S > R)$. When one or more loading variables are stochastic this also holds for the loading parameter and consequently the probability of failure can be determined from the probability density function of S according to:

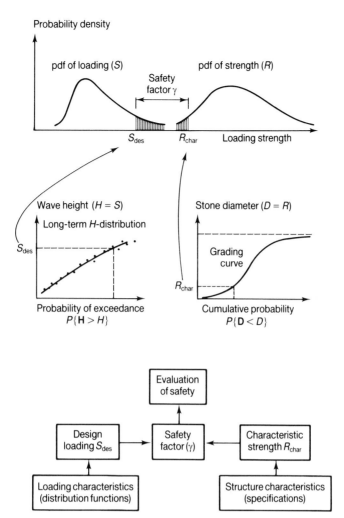

Figure 30 *Example of a deterministic approach*

$$P\{S>R\} \qquad (2.11)$$

In the design an appropriate (distribution of) strength (R) must be chosen to obtain the desired reduction of this probability (safety) at a given distribution of the loading(s). The traditional deterministic and the presently available probabilistic methods differ in the way in which safety is provided for in the structural design.

2.3.1 DETERMINISTIC APPROACH

The traditional design is based upon the deterministic approach. In this approach a limit state condition (ULS or SLS) is chosen with respect to the accepted loading of the structure (see Section 2.1.3). This limit state usually corresponds to a certain strength value or the characteristic strength. Exceedance of the limit state condition ('failure') is not accepted, except for a certain probability (P_F). This probability is usually based upon a risk analysis (Section 2.2.4). To the accepted probability P_F there corresponds a certain loading (S_D, e.g. the design wave height) that is obtained from a statistical (wave) analysis. In practice, the (small) value of P_F is usually expressed as an (average) probability per year that the structure fails or as the reciprocal value of the latter: the (average) return period ($T_R = 1/P_F$). The return period should be interpreted as the expectation or expected value of the time between two occurrences of exceedance of the design loading (wave height).

An example is shown in Figure 30. A long-term wave distribution curve (Section

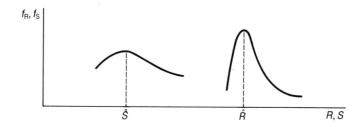

Figure 31 *Load and strength distribution*

4.2.2.3) is used to adjust the design wave, which is subsequently used in the stability condition (equation (5.44) in Section 5.1.3.1) to find an armourstone size, which answers the given probability of hydraulic stability.

An important limitation of a deterministic approach is that once S_D has been chosen, no account is taken of loadings below or exceeding S_D and contributions of these loadings to the expected damage is neglected. Loadings exceeding S_D (with probability $<P_F$) may contribute significantly to the expected damage. More frequent loadings below the design level S_D might also make a contribution to damage. This is a serious shortcoming when future damage must be estimated and quantified for maintenance assessment. In general, strength and loading will exhibit distribution functions as shown in Figure 31. Load and strength distribution functions usually have a shape comparable to the examples given in this figure and can be characterised by:

1. A maximum, corresponding to the 'average' values $\hat{S}$ and $\hat{R}$; and
2. A decreasing magnitude with increasing horizontal distance from these 'average' values.

The principle of the deterministic approach is to account for the uncertainties in S and R, reflected by f_S and f_R, by imposing a suitable safety factor F. This safety factor is defined as the ratio of the characteristic strength (R_D) and the design loading (S_D):

$$F = R_C / S_D \qquad (2.12)$$

It will be clear that F should always be greater than 1.

By choosing a characteristic strength value (R_C) that exceeds the design load (S_D) sufficiently, the resulting failure probability ($S>R$) is kept low. (This probability is shown in Figure 31 as the surface determined by $S>R$.) Thus through the choice of F a certain safety margin is maintained. The stochastic character of the strength and loading variables is accounted for in a simple way by composing F from several partial safety factors. Accumulation of partial safety factors may, however, result in a conservative design. General European or national standards for the choice of F are not available at present. Separate standards do exist for the application of specific materials used in industry and civil engineering. In the recent Dutch TGB ('Technical Foundations for Construction Standards') values for F are prescribed.

2.3.2 PROBABILISTIC APPROACH

Probabilistic methods are in fact a logical extension of the traditional methods. The problem of the choice of the appropriate safety factor is solved systematically by using statistical tools to describe the stochastic properties of both strength and loading variables.

In this approach a safety margin is obtained implicitly by taking into account all the uncertainties of the loading and strength variables. By accepting a certain probability of failure the designer adjusts the safety margin in a rational process.

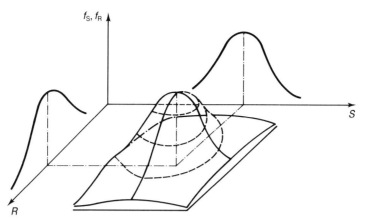

Figure 32 *Joint distribution of loading and strength*

An important support for a probabilistic approach has been the experience that this approach may offer the opportunity to locate unnecessary conservatism in the traditional deterministic design. It will be clear that in these cases the effort of a probabilistic approach will be rewarded in cost savings. On the other hand, by an improved appreciation of the stochastic elements in both loadings and strength, the approach also may offer the means to assess the damage to be expected during the lifetime of a structure. Summarising, the main advantages of the probabilistic approach compared to the traditional deterministic approach are:

- Better appreciation of strength and loading statistics;
- Prevention of unnecessary conservatism, leading to cost saving;
- Provision of means for maintenance assessment and data for management.

According to international conventions, probabilistic methods can be divided into four levels, which are listed here in the order of decreasing complexity:

Level III: These methods take account of the real probability distributions of the strength and loading variables;

Level II: These methods use schematised distribution functions for the strength and loading variables;

Level I: These are in fact quasi-probabilistic methods that use partial safety factors for the variables;

Level 0: Deterministic methods that use one safety factor F between characteristic strength and loading are often described as level 0 probabilistic methods.

Basic theory of probabilistic methods

The basic extension of deterministic design introduced by probabilistic methods is that the two-dimensional or joint probability density function of strength and loading, $f_{R,S}(R, S)$, is taken into account rather than characteristic design values. As a result, probabilities for any combination of strength and loading can be considered (Figure 32). This function forms a more or less undulating surface above the two-dimensional horizontal domain, any point on which is described by the distances S and R along the horizontal co-ordinate axes.

The partial probability density functions f_R and f_S shown in the figure are built up from the content of slices R and S of the volume determined by $f_{R,S}$. These partial probability density functions are characterised by their respective maxima $R=\hat{R}$ and $S=\hat{S}$. Basic variables can be used instead of R and S to describe the joint probability density function, but as soon as more than two variables are involved the function can no longer be described as a surface and requires an imaginary multi-dimensional description.

Figure 33 is an illustrative example of this approach in which the contour lines of a joint probability density function are given in terms of only two basic variables,

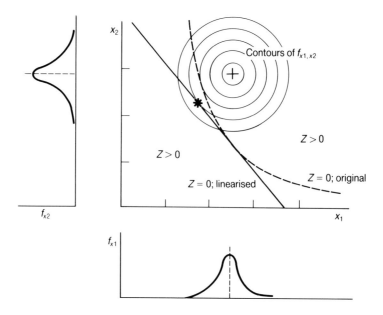

Figure 33 *Contours of constant $f_{x1,x2}$, partial distribution functions for x_1 and x_2 and limit state of failure*

x_1 and x_2, together with their corresponding partial distribution functions. In this example x_1 and x_2 are related to R and S, respectively, so $R = R(x_1)$ and $S = S(x_2)$.

By defining the reliability function Z as:

$$Z = R - S \qquad (2.13)$$

the state of failure of the structure under concern is determined by:

$$Z = 0 \qquad (2.14)$$

otherwise known as the limit state equation.

Using x_1 and x_2 as co-ordinates, the condition of equation (2.14) appears as the line $Z(x_1, x_2) = 0$, shown in the figure. The line $Z = 0$ separates the 'safe' and the 'unsafe' space determined by $(x_1, x_2, f_{x1,x2})$, giving:

$Z > 0$: safe

$Z = 0$: failure $\qquad (2.15)$

$Z < 0$: unsafe, structure has failed

As such, the line $Z = 0$ corresponds to the limit state of failure.

Generalising the above example for more than two basic variables by $R = R(x_1, x_2, \ldots, x_m)$ and $S = S(x_{m+1}, \ldots, x_n)$, Z becomes a function of n variables:

$$Z = Z(x_1, x_2, \ldots, x_m, x_{m+1}, \ldots, x_n)$$
$$= R(x_1, x_2, \ldots, x_m) - S(x_{m+1}, \ldots, x_n) \qquad (2.16)$$

Using the above basic theory, an outline of the Level III and Level II probabilistic methods will now be given in the following sections. All the methods are alternative ways of solving the limit state equation, having first described the joint probability density function of R and S. This outline will be illustrated with some examples using methods from Chapters 4 and 5.

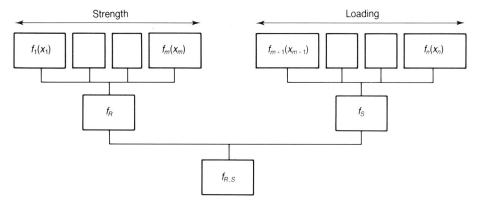

Figure 34 *Determination of joint probability density function*

2.3.2.1 Level III methods

The most extensive probabilistic methods use the exact probability density functions of load and strength: the Full distribution Approach (FA). The determination of the joint probability density function (jpdf) $f_{R,S}$ for R and S is ultimately based upon the partial probability density functions (ppdf) of the basic variables x_1 to x_n. This is shown schematically in Figure 34.

Where the loading (S) and strength (R) are independent (which is often a reasonable assumption) the joint probability density function (jpdf) for R and S can be determined from the partial probability density functions (ppdf) of R and S:

$$f_{R,S}(R,S) = f_R \cdot f_S \tag{2.17}$$

The probability of failure of a structure is expressed by the 'volume' between the 'surface' of the jpdf and the horizontal plane $f_{R,S} = 0$, under the condition $Z < 0$. (The latter means that the plane $Z = 0$ cuts off from the 'volume' contained between the jpdf and the horizontal plane the part for which $Z > 0$.) The resulting probability of failure can, with the assumption of independent loading $(x_m \ldots x_n)$ and strength $(x_1, x_2, \ldots, x_m)$ parameters, be written as:

$$\iiint \cdots \int_{Z>0} f_1(x'_1) \ldots f_n(x'_n) \, dx'_1 \ldots dx'_n \tag{2.18}$$

The functions $f_1(x_1) \ldots f_n(x_n)$ are the ppdf of the loading and strength parameters.

In general, the function $Z(x_1, x_2, \ldots, x_m, x_{m+1}, \ldots, x_n)$ may have a rather complicated form, thus causing practical problems in determining the condition $Z < 0$ in the above computation. In addition, f_1 to f_n are often not known to a satisfactory degree. This has led to a number of alternative (Level II) simplified methods, which will be introduced below.

However, where it is possible to remain with a Level III approach, the integrations of equation (2.18) can be done numerically (for example, using a Riemann procedure in which known or estimated distribution functions are substituted for $f_i(x_i)$). Alternatively, the well-known Monte Carlo procedure can be used.

2.3.2.2 Level II methods

Level II methods comprise the principal methods that are used to overcome the problems faced when using the Level III methods. Linearisation of Z is a common feature of Level II methods. The simplifications in Level II methods in comparison to those of Level III and the overview they give of the relative contributions of the various ppdf make them attractive to use. Three main Level II methods are used which simulate a chosen number of occurrences of combinations of loading strength in order to estimate the probability of failure. The method starts with the generation of a series of random numbers from which through inverse transformation, numbers comply with a chosen statistical distribution function. This process is repeated for each of the basic variables. For each value of Z, calculated with a set of variables, a failure ($Z<0$) or success ($Z>0$) is obtained. After a sufficient number of simulations the fraction of failures is an estimate of the probability of failure:

First-order mean value approach (FMA);
First-order design point approach (FDA);
Approximate full distribution method (AFDA).

First-order mean value approach (FMA)

In this method the reliability function Z is approximated by means of linearisation around the expected mean values of the basic variables x_1 to x_n. This is done by using Taylor series expansion, thereby neglecting all but the linear terms. Under the assumption of mutually independent and Gaussian-distributed variables x_1 to x_n, the linear approximation for Z is written as:

$$Z(x_1,\ldots,x_n) = Z(\mu_{x1},\ldots,\mu_{xn}) + \sum_{i=1}^{n} \{(x_i-\mu_{xi})\frac{\partial}{\partial x_i}[Z(\mu_{xi})]\} \qquad (2.19)$$

This linearised Z-function can be considered Gaussian distributed with a mean value and a standard deviation:

$$\mu_z = Z(\mu_{x1},\ldots,\mu_{xn}) \qquad (2.20)$$

$$\sigma_z = \{\sum (\partial Z/\partial x_i)^2 \sigma_{xi}^2\}^{0.5} \qquad (2.21)$$

With the mean value and the standard deviation of the function Z the reliability index β can be defined as:

$$\beta = \mu_z/\sigma_z \qquad (2.22)$$

The value of β is the relative distance from μ_z to the limit state or $Z=0$ (see Figure 35). As such, β is a measure of the probability of failure, $P\{Z<0\}$. The table in Figure 35 lists $P\{Z<0\}$ as a function of β. The factors $[\partial Z/\partial x_i \sigma_{xi}]^2$ are the respective 'influence factors' $(\alpha_{xi})^2$ for the final failure probability. When the partial derivatives are derived analytically the FMA method allows for a hand calculation.

In the case of a strongly non-linear reliability function (Z) the errors (over-estimates) in the computed failure probability $P\{Z<0\}$ can be significant. This is due to the unreliability of the value found for σ_z. In these cases the use of a Level III method will result in a better approximation.

First-order design point approach (FDA)

This method is similar to the FMA appoach in that the function Z is approximated by the same principle. However, a refinement is introduced whereby the expansion of the Taylor series is made around the *design point*. This point is defined as that on the failure envelope ($Z=0$) where the jpdf, $f(x_i,\ldots,x_n)$, has its maximum (see Figure 33).

This method requires an iterative procedure where the failure envelope, $Z(x_1,x_2,\ldots,x_m,x_{m+1},\ldots,x_n)=0$, is a non-linear function. This will, unfortunately, often be the case, and thus computer methods are essential here. As linearisation is carried out around a point $(x_1,\ldots,x_n)$ where relatively large contributions to the jpdf are found, the errors experienced with the FMA method are reduced.

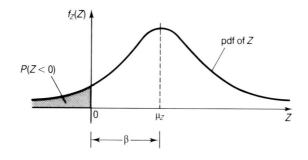

β	P{Z<0}	β	P{Z<0}
0.0	0.50		
0.1	0.46	3.1	$0.97 \cdot 10^{-3}$
0.2	0.42	3.2	$0.69 \cdot 10^{-3}$
0.3	0.38	3.3	$0.48 \cdot 10^{-3}$
0.4	0.34	3.4	$0.34 \cdot 10^{-3}$
0.5	0.31	3.5	$0.23 \cdot 10^{-3}$
0.6	0.27	3.6	$0.16 \cdot 10^{-3}$
0.7	0.24	3.7	$0.11 \cdot 10^{-3}$
0.8	0.21	3.8	$0.72 \cdot 10^{-4}$
0.9	0.18	3.9	$0.48 \cdot 10^{-4}$
1.0	0.16	4.0	$0.32 \cdot 10^{-4}$
1.1	0.14	4.1	$0.21 \cdot 10^{-4}$
1.2	0.12	4.2	$0.13 \cdot 10^{-4}$
1.3	0.10	4.3	$0.85 \cdot 10^{-5}$
1.4	$0.81 \cdot 10^{-1}$	4.4	$0.54 \cdot 10^{-5}$
1.5	$0.67 \cdot 10^{-1}$	4.5	$0.34 \cdot 10^{-5}$
1.6	$0.55 \cdot 10^{-1}$	4.6	$0.21 \cdot 10^{-5}$
1.7	$0.45 \cdot 10^{-1}$	4.7	$0.13 \cdot 10^{-5}$
1.8	$0.36 \cdot 10^{-1}$	4.8	$0.79 \cdot 10^{-6}$
1.9	$0.29 \cdot 10^{-1}$	4.9	$0.48 \cdot 10^{-6}$
2.0	$0.23 \cdot 10^{-1}$	5.0	$0.29 \cdot 10^{-6}$
2.1	$0.18 \cdot 10^{-1}$	5.1	$0.17 \cdot 10^{-6}$
2.2	$0.14 \cdot 10^{-1}$	5.2	$0.10 \cdot 10^{-6}$
2.3	$0.11 \cdot 10^{-1}$	5.3	$0.58 \cdot 10^{-7}$
2.4	$0.82 \cdot 10^{-2}$	5.4	$0.33 \cdot 10^{-7}$
2.5	$0.62 \cdot 10^{-2}$	5.5	$0.19 \cdot 10^{-7}$
2.6	$0.47 \cdot 10^{-2}$	5.6	$0.11 \cdot 10^{-7}$
2.7	$0.35 \cdot 10^{-2}$	5.7	$0.60 \cdot 10^{-8}$
2.8	$0.26 \cdot 10^{-2}$	5.8	$0.33 \cdot 10^{-8}$
2.9	$0.19 \cdot 10^{-2}$	5.9	$0.18 \cdot 10^{-8}$
3.0	$0.13 \cdot 10^{-2}$	6.0	$0.99 \cdot 10^{-9}$

Figure 35 *Normalised Gaussian pdf*

Approximate full distribution approach (AFDA)

This method, like the FDA discussed above, uses the design point for an expansion of the function Z. The additional refinement is to introduce simplified descriptions for the partial probability density functions (ppdf). The latter involve using the approximation of equivalent normal distribution functions around the design point. These functions are defined by a mean value, corresponding to the design point and the standard deviation. Like the FDA, this method generally requires computer numerical techniques.

2.3.2.3 Examples of probabilistic calculations

Some examples of the use of three of the above probabilistic methods will be given with regard to the stability of armourstone exposed to wave attack. According to the stability criterion for breaking waves, presented in Section 5.1.3.1, the loading function can be written as:

$$S = H^{0.75} T_z^{0.5} N^{0.1} \tag{2.23}$$

where H_s = significant wave height (Section 4.2.2.1),
T_z = zero crossing wave period,
N = number of waves,

are the basic loading variables. The corresponding strength function is:

$$R = 6.2 S_d^{0.2} (g/2\pi)^{-1/4} P^{0.18} \sqrt{\cot(\alpha)} \Delta D \qquad (2.24)$$

where S_d = damage parameter,
P = notional parameter of structure permeability,
α = slope angle of the armour layer,
Δ = relative density of armourstone,
D = stone diameter,

are the basic strength variables. In the final structural design the various elements of a structure have to be determined in detail (dimensions, weight). This implies that the strength variables $(x_1, \ldots, x_m = S_d, P, \alpha, \Delta, D)$ receive a value such that the strength R exceeds the design loading S to an acceptable degree and the probability of failure is acceptably small. This example will be elaborated using the methods introduced below.

Example of the results of a Level II FMA method

According to equation (2.13), Z can be written as:

$$Z = 6.2 S_d^{0.2} (g/2\pi)^{-1/4} P^{0.18} \sqrt{\cot(\alpha)} \Delta D - H^{0.75} T_z^{0.5} N^{0.1} \qquad (2.25)$$

When the partial derivatives are determined analytically a hand calculation can be made as follows. The partial derivatives $\partial Z/\partial x_i$ with respect to the stochastic variables can be written, respectively, as:

$$\partial Z/\partial S_d = 0.2 R/S_d$$

$$\partial Z/\partial H = -0.75 S/H$$

$$\partial Z/\partial T = -0.5 S_d/T$$

$$\partial Z/\partial D = R/D \qquad (2.26)$$

The mean values μ_{xi} are listed in Table 8. Multiplying the square of the partial derivatives in equation (2.26) by the corresponding σ_{xi}^2 according to equation (2.21) gives the influence factors α_{xi} of S_d, H, T, and D (see Table 8). Applying equation (2.21) gives $\sigma_Z = (28.5)^{0.5} = 5.3$, as shown in the table. A value of $\mu_Z = 13.7$ and application of equation (2.22) results in a value $\beta = 2.57$. Subsequent use of the normalised Gaussian distribution (Figure 35) then gives a $P\{Z<0\}$ of 0.0051.

Example of the results of a Level III Riemann calculation

Application of a Riemann integration, with $Z = R - S$ according to equation (2.25),

Table 8 Example of a FOMVA calculation

Variable x_i	Mean μ_{xi}	Stand. dev. σ_{xi}	Infl. factor α_{xi}^2	Rel. contr. $\alpha_{xi}^2/\Sigma\alpha_{xi}^2$
S_d	10	2	1.0	3%
H (m)	3.0	0.5	1.9	7%
T (s)	6.0	1.0	0.9	3%
D (m)	1.0	0.2	24.7	87%
$\cot g(\alpha)$	4	0	0	0
Δ	1.6	0	0	0
P	0.5	0	0	0
N	1000	0	0	0
Coeff.	6.2	0	0	0

$$\sigma_Z^2 = \Sigma \alpha_{xi}^2 = 28.5$$

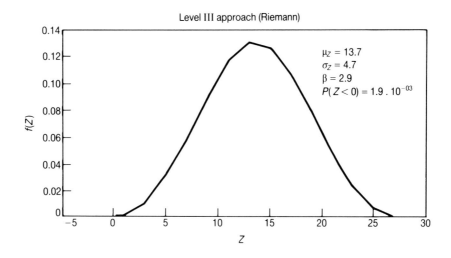

Figure 36 *Distribution of Z based on Gaussian distributed parameters*

Table 9 Example of an AFDA calculation

Variable x_i	Mean μ_{xi}	Stand. dev. σ_{xi}	Design point X^*	Rel. contr. $\alpha_{xi}^2/\Sigma\alpha_{xi}^2$
S_d	10	2	9.4	1.0%
H (m)	3.0	0.5	3.4	7.3%
T (s)	6.0	1.0	6.5	3.4%
D (m)	1.0	0.2	0.5	88.3%
cotg(α)	4	0	4	0
Δ	1.6	0	1.6	0
P	0.5	0	0.5	0
N	1000	0	1000	0
Coeff.	6.2	0	6.2	0

results in the distribution of Z shown in Figure 36. To obtain this distribution S_d, D, H and T were assumed to follow a Gaussian distribution (each with the μ and σ shown in Table 8). The remaining parameters have been assumed non-stochastic constants. The results are shown in Figure 36. From the resulting values of μ_Z and σ_Z and assuming that Z is also Gaussian distributed, the probability that values of Z become negative (failure) is $1.9 \cdot 10^{-3}$ (see the normalised Gaussian table in Figure 35).

Example of the results of a Level II AFDA calculation

Based on the same data as before and assuming the stochastic variables to be Gaussian distributed, a design point is found and a corresponding failure probability of $4.7 \cdot 10^{-3}$ (see Table 9 and Figure 35).

Comparison of results

Because of the limited simplifications involved in the Level II AFDA method, it will generally be the most reliable of the Level II methods discussed, as is seen by comparison of the various results for $P(Z<0)$. By introducing realistic distribution functions for the respective variables the reliability can be optimised.

The comparison also clearly shows the limitations of the FMA method, largely resulting from the linearisation of Z. Referring back to the example of Figure 33, where both a real and the linearised failure condition $Z=0$ were shown, it will be evident that, by using the linearised form, a considerable extra volume of $f_{x1,x2}$ may erroneously be included to contribute to the probability of failure. However, although this example demonstrates the limitations of the FMA method, it can be useful for obtaining a first estimate of the reliability of a design concept and for a structured sensitivity analysis of the parameters involved.

2.4 Environmental assessment

Environmental assessment (EA) is a procedure more commonly referred to outside the European Community as Environmental Impact Assessment (EIA), whereby the likely effects of a proposed development on the wider environment are considered as part of the planning and design of the project. The objectives of environmental assessment are:

- To make projects environmentally sensitive and less environmentally damaging;
- To provide suitable information to assist those organisations responsible for development control to make environmentally informed decisions.

2.4.1 REASONS FOR CARRYING OUT EA

Shoreline engineering works may affect (or be affected by) the natural and social environment in which they are situated. Typical aspects of the environment on which this type of project may impinge either during construction or in the long term are listed below and illustrated in Figure 37:

- Geomorphology, landscape and exciting physical processes (sediment transport, drainage, erosion, climate, etc.)
- Ecology (flora and fauna)
- Social and socio-economic aspects
- Human sensory aspects (visual intrusion, noise, odour, etc.)
- Palaeontological, archaeological, historical and cultural aspects
- Air, water and soil quantity (pollution, contamination, etc.)

The developer may perceive environmental assessment as an additional financial burden. However, the cost of the environmental assessment will be a small proportion (usually less than 0.5–1.0%) of overall project value and the procedure has a number of advantages from the developer's point of view:

- Identification of environmental impacts during the planning and design stage will enable the most cost-effective inclusion of measures to mitigate adverse impacts.
- The liaison and consultation exercise, which is normally an integral part of the assessment, will reduce the risk of an unexpected refusal of development consent at a late stage.
- EA should identify probable areas of objection to a project at an early stage to enable sensitive matters to be addressed and project delays minimised.
- EA provides an opportunity for public consultation and resulting greater acceptance of a project in the affected community.
- Unforeseen adverse environmental impacts may incur considerable future financial liabilities.

Environmental assessment may be voluntary or a specific legal requirement, depending on the state or other authority within whose jurisdiction the project falls. Advice should be sought from the relevant planning or other authority in this regard. Regulations affecting environmental assessment within the European Community are discussed in Box 13 (page 58). However, for the reasons just discussed, even where an EA is not formally required, some form of environmental study will prove valuable in project planning.

2.4.2 EA AS AN INTEGRAL PART OF THE DESIGN PROCESS

It is most beneficial for the environmental assessment to be commenced at the

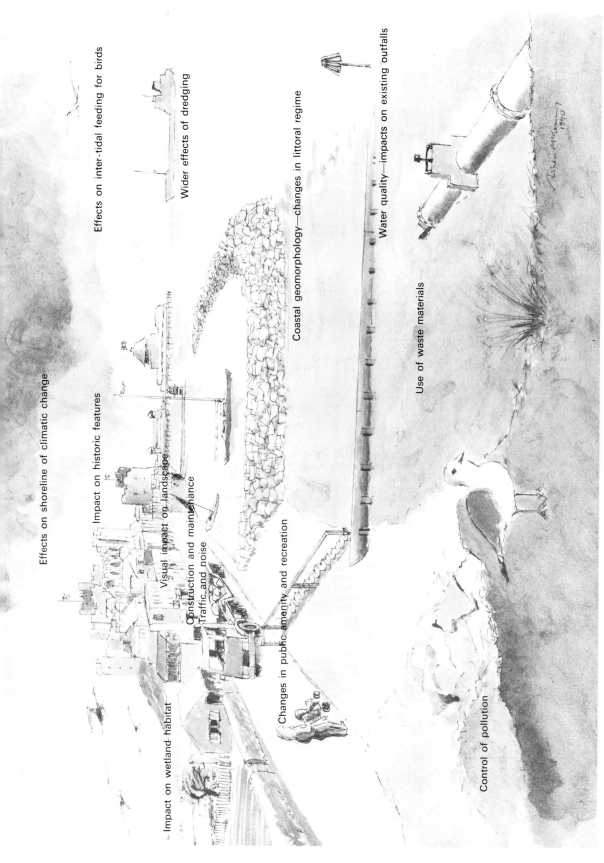

Figure 37 *Typical coastal environmental impacts*

Box 13 European Community (EC) environmental assessment regulations

Legal requirements

In mid-1988 the requirements for environmental assessment in the European Community became formalised with the implementation of EC Directive 85/337 (Council Directive of 27 June 1985 'on the assessment of the effects of certain public and private projects on the environment' refers). Relevant statutory instruments, regulations and guidelines are listed in Appendix 7. The EA regulations, recommendations and publications are continually being updated. If in doubt, reference should be made to the appropriate authorities before undertaking a formal assessment.

EC Directive 85/337 states that environmental assessment shall be a mandatory part of the planning process for project types listed in Annex I of the directive. For projects listed in Annex II, which covers most typical shoreline engineering works using rock; the requirement for environmental assessment is at the discretion of the planning authority or other body responsible for granting consent for the development to proceed. (If an environmental assessment is required as part of an application for planning consent an increased period may be allowed for the planning authority to determine the consent.)

Scoping; aspects to be addressed

Article 3 of EC Directive 85/337 gives a particular description of the aspects to be covered in any EA:

1. Human beings, fauna and flora;
2. Soil, water, air, climate and the landscape;
3. Interaction between the factors mentioned in (1) and (2) above;
4. Material assets and the cultural heritage.

Requirements for the Environmental Statement

There is no precise definition with EC Directive 85/337 as to the form of the Environmental Statement. The only requirement is that it contains the information specified in Articles 3 and 5 and Annex III of the Directive and includes a 'non-technical summary'. More detailed formats for the ES may be specified in national regulations.

Statutory consultees

In addition to the issuance of public notices as required by regulations, the proponent of the scheme may be under a statutory obligation to notify or consult certain authorities and organisations specified in EA or planning regulations as part of the formal environmental assessment process.

(Continued opposite)

	Human beings	Fauna	Flora	Soil	Water	Air	Climate	Landscape
Human beings	(cd)		(b)	(b)	(cf)	(cf)	(cf)	(ca)	(da)
Fauna	(b)	(b)		(b)	(bf)	(bf)	(bf)	(ba)	(ba)
Flora	(b)	(b)	(b)		(bf)	(bf)	(bf)	(ba)	(ba)
Soil	(f)	(fc)	(fb)	(fb)		(f)	(f)	(fa)	(fa)
Water	(f)	(fc)	(fb)	(fb)	(f)		(f)	(fa)	(fa)
Air	(f)	(fc)	(fb)	(fb)	(f)	(f)		(fa)	(fa)
Climate	(a)	(ac)	(ab)	(ab)	(af)	(af)	(af)		(a)
Landscape	(a)	(ad)	(ab)	(ab)	(af)	(af)	(af)	(a)	
Material assets	(ce)
Cultural heritage	(e)

Note: Letters in the above matrix refer to the discipline-based environmental aspects discussed in Section 2.4.1

Figure 38 *EC Directive 85/337 aspect and aspect interaction scoping chart*

> **Box 13** (Continued)
>
> Consent from other bodies may be required whether or not a development is outside the jurisdiction of the local planning authority. For example, shoreline works may require a licence for placing material on the seabed or approval for navigation. These bodies may also require an environmental assessment as part of the application for their consent, licence or approval. Statutory consultees for the UK are listed in Appendix 7.
>
> *Sensitive locations*
>
> Projects which impinge on particular sensitive locations (such as those listed in Appendix 7) will require special attention. Development directly or indirectly affecting such sites is likely to make the requirement for environmental assessment mandatory.
>
> This categorisation of environmental aspects is peculiar to the EC directive, but it is compatible with a more conventional discipline-based aspect listing such as that given in Section 2.4.1. Figure 40 indicates in each of the matrix boxes by letter (referring to the list in Section 2.4.1) the discipline-based aspects which are relevant to each of the EC environmental categories.

conceptual stage of a project. The simplified flow diagram of Figure 39 illustrates the integration of EA into project planning and design.

2.4.3 METHOD OF ASSESSMENT

Phases of an assessment

An environmental assessment will normally comprise three phases:

1. The baseline survey, to define the existing environment;
2. The projection of the proposed project onto the existing environment and the assessment of the probable impacts (beneficial and adverse);
3. Where impacts are found to be adverse, the investigation of measures to mitigate those impacts and the possible incorporation of those measures into the design.

Scoping

The aspects of the environment on which rock coastal and shoreline engineering works may have significant effects will include those listed in Section 2.4.1 and the various interactions between them. For an EA to be cost-effective, the distribution of time and effort must match the relative needs of the different aspects. At the start of an assessment it will be necessary to decide on the probable scope of the investigations and the appropriate level of detail required from the results.

Where the assessment is to be carried out under European Community Directive 85/337 or its derivative national regulations, a particular presentation of the aspects is required. This is compatible with the aspects as listed in Section 2.4.1 and is detailed in the scoping diagram in Box 13.

2.4.4 ADDITIONAL SOURCES OF INFORMATION

In addition to the normal investigation and survey process, environmental data for use in the assessment will come from two sources: consultation and monitoring.

Consultation

It is impossible to over-emphasise the importance of widespread consultation as part of an environmental assessment. Initial consultations should be with the planning, coast protection, drainage and navigation authorities and other statutory bodies whose consent will eventually be required to allow the project to proceed. It is advisable to contact statutory conservation organisations (e.g. in the UK the

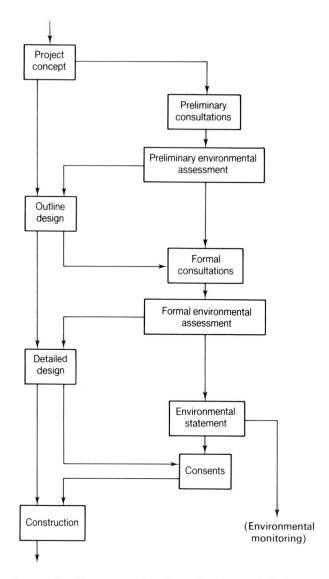

Figure 39 *Flow diagram for EA as part of project planning and design*

Nature Conservancy Council and the Countryside Commission) whether or not there is a legal obligation to do so. It is usually beneficial to consult all organisations whose interests are likely to be directly affected by the proposed works. As a general rule, consultation should be as widespread as practicable. It may be necessary to revise the scope of the assessment as a result of the consultations.

Monitoring

Environmental monitoring before construction of the works provides input data for calibration of numerical models and, after construction of the works, validation of predicted impacts. Monitoring after the works are in use may also provide early warning of unexpected impacts, allowing time for the implementation of remedial measures before the situation becomes too far advanced. The subjects, frequency and extent of the monitoring will be dictated by the site and the probable impacts of the project.

There is considerable scope for the development of more effective and informative environmental monitoring procedures, notably for impacts on sensitive/indicator species. Similarly, monitoring data collected at one site may prove invaluable on a number of later projects (see also Chapter 7).

2.4.5 ASSESSING THE IMPACTS OF THE PROPOSED WORKS

The assessment of the impacts of a proposed scheme will be described under the environmental headings listed in Section 2.4.1.

2.4.5.1 Geomorphology, landscape and existing physical processes

Rock-armoured rubble-mound breakwaters, groynes, training walls, etc. will usually have significant impacts on the existing littoral regime—this is often a primary function of the works! The most familiar problems arise from the effects of these types of works on the existing regime, typically; accretion updrift and erosion downdrift with variations caused by local changes in wave exposure, refraction, diffraction, etc. Similarly, works such as rock revetments and beach-nourishment programmes are likely to have some impact on local geomorphology. Such impacts can be assessed by using coastline simulation models, the most simple and well-known example being that described by Pelnard-Considère (1956). Further models on tidal flow, turbulence/diffusion/advection (e.g. Abbott, 1979; Celik and Rodi, 1985) and groundwater flow (e.g. van Ree *et al.*, 1991) should be used to assess turbulent diffusion of dissolved or suspended matter (chemicals, sediments, heat ...) in the water or in the subsoil respectively. Additionally physical modelling may also be necessary.

The effects on the geomorphological environment are of considerable interest to the engineer and will normally be investigated as part of engineering studies. However, the geographical limits of such investigations have, in the past, frequently been too narrow. The destruction of the village of Hallsands, south Devon, in the early part of this century was a result of the wider geomorphological impacts of offshore dredging.

It is generally assumed that stability and perhaps accretion are beneficial, whereas erosion is considered undesirable. However, this may not always reflect the view of conservation organisations.

Impacts on the landscape of shoreline works may be of particular significance. Where, for example, breakwaters are designed to project some distance above high water, the structure (particularly viewed from beach level) may represent a major impact on the landscape. The exposure of shoreline works allows few opportunities for effective screening or camouflage.

Impacts and effects relating to climate should be taken into account. For example, predicted changes in water levels and wave exposure during the design life of a shoreline structure may be very significant.

2.4.5.2 Ecology

The distribution of coastal marine flora and fauna is partly a function of tidal levels, coastal geomorphology and seasonal variations. For example, exposed rocky shores subject to persistent wave action and swell impose severe limits on the organisms present, but they nevertheless typically support considerable numbers of a limited range of species. Sheltered rocky shores provide a wide range of variation in the relative abundance of both plant and animals exploiting the diverse range of available microhabitats.

In the early stages of the assessment it is necessary to determine the ecological

Table 10 Possible social and socio-economic considerations in the environmental appraisal of rock structures

	Duration of effect	
Possible effects	Construction (see Section 2.4.6.3)	Long term
Changes in local employment	*	*
Effects on recreation	*	*
Change in access to shoreline	*	*
Immigration into small community	*	
Loss of safe/shallow areas for children's play		*
Effects on local commerce	*	
Change in vehicle access		*
Loss of boat trailer access		*
Impacts on sea view from houses		*
Improved sea angling locations		*
Relief of risk of flooded homes		*
Access for the disabled		*
Vulnerability to vandalism	*	*

significance of any future disturbance to both the immediate site and to any other locations or biological systems which may be indirectly affected. This is achieved by determining a series of ecological descriptors by which such disturbances can be assessed. These descriptors might typically include the following:

1. The presence of permanently resident species, relative abundance and distribution, significance locally, nationally and internationally;
2. Seasonal changes, with particular reference to migratory birds, breeding patterns/spawning seasons of marine organisms and the significance of these events as in (1) above;
3. Influence and level of human activity and the 'built' environment in determining ecological status of the site at the time of assessment (pre-project, post-project, etc.).

A monitoring programme should be established to run before, during and after construction to provide data for incorporation into design models and also to allow for predictions to be validated (see Section 2.4.4 above).

Generally, stable conditions provide the best opportunities for successful colonisation by plants and animals. Thus works designed to stabilise geomorphological factors may eventually have positive impacts on the local ecology. The benthos under the footprint of a rubble-mound breakwater may be lost, but its rock-armoured sides can provide suitable habitats for a variety of marine life. In areas where there are few natural rock outcrops and interstices on the seabed, a rock breakwater may increase local habitat and species diversity.

Ecological damage is more likely to occur due to project-related loss of sheltered inter-tidal zones and wetlands or the activities associated with construction of the engineering works, rather than the completed rock structure itself.

2.4.5.3 Social and socio-economic aspects

The coastline is a major zone of recreation and the impact of the works on the existing coastal amenity may be considerable (e.g. coastal revetments made up of large rocks may be unpleasant to traverse for beach users and may trap litter and other debris). The construction of the works themselves may provide much-needed employment or they may interfere with the livelihood of existing residents. Other social impacts may be due to the influx of an outside workforce during the construction of a major project in a lightly populated area. Some typical social and socio-economic considerations are summarised in Table 10.

The public perception of the advantages or disadvantages of a project needs to be addressed. Public consultation exercises will be of greatest value when there is an intention to allow those views to influence the design of the project. There are statutory requirements for public notification under various EA regulations (see, for example, Box 13).

Pedestrian access along rock structures can be difficult to provide. Public access along a particular stretch of shoreline may by inviolable. However, while members of the public would see access along a breakwater as advantageous, it may be cheaper and safer to actively prevent it. Positive socio-economic impact is the usual motivation behind development.

2.4.5.4 Human sensory aspects

Impacts on the human senses from a development may include visual intrusion. Beach-nourishment works should generally result in a visual improvement. If a much coarser material than the original is used to nourish a beach, stability may improve at the expense of appearance and amenity.

Once in place, rockfill structures are unlikely to cause any particular noise problems and are less likely to transmit wave-induced vibration than more rigid forms of construction. Impacts from beach-nourishment schemes may result where periodic recycling comprises transporting sediment up-drift in lorries.

Sediment pumped ashore from deep deposits may contain organic debris which gives off strong odours when exposed to air. This impact should only last as long as it takes the material to oxidise. It has been claimed that the interstices of rock revetments trap weed and other debris which may smell unpleasantly as they rot.

2.4.5.5 Palaeontological, archaeological, historical and cultural aspects

Information on the locations of sites of palaeontological or archaeological interest may be obtained from local authorities. In the case of fossil exposures, buildings and monuments which have been classified as of particular historic value, details may be obtained from statutory organisations charged with their preservation. It should be noted that many historically and culturally significant sites and buildings have no specific recognition or protection. The best advice is, if in doubt, consult an appropriate authority (e.g. a local historian may be able to advise). Where archaeological sites cannot be avoided, arrangements should be made to allow for their exploration in advance of the works.

It is always of value to prepare a concise history of the site for the following reasons:

- Historical investigation may help to identify areas of contamination at the site.
- Historical performance of shoreline regime may be of great significance to contemporary analysis and design.

2.4.5.6 Air, water and soil quality

Significant air and water quality impacts are unlikely to result specifically from the use of rock in shoreline works. (However, controls were placed on the supply of rock for coastal structures in Twofold Bay, New South Wales, Australia, due to the natural occurrence of asbestos in the rock joints.)

Structures which alter the coastal regime may adversely affect the performance of existing marine outfalls. Any increased risk of vessel damage or collision will mean greater probability of polluting incidents.

Changes in air and water quality are more likely to be the result of other activities related to the purpose for the works (e.g. pollution in harbours, etc.).

Where there are risks of marine pollution from oil spills measures to facilitate containment and clean-up may be usefully incorporated into the design. Works which bury or enclose pipelines or storage tanks may increase the risk of late discovery of leakage.

2.4.6 IMPACTS FROM THE CHOICE OF MATERIALS AND CONSTRUCTION

The most significant environmental impacts of a scheme may relate to the winning of materials and the construction methods, rather than the works themselves.

2.4.6.1 Sources of material

Rock and shingle for use in coastal and shoreline engineering works will come from two primary sources.

Onshore working from quarry or gravel pit

It is unusual in Europe for a project developer to consider opening a quarry for the sole purpose of supplying rock to the works. The quarry will probably not be under the developer's control; it may even be in another country. However, if opening a temporary dedicated quarry is considered, as may well be the case for major works outside Europe (see Section 3.4.2), it should be noted that the environmental impacts of quarrying may be major and will involve consideration of noise, vibration, dust, waste, effects on landscape, loadings on highways, etc. The environmental impacts of quarrying are beyond the scope of this manual, and advice on this subject should be sought from the relevant local planning authority and the appropriate government department.

Dredging from the seabed

Material may be dredged from the seabed specifically for beach-nourishment and reclamation projects or to provide depth for navigation. The environmental impacts of dredging may be less obvious than those of land excavation, as they are largely hidden from terrestrial view.

Dredging may have major impacts on seabed ecology (benthos) and geomorphology. The prime ecological impact will be the loss of seabed flora and less mobile fauna. Progressive studies of seabed geomorphology have shown sediment mobility at increasing depths, and it is generally accepted that the dredging of sediment in shallow seas has some effect beyond the immediate area of excavation. Considerable volumes of fine material may be lifted from the seabed and go into suspension during dredging, resulting in an increase in turbidity which reduces the penetration of sunlight into the sea, to the detriment of light-dependent organisms. Fine material may reduce the nourishment of 'filter-feeding' animals. Only where the ambient turbidity is already very high will the effect of dredging be negligible.

Toxins (notably metals) may be found in seabed sediments and traces are usually found in sea water. Sediments badly contaminated with heavy metals will not normally cause pollution during dredging, provided they are dredged and deposited in the same waterway. Heavy metals bind to seabed and estuarine sediments, and although the level of contamination may be high, the heavy metals remain stabilised in the sediment. However, if such material is pumped ashore and subjected to rainfall and drainage, the resulting change in redox potential and pH can result in the heavy metals being released from the sediment and becoming bio-available.

2.4.6.2 Use of waste materials in shoreline works

Possible additional environmental impacts may result from the use of waste products as substitutes for stone filling, e.g.:

WORKS ASPECTS \ ENVIRONMENTAL ASPECTS	Ecology	Social and socio-economic	Visual and landscape	Physical/ geomorphology	Air and water quality	Noise and vibration	Archaeological	Health and safety
CONSTRUCTION								
Access	o	XP	o	o	o	xP	o	xP
Dredging	XS	XP	o	o	o	XP	o	xM
Concreting	o	xP	o	o	o	x	o	xM
Placing of of stone etc.	o	o	o	o	x	xM	o	xM
Night/weekend work	o	XP	o	o	o	XP	o	o
Workforce	o	+x	o	o	o	o	o	o
PERMANENT WORKS								
Flood protection	oS	+X	o	+x	+x	o	o	+X
Siting	oS	X	X	oS	o	o	oS	o
Crest height	o	x	XP	o	o	o	o	o
Navigation	o	xP	XP	xS	o	o	o	xM
Drainage	o	o	xP	xS	o	o	o	xP
Finishes	o	xP	XM	o	o	o	o	o
Operation	oS	xP	o	xP	xM	o	o	xP
Revetments	xM	xM	XP	o	o	o	xM	o
Access	o	XP	o	o	o	xP	o	xP

Key:

x = minor effect   X = major effect
+ = benign effect   o = no significant effect
M = subject to mitigation
P = subject to partial mitigation
S = predictions to be subject to monitoring

Figure 40 *Example of an assessment matrix*

- Metalliferous slags
- Mine waste
- Baled domestic refuse (for reclamation fill).

The performance of some of these materials in coastal exposure is known (e.g. slag from iron and steel making at Workington, Cumbria, was deposited on the shore for many years and partially nourished the down-drift beaches). The history of the coastal disposal of colliery waste in north-east England is a similar case.

The risks of using waste materials include superficial and groundwater pollution, material instability and loss of amenity. The durability of certain waste materials, including some colliery shales, is known to be poor in the sea. Unless the chosen material has a confirmed history of successful use in the marine environment in similar exposures to that anticipated, the material should be tested and its performance evaluated before use.

Allowable levels of heavy metal pollution in coastal sediments are not generally defined except in terms of the impact on commercial shellfish or restrictions on their use in agriculture. However, this is a rapidly developing field of regulation, and guidance should be sought from the appropriate authorities before embarking on a design based on the use of waste materials.

2.4.6.3 Construction

The greatest environmental impacts from the use of rock in shoreline engineering may result from the construction of the works. As long as the site is adequately cleared and cleaned after construction is complete, these impacts should be limited to the duration of the construction. Impacts will result from the transport of material to the site—particularly if by road, storage at the site, placing in the works, any associated piling and similar operations, temporary loss of public amenity and various construction-related hazards. Work may be scheduled to minimise traffic nuisance, dust generation, noisy night working, loss of public access to beaches during the tourist season, etc. Site working should be organised to minimise the risks of oil spills and other pollution.

2.4.7 THE ENVIRONMENTAL STATEMENT (ES)

Where an environmental assessment has been carried out, the results are normally presented in a report called the Environmental Statement (ES) (or Environmental Impact Statement (EIS)). For mandatory assessments this report will form a principal part of the planning (or other) consent application.

The Environmental Statement should include the following:

1. A full description of the proposed scheme;
2. A description of the existing environment in which the scheme is to be situated and which may be affected by the proposed scheme.
3. A brief description of other options considered (including the option of doing nothing) and reasons for their rejection;
4. A statement of the predicted environmental impacts of the proposed scheme;
5. Where the predicted impacts are adverse, a description of the measures which will be adopted to mitigate those impacts.

The ES should include a concise summary written so that non-specialists may understand the results of the environmental assessment. The impacts of the proposed scheme may usefully be summarised in a matrix, as illustrated in Figure 40 on page 65.

3. Materials

The cost of production and transportation of the very large quantities of rock often required for coastal and shoreline structures is an important consideration when selecting a particular design solution. Thus it is important to establish the availability and quality of rock materials for a particular site at an early stage when considering design options.

The relation between the design process and that of materials assessment and evaluation is illustrated in the logic diagram of Figure 41. The various sections provide information concerning the geological materials that may be encountered, the materials properties which may be expected in relation to the design requirements and the test procedures which are usually applied to assess quality and durability of these materials before and during construction. Information is

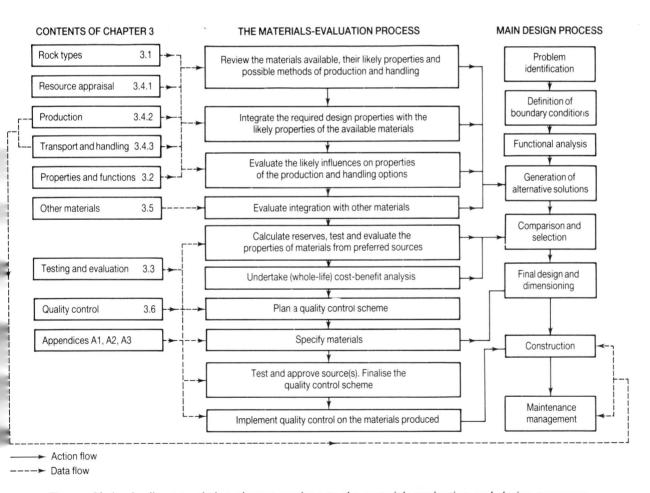

Figure 41 *Logic diagram relating chapter sections to the materials evaluation and design processes*

also provided relating to other material systems used whether as alternatives or in conjunction with the bulk geological materials. Methods of specifying materials are outlined in Appendix 1.

3.1 Rock Types

3.1.1 GENERAL CONSIDERATIONS

This section deals with the geological classification of rock materials and their general suitability for use in coastal and shoreline structures. The use of rock in such structures ranges from a few thousand to well over one million tonnes. Typically, the core of the structure uses most of the material (for example, a large breakwater in Iceland involved some 1 847 000 tonnes of rock of which 80% was core and 20% armour).

Maintenance, repair and the construction of small structures often makes use of materials stockpiled or obtained during the normal operations of an existing quarry. However, sizeable structures requiring large quantities of rock favour the extension or development of large quarries of an appropriate rock type which are dedicated partly or wholly to the production of materials, and which are conveniently sited adjacent to deep coastal waters. Section 3.4 groups quarries into three broad categories. Those supplying rock for coastal and shoreline structures should be capable of producing large blockstone by the appropriate choice of cutting, drilling or blasting technique. Typically, quarry outputs produce excess smaller gradings compared to the proportion of armour-sized blocks.

3.1.2 EVALUATION OF ROCK TYPES AT SOURCE

The selection of appropriate rock materials for use in a marine structure, whether from extension of an existing quarry or from the development of a new quarry source, must eventually be made on the detailed assessment of the rock's physical properties and the geological considerations at the quarry. However, a number of generalisations concerning the relevant properties of groups of rocks may be made which are helpful in the initial selection of potential sources. Once particular sources have been identified, it is, of course, essential for them to be investigated in detail, since the state of the rock weathering can give rise to considerable variations in properties determined by testing. In addition to weathering, many other reasons may give exceptions to the generalized values in Tables 11–14. The validity of these generalisations arise as a result of common mineralogies, textures and modes of formation of various groups of rocks, and can rapidly establish a framework of expected properties for different sources of rock materials.

Geologists divide all rock into three primary groups, depending on their mode of formation:

1. *Igneous rocks* formed by the crystallisation and solidification of a molten silicate magma;
2. *Sedimentary rocks* formed by the sedimentation and subsequent lithification of mineral grains, either under water or, more rarely, on an ancient land surface;
3. *Metamorphic rocks* formed by the effect of heat and pressure on igneous or sedimentary rocks for geological periods of time with the consequent development of new minerals and textures within the pre-existing rock.

These modes of origin result in a small number of typical discontinuity patterns within the rock masses, as illustrated in Figure 42. They arise from lamination or orientation of mineral grains, systematic variations in grain size, or from shrinkage due to drying (sediments) or cooling (igneous). These relatively simple patterns are often overprinted with additional discontinuities, resulting from regional faulting or folding of the earth's crust. However, examination of a geological map of the area

Blocky:
Typical of igneous rocks such as granites; also massive sandstones and limestones

Irregular:
Typical of igneous rocks and some hard sandstones and limestones

Flaggy:
Typical of bedded sedimentary rocks such as siltstones and sandstones

Bladed:
Typical of many metamorphic rocks such as phyllites and schists

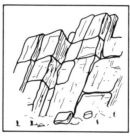
Slaty:
Slates and some shales

Columnar:
Typical of igneous rocks which occur as sheets such as basalt

Figure 42 *Idealized sketches of common natural outcrop forms*

will usually give a clear indication as to whether regional deformation is likely to introduce additional complexity to the discontinuity pattern. For example, superimposed periods of tight folding and/or extensive shearing and thrusting will leave distinctive outcrop patterns on the geological map. These patterns will indicate complex discontinuity or faulting patterns which seriously reduce the armour stone block sizes, particularly if the deformation style is brittle.

Within this three fold primary grouping, the rock types may be sub-divided into some 18 families of rocks, as indicated in Tables 11–13. Each family has essentially similar characteristics, so that the engineering properties of these rock types can be presented in general summary form as shown in Table 14. The figures in Table 14 refer only to fresh rock. Geological weathering, which is not always identifiable by simple visual inspection, can often seriously reduce the quality of the rock concerned.

Weathering

Rock weathering is brought about by the exposure of the rock during long periods of geological time to the climatic conditions at the earth's surface and involves mechanical disintegration and chemical decomposition acting together. The effects are most marked in humid, hot climates, but it must be remembered that climatic conditions pertaining in the geological past as well as present conditions may influence the weathered state of a given rock mass. The influence of climate on the weathering profiles of rock is illustrated in Box 14 for typical quarries in three different climates: north-western European, tropical hot-wet and hot desert.

The degree of weathering can often be assessed subjectively, though in some cases recourse has to be made to laboratory petrographic examination. A number of weathering classifications have been proposed, and the system shown in Table 15 is of general validity for field assessment and can be readily correlated with

Table 11 Igneous rocks: strong rocks with interlocking crystals

Rock type geological name	Typical grain size range (mm)	Visible voids	Typical texture	Typical rock mass appearance	Typical joint spacing (m)	Typical basic fragments shape*	Typical geological distribution
Granite	20–2	Common, small or microscopic	Isotropic, interlocking crystals	Blocky/irregular	0.5–10	Cubic	Mountain and shield areas, extensive
Diorite	3–1	Rare	Isotropic interlocking crystals	Blocky/irregular	0.2–10	Cubic	Localized areas
Gabbro	5–2	Very rare	Isotropic, interlocking crystals	Blocky/irregular	0.5–10	Cubic	Mountain areas localized
Rhyolite	Grains not visible to unaided eye	Rare	Locally variable microfractures common	Irregular	0.1–2	Cubic/prismatic	Localized areas
Andesite	Grains not visible to unaided eye	Rare, small and large	Isotropic uniform	Irregular/tabular	0.2–2	Cubic/prismatic/tabular	Extensive sheets
Basalt	Grains not visible to unaided eye	Common, large and small	Isotropic uniform	Irregular/blocky/columnar	0.2–5	Cubic/prismatic/elongate	Extensive sheets
Serpentinite	Grains not visible to unaided eye	None	Microfractures common	Irregular	0.05–1	Cubic	Mountain areas localized

*Typical fragment shapes when unmodified by external stress-induced fractures

Table 12 Sedimentary rocks: bedded rocks with grains cemented by interstitial material

Rock type name	Typical grain size range (mm)	Visible voids	Typical texture	Typical rock mass appearance	Interbedded or associated rocks	Typical bed thickness (m)	Typical basic fragment shape*	Typical geological distribution
Quartzite	2–0.2	Very rare	Narrow grain size range, compact uniform	Blocky/flaggy	Sandstones Siltstones Shales	0.1–5	Cubic/tabular	Localized areas
Sandstone	2–0.6	Uncommon	Narrow and wide grain size ranges, variable intergrain cement	Flaggy/blocky	Siltstones Shales	0.1–10	Cubic/tabular	Extensive areas
Siltstone	0.06–0.002	Very rare	Narrow size range, often laminated	Flaggy	Sandstone Shales Limestones	0.05–1	Tabular	Extensive areas
Shale	<0.002	Very rare	Narrow grain size, laminated	Flaggy/slaty	Sandstones Siltstones Limestones	0.005–0.01	Tabular	Extensive areas
Limestone	2–0.01	Common, large and small	Narrow grain size ranges or cemented fragments	Blocky/flaggy	Marls Shales Sandstones	0.5–1	Cubic/tabular	Extensive areas
Chalks	<0.01	Rare	Narrow grain size range	Blocky/flaggy	Limestones Marls Flinty soils	0.1–2	Cubic/tabular	Extensive areas

*See note to table 11

Table 13 Metamorphic rocks: crystals usually interlocking but grain alignments common

Rock type name	Typical grain size range (mm)	Visible voids	Typical texture	Typical appearance	Typical joint spacing (m)	Typical basic fragment shape*	Typical geological distribution
Slate	<0.01	Very rare	Narrow grain size range orientated grains	Slaty	0.002–0.1	Tabular	Localised areas
Phyllite	0.5–0.1	Rare parallel foliation	Narrow grain size range orientated grains	Bladed	0.01–0.2	Prismatic/tabular/elongate	Extensive areas
Schist	5–0.5	Rare parallel foliation	Orientated grains with wide size ranges	Bladed	0.01–1	Prismatic/tabular/elongate	Extensive areas
Gneiss	5–0.5	Vary rare	Orientated grains with wide size ranges	Blocky/Irregular	0.5–10	Cubic/tabular/elongate	Extensive areas
Marble	3–0.1	Very rare	Narrow interlocking grain size ranges	Blocky	1–10	Cubic	Localised areas

*See note to Table 11

Table 14 Some generalised engineering characteristics of unweathered common rocks

Rock group name	1 Rock mass density (t/m³)	2 Unconfined compressive strength (MPa) × 10^8	3 Water absorption (%)	4 Porosity (%)
Igneous				
Granite	2.8–2.5	260–160	0.2–2.0	0.4–2.4
Diorite	3.1–2.6	260–160		0.3–2.7
Gabbro	3.2–2.8	280–180	0.2–2.5	0.3–2.7
Rhyolite	2.8–2.3	260–100	0.2–5.0	0.4–6
Andesite	3.0–2.4	260–160	0.2–10	0.1–10
Basalt	3.1–2.5*	280–160	0.1–1	0.1–1
Sedimentary				
Quartzite	2.8–2.6	260–220	0.1–0.5	0.1–0.5
Sandstone	2.8–2.3	220–15	1.0–15	5–20
Siltstone	2.8–2.3	100–60	1.0–10	5–10
Shale	2.7–2.3	60–15	1.0–10	5–30
Limestone	2.7–2.3	120–30	0.2–5.0	0.5–20
Chalks	2.3–1.5	30–5	2.0–30	20–30
Metamorphic				
Slate	2.8–2.7	120–70	0.5–5	0.5–5
Phyllite	2.7–2.3	90–60	0.5–6.0	5–10
Schist	3.2–2.7	120–70	0.4–5.0	5–10
Gneiss	2.8–2.6	260–150	0.5–1.5	0.5–1.5
Marble	2.8–2.7	240–130	0.5–2.0	0.5–2.0

*Dense basalts, excluding vesicular basalts
Column 1: Typical range
Columns 2, 3 and 4: The typical ranges of values to be expected, data typically has a strong positive skew
Column 4: Porosity defined as the ratio pore volume: total volume for typical coherent individual blocks
Sources: Carmichael (1982); Bell (1983); Winkler (1973); Fookes and Poole (1981)

petrographic evaluations. As indicated in Table 15, rocks of weathering grade III or higher are generally unsuitable for the outer layers of a marine structure because they will have poor durability characteristics when subjected to marine conditions. They are also unlikely to be produced at the quarry as large unflawed blocks.

Selection of rock materials for marine structures must take account of the grade of geological weathering exhibited by the material, but it is also clear that some of the rock groups listed in Tables 11–13 are potentially more likely to be suitable than others, particularly for armouring and underlayers of a structure. An indication of suitability can be obtained by comparison of those rock properties listed in Tables 11–14 against the required idealised properties in Table 16.

However, under certain circumstances rock materials of lower quality may have to be selected. In such cases the satisfactory performance of the structure will depend on appropriate changes being incorporated into the design which will allow for the reduced quality of the materials used.

Box 14 Influence of climate on weathering of rock in quarries

Proper recognition (Fookes, 1980) should be taken of the weathering profiles existing within the rock face being excavated. In general, the igneous and metamorphic rocks, which were formed in conditions of high temperature and pressure not found at the earth's surface, show the greatest tendency for well-developed weathering profiles, with the breakdown of their mafic minerals in particular, to form secondary minerals which can be clearly seen under the microscope. The sedimentary rocks break down less easily because they were formed at conditions of temperature and pressure existing at the earth's surface, and therefore are probably not too much out of harmony with the climatic regime they experience at the present ground surface.

Figure 43 shows three quarries in an extremely simplified form; the right-hand one (No. 1) portrays an igneous rock quarry in north-western Europe. Immediately prior to the current geological processes operating on the earth's surface, the Pleistocene ice advances have planed off much of the existing residual soil and weakened weathered rock to leave fairly fresh rock generally containing only faintly to slightly weathering grades of material. Often this rock is shattered near the surface by freeze/thaw cycles and it may be covered with glacial or post-glacial debris. (Fookes, 1980.)

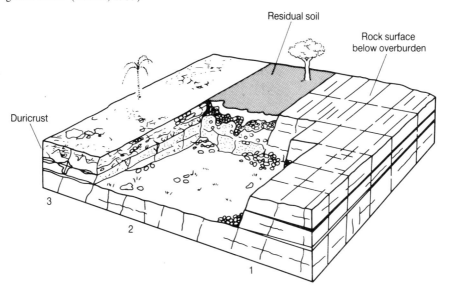

Figure 43 *Three idealized quarries showing different types of weathering*

The centre quarry (No. 2) shows the same rock in a tropical hot and wet climate. Here, physical planing by ice during the Ice Ages did not occur, and therefore a thick development of residual soil still exists sometimes lateritic. The soil passes down into highly weathered rock which in turn passes down into less weathered and finally into the fresh rock. In such a quarry the various grades of weathering can easily be seen and recognised.

The left-hand quarry (No. 3) portrays a limestone quarry in a hot desert climate. The rock is usually fairly porous and weakened some metres below the existing ground surface by leaching but with a thick hardened surface duricrust composed of calcrete (the specific type of duricrust formed by the upward leaching of a limestone bedrock).

Table 15 Weathering and alteration grades

Term	Grade*	Description†	Materials characteristics†
Fresh	IA	No visible sign of rock material weathering	Properties not influenced by weathering. Mineral constituents of rock are fresh and sound
Faintly weathered	IB	Discoloration on major discontinuity (e.g. joint) surfaces	Properties not significantly influenced by weathering. Mineral constituents sound
Slightly weathered (this grade is capable of further sub-division)	II	Discoloration on discontinuity surfaces indicates weathering of rock material. All the rock material may be discoloured by weathering and may be somewhat weaker than in its fresh condition	Properties may be significantly influenced by weathering. Strength and abrasion characteristics show some reduction. Some alteration of mineral constituents with microcracking
Moderately weathered	III	Less than half of the rock material is decomposed and/or disintegrated to a soil. Fresh or discoloured rock is present either as a continuous framework or as corestones	Properties will be significantly influenced by weathering. Soundness characteristics markedly affected. Alteration of mineral constituents common and much microcracking
Highly weathered	IV	More than half of the rock material is decomposed and/or disintegrated to a soil. Fresh or discoloured rock is present either as a discontinuous framework or as corestones	Not generally suitable for armour or underlayer materials—sometimes appropriate for core material if better material is not available
Completely weathered	V	All rock material is decomposed and/or disintegrated to soil. The original mass structure is still largely intact	Not suitable for armour or underlayers but may be suitable for core materials in some cases if other materials are not available
Residual soil	VI	All rock material is converted to soil. The mass structure and material fabric are destroyed. There is a large change in volume, but the soil still has not been significantly transported	May be suitable for random fill or core material under certain circumstances (e.g. in conjunction with geotextiles)

*After London Geological Society Working Party Report (1977). Also BS 5930
†For general engineering characteristics, see Fookes et al. (1971)

Table 16 Idealised typical rock quality parameter ranges for marine structures. Note additional test parameter values relevant to specification have been assembled in Section 3.4 and Appendix 1

Test or observation	Facings or armour	Underlayers	Core/fill
Weathering grade	I–II	I–II	I–II
Discontinuity spacing	1 m+	0.5 m+	0.2 m+
RQD (%)*	80–100	75–100	55–100
Porosity (%)	0–5	0–10	0–10
Water absorption (%)	<2.0	<2.5	<3.0
Unaxial compressive strength strength (σ_c) (MPa)	>100	>100	>50
Rock density (ρ_r) (kg/m^3)	>2600	>2600	>2000

*RQD: Rock Quality Designation: the ratio of intact rock core sections greater than 100 mm length to the total length drilled, expressed as a percentage (Wakeling, 1977; Fookes and Poole (1981); Poole et al., 1985; Allsop et al., 1985)

Many structures provide exceptions to the ranges of values indicated in Table 16, but the integration of Tables 11–14 with Table 16 allows Table 17 to be constructed, which may assist at the initial desk study stage of the assessment of potential construction materials. Once the expected properties of the various sources have been evaluated in the general way, the essential walk-over survey, detailed site investigation and laboratory testing programmes will establish the actual materials properties of the potential sources.

Table 17 Generalised evaluation of the use of fresh rock in marine structures

Rock group	Armour/ facings	Underlayers/ filters	Core/ fill	Comments
Igneous				
Granite	*	*	*	Good equant shapes. Beware of weathered rock
Diorite	*	*	*	Good equant shapes. Beware of weathered rock
Gabbro	*	*	*	Good equant shapes. Beware of weathered rock
Fresh rhyolite		*	*	Blocks typically angular, equant but small
Andesite	*	*	*	Block sizes sometimes small, beware of weathered rock
Basalt	*	*	*	Equant blocks sometimes small, beware of weathered or vesicular† rock
Serpentine	*	*	*	Often the blocks are angular and small
Sedimentary				
Pure quartzite	*	*	*	Sometimes poor tabular shapes. Abrasion resistance sometimes poor
Sandstone	*	*	*	Sometimes tabular and soft. Abrasion resistance sometimes poor
Siltstone			*	Usually tabular and of small size
Shale			*	Small tabular fragments, soft††
Pure limestone	*	*	*	Sometimes tabular, sometimes soft††
Chalks			*	Soft, easily eroded
Metamorphic				
Slate			*	Tabular shape, hard, has been used as armour
Phyllite			*	Elongate shapes, often soft
Schist	*	*	*	Elongate and tabular shapes common
Gneiss	*	*	*	Good equant shapes, hard; beware of weathered rock and micaceous planes
Marble	*	*	*	Usually good equant shapes

*Potentially of use
†Of which pumice is an extreme example
††When it is necessary to use these materials, consideration should be given to design options using geotextiles

3.1.3 ALLUVIAL, GLACIAL AND MARINE SOURCES OF ROCK MATERIALS

The previous section has considered the whole range of rock types available. Typically, these materials are obtained by conventional quarrying operations. However, there are also extensive areas where rock does not crop out at the surface and recourse has to be made to deposits of glacial till, river or marine sediments.

Large blocks of stone are not common in marine or river gravels, which will usually only contain materials below about 3 kg in weight with a high proportion of fine materials. Four exceptions to this generalisation will provide varying proportions (but usually small percentages) or larger material, usually in the 1–3 tonne range:

1. Glacial tills and boulder clays (glaciated regions of the world only);
2. Boulder river gravels to be found near the headwaters of rivers and sometimes in the outwash fans associated with rivers, particularly in desert regions subject to flash floods;
3. Residual blocks of lithified sedimentary material left after the erosion of weaker, less well-cemented material which surrounds them;
4. Duricrust materials. These rocks are common in desert regions where modern sediments have been lithified by salts carried to the surface in solution by capillary rise. Evaporation causes precipitation of the salt which cements the sedimentary grains. Care needs to be taken in selection to ensure that the cementing materials involved are carbonates or silica, rather than halite or gypsum, because the latter are soluble in water;

5. Concentrations. These are formed where a cementing material (e.g. calcite, silica) has concentrated and crystallised in otherwise soft deposits. They vary considerably in size, from flint modules in chalk up to the sarsen stones used in Stonehenge, UK.

The small percentage of large blocks present in such sources together with the typically high percentage of fines often requires appropriate methods of selecting, handling and transporting materials from an extensive source area, though in some situations the finer fractions have already been removed by erosion, leaving boulder fields of large blocks.

The rock types present in these types of deposit are usually durable and with the appropriate properties for satisfactory armouring, facing or underlayer materials. However, because of their transportation to the source area by environmental processes the boulder gravels and tills tend to have rounded shapes while the residual blocks and duricrust materials are often tabular.

Small materials from marine and river gravels are also typically durable, but the deposits can vary in both size grading and grade proportions over very short distances (even tens of metres) and typically usually contain a high proportion of fine sands, silts and clays. Thus materials from such sources may be appropriate as core material but rarely provide size ranges appropriate for armour or underlayers. The possible variability of such deposits requires that the quality and gradings of the material will need to be monitored carefully during the construction phase.

3.2 Properties and Functions

This section provides detailed information, supplementary to Section 3.1, that is needed to integrate the available rock properties with the required functions of the design solution (Figure 41). The following properties may be treated separately:

- Intrinsic properties of the rock mass (e.g. mainly material properties such as density);
- Properties affected by production (e.g. geometric properties such as shape);
- Execution-induced properties (e.g. bulk properties of granular media);
- Durability (i.e. resistance to degradation in service).

The flow diagram in Figure 44 shows those relationships between properties and functions which are of particular importance within the different structural elements, namely armour, underlayer and core. This section concentrates on the intrinsic and production affected properties shown in the left-hand side of the figure. The execution-induced properties are bulk properties which are mostly covered by the discussion of geotechnical parameters in Section 5.2.2.

In simplistic terms, the designer uses the system response approach (Section 2.2.2), in which the loadings described in Chapter 4 operate on the rock properties according to processes described and modelled in Chapter 5. Using the available tools and models, the structure can be designed to perform the functional requirements. An additional problem is that these functions will change with time in service because of rock-degradation processes. Therefore the designer's skill must also encompass consideration of durability and degradation processes. A provisional degradation model for armourstone which considered rock and environmental parameters is provided in Section 3.2.4 so that the whole-life consequences of using less than ideal materials may be considered at the design stage. As a supplement to the guidance in Figure 44, the important functional requirements of the three main elements of a rock structure in coastal and shoreline engineering are also worth briefly considering.

Armour layers and facings

The outer layers of a coastal or shoreline protection structure are subject to the

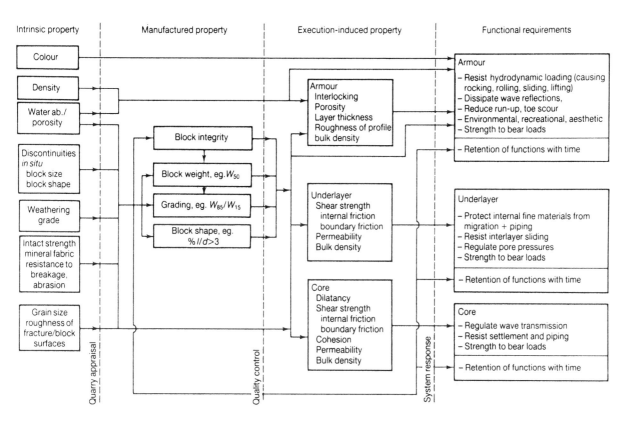

Figure 44 Flow diagram for rock properties and functions

most severe environmental conditions, and consequently require high strength and durability characteristics as well as stringent geometric constraints on the production of block size, weight, shape and grading. The main functional requirements are given in Figure 44. In the case of dynamically stable structures block movement occurs under wave attack, so that breakage, attrition, general wear and rounding is potentially more rapid and therefore durability potential is more important. There will also be implications for the design specification as the grading curve will modify with time.

Underlayers and filters

In a conventional rock breakwater or coastal protection structure the primary armour facing layers rest upon an underlayer of secondary armour which, in turn, rests on graded rock underlayers which act as a series of filters designed to prevent the net underlying material being drawn through them by the hydraulic forces induced by wave action. These filters rest on the core of the structure, thus preventing its removal by wave erosion.

The functional requirements of underlayers and filters (given in Figure 44) can be subdivided into hydraulic and structural requirements as for the armour layers. However, these requirements differ in important ways from those of armouring and the aesthetic aspects of the materials used are also no longer a factor.

The underlayer system will normally be required to contribute to the energy dissipation of the armour by turbulent flow through the void spaces. It will also need to be designed to prevent the hydraulic removal of particles from the lower layers, or 'piping', of core materials through the layer. The most important design considerations in this context are the grading curves, particle shape characteristics and rock density relative to the water. Layer packing is also important in relation to the porosity which, together with particle shape and grading, will control interlock between particles and, hence, the shear strength of particles within the pack and between layers.

Core materials

The principal requirements of core material, whether designed as a permeable or impermeable structure, is that it should have sufficient strength to support the overall structure, and that its compaction characteristics should be such as to preclude significant post-construction settlement. Material density, shear strength grading and particle shape will be the most important parameters for core materials but these materials should also be free of soluble salts or gypsum, which may be removed in solution.

3.2.1 INTRINSIC PROPERTIES

3.2.1.1 Colour

The aesthetic requirements of armour in amenity areas can make the correct choice of rock colour a vital consideration. Often, to help integrate the new structure with the landscape, the public preference will be for a rock with an appearance similar to that in local cliff and hill outcrops, especially if these are of lighter-coloured sedimentary rocks. Of the rock sources under consideration, the aesthetically preferred rock will not usually correspond with the one of greater durability, and therefore rock appearance considerations will sometimes influence the design process through Environmental Assessment (Section 2.4). Note that biological colonisation (by seaweed, barnacles, etc.) will rapidly alter the intertidal zone appearance and that colour can only be important for the visible upper armour layers.

3.2.1.2 Density

Determination of the rock density is crucial because of:

1. The correlation between density and durability;
2. The use of density in armour stability formulae and geotechnical equations;
3. The use of density in size–weight relations required for dimensioning of the design.

The rock density will generally correlate with the darkness of the rock's minerals and decrease with porosity and degree of weathering (see Tables 14 and 15). Density variation is a good indicator of quality variation and should be tested where this is suspected. Most geologists can estimate the density of rock to within 100 kg/m^3 from a hand specimen. In general, dealing with one type of rock in a quarry, the 90% exceedence value is not more than 100 kg/m^3 less than the average density.

There are several rock density parameters used in this manual (Chapter 5) and their definitions and relationships with the water-absorption parameter (for example, as given by BS 812:1975) depend on the method of sample mass determination, as follows:

Oven-dried relative density $= C/(A-B) = C/D$ $\qquad \rho_r/\rho$
Saturated surface dry relative density $= A/(A-B)$ $\qquad \rho_{ssd}/\rho$
Water absorption (%) $= 100(A-C)/C = A/D$ $\qquad w_{ab}$

Where $A =$ saturated and surface-dried mass,
$B =$ apparent mass when saturated and submerged,
$C =$ oven-dried mass,
$D =$ mass equivalent water volume.

The mass density of the rock equals the oven-dried relative density multiplied by the density of water used in the test (ρ). The oven-dried relative density

> **Box 15** Effect of water absorption on relative buoyant density
>
> Armour stability design formulae (e.g. equations (5.44) and (5.45)), employ the relative buoyant density (Δ) in relation to sea-water density on site (ρ_w) to indicate the effect of rock density and water absorption on stability. Elsewhere in this manual water absorption is neglected and the following definition is used:
>
> $$\Delta = (\rho_r/\rho_w) - 1$$
>
> Its precise definition, important for absorbent rocks, is written:
>
> $$\Delta = (\rho_{ssd}/\rho_w) - 1$$
> or $\quad \Delta = (\rho_r (1 + 0.01 w_{ab}) - \rho_w)/\rho_w$
> since $\quad \rho_{ssd} = \rho_r (1 + 0.01 w_{ab})$
>
> It is important therefore to use ρ_{ssd} (or ρ_r and w_{ab}) in calculating Δ for water-absorbing rocks, as ρ_{ssd} can depart significantly from ρ_r. Similarly, the density parameter being referred to in the specification must be stated clearly.

(dimensionless) is therefore numerically the same as the mass density (units of $1000 \, kg/m^3$) if the water density during testing is equal to $1000 \, kg/m^3$.

Special care should be taken to ensure correct use of density terms, as variation in meaning between internationally-used terms abounds. For example, for ρ_r/ρ, the terms bulk specific gravity (ASTM C 127, 1984) and 'apparent relative density' (PIANC) are used.

The consequences of using denser rock armour of the same nominal weight for the typical case of filling a given area with a double layer of armour can be considered in terms of:

Increased total weight material required
Increased total number of blocks required
Reduced layer thickness
Increased stability
Increased durability

For the same stability, a denser rock may allow substantial decreases in the size of armour and total weight of armour needed (see Box 16). The relative merits are often complex to evaluate, particularly for a system of standardised gradings, but these can be considered objectively using the available methods (see Section 3.2.3.4 and the example in Section 6.2.5). Stability is a function of relative buoyant density, Δ (see equations (5.44) and (5.45) in Chapter 5) and the effect of water absorption on Δ is discussed in Box 15. An advantage of using less dense material for core arises from its greater volume per unit weight of rock transported to site, since it is the bulk volume making up the design levels which is important provided the geotechnical properties are satisfactory.

3.2.1.3 Water absorption and rock porosity

Not to be confused with the porosity of a bulk granular material, the porosity of a piece of rock is the volume of voids per unit volume of the rock, and is therefore approximately equal to $w_{ab} \times \rho_r$. Water absorption at atmospheric pressure is the mass of absorbed water per unit dry mass of rock, which is lower than the absorption at full saturation of the voids (see Table 14).

Water absorption, which is easy to test, gives the most important single indication of in-service durability. It is particularly important to restrict the acceptable w_{ab} values for armourstone which is to undergo severe cycling stresses such as salt crystallisation and in impure limestones which may be susceptible to dissolution. However, the results can be misleading if taken in isolation, since rocks with w_{ab}

> **Box 16** Effects of rock density and water absorption on required armour block weight and total armour weight
>
> For identical stability according to the design equations (5.44) and (5.45) the median armour block weight W_{50} is proportional to ρ_r/Δ^3 and the total weight of armour in the armour layer is proportional to ρ_r/Δ. The effect of true (saturated surface dry) rock density, ρ_{ssd}, on block weight and total armour weight are shown in Figure 45 for rocks with water absorption, w_{ab} of 0%, 2% and 6%. For example, for rocks with 0% water absorption, Figure 45(a) shows that with ρ_r of 3.1 and 2.6, the W_{50} required for the denser rock is 0.38/0.71 times that of the dense rock. Similarly, Figure 45(b) indicates that 1.68/1.53 times more total weight of armour is required for the less dense rock. See also Box 29 (note, ρ_r is the oven-dry rock density).
>
>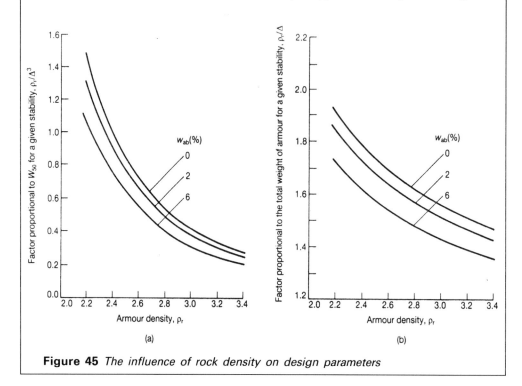
>
> **Figure 45** *The influence of rock density on design parameters*

greater than 4% which have a free-draining pore structure have been shown to perform well in service for certain applications.

For core material in which a high permeability through the structure is envisaged, solution losses can be considerable because of the relatively high rock surface area per unit volume occupied by these smaller particles. For porous limestones it may be necessary to specify higher-quality core than armour (e.g. as specified by Fookes and Thomas, 1986, who give details of a practical approach to materials selection used where only low-quality carbonate rocks were available). The evidence for significant dissolution of limestones in sea water is by no means clear and testing may be necessary.

3.2.1.4 Discontinuities

A distinction can be made between the intact rock material and the *in-situ* rock mass in the quarry. A discontinuity is defined as any surface in the rock along which cohesion is low or is lost between the intact rock on either side. The most common type is the joint. Others include weak bedding planes or schistosity planes, weakness zones and faults. Discontinuities provide essential information on:

In-situ block shapes
In-situ block sizes
Likely states and distributions of weathering.

The influence of rock type in association with discontinuity pattern upon the block shapes was summarised in Figure 41. The spatial distribution of discontinuities determines the *in-situ* block size distribution which is crucial in predicting the percentage of extracted blocks that can be used for armourstone and the yields of the various sizes.

The accurate prediction of the blasted block sizes is fraught with difficulties, and is the subject of ongoing research. The predicted fragmentation curve can be used to make a comparison of the initially proposed material design parameters (e.g. median block weight and grading of primary armour) with the likely source availability and rates of supply of such materials. Conversely, with this block size prediction and/or trial blast information it may be possible in a dedicated quarry to design the structure around the expected as-blasted block sizes and to aim for a 100% utilisation of blasted rock (see Vrijling and Nooy v.d. Kolff, 1990; Baird and Woodrow, 1988; and Section 6.1.6.1).

The prediction of fragmentation will require specialist expertise and involves the following stages:

1. Field techniques for assessing the distribution, orientation and frequency of the discontinuities:
   (a) From exposed discontinuities, use photogrammetric methods (Rengers *et al.*, 1988) and manual methods to give scanline and other spacing data (Brown, 1981) or detailed scanline mapping (Wang *et al.*, 1991; see Figure 46);
   (b) From unexposed discontinuities, use boreholes to give RQD (see Table 16), downhole techniques (e.g. closed-circuit television, CCTV), sound velocity profiling (LCPC, 1989)—both potentially very useful—and shallow seismics, which gives more general information.

2. Calculation of *in-situ* block sizes. Results based on volumetric joint count, block size index (Brown, 1981) and joint density (Rouleau and Gale, 1985) are often used to obtain average block sizes. More useful is the complete size distribution which can be assessed from mean joint spacings using new simple formulae (see Box 17).

3. Blast design and the prediction of the results of blasting together with trial blast data if available (Section 3.4.2.2). In addition to spatial information, the persistence, aperture, filling and water seepage of discontinuities (Brown, 1981) can give a general indication of rock quality, especially with respect to the state of weathering and therefore rock durability.

3.2.1.5 Weathering grade

The state of physical or chemical weathering or hydrothermal alteration of the bedrock near the earth's surface can be anything from minimal (weathering grade I), giving no signs of mineral discoloration or alteration, through to total (weathering grade VI) in which the rock is reduced to a soil (see Table 15 for grade definitions). The assessment of weathering grade which can be performed visually by an experienced geologist (see Lee and de Freitas, 1989, for a recent review) is useful for the rapid identification of sources or regions within a quarry which are unsuitable for rock armourstone exploitation. The cut-off for armourstone and underlayer acceptability usually falls within grade II. Little information exists on the rate of degradation of weathered materials in the core of the structure, but it will usually be acceptable to relax the quality requirement for core materials.

Certain tests (e.q. sulphate soundness, freeze/thaw, Methylene Blue absorption) have been developed to predict the degree to which geological weathering has affected the rock's ability to resist degradation in engineering time (see Box 18).

Figure 46 *Scan line mapping in progress at a quarry (courtesy J.-P. Latham)*

These provide quantitative support for the field classification given in Table 15 and, together with water absorption, form the basis of durability testing.

3.2.1.6 Intact fabric strength

The strength of a piece of rock which has no discontinuities or major flaws (i.e. intact rock) is dependent on the strength of the rock minerals, matrix cement, grain boundaries and pore structure weaknesses as well as the shape, size and degree of interlocking (i.e. texture) of the grains. This mineral fabric strength can be divided, according to the main types of loading in hydraulic structures, into resistance to (1) breakage and (2) abrasion. The importance of these properties and their specification for armour layers is dependent on site conditions such as the type of stability in the top layer (static or dynamic), foreshore attrition agents and wave energy, but even these strength properties may be overshadowed by the importance of resistance to weathering, particularly if the meteorological climate on-site is severe.

Box 17 Prediction of *in-situ* block size distribution

Discontinuities in most rock masses can be classified into sets, usually three, according to their orientations. Each set of discontinuities has a mean orientation and a mean spacing valued (Hudson and Priest, 1979). If it is assumed that all discontinuities in the rock mass are planar and persistent and can be divided into three sets, the *in-situ* block size distribution can be predicted by the following equation (Wang et al., 1990):

$$V_i = C_i * (\gamma_1 \gamma_2 \gamma_3)/[\cos(\theta)\cos(\phi)\cos(\alpha)] \quad i = 10, 20, \ldots 100$$

where V_i is the percentage passing volume (m³) on the cumulative distribution curve and C_i is a coefficient, where i corresponds to percentage. γ_1, γ_2 and γ_3 are the mean spacing values of the three sets, respectively, in metres. θ, ϕ and α are angles between mean orientations of the sets, which can be determined on a stereographic plot, as shown in Figure 47. The great circles on the plot represent the mean orientations of the discontinuity sets. The product $\cos(\theta)\cos(\phi)\cos(\alpha)$ will be 1 if the three sets are mutually perpendicular.

The coefficients C_i required to obtain the *in-situ* block volumes are listed in the table below. Their derivation is based on the further assumption that the discontinuity spacings in each set have a negative exponential distribution. The natural discontinuity spacings could be better fitted by lognormal distributions, but the given coefficients are still applicable.

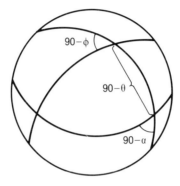

Figure 47 *Measurement of the angles of θ, φ and α between the mean orientations of rock mass discontinuities*

Coefficients and their 90% confidence intervals

V_i(m³)	Coefficient C_i	90% confidence intervals
V_{10}	0.322	±0.131
V_{20}	0.710	±0.249
V_{30}	1.207	±0.424
V_{40}	1.852	±0.645
V_{50}	2.708	±0.984
V_{60}	3.980	±1.550
V_{70}	5.867	±2.597
V_{80}	8.948	±4.515
V_{90}	15.332	±9.531
V_{100}	38.922	±23.734

Resistance to breakage

Experience of handling blocks in quarries suggests there are two types of breakage in armourstones:

Type 1: Breakage is into major pieces along flaws;
Type 2: Breakage is along new fractures usually resulting in edges and corners being knocked or sheared off.

In general, for large pieces of rock the possibility of major flaws is increased and so their overall strength to resist breakages is likely to be reduced, the intact strength being less significant. For armourstone blocks larger than 50 kg and 300 kg

> **Box 18** Weathering in geological and engineering time
>
> A distinction should be made between weathering in geological time (usually, many tens of thousands of years) and engineering time (typically, of perhaps 50 years for coastal structures). The general principle is that a weathered rock may still be initially strong but will usually degrade rapidly in service. For a given, often satisfactory, rock type such as granite, such geological weathering may have rendered the rock unsuitable for armour, not necessarily because it lacks overall strength at the time of excavation or winning from the ground but because it is likely to break down rapidly through spalling and abrasion when exposed to the rigours of the site location. This is because geologically weathered rock often contains an abundance of microcracks and deleterious minerals such as swelling clays, and is highly susceptible to rapid degradation by physical weathering and other disruptive forces in service such as impacts and abrasion. In some cases it may also be affected by chemical weathering if the blocks are in a hot and wet climatic regime.

Figure 48 *Type 1 breakage of gabbro block with many joints (courtesy G. J. Laan)*

(depending on the joint pattern in the quarry), Type 1 Breakage is more common. Flaws in armourstone blocks constitutes a major problem for quality control. Their presence is linked to production methods and their identification is described in Section 3.2.2.1. A typical example of Type 1 breakage of blocks is shown in Figure 48 for a gabbro.

Type 2 breakage in armourstone can result in rounding of blocks during handling, giving a reduced interlock and porosity on the armour layer. This effect will be minimised by using rocks with higher fracture toughness, tensile and compressive strengths. Test methods to determine uniaxial compressive strength, Brazilian indirect tensile strength, point load strength index, fracture toughness and aggregate impact value are given in Section 3.3. Fracture toughness, an unfamiliar concept in engineering geology, is elaborated upon in Box 19. The two extreme behaviours require that the fracture toughness and the block integrity should be assessed.

The French (LCPC, 1989) strongly recommended the measurement of Sound

Box 19 Fracture toughness and other intact strength measures

The importance of fracture toughness in assessing rock suitability in coastal engineering was first described in Dibb et al. (1983), and, because of its likely future role, the concept is elaborated upon here. For further information, Atkinson (1987) is a useful general reference.

The uniaxial compressive strength test is a widely used measure of intact rock strength. So too is the Brazil test, which is employed as an indirect measure of tensile strength. The point load test has often been reported as an indirect index of compressive or tensile strength, and its wide use in practice is due to the speed and simplicity of testing, the ease of specimen preparation and its field utility.

The above strength tests measure the stress level or load at which the particular specimens of rock will break. If the specimen contains hidden weaknesses due to flaws on a larger scale than the rock fabric, these will often be exploited. Results are therefore particularly dependent on the relation between specimen size (usually about 50 mm diameter cores) and inherent weakness sizes. A typically high test variability of these strength measures was reported by Gunsallus and Kulhawy (1984) for a suite of sedimentary rocks.

By contrast, the fracture toughness of a rock is experimentally derived by generating a new crack independent of dominant flaws in the test specimen. The result is a controlled simulation of sudden propagation from a crack tip or sharp flaw on the scale of the mineral fabric. The fracture toughness, K_{IC}, is defined as the characteristic level of stress intensity ahead of the crack tip that is required to propagate the new crack catastrophically through the mineral fabric of the rock. The stress intensity factor is dependent on the applied stress and the crack geometry. It is a fundamental material property that is independent of the discontinuities that may exist in large armour blocks, and reflects the weakness imposed by the pore structure together with the strength of the rock minerals, matrix cement and grain boundaries.

In the above-mentioned study, the fracture toughness was shown to have the least test variability compared with the three other strength tests referred to above, although it is not certain that this would be the case for all rock types and test facilities. Figure 49 indicates the K_{IC} results for fresh rocks of different mineralogies (data from Meredith, 1988).

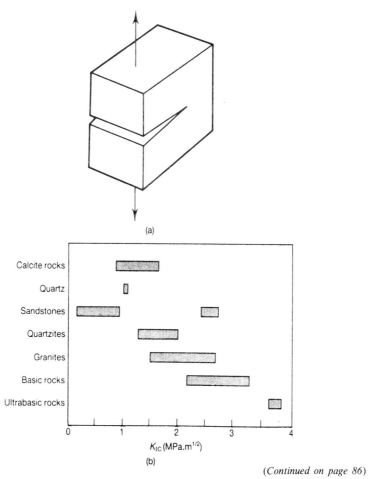

(Continued on page 86)

Figure 49 (a) Mode I crack geometry; (b) Synoptic diagram of the variations of future toughness of rocks with various mineralogical compositions

> **Box 19** (Continued)
>
> The fracture toughness test, which has recently been developed to a standard recommended method (ISRM, 1988), develops a crack in mode I, the opening or tensile crack mode. The fracture toughness values are called K_{IC} values for critical intensity factor, mode I. Modes II and III are crack sliding and crack tearing modes which both involve shearing. The mechanisms involving impacts of wind- and waterborne particles which cause small tensile cracks at the surface of blocks suggest that a good correlation between K_{IC} and resistance to abrasion by attrition can be expected. Data from Cost (1985) also suggested a tentative correlation between resistance to removal of material from corners of blocks during handling and transport, and K_{IC}.
>
> A considerable body of research data now exists on correlations between the traditional strength measures and K_{IC}. With the new ISRM standard method for K_{IC} further research data will be emerging in the near future. In dynamic fracture processes (e.g. impacts) K_{IC} is considered a promising strength parameter, particularly for use in detonics and fragmentation applications (Grady and Kipp, 1987).

Velocity as a means for detecting the intrinsic fabric strength. The index of continuity, I_c (NFP 18-556, 1980), is the ratio (%) of measured velocity to the theoretical velocity calculated using the known sound velocities of the individual minerals and their proportions obtained from petrographic analysis. The I_c value decreases with increasing prevalence of microcracks, rock porosity (n_r) and altered minerals. Moisture content in the pores can also increase I_c about its equivalent dry value. The French propose to avoid some of these problems by using a separate parameter, the degree of fissurisation $D_f = 100 - 1.4 n_r - I_c$.

The elastic moduli (e.g. Young's Modulus and the Shear Modulus) are fundamental rock properties that need not correlate well with strength but are important in seismic methods of site investigation, non-destructive materials assessment, and in various blast design considerations. Elasticity and strength properties are often anisotropic, as many rocks possess a preferred mineral orientation. Where it is necessary to prevent damage to underlayers and fine materials that must undergo compaction, the resistance to crushing may be specified using an aggregate crushing test that best simulates the conditions.

Resistance to abrasion

Materials handled in bulk (typically, core and underlayer materials of less than 300 kg) and/or loaded many times will undergo considerble mutual shearing, resulting in abrasive degradation with greater proportional weight losses for finer materials. Resistance to abrasion in service is most important for sites where the foreshore attrition agents such as shingle or sand can attack the armour. Also, for structures using dynamic design concepts, the increased risk of rocking, sliding and rolling of stone will require that only highly abrasion-resistant materials should be used. (see Section 3.2.4.2). Test methods are given in Section 3.3.

Grain size

Generally, rocks with a large grain size in any rock type will have higher fracture toughness (Meredith, 1988) and will exhibit fractures and therefore planar block surfaces with greater roughness and skin friction to resist sliding between blocks.

3.2.2 PRODUCTION-AFFECTED PROPERTIES

In addition to block breakage considerations, this section deals with the geometric properties of the materials prior to construction and the parameters necessary to describe, specify and thus control the bulk material behaviour in the structure.

3.2.2.1 Block integrity

Block integrity is affected by the persistence and frequency of fissures and planes

> **Box 20** Brittle influence on Type 1 breakage of large blocks
>
> It is more important to take steps to minimise Type 1 breakage of heavy blocks of brittle rocks than the less brittle ones such as porous shelly limestones and sandstones. The latter can give greater dampening of impact forces and stresses due to crushing (with a resultant loss of some edge material of minor importance). In contrast, for more brittle rocks such as pure fine-grained limestones and many igneous and metamorphic rocks, the stress from knocks is transmitted and reflected to produce the critical tensile strengths required to open and propagate the major flaws.

of potential weakness in the blocks. Block integrity or 'wholeness', which varies widely from one quarry to another, greatly influences the number of Type 1 breakages of armourstone during quarry and construction handling and also in service during exceptionally stormy sea states, particularly for dynamic structures. The number of Type 1 breakages is also affected by the intrinsic brittleness of the rock types (see Box 20).

Fissures in large blocks may be induced by blasting. These will generally be more common where high-fragmentation blasting techniques for aggregates production have been used giving low armourstone yields (often considered a waste product). The propagation of cracks after blasting (e.g. while rocks sit in stockpiles) has been cited (Lienhart and Stransky, 1981) and long 'curing' times of at least 30 days or 6 months have been recommended, so that cracking will not continue after blocks are in place on the structures. The extent of this problem is not well understood. Blasting during non-freezing weather will help to reduce the incidence of 'popping', which occurs when frozen pore water causes cracking because the rock is no longer confined.

For obvious economic reasons and to reduce breakages, a minimum of handling is recommended. However, there are many other considerations (e.g. availability of space and plant) that determine handling and stocking methods during the construction procedure, and these must be evaluated against the possible damage and breakages. For example, tipping to off-load armourstone at site may typically carry the penalty of 2–3% of Type 1 breakages, but, depending on block integrity, this figure could be as high as 10%. Experience with multiple rough-handling operations during execution of the Eastern Scheldt Stormsurge Barrier in Holland showed that between 10% and 20% reduction in W_{50} occurred for stone classes between 60 kg and 3000 kg. Thus refinements in grading can sometimes be necessary prior to construction. Alternatively, broken armour blocks may be reassigned as usable elsewhere on the structure.

Figure 50 shows the general case where size reduction during handling is implied. However, reselection and rejection of small pieces can even increase the W_{50} of the consignment arriving at site. Rejection of fissured blocks during quality control at the quarry (see Sections 3.3.7 and 3.6) will help to minimise these breakages, and the effects of handling can be monitored and revisions made early in the construction procedure. For the majority of armour sources, however, minimising the handling phase breakages does not constitute a significant problem. Methods for assessing or predicting Type 1 breakage and the overall effect on grading are described in Section 3.3.7.

3.2.2.2 Block shape

Axial ratio measurements are the most widely accepted simple measures of gross shape since, if clearly defined, they are amenable to objective measurement, allowing shape limits to be set by designers. Also, a system of axial dimension measurement can be consistent for both armour block materials (300–3000 mm) and smaller underlayer and core materials (5–300 mm).

In detail, several systems for defining gross shape using axial dimensions have been

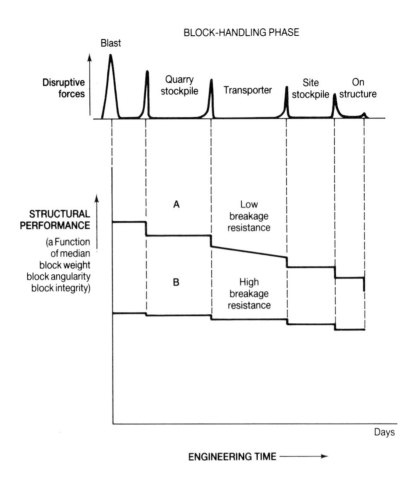

Figure 50 *Conceptual scheme for weight reduction of armour blocks during handling*

proposed or used. For example, the maximum (X), intermediate (Y) and minimum (Z) dimensions of the enclosing cuboid (Thompson and Shuttler, 1975) are often used, and their ratios can provide a fully three-dimensional system that will, for example, distinguish tabular from elongated shapes. The French shape description (LCPC, 1989) is also based on three axial dimensions, L, G and E (in decreasing order), which are combined into a shape factor (equal to $(L+G)/2E$). The difficulty with the French scheme is that L, G and E are defined differently for general irregular blocks and for parallelipipedic ones, which means that a vital distinction that must be made is, in practice, very difficult. A more practical scheme (because it is more amenable to objective measurement) has been proposed as follows:

z = Sieve size (i.e. the smallest square hole that a block can pass through with optimum orientation),
l = Maximum axial length (given by the maximum distance between two points on the block),
d = Thickness or minimum axial breadth (given by the minimum distance between two parallel straight lines between which the block could just pass).

The z/d ratio can also distinguish tabular from elongate blocks. In practice, this distinction is not of great significance for armourstone, and the aspect ration l/d alone can provide the essential degree of departure from the equant form. (Columnar jointing in some basalts which can give an exceptional preponderance of elongate blocks is the exception, but such basalts are normally only used in regular type block revetments now being simulated by proprietary systems.) Typical block aspect ratios for blocks of different rock type and size are given in Box 21.

Objective determination of z becomes impractical above laboratory sieve sizes of 250 mm. The control of z using screens and of d employing industrial grizzlies is

> **Box 21** Block aspect ratio data
>
> For a range of rock types it is found that the ratio l/d decreases as fragment size increases. The following average ratios were found for a range of rock types and sizes after transport and handling and for which a small degree of shape selection of armour sizes may have occurred (Laan, 1981).
>
> | Rock type | Weight range of sample (kg) | Average l/d | Average z/d |
> |---|---|---|---|
> | Dunite | 5000–10 000 | 1.9 | |
> | Dolerite | 60–300 | 2.1 | |
> | | 6000–10 000 | 2.0 | |
> | Granite | 60–300 | 1.8 | |
> | | 1000–3000 | 2.0 | |
> | Basalt | 0.9–2.5 | 2.5 | 1.37 |
> | | 2.5–9.0 | 2.2 | 1.39 |
> | | 750–3250 | 2.0 | |
> | | 6000–10 500 | 2.0 | |
> | Gabbro | 300–1100 | 2.0 | |
> | Limestone | 20–260 | 2.1 | |
> | | 0.4–7.0 | 2.2 | 1.32 |
> | Sandstone/quartzite | 0.7–7 | 3.0 | 1.70 |
> | Summary | $l/d = 2.0$ (armour size) | $l/d = 2.2$ (0.5–10 kg, i.e. fine sizes) | |
> | | | $z/d = 1.33$ (fine sizes) | |
>
> Data are not available for a wider range of bedded sedimentary and foliated (i.e. anisotropic) metamorphic rocks, but experience suggests that for these rock types, armourstones in the blast pile may have an average value of l/d between 2.2 and 2.5 or higher, depending on rock anisotropy and natural jointing in the quarry.

inefficient, and at present cannot be achieved using bar separations of greater than 200 and 400 mm, respectively for any routine production purposes, although grizzly separations of up to 750 mm have been used.

Shape specification

Specification of armourstone shape is important because of its influence on the following:

1. Ease of handling and placement—equant shapes are least problematic;
2. Number of breakages—equant shapes are generally stronger;
3. Range of layer thicknesses achievable in a double layer by different construction methods—equant: small range, tabular: large range (see Table 20, in Section 3.2.3.4);
4. Range of porosity—rounded: lower void content;
5. Static hydraulic stability—experimental evidence for a shape effect on stability of armour under certain construction and design conditions has been reported (Bradbury et al., 1990). The tentative conclusion of that flume study applies to a limited set of conditions but, perhaps surprisingly, indicates that tabular material is more stable than equant, which in turn is much more stable than rounded material (see Section 5.1.3.1 for details).

Because of the large proportion of armour size blocks with l/d greater than 2, specification on block shape should generally quote a limitation on the percentage (e.g. 3% or 5%) of blocks permissible with l/d greater than 3 (rather than 2). If, however, a tight specification on shape is required (e.g. for narrow gradings of heavy classes of armourstones where relatively equant shape source material is realistic) then a 50% or similar restriction on blocks with l/d greater than 2 with a proviso of no blocks greater than 3 is a workable alternative specification. This corresponds to a practical quality control procedure of rejecting blocks judged by eye to be clearly greater than 2:1 while retaining borderline 2:1 blocks. These criteria form the basis of the shape requirements for the model specification (Appendix 1). Designers should note that a previously commonly adopted criterion

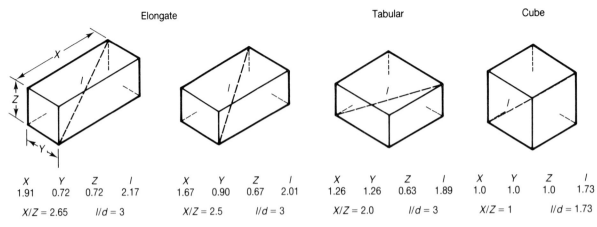

Figure 51 *Illustration of blocks with equal volume and l/d=3, compared with a cube with l/d=1.73*

of X/Z generally less than 2 for equant rock using the cuboidal system represents a criterion of l/d generally less than 2.7. Idealised blocks with l/d ratios of 3 are shown in Figure 51, where they are compared with a cube (having an l/d of 1.73).

As with most specification clauses, the economics of producing to a given specification must be evaluated with respect to achieving the functional requirements without excessive reserves of safety. Ultimately, the client must bear the cost of production and selection of both acceptable and reject materials (together with the subsequent disposal of the latter). Handling and disposal costs of reject material are reduced if this can be done at the production site, where the necessary equipment for economic disposal (by secondary breaking and crushing) is usually available and transport costs have not been incurred. For processed rather than individually selected rock gradings with an average mean weight less than about 300 kg, the shape of material is a function of the source rock and cannot usually be influenced economically during processing.

The l, d, z system can be used to specify the most general gross shape characteristics of blocks, but may require additional qualifying information if it is to discriminate between certain block shapes (for example, between rounded and angular or between equant and pyramidal shapes). It is also unable to indicate features of surface texture without additional data. The shape specification should also include a clause designed to eliminate blocks that are too rounded or spherical for stability, such as stating that the blocks should be predominantly angular. To give this clause more objectivity is difficult other than by visual comparison with blocks of unacceptable roundness, although elaborate image analysis techniques are available.

Shape classification

An introductory review of the problems of shape description is given by Barret (1980). Image analysis techniques which use Fourier and Fractal descriptors are described briefly in Latham and Poole (1987).

The proposed shape classification can be applied to individual blocks and to samples of blocks, as shown in Table 18 and Figure 52. The five shape classes of this classification correspond closely with the shapes of the five samples used in flume tests (Bradbury et al., 1988). Table 18 includes details of the shape data from the samples from which provisional flume test results for hydraulic stability were obtained. The Fourier asperity roughness parameter, P_R is the only parameter which conveniently and systematically classifies the block populations into the five suggested shape classes and which appears from that study to have a systematic relationship with armour stability. It can be obtained from random projections of block outlines using image analysis (Latham and Poole, 1987), or can be estimated by visual comparison with Figure 52 (see Box 22).

Table 18 Shape classification for armourstone

	ANGULAR			ROUNDED	
	Blocks with surfaces bounded by sharp edges and corners			**Most corners and edges show clear signs of wear or crushing**	
Shape class	Elongate + tabular (ET)	Irregular (IR)	Equant (EQ)	Semi-round (SR)	Very round (VR)
Typical sources	Columnar jointed basalts, bedded sedimentary and metamorphic rocks	Massive sedimentary and igneous rocks	Softer sedimentary rocks, rounded during handling, blocks already in service	Dredged sea stones, glacial and river boulders, blocks already in service	
Individual block classification					
Aspect ratio l/d	>3.0	2.0–3.0	1.5–2.0	—	—
Typical weight loss by wear (%)	0–2	0–2	0–2	5–10	10–30
Most corners and edges worn	No	No	No	Yes	Yes
Most of surface smooth	No	No	No	No	Yes
Visual comparison with photograph	(a)	(b)	(c)	(d)	(e)
Block population classification					
Mean aspect ratio l/d	>3.0	2.0–3.0	1.5–2.0	—	—
Percentage* of stones with $l/d>3$	>35	5–35	<5	—	—
Fourier asperity roughness, $\bar{P}_R$†	>0.015	0.013–0.015	0.011–0.013	0.009–0.011	<0.009
Shape data from flume-tested samples (Latham et al., 1988)					
Mean aspect ratio l/d	3.2	2.2	2.0	2.3	2.0
Mean aspect ratio z/d	1.59	1.27	1.30	1.31	1.25
Percentage* of stones with $l/d>3$	66	6	0	15	2
Percentage* of stones with $l/d>2$	100	64	56	67	50
Roughness, $\bar{P}_R$†	0.017	0.014	0.012	0.010	0.005
Range of P_R for 70% of stones	0.012–0.032	0.011–0.022	0.009–0.017	0.008–0.015	0.003–0.009

*The percentage is given here by number as percentage by weight is approximately the same for narrow gradings of the type used in armour layers.
†$\bar{P}_R$ is the log mean of the individual values obtained for P_R

(a) ELONGATE/TABULAR (ET)
$P_R > 0.015$

(b) IRREGULAR (IR)
$P_R = 0.013–0.015$

(c) EQUANT (EQ)
$P_R = 0.011–0.013$

(d) SEMI-ROUND (SR)
$P_R = 0.009–0.011$

(e) VERY ROUND (VR)
$P_R < 0.009$

Figure 52 *Visual comparison of block shapes (photographs courtesy HR Wallingford)*

Box 22 The Fourier asperity roughness parameter, P_R

This new parameter is based on the analysis of the outline of a random projection of a particle. The outline is described by a series of independent uncorrelated harmonics of the form

$$C_n \cos(n\theta - A_n)$$

where C_n is the amplitude coefficient of the nth harmonic, θ the polar angle and A_n the phase angle (Ehrlich and Weinberg, 1970). Each harmonic represents a specific geometric contribution to total particle shape. Gross form contributes to lower harmonic amplitudes while higher orders give fine-scale texture. The coefficient Q, which is size independent, provides a flexible quantitative index which can be computed over all or a chosen range of harmonics, and is defined as

$$Q = (0.5 \sum C_n^2)^{1/2}$$

where the Fourier asperity roughness factor $P_R = Q$ for n summed from the eleventh to twentieth harmonic (Figure 53).

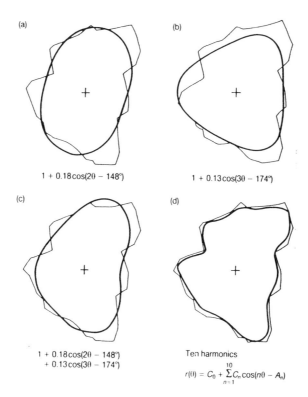

The thin line represents the digitized co-ordinates: (a) and (b) show the contribution of the second and third harmonics respectively. In (c), their combined contributions are shown and (d) which includes the first 10 harmonics gives quite a good approximation.

Harmonic contributions to shape, after Ehrlich & Weinberg (1970)

Figure 53 *Harmonic contributions to shape*

As P_R is based on a random projection outline, this parameter is measured for classification purposes when a large number of particles (at least 30) can be considered to be in random orientation. The maximum projection gives a lower value.

The importance of P_R is that it appears to have a relationship with hydraulic stability for certain conditions tested (Bradbury *et al.*, 1990) and is ideal for quantifying wear processes (Latham and Poole, 1988).

3.2.2.3 Block weight and size

The relationships between size and weight of individual blocks may be defined in terms of the equivalent-volume cube (side D_n) or the equivalent volume sphere (diameter D_s) which, with weight density γ_a and block weight W, give the following relationships:

$$D_n = 1.0(W/\gamma_a)^{1/3}$$

$$D_s = 1.24(W/\gamma_a)^{1/3}$$

with $D_n = 0.806 D_s$

As indicated in the previous section, where graded rock materials are used, size and weight relations refer to medians or averages.

Where the particles are small enough (less than 200 mm) the statistical values are most conveniently derived from sieve analyses. The median sieve size (square openings) (D_{50}) on the percentage passing cumulative curve can be related to the median weight (W_{50}) on the percentage lighter by weight cumulative curve by a simple conversion factor and γ_a or ρ_a. (This dimension D_{50} is the same as the median value of the gross shape dimension z.)

The 50% passing nominal diameter D_{n50} is the size of the cube with equivalent volume to the block with median weight, and is given by

$$D_{n50} = (W_{50}/\gamma_a)^{1/3}$$

The conversion factors relating D_{50} to D_{n50} or W_{50} have been determined experimentally by various workers. The most extensive study is that by Laan (1981), for which the following summary value was obtained and recommended for future use:

$$D_{n50}/D_{50} = 0.84 \quad \text{or} \quad F_s = W_{50}/(\gamma_a D_{50}^3) = 0.60$$

where F_s is the shape factor. As a cautionary note, Laan's study used several different rock types and sizes of stones and found that the value of F_s varied from 0.34 to 0.72, while in the study using different shape classes of limestone fragments (Latham et al., 1988), F_s for all five classes fell between 0.66 and 0.70.

Samples containing blocks larger than 100 mm are difficult to analyse by sieving techniques and direct measurement becomes a more appropriate means of determining size distribution. The relationship between size and weight distributions have been noted above. With large block sizes (i.e. larger than sieve sizes of 250 mm) measurement of weight may prove more practicable than measurement of dimensions, and the size–weight conversion factor such as F_s can be used to determine the median sieve size and other geometric design parameters that may be expressed in terms of sieve sizes.

W_{50} (or M_{50}) is the most important structural parameter in the rock armour design being related to the design parameter D_{n50}, as described above. The controls on block weights were discussed under 'Discontinuities' (Section 3.2.1.4) and the changes in W_{50} from quarry to site were discussed under 'Block integrity' in Section 3.2.2.1. Gradings of rock fulfilling the class limit specification described in the following section may be expected to have standard deviations in D_{n50}, varying from 1% for heavy gradings to 7% for wide light gradings.

3.2.2.4 Grading

In a sample of natural quarry blocks there will be a range of block weights, and in this sense, all rock materials is, to some extent, graded. The particle weight distribution is most conveniently presented in a percentage lighter by weight

Box 23 Useful theoretical equations for stone gradations and fragmentation curves

These are most conveniently presented on a log-linear plot of block weight versus percentage lighter by weight. Defining W_y as the weight for which the fraction y is lighter on the cumulative curve, three useful equations are given below and compared in Figure 54.

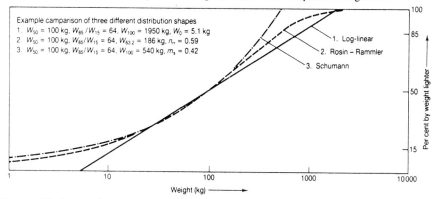

Figure 54 *Comparison of log-linear, Rosin Rammler and Schumann equations for stone gradations and fragmentation*

1. Log-linear equation

This equation is useful for describing narrow and possibly some wide gradations. Van der Meer's (1988a) experimental gradings were linear on a log-linear plot:

$$y = \frac{\log W_y - \log W_0}{\log W_{100} - \log W_0}$$

$$W_y = W_0 (W_{100}/W_0)^y$$

A useful form of this equation is expressed using the common grading parameters W_{85}/W_{15} and W_{50}. The equation readily gives all weight values W_y for a given $y\%$ which can be very useful:

$$W_y = W_{50}(W_{85}/W_{15})^{[(y-0.5)/0.7]}$$

2. Rosin–Rammler equation

This is equivalent to the Weibull distribution with a location parameter at the 63.2 percentile. It is useful for describing *in-situ* and as-blasted fragmentation curves, i.e. very wide gradations (see Section 3.4).

$$y = 1 - \exp[-(W_y/W_{63.2})^{n_{rr}}]$$

or, in terms of W_{50} and W_{85}/W_{15}:

$$W_y = W_{50}\left[\frac{\log(1-y)}{\log 0.5}\right]^{\left(\frac{\log(W_{85}/W_{15})}{\log 11.67}\right)}$$

3. Schumann equation

This equation is also used to describe *in-situ* and as-blasted fragmentation curves. It is useful for modelling wide gradations where oversized material has been removed:

$$y = \left(\frac{W_y}{W_{100}}\right)^{m_s}$$

or, in term of W_{50} and W_{85}/W_{15}:

$$W_y = W_{50}\left(\frac{y}{0.5}\right)^{\left(\frac{\log(W_{85}/W_{15})}{\log 5.67}\right)}$$

Example

The above equations may be applied to the problem of ensuring that the largest block, D_{100}, does not protude above a $2D_{n50}$ thick rip-rap layer. The approach to solving the problem is to limit the grading width to a maximum value. Depending on the form of the gradation equation, the limiting values given by $W_{max}/W_{50} < 8$ are:

1. $W_{85}/W_{15} < 18$
2. $W_{85}/W_{15} < 33$ (assumes $D_{max} = D_{95}$)
3. $W_{85}/W_{15} < 180$

Although wide gradings typical of rip-rap will generally be curved on a log-linear plot, a criterion based on the above log-linear equation which is a better fit for the upper half of the grading curve offers a safer approach. It is also sound advice to require that all very large blocks be laid with their shortest dimensions normal to the rock slope. The use of very wide gradings ($W_{85}/W_{15} > 16$) for rip-rap is not advised (see Box 53).

cumulative curve, where W_{50} expresses the block weight for which 50% of the total sample weight is of lighter blocks (i.e. the median weight) and W_{85} and W_{15} are similarly defined. The overall steepness of the curve indicates the grading width, and a popular quantitative indication of grading width is the W_{85}/W_{15} ratio or its cube root, which is equivalent to the D_{85}/D_{15} ratio determined from the cumulative curve of the equivalent cube or sieve diameters of the sample. The following ranges are recommended for describing the grading widths:

	D_{85}/D_{15} or $(W_{85}/W_{15})^{1/3}$	W_{85}/W_{15}
Narrow or 'single-sized' gradation	Less than 1.5	1.7–2.7
Wide gradation	1.5–2.5	2.7–16.0
Very wide or 'quarry run' gradation	2.5–5.0+	16.0–125+

The term 'rip-rap' usually applies to armouring stones of wide gradation which are generally bulk placed and used in revetments. The phrase 'well graded' should generally be avoided when describing grading width. It merely implies that there are no significant 'gaps' in material sizes over the total width of the grading.

Standardisation of gradings

There are many advantages in introducing standard grading classes. These mostly concern the economics of production, selection, stockpiling and quality control from the producer's viewpoint. With only a few specified grading classes, the producer is encouraged to produce and stock the graded products, knowing that designers are more likely to provide them the market by referring to these standards wherever possible. The proposed standard gradings for armour are relatively narrow. This can result in increased selection costs, but these will often be completely offset by the possibility of using thinner layers to achieve the same design function. Standard gradings are not needed for temporary dedicated quarries supplying single projects where maximised utilisation of the blasted rock is required.

It is convenient to divide graded rock into:

'Heavy gradings' for larger sizes appropriate to armour layers and which are normally handled individually;
'Light gradings' appropriate to sheltered cover applications, underlayers and filterlayers that are produced in bulk, usually by crusher opening and grid bar separation adjustments;
'Fine gradings' that are of such a size that all pieces can be processed by production screens with square openings (i.e. less than 200 mm). For fine gradings, the sieve size ratio from the cumulative sieve curve ($D_{60}/D_{10} = U_c$, the uniformity coefficient) is often used to characterise the width of grading in a sample.

Standard gradings are more or less essential for fine and light gradings. However, for heavy gradings it is not difficult because of individual handling to define and produce gradings other than standard (see below). For example, if the 1–3 t grading is (just) too small for a particular application, choice of the first safe standard grading of 3–6 t will lead to an excessive layer thickness and weight of stone, and here use of a non-standard grading may well be appropriate. Again, ceiling sizes of stones in quarries arising from geological constraints may dictate an upper limit.

A consistent scheme for defining grading requirements for the four suggested heavy standard grading classes was proposed in the draft Dutch standard NEN5180, and these have been adopted, as indicated in Appendix 1, Figure 56 and Table 19. Box 24 explains the scheme in detail and in Box 25, the effective mean, W_{em}, which is important for rapid assessment of grading is described.

Three light grading classes have been defined by weight in an identical way to the heavy gradings, as shown in Table 19. Corresponding requirements have been proposed when these gradings are specified using 'sieve' sizes and have been designated 200/350 mm, 350/550 mm and 200/500 mm classes. The test verification in

Table 19 Requirements and supplementary information for standard gradings

	Class designation (kg)	ELCL ($y<2$)	LCL $0<y<10$	UCL $70<y<100$	EUCL $97<y$	Requirements for range of effective mean weight, i.e. excluding pieces less than ELCL		Additional information Expected range for compliance with standard gradings	
		where y is the percent by weight lighter on the cumulative plot						W_{50}	W_{85}/W_{15}
	Class limit definition by weight, W_y (kg)								
Heavy gradings	300–1000 kg	200	300	1000	1500	540–690 kg		595–760 kg	2.3–3.8
	1000–3000 kg	650	1000	3000	4500	1700–2100 kg		1800–2200 kg	2.2–3.6
	3000–6000 kg	2000	3000	6000	9000	4200–4800 kg		4200–4800 kg	1.6–2.2
	6000–10000 kg	4000	6000	10000	15000	7500–8500 kg		7500–8500 kg	1.4–1.8
Light gradings	10–60 kg	2	10	60	120	20–35 kg		26–46 kg	3.2–7.7
	60–300 kg	30	60	300	450	130–190 kg		150–220 kg	2.8–6.0
	10–200 kg (wide)	2	10	200	300	30–90 kg		70–130 kg	5.0–11.4

Average weight retained on L hole, $\bar{W}_L$ (kg) for rock density, ρ_r (t/m³)

	Class limit definition by square hole (mm)	<2.5		2.5–2.9		>2.9	
		min.	max.	min.	max.	min.	max.
	200/350 mm	100	200	(350)	400	20	40	25	45	25	50
	350/550 mm	250	350	(550)	650	115	180	130	200	145	240
	200/500 mm (wide)	100	200	(500)	550	45	80	50	90	55	100

Note
The 10–60 kg, 60–300 kg and 10–200 kg classes are approximately equivalent to the 200/350 mm, 350/550 mm and 200/500 mm classes, respectively. The sizes given in brackets are those of the equivalent UCL or U hole and are not requirements but for consistency are used for designating the class of grading.

Box 24 Explanation of class limit system of standard gradings

Rather than using envelopes drawn on a cumulative plot to define the limits of a standard grading, the proposed system refers to either:

1. A series of weights of stones and their corresponding ranges of the cumulative percentage by weight lighter than values which are acceptable; or
2. A series of 'sieve' size of stones and their corresponding ranges of cumulative percentage by weight passing values, together with the average weight of stones in the grading.

Each weight-standardised grading class is designated by referring to the weights of its lower class limit (LCL) and its upper class limit (UCL). In order to further define the grading requirements realistically and to ensure that there are not too many undersized or oversized blocks in a given grading class, it is necessary to set two further limits, the extreme lower class limit (ELCL) and the extreme upper class limit (EUCL). The standard grading scheme then uses percentage by weight lighter than values, y (equivalent to percentage by weight passing for aggregated), 0, 2, 10, 70, 97 and 100 to set the maximum and minimum percentiles corresponding with each of the four weight values given by ELCL, LCL, UCL, and EUCL (see below and Figure 51, Table 19). Note that although straight-line segments have been drawn in Figure 51 to indicate an envelope, grading curves can go outside these straight-section envelopes and still fulfil the requirements given by the four class limits. However, the further requirement specifying that the effective mean weight, W_{em}, should fall within a set range will help to reassure the designer that there is an appropriate range of W_{50} values.

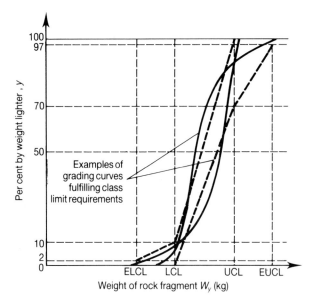

Figure 55 *Explanation of the grading class limits of a standard grading*

Size and average weight standardised gradings

Each size-standarized grading class is defined with reference to the 'sieve' sizes of its ELCL, LCL and EUCL as explained above (no UCL specified) together with the average weight, $\bar{W}_L$, of all rocks not passing the LCL hole. However, for consistency they are designated i.e. named using the LCL and UCL hole sizes. Apart from bulk weighing to determine $\bar{W}_L$, the test verification of size is limited to gauging of blocks using three square guage holes of sizes as defined in Table 19 and imposing the requirment consistent with the weight-grading scheme that:

1. Less than 2% by weight shall pass the ELCL hole, known as the EL hole;
2. Less than 10% by weight shall pass the LCL hole, known as the L hole;
3. At least 97% by weight shall pass the EUCL hole, known as the EU hole.

the latter case requires both gauging of blocks and average weight determination and is given in Appendix 2, Section A2.2: Part 2. Both specifications for the light gradings are intended to produce approximately similar graded stone products. The size-limit specification of light gradings is only appropriate for underlayers applications whereas the weight-limit specification is always necessary for cover layers. Any of the three standard light gradings specified by size limits but tested

> **Box 25** Median and effective mean weight
>
> A number of parameters in addition to the designated class limits may be required for design and specification purpose. An acceptance range for the important design parameter, W_{50} (median weight), is also useful for each grading class. However, in some circumstances an effective mean weight W_{em} for a particular consignment may be more easily obtained simply by bulk weighing and counting. In order to avoid including fragments and splinters in the distribution, W_{em} is defined as the arithmetic average weight of all blocks in the consignment or sample, excluding those which fall below the ELCL weight for the grading class. An empirical conversion factor relating W_{em} to W_{50} allows an estimate of W_{50} without the necessity of weighing each piece to obtain a weight-distribution curve. As the grading becomes wider, so will W_{50} depart more from W_{em}. Approximate relations are as follows:
>
> | | |
> |---|---|
> | 10–60 kg | $W_{50}/W_{em} = 1.3$ |
> | 60–300 kg | $W_{50}/W_{em} = 1.15$ |
> | 300–1000 kg | $W_{50}/W_{em} = 1.10$ |
> | 1000–3000 kg | $W_{50}/W_{em} = 1.05$ |
> | 3000–6000 kg and 6000–10 000 kg | $W_{50}/W_{em} = 1.00$ |

and certified by their equivalent weight designated grading class would be deemed acceptable for all uses as the weight definition is taken as primary.

The weight envelopes for the standard light and heavy gradings are shown in Figure 56 with appropriate size-weight conversion. In this figure, rocks of any density and weight can be related to the correponding nominal diameter, D_n, sieve size, z, and minimum rock thickness, d, using the alternative horizontal scales on this chart. The 'sieve' size, z, enveloped for the standard fine and light gradings are shown in Figure 57, again with conversions to D_n and d.

The standard fine and light gradings are produced by screens and grids and sometimes with eye selection to remove oversize material in the 60–300 kg and 10–200 kg classes. The poor screening efficiency that occurs in practice means that a correction factor would be needed in addition to the given theoretical relationships for sieve size, z, and minimum stone thicknesses, d, should a producer wish to use Figures 56 and 57 to indicate the combination of screens and grids that would yield the standard gradings. For the 10–200 kg more widely graded class, experience in the UK quarries using simple grizzlies to remove undersize material and eye selection to remove oversize material suggests that for rock of average (2.7 t/m³) density, setting the grizzly clear spacing at 225 mm and using eye selection of material above 450 mm will be a good starting point. This advice must be tempered, however, by individual quarry managers' own experience at their particular quarry.

The grading envelopes become progressively narrower in the 'heavy grading' classes, consistent with design requirements and the geological constraints on producing large sizes of blocks. However, projects requiring blocks larger than 10 t should not make the (non-standard) grading class excessively narrow because of the producer's extreme difficulty in selecting accurately and the wastage from oversize block production. For example at $\rho_r = 2.7$ t/m³, D_n for a 15 t block = 1.7 m, and D_n for a 20 t block = 1.95 m, a difference of about 10%, which is very difficult to select precisely except by individual weighing.

Wide or quarry run gradings may be economic to produce and will certainly be used for core. The amount of fine material may be difficult to control. One rapid method is to use a loader with a meshed bucket base such that material of less than 100 mm passes through. The operator's success in removing fines will depend on not overfilling the bucket, the type and dryness of the rock material. Other possiblities using grizzlies and divergators can give greater control of fines. Screening is undoubtedly necessary in the production of permeable core materials when a high percentage of fines are present in the blast pile.

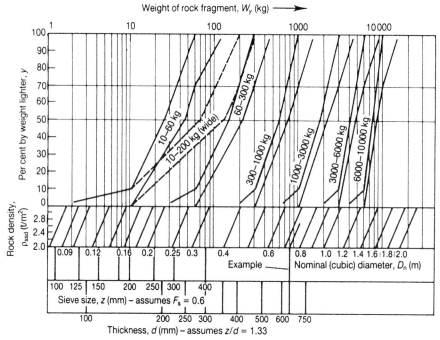

Figure 56 *Weight gradings and size relationships for the standard light and heavy grading classes*

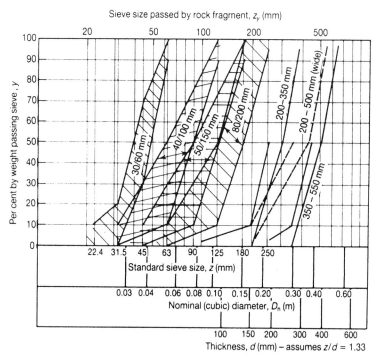

Figure 57 *Size gradings and relationships for the standard fine and light grading classes*

Derivation of non-standard gradings

It may be necessary or desirable to define and use non-standard gradings for a number of situations, the following being the most common:

1. For extremely heavy rock armour greater than 10 t;
2. For heavy rock armour gradings in the range of 0.3–10 t where the calculated design W_{50} falls inconveniently in relation to the standard gradings and where

the need to use heavier rocks than those theoretically required has a definite cost penalty, even when the reduced number of rock pieces required is taken into account;

3. For wide gradings that can be produced at considerable economic saving compared with standard gradings and which are acceptable to the designer in terms of their function;

4. Where a quarry is being developed for the specific breakwater, dam or coastal protection project (temporary dedicated quarry) and economics dictate minimum wastage of quarried materials.

In order to apply the same grading test methods described for the standard gradings in Appendix 2, Section A2.2, Parts 1 and 2 to non-standard gradings it is clearly necesssary to describe any non-standard gradings in a similiar way. Specifically, given a desired W_{50} and W_{85}/W_{15} (e.g. from hydraulic stability considerations in Section 5.1.3), it is necessary to define a method of deriving the ELCL, LCL, UCL and ELCL class limits of weight or size.

Boxes 26 and 27 provide this required guidance, for heavy narrow non-standard gradings and for light and light/heavy wide non-standard gradings, respectively. By inspection, the reader will note that the standard gradings described are consistent with this guidance for non-standard gradings. The non-standard narrow heavy gradings might be required for berm breakwaters or coastal and shoreline armour layers. The non-standard light and light/heavy wide gradings will be appropriate for situations such as core, underlayers and armour layers in moderate wave climates where 'rip-rap' has traditionally been used.

Box 26 Derivation of non-standard specification for narrow heavy gradings

Conventional (i.e. narrow) gradings, for berm breakwaters shoreline and coastal armour layers:
1. In the range of W_{50} between 0.5 and 3t, for a given design W_{50}, set:

$EUCL = 2.25 W_{50}$
$UCL = 1.5 W_{50}$
$LCL = 0.45 W_{50}$
$ELCL = 0.30 W_{50}$
W_{em} range $= 0.8 W_{50} - 1.0 W_{50}$

Further information on expected ranges for compliance with the above requirements:

W_{85}/W_{15} range $= 2.0 - 4.0$
W_{50} range $= 0.9 W_{50} - 1.1 W_{50}$

2. In the range W_{50} between 3 and 15t or greater

$EUCL = 2.1 W_{50}$
$UCL = 1.4 W_{50}$
$LCL = 0.7 W_{50}$
$ELCL = 0.47 W_{50}$
W_{em} range $= 0.95 W_{50} - 1.1 W_{50}$

Further information on expected ranges for compliance with the above requirements:

W_{85}/W_{15} range $= 1.5 - 2.5$
W_{50} range $= 0.95 W_{50} - 1.1 W_{50}$

Notes:
These theoretically derived grading specifications are based on consideration of log-linear distributions with appropriate limiting values of W_{85}/W_{15}. The W_{em} range requirements are based on similar rules for the standard gradings of equivalent grading width. The designer should numerically round the EUCL, UCL, LCL, ELCL and W_{em} figures and should be aware of the implications of such rounding.

3.2.3 EXECUTION-INDUCED PROPERTIES

When the armour blocks are placed together, the destructive wave energy is mostly dissipated through the turbulence generated within the void space between the

Box 27 Derivation of non-standard specification for wide, light and light/heavy gradings

This system for deriving a grading specification assumes that a design W_{50} is set. This value must be between 0.05 and 1.0 t. It also assumes that W_{85}/W_{15} is to lie approximately in the range 5–11, Typical of wide gradings. The upper limit of W_{85}/W_{15} of 11.4 corresponds to a value which for a log-linear size distribution was shown to be equally stable for coverlayers, as were narrow gradings with W_{85}/W_{15} of 2.0 (van der Meer, 1988a). The production of these gradings involves settings for screens, grids, crusher openings, etc. and possibly visual selection to remove under- and oversize material.

Weight designation

This may be simply calculated from the design W_{50} as follows:

Requirements:

EUCL $= 3W_{50}$
UCL $= 2W_{50}$
LCL $= 0.1W_{50}$
ELCL $= 0.02W_{50}$

Rock density ρ_r (t/m³)	W_{em} required	W_{50} expected
>2.9	$0.35W_{50}$–$1.0W_{50}$	$0.8W_{50}$–$1.45W_{50}$
2.5–2.9	$0.3W_{50}$–$0.9W_{50}$	$0.7W_{50}$–$1.3W_{50}$
<2.5	$0.25W_{50}$–$0.8W_{50}$	$0.6W_{50}$–$1.15W_{50}$

Size and average weight designation

1. Obtain the weight designation class limits as above;
2. Convert these to equivalent sieve sizes using $\rho_r = 2.7\,t/m^3$, $D_n/D_z = 0.84$.

Requirements: EUCL gives EU hole dimensions
LCL gives L hole dimensions
ELCL gives EL hole dimensions

e.g. for $W_{50} = 500$ kg, length of sides of L hole =

$$(0.1 \times 0.5/2.7)^{1/2}/0.84 = 0.315\,m$$

3. Calculate from the design W_{50} the requirement for the average weight of pieces not passing the L hole, $\bar{W}_L$:

Rock density, ρ_r (t/m³)	$\bar{W}_L$ required
<2.9	$0.55W_{50}$–$1.0W_{50}$
2.5–2.9	$0.5W_{50}$–$0.9W_{50}$
<2.5	$0.45W_{50}$–$0.8W_{50}$

Notes

For these gradings, a log-linear size distribution cannot usually be produced economically. These theoretically derived weights and size limit designations are based on considerations of the Schumann distribution below W_{50} and a log-linear distribution above W_{50} with appropriate limiting values of W_{85}/W_{15}. The designer should numerically round the EL, L, EU and $\bar{W}_L$ figures to the nearest convenient values and be aware of the implications of such rounding.

The producer in a quarry has the production equipment and rock types experience to carefully select settings to give, as near as possible, the W_{50} required for the design (this can be inferred from calculating $10 \times$ LCL, where LCL is given from the L hole dimensions). The producer is responsible for meeting all the requirements but will need to concentrated on the L hole, which defines the size hole through which no more than 10% by weight shall pass, and the average weight, $\bar{W}_L$, of pieces not passing the L hole.

Examples of such gradings produced by grizzly and eye selection suggest that to fulfil grading requirements tested according to Appendix 2, Section A2.2, Part 2, the following rules of thumb may be useful for rocks in the range $2.5 < \rho_r < 2.9$:

1. Set grizzly ≈ 1.2 times L hole to remove undersize;
2. Eye select $D_z \approx 0.8$ times EU hole to remove oversize.

For $W_{50} < 100$ kg the required multiples in (1) and (2) will probably need increasing to 1.3 and 0.9 times, respectively.

(*Continued on page 103*)

> **Box 27** (Continued)
>
> With the size and average weight definition of gradings, EL, L, EU and $\bar{W}_L$ must be specified. In keeping the test easy to perform, the grading curve position is only located near the light end using L. To ensure that the heavy end is correctly represented, the test relies on $\bar{W}_L$, which cannot go below 0.5 times the design W_{50}. Inaccurate oversize selection with two few blocks just below EU size is usually the cause of $\bar{W}_L$ falling below 0.5 times W_{50} design. The true value of W_{50} in a grading which fulfils the $\bar{W}_L$ and EL, L and EU requirements may range from about $0.7W_{50}$ to $1.0W_{50}$.
>
> A rock density of 2.7 t/m^3 has been assumed throughout, but the producer will be reponsible for making adjustments to ensure that $\bar{W}_L$ as well as the hole requirements are met, provided ρ_r is between 2.5 and 2.9 t/m^3.
>
> Either the weight designation or size and average weight designation may be specified corresponding to tests in Parts 1 and 2 of Section A2.2 of Appendix 2, respectively. The latter will give a simpler test to perform but with less control on the shape of the grading curve at the heavy end. In many cases where narrow and wide gradings are used for underlayers, the specification by size and average weight will give sufficient control.

seaward-facing outer blocks. A high porosity in the armour layer (as well as high core permeability) helps to achieve this energy dissipation. To avoid blocks being plucked out and rolling down (or up) slope, a combination of the following is generally necessary:

1. Sufficient block weight and density;
2. Sufficient porosity;
3. Sufficient degree of interlocking.

3.2.3.1 Interlocking and armour layer porosity

The term 'interlock' often applied to blocks in armour layers is a general and relative term, understood to apply to a pack of blocks and to reflect the degree to which individual particles are constrained by their neighbours from rotating or sliding relative to each other. It depends mainly on placement methods. A precise interlock parameter is difficult to define (Box 28).

Improved interlocking of armour does not accompany increased porosity. For example, an instruction to produce a tightly packed arrangement of rectangular sectioned blocks can lead to high interlocking but at the expense of a lower layer porosity and poorer energy dissipation. Both rounded block corners and wider gradings tend to produce a lower armour layer porosity. Rounded blocks tend to reduce interlocking, although this effect is less well established.

Both block shape and grading will influence the ranges of porosity and interlock

> **Box 28** Interlock
>
> Interlock between identical precast concrete units can be precisely defined and experimentally determined. It is assumed that the natural angle of repose can be expressed as the sum of two parts representing the angle of contact friction (e.g. for the surface of precast concrete) and the angle of interlocking. Under a 'typical' packing arrangement, the interlock angle appears to be constant for a given unit shape. This definition is less workable for natural rock armour because of (1) surface friction roughness relating to mineral grain size and the gross shape roughness of the rock (stone) form the two ends of the roughness spectrum so that the distinction between friction and interlock is blurred, (2) highly variable packing, (3) the effect of grading and (4) strength scale effects with corner breakages.
>
> A more practical prototype measure indicating interlocking is one based on the average co-ordination number or average number of point contacts that each block makes with its neighbours. A special definition would apply to a surface row of blocks.

that are possible in armour layers. Placement technique will be the final adjusting factor determining the layer porosity and interlock as well as the final layer thickness that is achieved upon construction.

3.2.3.2 Internal friction

For particulate materials, the internal friction angle, ϕ, depends on particle size distribution, shape and surface roughness, applied strain field (triaxial, plane) and strain rate, intact particle strengths and stress field. The role of this parameter in geotechnical models of the deformation behaviour are given in Section 5.2.4.

3.2.3.3 Roughness of layer topography

The roughness of the armour layer surface and the porosity of the armour layer will influence the wave reflections and run-up/run-down on a given slope (Section 5.2.1.1). Blocks having an approximately rectangular cross-section, when used in smaller revetment structures, will tend to facilitate a placement operation, resulting in a 'dry stone wall' final appearance. The effect of having flat block faces lying nearly parallel to the seaward slope is to reduce the incidence of locally high dynamic loads during wave breaking and to increase interlock. Another benefit of such smooth outer profiles in amenity areas could be one of safety to people and encouragement of recreation. Potentially dangerous inter-block cavities could be reduced to a minimum using skilfully crafted placement. However, this can only be achieved in a reasonable construction time if sizes, shapes and grading of blocks are appropriate. Wave run-up will be increased by such a smoothing of the profile (Section 5.2.1.1) and for the case of tabular blocks laid flat, damage may progress very suddenly.

The topographic roughness has implications for the accuracy and interpretation of profile survey methods. Angular equant blocks with typical placement and therefore roughness can be expected to give average profile heights of about 0.1–$0.2 D_{n50}$ higher for a survey based on the highest point on each block compared with one based on the hemispherical probing at regular intervals (as required in Appendix 1).

Interlayer friction is normally considered sufficient if size and grading relations between layers have been set by filter rules. However, it has been suggested that there are considerable stability advantages of using a $W_{50}(\text{above})/W_{50}(\text{below})$ ratio of 10 to 15 (see Section 5.1.3.1).

3.2.3.4 Implications for porosity, density and layer thickness

Porosity and layer thickness of armour layers and underlayers need to be calculated in order to estimate the volume or weight requirements of rock to cover a given area of a structure. Volumes and weights are clearly important in the billing of quantities and for meeting specification. Also, because the notional permeability factor P for the total structure (Section 5.1.1.2) as used with the van der Meer design equations will, in some cases, be sensitive to differences in layer thickness. Results reported in Bradbury *et al.* (1990) suggest that layer thicknesses in double armour layers can often be as low as $1.5 D_{n50}$, with potentially significant effects on the armour stability.

The perpendicular layer thickness and porosity of a layered system of armouring will depend on the following factors:

1. Weight and grading of rock (W_{50} and W_{85}/W_{15});
2. Weight density of rock (γ_a);
3. Method of placement;
4. Block shapes;
5. Definition of the surface and the survey method of measurement.

The last three factors provide the greatest areas of uncertainty for design and construction.

For narrow gradings, good approximation of armour layer thickness (t_a) and porosity (n_v) are given by the equations

$$n_v = 1 - (\gamma_b / \gamma_a)$$

$$t_a = n k_t D_{n50}$$

$$N_a = n A k_t (1 - n_v) D_{n50}^{-2}$$

where γ_b = bulk weight density as laid,
n = number of layers making up the total thickness of armour,
k_t = layer thickness coefficient,
N_a = total number of blocks in area considered,
A = area considered.

The porosity for double layers surveyed by a method similar to that given in Appendix 1 ranges from 35% to 42% in narrow graded armour layers. If these same stones were placed in a bulk-filling operation, the appropriate porosity values are usually about an additional 5% higher, although it is difficult to generalize about such boundary effects on porosity determinations. Note also that if $W_{50}/W_{em} > 1$ (see Box 25) then N_a in the above equation should be increased by a factor approximately given by $(W_{50}/W_{em})^{2/3}$.

Table 20 Thickness and porosity in narrow gradation armour layers

Shape spec. or description	Placement	Above/below water	Layer coeff. k_t	Porosity (%) n_v
Irregular	(b)+(d)	Above	1.20	39
	(b)+(e)	Above	1.05	39
	(c)+(f)	Above or below	0.75	40
Semi-round	(b)+(d)	Above	1.25	36
	(b)+(e)	Above	1.10	36
	(c)+(f)	Above or below	0.75	37
Equant	(b)+(d)	Above	1.15	37
	(b)+(e)	Above	1.00	37
	(c)+(f)	Above or below	0.80	38
Very round	(b)+(d)	Above	1.20	35
	(b)+(e)	Above	1.05	35
	(c)+(f)	Above or below	0.80	36

Notes

1. Shape descriptions refer to the block population classifications of Table 11
2. Placement techniques and/or their implied effects
   (b) Section 12.3 of Appendix 1
   (c) Random placement to achieve a high porosity for wave energy dissipation
   (d) Long axes of elongate, tabular and irregular blocks to be placed normal to slope
   (e) Long axes of elongate, tabular and irregular blocks to be placed upslope with short axes along slope
   (f) Long axes of elongate, tabular and irregular blocks may take any orientation

The choice of placement technique is very dependent on whether placement is above or below water and the availability of plant. Careful packing and block-orientation placement can only be achieved within the range where hydraulic grabs and buckets can operate. Placement by crane (with or without eye bolts) is more compatible with (f) and the resulting layer thickness is reduced (for typical eyebolt positioning on the top face of stockpiled blocks). Underwater placement to the correctly gridded system should yield on average about the same k_t and n_v results, although laying of discrete layers is not normally possible.

Tolerances (not given in the table) must reflect the uncertainty of placement and the lower precision of the profile checking survey method

Recent experience with numerous contracts in the UK indicates that the selection of k_t and n_v from values tabulated in the *SPM* (1984), together with lack of clarity with profile survey methods, has often been the cause of later contractual disputes between designer and contractor. Usually, the *SPM* (1984) values of k_t are higher than those given by the thickness of double armour layers constructed with normal satisfactory methods. It is therefore recommended that a verification test panel (see Box 39) be constructed to resolve such problems for any project where layer thickness coefficients have not been established from prior practical experience.

Improved guidance in the selection of appropriate values for k_t and n_v is clearly now required. Reported values to date for layer thickness coefficients have varied considerably, at least in part due to inconsistencies in profile survey technique. The values set out in Table 20, which refer only to armour placed in discrete layers, are first estimates based on work by Latham *et al.* (1988) on static models and flume-tested sections using the profile measuring techniques recommended in this manual (Appendix 1, Clause 13). Additional difficulties arise in that for wide gradations, it becomes more impractical to consider separate layers, so that a different approach to thickness specification is required. In Box 29 the effects of differing rock density on stability and geometric properties are explored using the above equations.

Box 29 Consequences of using different rock density

	A High ρ_r Low w_{ab}	B Low ρ_r Low w_{ab}	C Very low ρ_r High w_{ab}	D Low ρ_r Low w_{ab}	E Very low ρ_r High w_{ab}
$A\,(m^2)$	100	100	100	100	100
k_t	0.9	0.9	0.9	0.9	0.9
n_v	0.4	0.4	0.4	0.4	0.4
$\rho_w\,(t/m^3)$	1.03	1.03	1.03	1.03	1.03
$\gamma_r, \rho_r\,(t/m^3)$	3.1	2.6	2.4	2.6	2.4
w_{ab}	0.5	0.5	4.0	0.5	4.0
ρ_{ssd}	3.12	2.61	2.50	2.61	2.50
Δ	2.02	1.54	1.42	1.54	1.42
$D_{n50}\,(m)$	1.17	1.24	1.28	1.54	1.67
$W_{50}\,(t)$	5	5	5	9.52	11.21
$\Delta D_{n50}\,(m)$	2.37	1.92	1.81	2.37	2.37
$\gamma_b, \rho_b\,(t/m^3)$	1.86	1.56	1.44	1.56	1.44
$t_a\,(m)$ Double layer	2.11	2.24	2.30	2.77	3.01
N_a	78.5	69.8	66.2	45.5	38.7
Total weight	393	349	331	434	434

As damage is a function of ΔD_{n50}, the armour layers in columns A, D and E have the same stability but D and E require heavier armour and a greater total weight, although fewer pieces need handling. The variation in ΔD_{n50} indicates, when compared with equation (5.44) in Section 5.1.3.1, that for plunging wave attack, armourstone with the same W_{50} but lower density gives an armour layer likely to damage 3.9 times more (column C) or 2.9 times more (column B) than the armourstone of high-density rock in column A.

3.2.4 DURABILITY AND ARMOURSTONE DEGRADATION RATES

Durability is defined as the ability of a rock to retain its physical and mechanical properties (i.e. resist degradation) in engineering service. It is therefore a function of the rock properties and the engineering environment or site conditions. For elements in a structure durability reflects the rate at which the structural performance is diminished during service conditions. The structural performance of armour blocks is a catch-all term, which is largely a function of median block weight. However, block angularity and integrity will also contribute to it. The structural performance refers to a given time or stage in the engineering life of the blocks and is a concept distinct from durability. Resistance to degradation is essential for the retention with time of the main functional requirements (Figure 44).

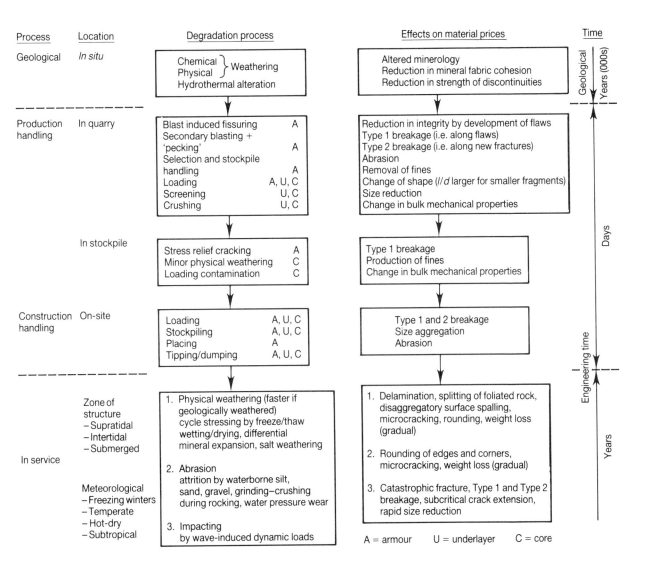

Figure 58 *Factors affecting degradation of rock materials*

3.2.4.1 Design implications

Degradation has the effect of reducing block integrity, average weight and angularity, so modifying the weight distribution and interlocking properties of the armour and thus the expected damage. When rapid, the degradation processes can significantly impair the performance of the armour layer during the design life of the structure and the rapid onset of failure in only moderate storm conditions could result. The designer will therefore need to make some prediction as to the rock's performance in a given site and design situation.

If, for example, using Section 3.1, it is concluded that local rock sources and environmental site conditions are likely to give rapid degradation of armourstone, then it will be necesssary to assess certain design options such as:

1. High frequency of maintenance and repairs;
2. Over-dimensioning of armourstone;
3. Gentler seaward design slope and greater material volumes;
4. The use of higher-quality rock from a more remote source.

A generalised scheme of how the various degradation processes are likely to affect the structural performance of armour, underlayer and core materials is set out in Figure 58. The relationships between structural performance of armourstones and engineering time for the in-service phase of the works are shown in Figure 59.

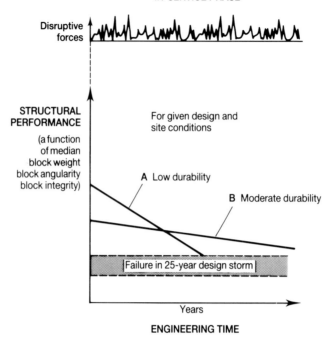

Figure 59 *Conceptual scheme for structural performance of armour blocks in service*

Figure 60 *Prototype production model of QMW abrasion mill apparatus (courtesy Queen Mary & Westfield College)*

Table 21 Guide to rock durability from test results

Test	Excellent	Good	Marginal	Poor	Comments
Rock density (oven dry) Appendix 2, Section A2.6 ρ_r(t/m^3)	>2.9	2.6–2.9	2.3–2.6	<2.3	Physical property affecting hydraulic stability. Good indicator of durability except dense but weathered basic rocks
Water absorption Appendix 2, Section A2.7 W_{ab} (%)	<0.5	0.5–2.0	2.0–6.0	>6.0	Single most important indicator of resistance to degradation. Good indicator of weathering resistance. Often misleading for porous limestones with large free-draining pores
Magnesium sulphate soundness BS 812/BS 6349 MSS(%)	<2	2–12	12–30	>30	Indicates resistance to weathering. Important test for porous sedimentary rocks for use in hot dry climate. Good correlation with water absorption
Freeze/thaw Appendix 2, Section A2.4 FT(%)	<0.1	0.1–0.5	0.5–2.0	>2.0	Important test for freezing winter climates (especially reservoir dams). Good correlation with water absorption
Methylene blue absorption Appendix 2, Section A2.10 MBA (g/100 g)	<0.4	0.4–0.7	0.7–1.0	>1.0	Indicates presence of deleterious clay minerals
Fracture toughness ISRM (1988) K_{IC}(MP$_a$.m$^{1/2}$)	>2.2	1.4–2.2	0.8–1.4	<0.8	Indicates resistance to Type 2 breakage, degree of weathering (geological). Good correlation with abrasion resistance. Can be misleading for impact resistance of large blocks
Point load index ISRM (1985) $I_{s(50)}$(MP$_a$)	>8.0	4.0–8.0	1.5–4.0	<1.5	Indicates resistance to Type 2 breakage. Quick test with high test variability. Can be misleading for impact resistance of large blocks
Wet dynamic crushing value Appendix 2, Section A2.5 WDCV (%)	<12.0	12–20	20–30	>30	Indicates resistance to Type 2 breakage. Quick test
Mill abrasion resistance index Appendix 2, Section A2.9 k_s (loss per 1000 revs)	<0.002	0.002–0.004	0.004–0.015	>0.015	Indicates resistance to abrasion by mutual grinding of saturated rock surfaces. Test result may be used with rock-degradation model (Section 3.3.4.2)
Block integrity drop test Section 3.4.8 I_d(%)	<2	2–5	5–15	>15	Indicates resistance to Type 1 breakage of large blocks

A model to predict weight loss with time in different design and site conditions is proposed in Section 3.2.4.2. General guidance for the durability of rocks may be obtained from the tests as indicated in Table 21 and test methods described in Section 3.3 and Appendix 2.

3.2.4.2 Rock-degradation model for armourstone

This model (a fuller description is given in Latham, 1991) is a design tool which may be used for guidance in the assessment of weight loss with time in service on rock-armoured structures. It must be emphasised that the model has not undergone thorough evaluation, and the recommended input parameters are only the best estimates from currently available data. The model is designed to introduce a systematic approach to the forecasting of degradation of rock with time. The designer is then in a position to incorporate these predictions into the design together with engineering experience and judgement such as may be obtained following an examination of degradation and block-rounding processes near the proposed site (e.g. see Fookes and Thomas, 1986). The model is no substitute for the expertise of a qualified and experienced engineering geologist's judgement, though such a person would be encouraged to consult the model.

Factors affecting the degradation rate are divided into intrinsic material properties of the rock source, the production influenced geometric properties of the armourstone, the environmental boundary conditions at the coastal site, and the armour layer design concepts used. These give a total of ten parameters.

To apply the model to a given rock source, a sample of the material is tested in an abrasion mill (see Figure 60) simulation of the wear process (test method given in Appendix 2, Section A2.9). This gives a fractional weight remaining versus laboratory time plot. This weight loss versus time plot is the intrinsic rock property data summarised by the mill abrasion index k_s, the first of ten parameters. A zonation of typical plots that may be expected is given in Figure 61 and corresponding typical k_s values in Table 22.

Having established the weight loss versus time curve, laboratory time is then converted to years on-site using the *equivalent wear time factor*, X (Figure 62), which is derived from the product of nine weighted parameter ratings. The effects of the principal degradation mechanisms, excluding abrasion (i.e. fracturing and spalling), are also included. For example, using parameter estimates that couple meteorological climatic conditions with particular rock type susceptibilities, the influence of physicochemical degradation can be included (e.g. X_6). Similarly, breakages can be included through a wave energy parameter (X_4).
The parameter ratings are set out in Table 23. A summary of model calibration procedures is given in Box 30 and applications of the model are described in Box 31.

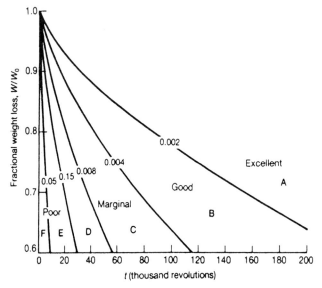

Figure 61 *QMW abrasion test fractional weight loss plot zonations for different rock types tested (see also Table 22)*

Table 22 Typical QMW mill abrasion results for rocks of different durability zonings. (see also Figure 59)

Zone	description	Example rock type	Typical k_s value (per 1000 revs)
A	Excellent	Fresh igneous and metamorphic, chert	0.002
B	Good	Igneous and metamorphis, quartzite, hardest limestone	0.002–0.004
C	Marginal	Igneous and metamorphic weathered (grade II), pure limestone (sparite), hard sandstone	0.004–0.008
D	Marginal	Oolitic and impure limestone, sandstone, shale, siltstone, weathered rocks	0.008–0.015
E	Poor	Oolitic limestone, hard chalk, poorly cemented sandstone, weathered rocks	0.015–0.050
F	Poor	Soft chalks, weathered rocks	>0.05

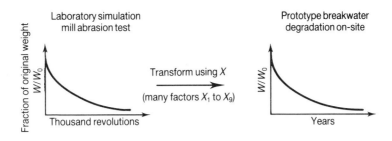

Equivalent wear time factor $X = \dfrac{\text{Number of years in service}}{t,\text{ equivalent number of thousands of revolutions}}$

Problem: to determine X as a function of the nine factors
Solution: X = product of all ratings $X_1, X_2, ..., X_9$ given in Table 23

Figure 62 *Degradation model principles*

Table 23 Ratings estimates for factors in degradation model

Parameter	Rating estimates						Parameter influence X_{max}/X_{min}	Quality of calibration
(k_s)	**Rock fabric strength** Use mill abrasion test to plot W/W_0 versus revolutions, or select from plots of similar material tested in mill						>20	Very good
X_1	**Size** Effect given by $0.5\,(W_{50})^{1/3}$ for W_{50} in tonnes						~10 (~2 armour)	Good
	W_{50}	15	8	1	0.1	0.01		
	Rating	**1.23**	**1.0**	**0.5**	**0.23**	**0.11**		
X_2	**Grading**						~2.5	Fair
	$(W_{85}/W_{15})^{1/3}$	1.1–1.4	1.4–2.5	2.5–4.0				
	Rating	**1.2**	**1.0**	**0.5**				
X_3	**Initial shape**						~2.0	Fair
	P_R (asperity roughness)	>0.013 (irregular, tabular, elongate)	0.013–0.010 (equant)	<0.010 (semi-rounded, rounded)				
	Rating	**1.0**	**1.5**	**2.0**				
X_4	**Incident wave energy** Wave height						~10	Poor
	H_s (m)	>8		4–8		<4		
	Integrity of blocks	Good Poor		Good Poor		Good Poor		
	Rating	**1** **0.3**		**2** **1**		**3** **2**		
X_5	**Zone of structure**						~10	Good
		Supra-tidal Hot temperate		Inter-tidal	Submerged			
	Rating	**2.5** **8**		**1**	**10**			
X_6	**Meteorological climate**—effects on specific rock types and of water absorption $(W_{ab}\%)$						~5	Fair
	Hot + dry	Hot + humid	Freezing winters	Temperate				
	$W_{ab}>2$ $W_{ab}<2$	Basic Acidic	$W_{ab}>2$ $W_{ab}<2$	ALL				
	0.2 0.5	**0.2 0.8**	**0.5 1.0**	**1.0**				
X_7	**Waterborne attrition agents**						~7.5	Poor
		Shingle	Gravel	Sand	Silt	None		
	Rating	**0.2**	**0.5**	**1.0**	**1.2**	**1.5**		
X_8	**Concentration of wave attack**						~2.0	Fair
	Tidal range (m)	<2		2–6		>6		
	Seaward slope (cot α)	<2.5 >3		<2.5 >3		<2.5 >3		
	Rating	**1** **1.5**		**1.2** **1.8**		**1.5** **2.0**		
X_9	**Mobility of armour in design concept**						~4(~10)	Fair
	$H_s/\Delta D_{n50}$	1–3	4–6	6–20	(20–500)			
	Rating	**2**	**1**	**0.5**	**0.2**			

Box 30 Calibration of degradation model

The three methods and various data sources which have been used to calibrate the model are described below:

Method 1: Block shape measurement
Measure prototype rounding for years in service
Convert rounding to weight loss (laboratory relationship)
Abrasion mill test to obtain k_s
Estimate X_1–X_9 for prototype situation, carry out sensitivity analysis

Data source 1: Dibb *et al.* (1983)
              Middle East—limestone
              Australia—granite

Data source 2: Latham and Poole (1988)
              Buckhaven, Scotland—dolerite

Method 2: Block surface profile monitoring
Measure skin thickness removed between reference pins
Convert thickness change—weight loss (area/volume relationship)
Abrasion mill test to obtain k_s
Estimate X_1–X_9 for prototype situation, etc.

Data source 1: Clark (1988)
              West Bay, Dorset—Portland limestone

Data source 2: Fookes *et al.* (1988)
              Review of wear rates from tombstones, etc.

Method 3: Direct weighing
Remove blocks, weigh and return them to structure

Date source: Allison and Savage (1976)
              North Carolina—'New Bern' limestone

Future upgrading and expansion of the rating system will require detailed monitoring using either of the calibration methods during and after construction. Poor and fair calibrations for parameters with an apparently large influence such as attrition agents, block mobility and meterological effects should be further investigated for different site conditions in which other parameters are well defined or kept constant.

Box 31 Application of degradation model

The examples are based on calibration data and are therefore not independent of the model development but will illustrate the approach. The rock types are tested in the QMW abrasion mill giving plots A and D in the figure below. Note that this figure gives the typical loss in asperity roughness observed for milling tests, which can result in greater loss of armour stability than the weight loss itself. The site situations are summarised below, with ratings in bold type indicating changes.

	Parameter	Ratings			
		Example D		Example A	
		(i)	(ii)	(i)	(ii)
X_1	Size	**1.0**	**0.63**	0.72	0.84
X_2	Grading	1.2	1.2	**1.0**	**1.2**
X_3	Slope	1.0	1.0	1.5	1.5
X_4	Wave energy	**1.0**	**3.0**	2.0	2.0
X_5	Zone	1.0	1.0	1.0	1.0
X_6	Climate	1.0	1.0	**0.2**	**1.0**
X_7	Attrition	**0.2**	**1.0**	1.0	1.0
X_8	Conc. of attack	1.5	1.5	**1.5**	**1.0**
X_9	Block mobility	2.0	2.0	**1.0**	**2.0**
	$X =$	0.7	6.8	0.6	6.0

Example D

(i) Eight tonne, narrow grading, static design, shingle attrition, etc.
(ii) Two tonne, low wave energy, sand attrition, etc.

Example A

(i) Three tonne, basalt in tropical climate, medium grading, dynamic design, etc.
(ii) Four and a half tonne, temperate climate, narrow grading, static design, etc.

Design implications
Rock type D

Situation (i) $X = 0.7$, $W_0 = 8$ tonnes (design requirement).
After 25 years ($t = 25/X = 35\,000$ revolutions in mill), from graph $W/W_0 = 0.7$ and so $W = 5.6$ tonnes, 30% reduction, totally unsatisfactory. *If overdimension*, require $W_0 = 11.5$ tonnes to obtain $W = 8$ tonnes after 25 years' degradation.

Situation (ii) $X = 6.8$, $W_0 = 2$ tonnes (design requirement).
After 50 years ($t = 7000$ revolutions in mill)
$W = 1.85$ tonnes.

Rock type A

It is economic for the local temporary quarry dedicated to a particular project to produce a medium width grading with $W_{50} = 3$ tonnes but the wave climate is such that a dynamic design concept must be used to exploit this source of relatively small armour. The rock is a basalt and the climate is tropical. What is the predicted weight reduction after 50 years?

Situation (i) $X = 0.6$, $W_0 = 3$ tonnes
After 50 years ($t = 80\,000$ revolutions in mill)
$W = 2.6$ tonnes
13% reduction

Suppose the same rock was produced in a non-dedicated North European quarry in stockpiles of 3–6 tonne armourstone ($W_{50} = 4.5$ tonnes) for such a temperate climate sited structure, where a single-size static armour design was considered, but with a more concentrated wave attack:

Situation (ii) $X = 6.0$, $W_0 = 4.5$ tonnes
After 50 years ($t = 8000$ revolutions in mill)
$W = 4.4$ tonnes
2% reduction

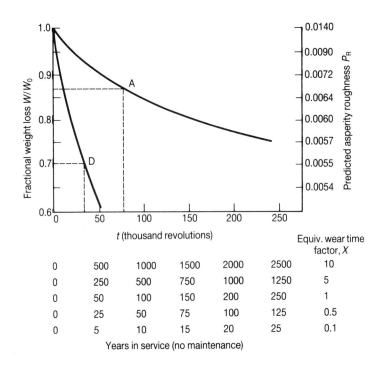

Figure 63 *Application of abrasion mill test results to the degradation model*

3.3 Testing and Evaluation

General reasons for testing rock and aggregate are to:

- Assess the size, quality and usefulness of a new source of stone;
- Ascertain whether the rock from a given source is changing or constant;
- Compare the size and quality of stone from different sources;
- Assess sample variabilities from within one source;
- Predict the performance of material in service;
- Ascertain that the rock characteristics satisfy a specification.

No single test can fulfil all these functions. The possible criteria for selecting a suitable suite of tests for material source evaluation and contract specification include:

1. Time and effort taken to collect sample, prepare and perform test, and report test result, relative to total amount of materials and project value;
2. Sophistication of preparation and testing equipment. Can results be obtained in the field, at temporary site laboratory, at advanced laboratory?
3. Is the rock suspected of being marginal in quality (e.g. from field hammer 'ring', earthy odour on breakage, ease of breakage with hammer, grainy appearance, feels light)? If so, are there any susceptibilities to degradation associated with this rock type (e.g. water-absorbing secondary minerals in geologically weathered basalts, microporous sandstones)?
4. Is the in-service application known to involve high abrasive forces (e.g. from continuous shingle attack and block mobility)? High impact loads due to aggressive storm conditions? Freezing conditions? High temperature and humid conditions?

Generally, a test for each specific property is more meaningful than a simulation that attempts to combine all loading types into one test. Usually, tests should be chosen from Table 21. The materials evaluation process following the desk study (Section 3.1) can be represented by a flow chart (Figure 64) which was originally designed for aggregates (Stapel and Verhoef, 1989) but has been modified appropriately. The petrographic examination (e.g. Section 3.3.4, International Society for Rock Mechanics (ISRM) method in Brown, 1981, and suggestions in Latham *et al.*, 1990) should always be undertaken at the earliest possible stage and may sometimes be run in parallel with the quick physical tests that can also be used to obtain rock-durability indicators (Box 32) and assessment of shape (Section 3.3.2), size and weight gradations (Section 3.3.1).

The follow-up tests such as freeze–thaw (Section 3.3.7.2), sulphate soundness (Section 3.3.7.1), Methylene Blue absorption (MBA) (Section 3.3.7.3), fracture toughness (Section 3.3.6.1) and abrasion resistance (Section 3.3.6.4), that are more involved and may take weeks to report on, can be used to give final guidance on the more marginal rocks and will not be required as frequently in a quality control programme. The importance of representative sampling cannot be overstated.

The following sections identify the standard and non-standard methods for rock materials size and quality assessment quoting the ISRM suggested methods (many are published together in a volume edited by Brown, 1981) as well as the British (BS), German (DIN), Dutch (NEN) and American (ASTM) reference standards and, where appropriate, a brief comment and recommendation as to the preferred method for this manual is also included.

The British Standard Code for Maritime Structures, BS 6349, Part 1: 1984 makes recommendations on rock quality for use as armourstone based on BS 812 (Sampling and testing of mineral aggregates, sands and fillers) which, because they are based on requirements for aggregates, are not ideally suited to sampling and testing methods appropriate for large blocks in coastal and shoreline protection. The present Dutch draft (November 1988) of NEN 5180 (Quarried rock for hydraulic engineering purposes) has therefore concerned itself particularly with sampling and testing methods and requirements for large blocks. Many of the recommended test procedures in this manual follow from the associated series of tests, NEN 5181–5188, which are currently also in draft form. Therefore a number of the draft NEN standard tests have been translated as part of the CIRIA/CUR project and reproduced, with some modifications agreed during the project, as Appendix 2 of this manual.

Acceptance values for the results from the various test procedures described in this section are summarised in Table 21 and in Appendix 1, Part B.

3.3.1 SIZE AND WEIGHT GRADATIONS

3.3.1.1 Particle size distribution

Recommended method: Appendix 2, Section A2.1; Essential test

Appendix 2, Section A2.1 (based on NEN 5181) offers the best test for the size distribution of coarse materials. This test is used to verify (see Figure 65) requirements for fine gradings as defined in Appendix 1 and is recommended in both the draft NEN 5180 and this manual for gradings between nominal stone diameters 30 mm and 200 mm. For finer materials usually associated with aggregates (<75 mm) the BS 812 test is recommended. This test method is required for control of the parameters D_{50}, D_{15} and D_{85} for underlayers and filter layers. The values specified will depend on these design requirements and on producibility and/or availability.

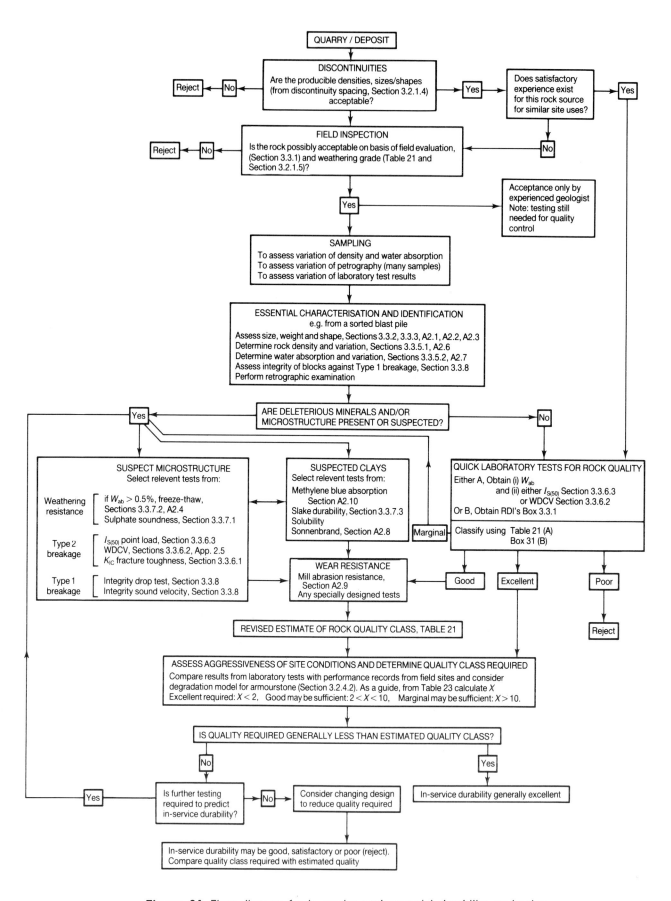

Figure 64 *Flow diagram for in-service rock material durability evaluation*

Figure 66(e)

Figure 66 *Weighing devices being employed in grading tests for light/heavy gradings. (a) Grading test with a mechanical type potato scale; (b) and (c) grading tests, spring-type scales; (d) grading test with a weigh table with built-in pressure cells and concrete calibration test blocks; (e) weighing with a spring-type device (PIAB Dynamometer) (courtesy G. J. Laan)*

Comments on test methods

The UK and Germany do not correct for density variation of test water, expressing results as relative density. Distilled, de-ionised or boiled tap water is specified in BS 812. The accuracy for the saturated and surface-dried condition decreases as sample size is reduced. The testing procedure is invariably coupled with that for the water-absorption test. Variation in test method between the above standards has almost negligible effects on results for design purposes, with the possible exception of saturated surface dry density results (see Section 3.2.1).

Comments on requirements

A minimum average density is often specified because of the relation between durability and density. Usual minimum requirements in the UK (BS6349), 2600 kg/m^3 (for apparent relative density), Holland (draft NEN 5180), 2500 kg/m^3, and Germany (DIN 5202, DIN 52106), 2300 kg/m^3, depend upon the policy with regard to the use of marginal materials.

3.3.4.2 Water absorption, W_{ab} at atmospheric pressure

Recommended method: Appendix 2, Section A2.7 (based on draft NEN 5187). Essential test.
Other standards: BS 812, ASTM C127–84

Comment on test methods

The standard test method for absorption at atmospheric pressures is principally used to indicate durability. Water absorption also influences the relative buoyant density Δ on-site (see Box 15). The influence is negligible for low absorption. This 'on-site' absorption depends on the rock's position in the zone of the structure (i.e. supratidal, intertidal or submerged). The maximum absorption in the submerged zone is attained after a long immersion time. This water content probably equals the 'maximum saturation' given by tests based on boiling or pressure variation provided specimens are no larger than about 350 g. The relative difference between atmospheric and 'maximum' water absorption varies between 15% and 60%. For blocks in the intertidal zone w_{ab} at atmospheric pressure given in Appendix 2, Section A2.7 probably gives the best value for estimating Δ values.

Comments on requirements

The test is to indicate durability. If w_{ab} is less than 0.5% then it will not be necessary to perform durability tests for resistance to weathering. For w_{ab} between 0.5% and 2%, the rock is likely to perform adequately in service but a weathering resistance durability test should be used. For w_{ab} greater than 2%, the material should be considered marginal until durability tests suggest otherwise. The proportion of small pore spaces (i.e. the size distribution of pores in the rock) is often most critical for durability. The resistance to impacts and abrasion in service may be unacceptable for material which is otherwise satisfactory against weathering.

3.3.5 INTACT STRENGTH

Many rock types show a pronounced weakening upon water saturation and testing for breakage resistance of the rock fabric should be performed on saturated samples to simulate in-service conditions. Argillaceous sedimentary rocks and some quartz-rich igneous rocks generally show the most pronounced water-weakening effects. Modern portable equipment for taking rock cores in potential and existing quarries are becoming increasingly important aids to sampling. Tests using core geometry will be favoured.

3.3.5.1 Fracture toughness (K_{IC})

Recommended method: ISRM (Short rod or chevron bend specimen). Recommended test for detailed appraisal of intact rock strength (see *Dibb et al.*, 1983).
Standard: ISRM (1988)

Comments on test methods

The two ISRM recommended test methods are based on the chevron bend and the short rod specimen. The significant advantages of these methods compared with other specimen geometries are given by Meredith (1988) as follows:

1. It can be fairly easily machined from rock cores.
2. No pre-cracking is necessary because crack propagation is initially stable.
3. No crack-length measurement is required because the load passes through a smooth maximum at a constant material independent crack length.
4. The testing technique can be adapted to correct the measured fracture toughness for non-linear effects. In addition, the short rod uses less rock material than the chevron bend specimen but the latter is slightly easier to prepare for testing. Each test may be performed at two levels of sophistication. The use of the fracture toughness test at level I, where the test can be made with portable equipment (Figure 67), requires only the measurement of peak loads and is sufficient for classifying fracture resistance of a rock fabric for the purposes of this manual. The non-linear corrections applied to level II testing may have the effect of increasing the K_{IC} value by a factor of between 0% and 30%, but it should be noted that the larger corrections apply to the stronger rocks, which are unlikely to be of marginal durability, and that level I will always be a conservative value.

For the short-rod specimen, a piece of core is cut to the proper length with roughly parallel faces and a diametral notch is cut using a saw and jig. After notching, plates for load transfer are epoxied to the slotted end of the specimen, parallel and centred to the cut using a guide plate during adhesion. A splitting line load is applied horizontally to the end of the specimen perpendicular to the cut diameter of the core. This may be done by manually turning a handle on a threaded shaft at a constant rate and reading the peak load from a dial (e.g. Gunsallus and Kulhawy, 1984). For level I testing, field apparatus operated from a mobile laboratory with sawing facilities is a practical possibility for the near future. Standard laboratory materials testing rigs (e.g. Instron, Schenck, etc.) can also be used for the test.

Comments on requirements

The bar chart from Meredith (1988) given earlier in Figure 49 shows the ranges of K_{IC} values to be expected in rock types of differing mineralogy that have not undergone any significant geological weathering. Rocks that are found to have K_{IC} values below the appropriate ranges shown in the bar chart are likely to be geologically weathered and should be considered carefully with other tests that would indicate low durability due to physical weathering. In general, and noting the preceding comment, K_{IC} values of less than $1.1\,\text{MPa.m}^{1/2}$ would indicate unsatisfactory fracture toughness, and may be taken as an indication of unsatisfactory impact resistance. However, K_{IC} values as low as 0.8 may be acceptable for mild conditions while for very aggressive conditions K_{IC} should be greater than $1.4\,\text{MPa.m}^{1/2}$. Note that the acceptance of $0.7\,\text{MPa.m}^{1/2}$ in Allsop *et al.* (1985) is too low, and is based on results using a different test method.

3.3.5.2 Impact resistance

Recommended method: Dynamic crushing value determination according to Appendix 2, Section A2.5 (based on draft NEN 5185 but modified to use a saturated sample to allow for the water-weakening effect noted for many rocks). Useful quick test

Other standards: Aggregate impact value (AIV) BS 812. See also Los Angeles abrasion test (Section 3.3.6.4)

Other methods: Modified AIV (MAIV) (Hosking and Tubey, 1969)

Comments on testing methods

In all the above methods, a fixed impulsive force is repeatedly applied (using a standard apparatus) to the surface of an enclosed aggregate specimen and the weight loss of a certain sieve fraction is measured. During impacting, the broken finer material tends to build up, cushioning the blows and causing poorer results discrimination between the weaker materials.

Figure 67(a)

Figure 67(b)

Figure 67 *Fracture toughness test. (a) Test rig; (b) 2-inch diameter short rod specimen tested for fracture toughness (courtesy Queen Mary & Westfield College)*

The MAIV (required for rapid dynamic RDI determination, see Box 32) overcomes the problem of cushioning by fines by repeating the test and ensuring that only 5–20% fines are produced for the assessment of the MAIV and the number of blows is taken into account. The MAIV also requires a saturated test specimen.

The procedure in Appendix 2, Secton A2.5 makes two useful modifications to previous approaches to specimen preparation. The crushing procedure to obtain the test aggregate is standardised, and is then shape sorted through slotted screens prior to testing. Thus, high proportions of flaky aggregate cannot raise the test result value in a manner which would be misleading for applications other than roadstone usage.

Impact resistance tests are quick to performed and the testing machine is in widespread use, but its portability to field sites is more in doubt as there is a need to bolt the machine down onto a solid substrate.

Comments on requirement

For the standard AIV (BS 812), a maximum value of 30% (BS 6349) is usually quoted in the UK for average site conditions. With the Appendix 2 method, an absolute maximum of 40% is recommended, based on Dutch experience. This high value is because the coarser grained igneous rocks can give anomalously high values that do not reflect the real resistance to impacts of pieces larger than 100 mm. More generally, guidance given in Table 21 is applicable which would reject materials with values above 30.

3.3.5.3 Crushing resistance

Recommended method: ISRM Point Load Strength index
Standards: 10% fines BS 812 (1975)
Point load test index $I_{s(50)}$ ISRM (1985)
Uniaxial compressive strength (σ_c) ISRM (Brown, 1981)
Uniaxial compressive strength (σ_c) ASTM (D2938-71a)
Brazil indirect tensile strength (BTS) ISRM (Brown, 1981)

Comments on testing methods

The 10% fines test is a laboratory test which uses an aggregate specimen and is a variation on the aggregate crushing test (ACV, BS 812). A preliminary test is used to ensure that the load is recorded when between 7.5% and 12.5% fines are produced in the aggregate specimen by compression in a standard machine at a displacement rate of 20 mm in 10 min. The load (kN) to produce the 10% fines value is calculated using a simple formula. It gives a measure of interparticle shear strength of the bulk material as well as the crushing strength (usually by tensile failure) of individual rock fragments under relatively fast but essentially static loading.

The point load test index, $I_{s(50)}$, is a field test which is sometimes known as the Franklin point load test. It has some advantages over the other two intact strength testing methods and is used in the static RDI (see Box 32). The load is applied through round-end conical plattens using a hand-operated hydraulic pump and the rock specimen breaks in tension (see Figure 68). Although the latest ISRM suggested method shows clearly how irregular lumps can be tested and corrected to an equivalent $I_{s(50)}$ value, the variability of results obtained in such a manner is too high to form the basis of an accept/reject criterion, though useful for an initial rapid quarry assessment. Only when lump sides are trimmed on a saw or if cores are used will $I_{s(50)}$ values be valid for specification purposes. For weaker rocks, the distribution of results is often skewed towards lower values, so that a mean and standard deviation quoted will be difficult to interpret correctly.

To perform the σ_c or BTS test to the standards quoted, a sophisticated laboratory apparatus is required. Thus if σ_c is needed for use in determining Barton's equivalent strength parameter for particulate rock (see Section 5.2.2.4) or for other purposes it may be sufficient to quote that the value is 'compressive strength estimated from point load tests'. $\sigma_c = 22 I_{s(50)}$ is the best established relationship (Brook, 1985) and the one to be recommended. However, the proportionality constant does vary, possibly due to differences between rock types, weak rocks particularly affecting the value.

Comments on requirements

The requirement that 10% fines should not be less than 100 kN (BS 6349) will be fulfilled by many marginal materials that would probably fail on other test criteria. σ_c should be greater than 100 MPa or $I_{s(50)}$ should be greater than

Figure 68 *Point load test machine (courtesy Queen Mary & Westfield College)*

4 MPa are the most commonly quoted and therefore recommended requirements for armourstone in average site conditions. These values correlate with a BTS of about 10 MPa.

3.3.5.4 Abrasion resistance

Recommended method: Abrasion mill test according to Appendix 2, Section A2.9. If insufficient time, use BS 812 AAV test or sandblast test (Verhoef, 1987).
Other standards: Los Angeles test ASTM C 131-81
Aggregate abrasion test (AAV) BS 812: 1975
Sandblast test ASTM C 418-8, NEN 7014
Wet attrition test BS 812: 1951
Wet Deval test Norme NFP 18-577 (1979)
Other methods: Sandblast test (Verhoef, 1987)

Comments on test methods

The Los Angeles test (LAV) uses a steel drum containing a specified number of steel balls. When rotating with the specified sample this applies a combination of attrition due to wear between rock particles and impacts from the charge of steel balls which may be sufficient to cause whole-lump fracture. It was designed for testing highway materials, and the simulation of loading combines fracture with abrasion in a manner which bears little relation to processes in coastal structures. Varying degrees of cracking caused during sample preparation can influence results. However, it remains popular as the only standardised indicator of wear or impact resistance used in US and Canadian specifications. The machines are widely available in the USA, Canada and parts of Europe.

The aggregate abrasion test (AAV) is where a number of rock fragments are embedded in resin and then pressed against a rotating steel disc. A silica sand is fed between the rock and steel surface and acts as an abrasive medium. The test was designed for highway materials and is of little value compared wth other abrasion tests, with the possible exception of when there is a substantial amount of beach sand in the water or wind.

The French (LCPC, 1989) recommend the use of the wet Deval test for assessing wear resistance of armourstone. The wet attrition test and the wet Deval test are practically identical and give good simulations of rock-to-rock surface grinding, but their procedures, though simple, are often criticised as having poor reproducibility and discrimination, problems which have now been overcome with the QMW mill-abrasion test.

The sandblast test has been researched (Verhoef, 1987), specifically with abrasive wear of rocks in coastal protection works in mind. The surface of the test sample is abraded by an air blast containing silica sand under specified conditions. The result is expressed as an index which is the ratio of material volumes lost by the test rock and by soda-lime glass (as a standard reference) under the same sandblasting conditions. It is therefore easy to interpret in terms of on-site performance with time. The index has a very non-linear relationship with the traditional strength measures such as Brazilian tensile strength. It appears particularly suitable for discriminating materials that have BTS of less than 10 MPa but is probably of little value for discriminating tough rocks. Data from further tests on marginal rocks with some in-service performance ratings would help to establish the value of the new test method.

QMW mill abrasion test

A full description of this test procedure is given in Appendix 2, Section A2.9 with suggested requirements in Table 21, Section 3.2.4 and Appendix 1.

Comments on other abrasion test requirements

For LAV the maximum acceptable loss for average site conditions in the USA and Canada is usually about 35% for 500 revolutions. For AAV the maximum acceptable loss often quoted is 15% (B3 6349)

3.3.5.5 Deformability, hardness and sonic velocity

Elastic moduli

The elastic Young's Modulus, E, and Poisson's ratio, v, in uniaxial compression (ISRM, Brown, 1981) are obtained from uniaxial compression testing on intact cylinders of rock. In addition to σ_c the E (and v) test results can be used for classification of intact rock. The laboratory equipment and procedure required is relatively complex. Young's Modulus is the ratio of the axial stress change to the axial strain produced by the stress, and is calculated from the slope of the stress–strain curve. Poisson's ratio, μ, is the ratio of the lateral expansion to the axial contraction, and is lower in rocks with high porosity. Both moduli are often used in blast design calculations.

Hardness

Hardness of rocks is measured by indentation, rebound or scratch tests (see Brown,

1981). The most practical (because of its simple field use) is the Schmidt impact hammer test, used to determine the Schmidt Rebound Hardness (ISRM, Brown, 1981). The Schmidt hammer is a quantitative extension of the qualitative impression gained from the sound of the geologist's hammer striking a rock, in terms of toughness, elasticity and state of freshness. The results of the Schmidt hammer test (see Figure 69) are strongly influenced by methodological factors such as surface preparation for the piston and the unsuspected presence of open, face-parallel cracks close behind the rock surface. The instrument is held tightly against the rock and a spring-loaded hammer strikes the rock with a known energy input. The rebound is recorded on a scale as a percentage of forward travel and is influenced by the elasticity of the rock and the amount of permanent deformation of the mineral fabric. Different hammer types releasing varying amounts of energy are available. For the common test hammer, type L, fresh igneous rocks give a value of 50 or greater. A number of studies provide examples of the use of this technique on rocks (Duncan, 1969; *Geol. Soc. Rep.*, 1977) and correlations with σ_c.

Sound velocity

The sound velocity determination (ISRM, Brown, 1981) measures the velocity of propagation of elastic waves for different sonic frequency pulses. Dearman (1987) describes how zones of altered mineralogy and porosity in a dolerite were discriminated using the sonic velocity method and velocities from 3000 to 4200 m/s were correlated with porosity from 4% to 0.6% and oven-dry density of 2.82–2.89 t/m^3. A disadvantage is that the method is sensitive to the degree of saturation of the rock. It is a possible future test method because of its potential speed, but accurate results will require careful preparation of specimens. In France (LCPC, 1989) the index of continuity, I_c (NFP 18-593, 1980), based on sound velocity determination, is considered highly significant (see Section 3.2.1.6) for assessing microfractures.

3.3.6. RESISTANCE TO WEATHERING

Ideally, cyclic stressing simulation tests should be carried out on sample blocks the same size as the blocks to be used on site, but this is impractical and a compromise for routine testing is necessary. However, specially designed test procedures using large pieces (e.g. greater than 20 kg) may sometimes be incorporated into the initial material evaluation process. The standard freeze/thaw test should use 10–20 kg pieces. It is suggested that the aggregate sample size for the magnesium sulphate soundness test should be chosen from two alternatives, depending on whether microcracks are visible in hand-sized specimens. It has been estimated that five cycles of sodium sulphate is equivalent to about 200 freeze/thaw cycles under normal conditions.

Recommended weathering resistance tests

The magnesium sulphate soundness test (BS 6349) using a 7 kg sample of 63–125 mm sieved pieces with a 50 mm sieve to determine the losses is recommended. The test is important for rocks showing significant water absorption due to microcracks and/or if a small pore size encouraging capillary action is suspected. For site conditions with high temperatures and salt precipitation, it will be an indicative test for breakdown by salt action.

The freeze/thaw test (draft NEN 5184, see Appendix 2, Section A2.4) is useful for the same rock susceptibilities as above but applies particularly to cold climates with freezing winters, especially for inland shore protection. A small pore size together with high absorption is an indication of susceptibility to freeze/thaw.

The Methylene Blue absorption test (see Appendix 2, Section A2.10) may be useful for argillaceous sandstones and impure limestones as well as igneous and metamorphic rocks with suspected secondary (i.e. altered and weathered) minerals which may behave like swelling clays. The test has been known to result in rejection of perfectly sound dolerites and the most reliable results will be from

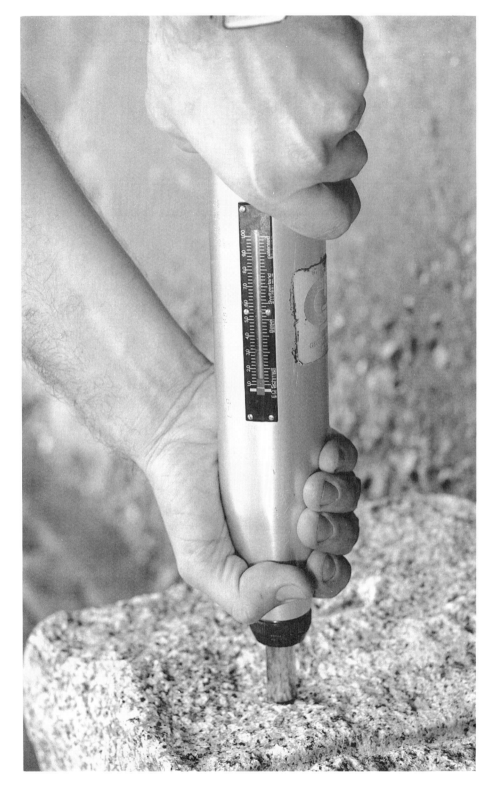

Figure 69 Schmidt hammer (courtesy Queen Mary & Westfield College)

samples of basalt. The 'Sonnenbrand' (draft NEN 5188) test (see Figure 70) has become important for testing basalts with possible deleterious minerals from a particular area near Eifel in Germany and may be important for basalts in other areas if there are no service records indicating the rock's durability. For a review of the subject of weathering and testing the resistance to weathering, see Fookes et al. (1988).

3.3.6.1 Sulphate soundness test

Standards: ASTM C88
                 BS 6349: 1984, Appendix B
                 BS 812, Part 121: 1989
Other methods: Property Services Agency (PSA) (1979)
                 Hosking and Tubey (1969)

Comments on test methods

The ASTM C88 test is the one most widely chosen internationally. It determines the resistance of an aggregate sample to disintegration by saturated solution of magnesium or sodium sulphate. The test results are thought to be a qualitative indicator of the soundness of an aggregate subject to weathering action, freeze and thaw conditions or salt crystallisation deterioration (i.e. the physicochemical cyclic stressing degradation mechanisms). In this test, samples of representatively graded aggregate (which can include material up to 125 mm) are subjected to alternate cycles of immersion in magnesium or sodium sulphate and oven drying. Crystallisation in the pores and microcracks may set up bursting pressures within the rock. The weight loss of material after five cycles denotes the soundness. Careful attention should be given to the qualitative description of the break-up of pieces larger than 20 mm. The test takes about 3 weeks.

The BS 6349 test is almost identical to that of the ASTM and was adapted to British practice. The PSA test uses a single aggregate size sample and forms the basis of an improved test method given in the new BS 812, Part 121. This new BS test introduces stricter procedural measures and uses only magnesium sulphate solution (with a 400–420 g sample of 10–14 mm pieces in the definitive method), but it also takes 3 weeks. Unfortunately, the test result is expressed as a percentage of original weight retained on the sieve in contrast to all other soundness test methods. Hosking and Tubey (1969) modified the test to a quick test taking 5 days for 5 cycles using 60 chippings in the 12.5–19 mm size range. This is the quick method required in the determination of the static RDI (Box 32).

The test may become prohibitively expensive if a widely graded sample is tested (e.g. using the full grading range in BS 6349). The test results depend on particle sizes and shapes tested, with increasing losses for angular particles. The chemical reaction which occurs between carbonate and sulphate in solution is another possible problem. The test may not be suitable for all rock types and reservations have been expressed elsewhere in respect of some carbonate aggregates and some aggregates having a high proportion of magnesium bearing minerals or of cryptocrystalline quartz. There is therefore always a high variability associated with the sulphate soundness test and usually at least a 3-week reporting time. When the strength of visible microcracks is in question, the visual examination by an experienced geologist of cored specimens subjected to a non-standard, five-cycle sulphate soundness test can be more useful than the numerical result of the standard test.

Comments on test requirements

The numerical value of sulphate soundness for acceptance as armourstone should be the same or lower than that for underlayer and core which are relatively protected from in-service weathering. BS 6349 proposes a maximum of 18% for magnesium sulphate soundness value. When water absorption into the rock is judged to be enhanced by visible microfractures or if the rock is anisotropic, the BS 6349 test with the 63–125 mm pieces is to be recommended, otherwise,

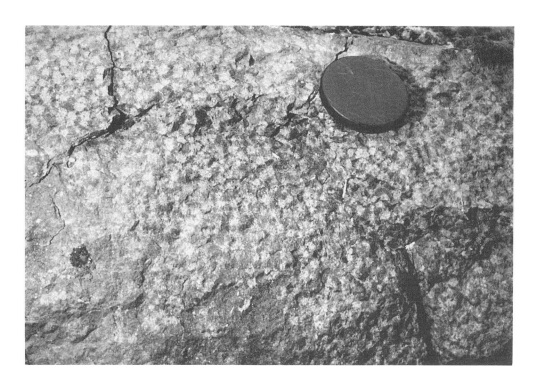

Figure 70(a)

Figure 70(b)

Figure 70(c)

Figure 70 *Rocks affected by the "Sonnenbrand" effect. (a) Star-shaped spots on surface of basalt block, indicating the Sonnenbrand effect due to deleterious minerals and leading to severe failure; (b) and (c) considerable cracking interpreted as being caused primarily by Sonnenbrand effects (courtesy G. J. Laan)*

BS 6349 with a 10–20 mm aggregate sample size is to be used. In the absence of research data, the level of acceptable loss is assumed equivalent for each sample size.

3.3.6.2 Freeze/thaw test

Recommended method: Appendix 2, Section A2.4 (based on draft NEN 5184)
Other standards: ASTM C-606
AASHTO T103-78

Comments on test methods and requirements

The test in Appendix 2, Section A2.4 will take a minimum of 1 month to report. That for stones heavier than 20 kg subjects pieces of at least 10 kg to 25 freeze/thaw cycles. Twenty pieces sampled at random are tested and 0.5% weight loss in more than one piece is taken as a failure to resist freeze/thaw according to NEN 5180. The values given in Table 21 suggest ten stringent criteria may sometimes be acceptable.

3.3.6.3. Breakdown due to clay minerals

Recommended method: Methylene Blue Absorption as described in Appendix 2, Section A2.10
'Sonnenbrand' effect test for basalts as described in Appendix 2, Section A2.8
Other methods: Slake durability index (Franklin, 1970 and ISRM, Brown, 1981).

Table 24 MBA values (g/100 g) of some clay minerals and rocks

Montmorillonite	19–36
Bentonites	5–23
Chlorite	0.6
Kaolinite	2–5
Illite	1.8–2.5
Halloysite	1.3
Palygorskite	14.6
Ferrihydrite	<0.6
Serpentine	1.2
Basalt	0.11–0.4
Granite	0.11–0.4
Weathered basalt	18.7
Bauxite	1.2
Organic limestone	0.9
Marble	<0.6

Data from Stapel and Verhoef (1989)

Wetting–drying (Lienhart and Stransky, 1984)
Washington degradation test (Department of Main Roads, T214, 1974)

Comments on test methods

The Methylene Blue Absorption (MBA) test has been recently revised (Stapel and Verhoef, 1989) and a full description of the method is given in Appendix 2, Section A2.10. Limiting Methylene Blue absorption values are given in Table 21 and may be compared with typical MBA values of clay minerals and rocks listed in Table 24.

The slake durability index measures the resistance to mixing with water and is an index test principally for very weak mud rocks. A number of small specimens of known weight are placed in a wire-mesh drum and immersed in water. The drum is rotated for 10 min, causing the specimens to rub and abrade, and after 10 min they are dried and weighed. The slake durability index expresses the weight loss as a percentage of initial weight. Two cycles are usually used.

The test for 'Sonnenbrand' effect is a curiously simple test for a phenomenon which is not well understood but known to indicate deleterious minerals in basalt. The test is given in Appendix 2, Section A2.8.

The wetting-drying test (Lienhart and Stransky, 1984) is designed to simulate summer conditions of alternating rainfall and drying by summer sun on armourstone and facings, and is an effective test for argillaceous rocks. For armourstone the test consists of soaking 65 mm thick slabs, cut (normal to bedding) from the sample, in water at 10°C for six hours followed by infra-red heating for six hours. The normal test of 80 cycles therefore takes at least 40 days, although the number of cycles may be based on experience and climatic conditions at the site.

The Washington degradation factor is a measure of fines produced by rock particles when they are abraded in the wet. This is measured by the height of a sediment column in a cylinder which has had all the abrasion products removed except for the material passing a 75 mm sieve.

3.3.7 BLOCK INTEGRITY

Recommended method: Drop test breakage index according to Appendix 2, Section A2.11
Test delivery
Other methods: Visual examination
Sounding bar
Sonic velocity
Infra-red photography.

The methods described should be useful for two main applications:

1. To predict breakage during handling and in service;
2. To apply accept or reject criteria for individual or consignments of blocks, once a source of adequate-sized armour has been located.

The limited development of the methods arises because of the difficulty in assessing block integrity.

The suggested drop test method applies to standard heavy gradings or non-standard gradings defined by the class limit and effective mean weight system. Experience with this form of the drop test is very limited and hence few data can be presented.

Recommendations for acceptances values of I_d are difficult to make, as this method is a relatively new suggestion. However, those values of I_d given for the descriptive terms poor, good etc. in Table 21 are probably reasonable based on current experience and are adopted in Table 22 for the degradation model ratings.

The test delivery index methods are identical to the drop test methods except that final weighings are obtained after damage by transport and handling impacts upon arrival and unloading at the construction site. If breakage is excessive, a secondary selection is carried out in order to meet the requirements for the armour layer.

A subjective but useful semi-quantitative method to assess the breakage resistance of the armour-sized blast products is by visual inspection. The number of significantly flawed blocks which are considered likely to break during rough handling are expressed as a percentage of the total. Examples of flaws are shale layers, stylolite seams, laminations, unit contacts, veins, joints, cleavage and foliation planes and cracks. The problem with this method is that even geologists may not agree on which blocks are significantly flawed. In this respect, two points may be useful. Veining, jointing and foliation planes in blocks are usually insignificant if the block surfaces are fresh and tend not to follow such flaws, although it is more common for block sides to follow flaws. Partial fissuring in freshly extracted rock is usually significant, as these cracks will invariably propagate.

Another qualitative technique used successfully in the dimension stone industry is to strike the blocks in various places with a 'sounding bar' and to listen to the ringing sound for an indication of whether serious undetected flaws are present.

To assess internal and otherwise invisible block discontinuities and distinguish whether they are a serious strength hazard to the block, non-destructive sonic velocity methods are currently being researched at the Delft University of Technology (Niese et al., 1990) and in France (LCPC, 1989). The Dutch method involves a statistical treatment of at least 20 acoustic velocity measurements per block which aims to identify serious discontinuities by a high incidence of poor acoustic wave transmission through the block relative to the maximum or median velocity measured. The French method uses the Index of Continuity I_c, with acoustic velocities measured in three orthogonal directions in each block (Section 3.2.1.6). A theoretical velocity for the rock type is then applied in order to normalise the measured velocity and to determine the significance of any low measured velocities. Although both these methods are operational, they are time consuming to perform and require further validation. They are most useful when calibrated with drop tests, as described in Appendix 2, Section A2.11. Infra-red photography, which can detect surface cracks by heat loss anomalies, has been used in concrete technology and may be worth future research.

3.4 Production and Handling

Production and handling of rock is described in this section under three main headings: resource appraisal (terrestrial and marine), production (quarry types, blast design, stone selection, handling, stocking, loading and costs) and transportation.

3.4.1 RESOURCE APPRAISAL

3.4.1.1 Terrestrial sources

Access to the contract site is an important consideration. Sea access allows materials to be obtained economically from greater distances, but if road transportation is the preferred option, the following steps in the appraisal of resources in the local area should be considered:

1. Study the available geological data of the area (see Section 3.1 for guidance).
2. Establish a priority list of likely locations from these data.
3. Carry out a visual inspection of the sites (see Section 3.2.1 for guidance on intrinsic properties).
4. Ascertain and take appropriate action on site ownership.
5. Comply with any statutory requirements to commence site investigations.
6. Consider environmental and public relation aspects.
7. Assess each site for materials quality and reserves (see Sections 3.2 and 3.3).
8. Carry out preliminary cost analysis.
9. Acquire land or mineral rights and planning consent for selected site(s).
10. Carry out a detailed drilling programme so that:
    (a) Material reserves (quantities available for production) may be established; and
    (b) Areas of suitable and unsuitable material may be identified.
11. Establish capital cost requirements.
12. Draw up production schedules.

Site investigations may involve a range of different techniques, ranging from direct geological mapping or drilling to indirect electrical, magnetic and seismic methods. The choice of techniques used must be related to the nature of the particular site, the availability and cost-effectiveness of the methods chosen, and must address the problems of materials quality, variability and spatial extent.

3.4.1.2 Marine sources

For coastal and shoreline structures, marine gravels are logically a suitable source of materials. In 1984, 1 250 000 tonnes of as-dredged materials was used for reclamation contracts in the UK.

Again, existing maps and charts should be used as a starting point to locate gravels and other materials. The interpretation of geomorphological processes such as river-sediment transport together with geological information relating to ancient river courses, glacier movements and old land surfaces can assist in the location of possible source areas. Ships with sub-bottom profile equipment can give a rough guide to the geological content of the bottom but little indication of bed thicknesses.

Precise location of position is difficult, and special navigational systems may have to be set up. Reserve evaluation is most simply achieved by sampling on a grid pattern with grab sampling and borehole samples obtained as appropriate. In selecting a particular site, consideration should be given to the following:

1. The ratio gravel/sand/silt is economically important. (Note that sand and silt are typically washed out from the top layers of a beach and are deposited elsewhere.)
2. A wide area a few metres thick is most economically dredged because each load can be dredged without much manoeuvring, but conflicts with ecological requirements to minimise the area of seabed destroyed.
3. Optimum depths of deposit below the water surface lie between 10 and 30 m. Depths of 40 m are possible for some dredgers but are not preferred, as production is low and costs are high.

4. Well-rounded shingle is most appropriate for recreational purposes and also produces least wear on dredger pumps and pipelines.

It is usually necessary to apply to the appropriate government agency (e.g. Crown Estates in the UK) for licences to explore and dredge offshore materials. Such licences are often difficult to obtain because of concern over environmental effects and coastal erosion. For large contracts it may be possible to gain a new extraction licence. For small ones the use of an existing licence is most likely.

3.4.2 PRODUCTION

3.4.2.1 Types of quarry

The majority of working quarries produce aggregates for roadstone and concrete; very few are dedicated or specialise in the production of stone for coastal and shoreline engineering. In general, quarries fall into three categories.

Aggregate quarries

In aggregate quarries, blasting and other processes are aimed at achieving the optimum production of material of around 40 mm and finer. Blast design will be aimed at achieving maximum reduction in size of material and minimal production of sizes suitable for armourstone. However, there will be a percentage of material from a blast that is too large for the primary crusher, and will either require further processing before crushing or be sold. Its geological characteristics will determine its potential for armourstone (see Section 3.1). In most aggregate quarries the production of blocks over 3 tonnes is unlikely to exceed 5% and may be as low as 1% or 2%.

The largest material available from routine production is likely to be the primary stockpile, i.e. material that has passed once through a crusher. Dependent on the type of crusher, the top size is unlikely to exceed 300 mm and may be as small as 150 mm. It is likely to be randomly graded down to dust or 40 mm.

Quarries may have, or may install, a separate production plant for larger materials required in coastal and shoreline engineering. It is unlikely that significant stocks of large material will be held and long lead times are likely to be required.

Dimension stone quarries

These quarries produce large blocks of stone for masonry or monumental works. Production methods may exclude blasting. There is usually a high incidence of rock unsuitable for its intended purpose and which appears ideal for armourstone. These 'off-cuts' can be of any size, are likely to be of irregular shape and may be available at a low price. However, by their very nature, it is likely that any blocks of a regular shape are likely to be flawed. Thus, careful selection procedures will be required if these blocks are to be used in coastal engineering works. The production of medium and lighter gradings, if produced, will be as for aggregate quarries. However, significant quantities of large stones may be available at short notice.

Dedicated armourstone quarries

There are a few permanent quarries that specialise in the production of stone for sea defences. Temporary dedicated quarries may be required for some contracts. In either case there is a need to maximise fractions of usable material in order to minimise costs and wastage.

For quarries serving more than one contract the standardisation of specifications will aid the maintenance of stockpiles. For the permanent quarries, stockpiles are likely to exist and lead times could be low. To open a temporary quarry could take years (see Section 3.4.1), though some dedicated quarries have been opened in only 6 months.

3.4.2.2 Armourstone blast design

Blast design is a significant factor in securing desired fragmentation. However, it is unlikely that an aggregates quarry will consider changing its blasting pattern to suit a particular contract unless a significant proportion of the blasted material is required. Quarries may be prepared to set aside one particular area of the quarry for special materials. (Figure 71).

Blast design largely depends on the skill and experience of the people involved. However, the sizes and proportions of rock particles produced will depend on the following factors:

1. The geological characteristics of the rock (see Section 3.1.2), which include the following:
   (a) Discontinuity spacings and orientation (bedding, joints, faults, cohesion across planes). *In-situ* block size distributions can be assessed using the techniques described in Box 17 in Section 3.2;
   (b) Strength and elasticity (rock type, weathering characteristics);
   (c) Density, porosity, permeability.
2. The properties and detonation methods of the explosives used;
3. The blast design (configuration and drilling pattern).

Generally, the proportion of armour-sized blocks increase with increasing tensile strength, decreasing Young's Modulus and increase in the discontinuity spacing. Discontinuity spacing is a primary factor in that it controls the maximum size of blocks obtainable from a rock mass.

In contrast to a high-fragmentation blast, greater percentages of armourstone can be produced if the blast design involves the following factors (see Figure 73):

1. A low specific charge. Generally, a specific charge as low as $0.2\,kg/m^3$ can be used. If possible, the explosive used should have lower shock energy higher gas energy.
2. A large burden and small hole placing (Figure 73). Initially, the burden may be chosen as 60 times blasthole diameter and should be larger than the discontinuity spacing in a jointed rock mass. The spacing to burden ratio should generally be less than or equal to 1 and may be as low as 0.5.
3. If the bench height (Figure 73) is either too high or too low, armourstone production will be poor. For an initial estimate, bench height could be selected as two to three times the burden. In planning bench levels, the rock mass from which most armourstones might be produced, such as thick bedded layers, should be located nearly at the top of the bench alongside the stemming section of the holes.
4. A large stemming length (Figure 73) is better than a small one. Usually, stemming length is selected to be larger than burden.
5. A small blasthole diameter. A diameter of less than 100 mm (normally between 50 and 75 mm) is recommended.
6. One row of holes is found to be better than multi-rows.
7. Holes should be fired instantaneously rather than using inter-hole delaying (this may cause high ground vibration).
8. A bottom charge of high concentration with a column charge of low concentration. A decking charge will be necessary in most situations. The material for decking can be either air or aggregates.
9. A small primer. This should be located either centrally or multipally distributed.

Prediction of the result of blasting

Prediction of the size distribution yield curve is the subject of current research, but accuracy in detail is limited because the geological conditions cannot easily be determined for every blast and there are limitations arising from the practical implementation of the blast design. Various equations have been proposed for predicting the blast fragmentation size distributions. The most widely used are

Figure 71 (a)

Figure 71 (b)

Figure 71 (a) *Blasting in a limestone quarry near Namur, Belgium (courtesy G. J. Laan);* (b) *blasting at Jebel Berri Jubail, November 1982 (courtesy Aveco Infrastructure Consultants)*

Figure 72 *Drilling equipment used in drilling blast holes (courtesy G. J. Laan)*

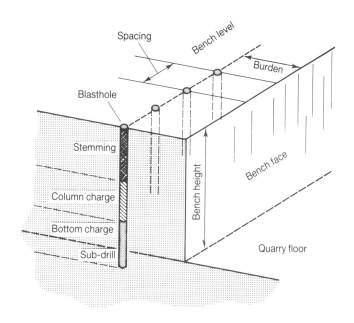

Figure 73 *Schematic illustration of definition of the bench blasting terms*

Bond's third theory of comminution and Kuznetsov's equation. The latter was established empirically using results of experimental and production blasting for high fragmentation, and is not discussed here. The application of Bond's third theory, together with the Rosin–Rammler equation, is often the most convenient means of predicting block size distributions (see Boxes 33 and 34).

These equations allow prediction of the mass fraction of blocks over a particular size and estimation of the proportions in the different size grades. It is unlikely

> **Box 33** Bond's third theory equation (Bond, 1959)
>
> The equation given below is used to determine the required blast product size parameters for a given *in-situ* size parameter:
>
> $$E_f = 10 * E_c * \left(\frac{1}{\sqrt{P_{80}}} - \frac{1}{\sqrt{F_{80}}}\right)$$
>
> where E_f is the required energy for fragmentation in kwh per ton of processed material, E_c is Bond's work index and P_{80} and F_{80} are the product and feed sizes, respectively. Here, 80 means that the particle size is equivalent to the sieve opening through which 80% of the materials pass (expressed in microns).
>
> In this equation, the parameter F_{80} can be estimated by the *in situ* block size distribution analysis techniques (Section 3.2.1.4). The required energy (E_f) can be taken as input energy, which can be determined from the type of explosive and the specific charge as follows:
>
> $$E_f = \frac{0.00365 * E_s * S_e}{\rho_r}$$
>
> where E_s is the weight strength of the chosen explosive in weight (%), S_e is the specific charge (kg/m^3) and ρ_r is the density of the rock mass (t/m^3).
>
> The Bond work index, E_c, which is defined as the energy required to crush a solid of infinite size to a product for which 80% passes a size of 100 μm, represents the rock's resistance to crushing, grinding and breaking by blasting. The accuracy of applying this equation lies in the correct determination of this index. Da Gamma (1983) presented an empirical equation which might be adopted to estimate E_c:
>
> $$E_c = 15.42 + 27.35 * F_{50}/B$$
>
> where F_{50} is the feed size which 50% of rock will pass and B is the burden.
>
> The product size of the blasted rock given by P_{80} can then be estimated by Bond's theory equation. The outstanding advantage of this theory is that the prediction of the blasted results is related to the *in situ* block size and the energy. This equation has been successfully applied to the size analysis of rock blasting for high fragmentation.

that blasting predictions will become totally accurate under normal conditions, but some useful and reasonably accurate generalisations can be made.

3.4.2.3 Selection in the quarry

Selection processes will differ, depending on whether material is selected from a blast pile or from a stock of oversize stones. Oversize blocks may be reduced by wedging along a row of small boreholes, by hydraulic 'pecker' (Figure 74), expansive grout techniques, or by a high-pressure water jet in a borehole. Blast pile selection is likely to be made using large machinery. The precise selection is made difficult by most of the stone being obscured by other material. Selection responsibility will, in the fist instance, lie with the machine operator, who will be in the cab of his machine some distance from the actual stone (Figure 75). Simple rules of thumb may be given or calculated. Generally, the selection is done by eye against 'example blocks'. Therefore, it is best to specify a wide range in weight for blocks, ideally following the class limit grading system suggested in Section 3.2.2.4.

Selection from 'stock' is made easier by the absence of obscuring material and the ability to get another person close to the material to indicate suitable stones. When individual stones are being considered, the operative on the ground may be given some method of sizing stones to the specification. Weighing devices on handling machines can also be used (see Section 3.6).

The stones in excess of 300–500 kg have to be selected individually and the rate

> **Box 34** Rosin–Rammler equation
>
> This equation is used to generate the entire fragmentation curve from the P_{80} value and the uniformity coefficient of the size distribution determined empirically from the blast design parameters, as explained below. It has the form:
>
> $$x = 1 - \exp[-(S_x/S_{63.2})^n]$$
>
> where S_x is the screen size for which the fraction x pass on the cumulative curve, $S_{63.2}$ is the characteristic rock size and n is an index of uniformity. For a particular sample of rock with a known density, this size equation can be transformed from the Rosin–Rammler equation in Box 23. However, n needs to be changed, since $n_{rr} = n/3$.
>
> The index of uniformity or Rosin–Rammler exponent, n, determines the shape of the curve. Larger values of n give a narrower range of fragment sizes, fewer fines and fewer large blocks.
>
> The value of n can be estimated by the modified algorithm (Cunningham, 1987):
>
> $$n = (2.2 - 14.0 * d/B) * \{0.5(1 + S/B)\}^{0.5} * (1 - W/B) * (\text{abs}(BCL - CCL)/L + 0.1)^{0.1} * L/H$$
>
> where
> - d = hole diameter (mm),
> - B = burden (m),
> - S = spacing (m),
> - BCL = bottom charge length (m),
> - CCL = column charge length (m),
> - L = total charge length (m),
> - W = standard deviation of drilling accuracy (m),
> - H = bench height (m).
>
> Characteristic size $S_{63.2}$ is a mathematical point of no great practical importance. It merely fixes the position of the curve where $S = S_{63.2}$. so that
>
> $$x = 1 - \exp(-1) = 0.632 = 63.2\%$$
>
> This is a point with 63.2% of rock passing the mesh size $S_{63.2}$.
>
> The whole distribution of the blasted block size can be approximated by applying the third theory with the Rosin–Rammler equation. The characteristic size, $S_{63.2}$, in the Rosin–Rammler equation can be determined by inputting $S_x = P_{80}$ and $x = 0.8$ (80%) if n is already known.

of output depends on the hoisting and rotation speed of the selection machine, not on the weight of the block. Consequently the smaller size category (just over 300 kg) is normally the most expensive in production.

For lighter gradings material may be selected by the use of grizzlies (passive grids) or mechanical grids. It is possible to pass as-blasted material over a selection screen comprising iron rails with gaps of up to 500 mm maximum. Grids may be used to a maximum of 200 mm. A combination of grids and screens may give the required material (Figure 76). However, some degradation of material will occur with these methods (see Section 3.2.4). The fixed cost of establishing large grids is such that the production of small quantities by this method may be prohibitively expensive. Tyning buckets may be used to select out material from a blastpile or stockpile.

3.4.2.4 Handling

A variety of handling equipment exists, and its availability will depend on type of production unit, location and economic factors. The principal types include the following.

Wheeled loaders

The conventional wheeled loader is suitable for handling lighter gradings but is not ideal for individual stones. They may give rise to contamination with surface material.

Figure 74 *Splitting oversize blocks with a hydraulic 'pecker'* (*courtesy G. J. Laan*)

Figure 75 *Selection by machine operator of 60–300 kg and 300–1000 kg basalt* (*courtesy G. J. Laan*)

Figure 76 *Production screen for 10–60 kg (falling through) and greater than 300 kg (sliding over)*

Hydraulic excavator

The hydraulic excavator may be in either the back hoe or front shovel configuration. Depending on the nature of the material pile, they can be used to good effect on either individual stones or mass gradings. Different types of bucket or forks can be fitted to suit the intended purpose.

Grabs

Grabs may be fitted to either cranes or hydraulic machines. The finger grab is suitable for cranes and can vertically lift and lower large blocks successfully. The hydraulic grab is more suited to picking up and placing stones and can work in a variety of planes. Normally, a hydraulic grab will have one fixed arm and pressure will be exerted by the moveable arm. There are a number of different designs of grabs or grapples.

Cranes

Cranes can be used with a variety of attachments including grabs. The relative speed of operation using a crane is likely to be slower than when using a

Figure 77 *Handling rock with chains (courtesy G. J. Laan)*

hydraulic machine. There are a number of options for lifting armourstones which will be dependent on the stone itself and the manoeuvre required. Some stones have a natural softness such that chains (see Figure 77), dogs, etc. can grip them. Harder stones cannot be gripped in this way and support must be given on the underside to pick them up. The use of chain in slings is continued because of its flexibility and superior shock-absorbing properties. It can stand rough usage and has a relative longevity in use and storage. Safe working loads must be marked on chains.

A wide variety of wire rope slings exist, and the suppliers should be consulted to obtain the most suitable rope for a particular purpose. Wire ropes are designated by size, strength and construction. Fibre ropes may be used but reference to the manufacturer is strongly recommended.

Eyebolts may be used in conjunction with a lifting machine. A hole is drilled in the rock and epoxy grout added. A bolt is then placed in the hole and, after a short curing time, a re-usable eyebolt can be attached. The safe working load of the bolts and epoxy grout needs to be checked and the positioning of the hole is critical for safe and efficient lifting. Epoxy-grouted hooks and ordinary bolts (Figure 78) are an alternative to eyebolts.

Figure 78 *Epoxy grouted transport bolt (courtesy G. J. Laan)*

3.4.2.5 Stocking

In general, stocking and rehandling of armourstones is considered to have a high cost. However, there are benefits in placing like sizes together and stocking them in a manner to aid re-handling (Figure 79).

3.4.2.6 Loading of vehicles

The same methods as for handling are generally available. However, there is an extra consideration in that damage to vehicles must be avoided (Figure 80). Unloading methods should also be considered in order to achieve minimum overall cost.

3.4.2.7 Gravel production

For gravel production two principal types of dredger are used: cutter suction dredgers (CSD) and trailer hopper suction dredgers (THSD). Sea-going suction hopper trailer dredgers have been built up to 20 000-tonne capacity and are suited to high-volume contracts.

3.4.2.8 Cost implications of production

In common with other processes, quarry production has costs that can be divided into fixed and variable elements. The importance of fixed costs will be relative to the type of quarry. For example, a dimension stone quarry might cost no fixed elements to armourstone, whereas a quarry opened up especially for a contract will need to recoup all costs.

Figure 79(a)

Figure 79(b)

Figure 79(c)

Figure 79 *Stockpiles of (a) 300–1000 kg and 1–3 tonnes standard gradings, (b) 300–1000 kg, basalt and (c) 1–3 tonnes, basalt (courtesy G. J. Laan)*

Figure 80 *Loading rock using a front-end loader (courtesy HR Wallingford)*

As an example, the cost of a dedicated quarry can be separated into fixed and running costs. Fixed costs are those of overburden removal, test blasts, establishment of a production face, mobilisation of equipment and construction of access roads. Running costs are proportional to the volume of material produced.

The investment cost in equipment for an average quarry can be in the range of $3–5 million. The time required for the preparatory activities for such a quarry is in the order of 3 months from the time appropriate plant is on-site.

Production costs are generally in the range of $5–8/ton (including capital investment cost but excluding mobilisation and demobilisation), with the following subdivision:

20–30% for drilling and blasting and overburden removal;
30–40% for loading and selecting at the face;
20%   for transport to the stockpile;
10%   for screening or separation;
10%   for loading at the stockpile before transport.

The costs may also have to be increased to cover royalties.

Production costs in total will vary considerably, with the type of stone quarried and the annual output. Smaller sizes in a large limestone quarry could be produced for less than $5 per tonne, whereas a small igneous quarry could have costs of $15 or more per tonne. If specialist equipment is required, the producer will wish to stipulate a minimum throughput per day for a contract, in order to minimise per tonne costs (see also Section 6.1.6). For gravel extraction, production costs are commonly very low and the more significant cost component is transportation to site (see Section 3.4.3).

3.4.3. TRANSPORTATION

In many coastal and shoreline projects, transportation over the last few metres to site is the critical factor in selecting likely rock sources. Coastal sites may be either unreachable by suitable road or difficult by sea due to navigational hazards. Transhipping at any point is usually an uneconomic process. Access by rail to a site is unusual and only the largest of contracts could justify new infrastructure of this kind.

Broad types of transport options are given below, but it should be noted that legal and safety requirements may vary from one country to another. The discussion of transport options is followed by a brief consideration of their cost implications. Gravel transport is discussed more fully in Section 6.4.

3.4.3.1 Road transport

The following vehicles are commonly used for road transport of quarried rock.

Flat-bed lorries

These may carry individual stones but each stone will need securing to avoid loss in transit. Maximum net weight in the UK is approximately 24.5 tonnes. Unloading must be arranged by the site. If articulated vehicles are used, pre-loading of trailers can take place.

Steel-bodied tippers and on/off-road dump trucks

These may be used for all materials but care must be taken, depending on the size involved. Pay loads may be restricted due to weight of body. Unloading is by tipping either to the stockpile or directly into the structure. Rigid vehicles with sub-frames may be used for hauling filled rock trays.

The possibility of damage to steel bodies must be considered. Figure 81 shows, for example, a purpose-built steel body truck less than 6 months old already severely damaged whilst transporting limestone armour up to 7 tonnes in weight.

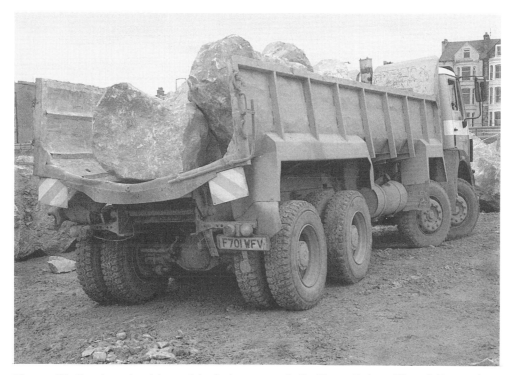

Figure 81 *Road truck with steel body (courtesy J. D. Simm, Robert West & Partners)*

The damage necessitated subsequent reconstruction of the body and strengthening using hardened steel.

Conventional aluminium alloy tippers (up to 20 tonnes net capacity)

The majority of tippers used in the UK now have alloy bodies. It is unlikely that material over 400 mm nominal size can be carried out without damage to the vehicle. This risk of damage can be reduced if a back hoe excavator is used to load materials.

On-site vehicles

These can be of any size or configuration according to the relevant conditions. In all but the worst conditions vehicles similar to the above can be matched to the demand. Stability of articulated rough-terrain vehicles will need to be considered when carrying single large armourstones. In general, for road vehicles the optimisation of payloads is more difficult for the larger armour blocks, particularly if a narrow grading is specified.

3.4.3.2 Rail transport

UK railways generally allow wagons of 70–75 tonnes nett or 35–37 tonnes nett. Other railways may have different capacity wagons related to axle weight limitations. The following wagon types are used.

Flat wagons

These have been employed for the transportation of individual stones. The use of low-level sides obviated the need for individual securing steps (Figure 82).

Box wagons

The use of conventional box wagons is usually restricted to material up to

Figure 82 Rail transport of rock at buffer stockpile, Europoort (courtesy Aveco Infrastructure Consultants)

Figure 83 Loading 10–15 tonne rock into a 2000-tonne barge using epoxy-grouted hooks (courtesy G. J. Laan)

Figure 84 *A split box for loading a ship* (*courtesy G. J. Laan*)

100 mm in size due to difficulties of unloading. Side tipping and tipper wagons are possibilities for larger stone.

3.4.3.3 Sea transport

Conventional coasters or barges of between 2000 and 3000 tonnes and draught 3–5 m may carry stones up to approximately 6 tonnes in weight without undue problem provided a safe loading and unloading system is used. The use of eyebolts or hooks is sometimes preferred for this reason (Figure 83). A split box may also be useful (Figure 84). Unloading requires a crane-mounted grab and for this reason a quay or wharf is usually required (Figure 85). The choice of capacity is wide, and the ideal vessel will depend on availability, rate of loading and unloading, draught and other factors.

'Drive on and off' pontoons can be used for carrying all sizes of stone (Figure 86). Protection to the deck may be necessary for larger material and 'fences' are required to avoid loss in transit and damage to the pontoon. Unloading may be achieved by on-board machines (Figure 87) simply pushing material over the side of the vessel (Figure 88). A variety of pontoons are available, including sea-going versions, and some are self-propelled. Specially strengthened pontoon barges can be used of up to 20 000 tonnes, with a draught of up to 8 m.

A variety of specialised vessels exist for carrying stone. These include coasters mounted with excavators for self-loading and unloading as well as side- and bottom-dumping barges. Generally, their capacity is lower than for pontoons and may only be 500–800 tonnes.

Sometimes the rock barges can bring the material directly to the site for dumping or placing by crane. Often the transport capacities do not match the placing

Figure 85 (a)

Figure 85 (b)

Figure 85 Loading rock (a) rail to sea and (b) into a barge (courtesy G. J. Laan/Aveco Infrastructure Consultants)

Figure 86 *A 20 000 tonne pontoon (courtesy G. J. Laan)*

capacities, so trans-shipment and stockpiling is required, even if the breakwater is built with seaborne equipment.

3.4.3.4 Cost implications of transportation

Road transport of material has the advantage of flexibility where delivery rate can be matched to the site requirements. Thus, while costs may be high per load, there can be saving in stocking and handling on-site.

The cost is generally higher than other options. For example, a recent cost calculation for a West European breakwater showed typically $0.10–0.15/ton km for truck transport against $0.02–0.03/ton km for ship transport.

If the distance between quarry and construction site is more than 25 km, it is generally economic to provide for buffer stockpiles at both the quarry and the construction site, since a perfect match of production and transport capacities is difficult. In this way, unnecessary production or placing delays are eliminated. In general, the logistics of the whole system should be considered to balance production, transport, stockpiling and placing capacities.

Shipping costs can be very low compared to road haulage costs, commonly of the order of 80% less. However, the likely high capacity of vessels (such as barges) may increase costs at the site. Bad weather can cause unexpected delays and demurrage is commonly charged where a vessel takes more than 24 hours to unload. Typical costs of loading and unloading are $0.5 to 1.0/tonne and typical capacity 200 tonne/hour. Rail rates are economic over long distances and for high-tonnage contracts, but double handling is likely to be required and the costs of the necessary infrastructure need to be considered.

For gravel transport, whether delivery is via a wharf or direct to site will have a large bearing on costs. Pump-ashore costs will be relative to the distances involved.

Figure 87 *Unloading 1–3 tonne rock from a 20 000 tonne pontoon. The fines and fragments are pushed aside using a 'bob cat' (courtesy G. J. Laan)*

Figure 88 *Unloading a 10 000 tonne pontoon directly onto a beach at Fairlight Cove, UK (courtesy C. P. Orbell-Durrant, Shephard Hill & Co. Ltd.)*

Wear costs are also significant for pumping and will be relative to mineral content, shape and size of material (see Section 6.4 for further information on gravel transport).

3.5 Other Materials

3.5.1 CONCRETE

The large proportions of the finer grade sizes produced in the quarry or present in sand and gravel deposits are sometimes used in the manufacture of concrete units for armouring marine structures. At some localities where there is no economic source of durable rock obtainable as large blocks and concrete armour, units have to be used. Concrete units are usually the appropriate choice when heavy armouring is required, since sources of large stone blocks in excess of 15 tonnes are rare. Concrete units also have higher stability coefficients (K_D) than stone blocks, though the handling and placing of units over 20 tonnes becomes specialised and costly. Concrete is also used extensively as small blocks or panels for the facing of revetments in coastal and river bank protection.

Concrete armoured units may be constructed as simple cubes but many designs of more complex shape are widely used. Units include the widely adopted tetrapod and dolos, the latest generation of randomly placed units (including Accropode and Haro) and single-layer regularly placed units such as Cob, Shed and Diode. Further details on such units are widely published (*SPM*, 1984; Thomas and Hall, 1991).

Typically, local gravel or crushed rock aggregates are used together with ordinary Portland cement to manufacture unreinforced units. Greater resistance to sea water can be achieved by using a sulphate-resistant cement or a blast furnace slag cement replacement. Fresh mix water should normally be used. (Sea water has been used successfully in unreinforced units, though there is a tendency for dampness and efflorescences to develop on the surface. It should not be used in steel reinforced concrete because of the risk of corrosion to the steel.) There has been some limited use of chopped fibres of steel or polyethylene and polypropylene in attempts to increase the tensile strength of the concrete.

Reinforced concrete demolition rubble is occasionally used for armouring. Depending on shape (P_R—see Section 5.1.3.1) such rubble may be more or less stable than equant rock. However, chloride-related corrosion of reinforcement causing spalling and concrete break-up and initiated via ingress of chlorides at sites of exposed bar ends may mean that tabular pieces, even if more stable, may have a relatively short lifetime (Simm and Fookes, 1989).

Concrete attacked by sulphates develops gypsum and ettringite crystals which can cause expansion and spalling of the concrete. However, these sulphate minerals are more soluble in a chloride solution, so that sulphates in sea water do not normally cause expansion. Instead, a slow increase in porosity due to leaching and a consequent loss in strength will occur. Salts crystallising in the surface of concrete units in the intertidal and supratidal zones of a structure subject to wetting and drying will gradually cause the disruption and disintegration of the surface. Attrition from waveborne sand and gravel can also cause rapid wear of armour units, particularly if low-quality porous concretes are used.

The ability of concrete units to carry out their function for their design life is broadly dependent on the concrete quality and the severity of the particular environment. The quality of the concrete is usually assessed in terms of its compressive strength and its porosity. (In fact, concrete armour units are generally unreinforced and therefore rely on tensile strength for their integrity, but tensile strength can be tentatively correlated with compressive strength.) High-strength, low-porosity concretes are achieved by careful mix design, appropriate aggregate gradings, water/cement ratio and admixtures.

3.5.2 INDUSTRIAL BY-PRODUCTS

Current European policies aim to increase the use of waste materials of all kinds and to find economic, satisfactory and safe means of their disposal. The use of waste materials in coastal and shoreline structures is limited by their particle size distributions, mechanical and chemical stabilities and the need to avoid materials which present an actual or potential toxic hazard. The principal groups of waste materials are listed in Table 25 together with notes concerning their possible use in coastal and shoreline structures.

There is an increasing use of crushed demolition rubble as core material. This is usually inert and does not normally contain soluble or biodegradable materials. This is also true of many other bulky industrial by-products such as slags and colliery waste. However, care must be taken to evaluate the possible environmental hazards associated with waste products of this type. Slags associated with copper smelting, for example, may contain dangerous levels of copper arsenic and other poisonous substances, while shales may contain pyrites which will degrade in water to produce undesirable and environmentally damaging solutions and gases.

The leaching or formation of undesirable or toxic compounds is perhaps the most important potential hazard associated with the use of many of these waste materials. If often proves difficult to relate laboratory test results to the actual *in-situ* conditions, leading, in turn, to difficulties in developing satsifactory models of the processes involved. These considerations have implications for the planning of quality control and quality assurance.

Although careful consideration must be given to the potential hazard associated with mineral and industrial wastes, many have little or no contamination and may be used in structures providing their physical properties are appropriate. Examples of materials which typically have high hazard potential are copper and lead slags.

Table 25 Industrial by-products and wastes and their potential for use in marine structures

By-product waste	Type of Material	Typical characteristics	Deleterious components	Potential use in coastal defences
Colliery and open-cast mine wastes Quarry wastes	Shales Sandstone Clay 'silex'	Flaky, friable small particle sizes, large percentage fines	Sulphides Coal (arsenic) Clay	Possible core, fill or filter material
Ferrous metal Extraction wastes	Slags Mine wastes Fly ash	Hard, metastable glassy material, small particle sizes	Sulphides Sulphates (Heavy metals)	Possible core, fill or filter material
Non-ferrous metal Extraction wastes	Metal slags Phosphorous slags Mine wastes Gangue minerals Fly ash	Brittle glassy material and crushed rock	Sulphides Sulphates Heavy metals Fluorides	If innocuous possible core, fill or filter material
Domestic wastes	Variable large paper plastics organic content	High biodegradable component	Inflammable and toxic gases and liquids	Unsuitable
Coal and incineration ash	Cinders, slags, fly ash	Friable, porous granular materials	—	Possible fill
Construction and demolition wastes	Concrete brick and building materials	Rubble	Timber Steel	Possible core or fill
Oil and chemical industry wastes	Wide range of chemical by-products	Highly variable fine grained or solidified	Toxic liquids and gases	Unsuitable
Clay industry wastes	Quartz, mica, feldspar	Sand–silt grade materials	—	Possible core or fill
Dredging sludge	Silts, clays	Silty clay	Possible organic and metal compounds	Possible core or embankment fill

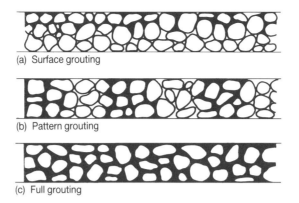

(a) Surface grouting

(b) Pattern grouting

(c) Full grouting

Figure 89 *Grouting methods*

Possible hazards from heavy metal contamination may arise in steel and silicomanganese slags, while certain mine wastes sometimes contain leachable arsenic and strontium compounds. Other waste materials such as silex, quarry wastes, dredging sludge and many minestone wastes have little or no hazardous contamination.

The engineering properties of many waste materials are often comparable or better than traditional materials. Slags have good frictional properties due to their angularity and roughness and typically have high density. Mine wastes sometimes have poor weathering characteristics, but are usually inert and have satisfactory grading for deep dills. The fine materials such as fly ashes and ground slags are already in general use as cement replacement and fillers. Good quality control, not only for limiting the potential for toxic hazard, but also of the mechanical properties of waste materials can considerably increase the use of such low-cost materials in appropriately designed coastal and shoreline structures.

3.5.3 GROUTS

The stability of an installed layer of loose stone pitching or open revetment blockwork (Section 6.2.1.8) can be improved by grouting (see Section 5.1.3.9). The grouting can be executed with a colloidal cement mortar or a stone asphalt grout. In general, three methods can be distinguished (Figure 89):

1. *Surface grouting.* Stones in the top of a layer are stabilised by spreading an even layer of mortar over them, and involves completely filling the quarried stone voids over at least the top one-third of the armour layer.
2. *Pattern grouting.* The stone layer is penetrated to the full depth according to a predetermined pattern in which 50–80% of the voids are filled.
3. *Full grouting.* The intergranular pore spaces are completely filled with grout so that a coherent layer is created.

3.5.3.1 Asphalt grout

Mix design

Asphalt grout is a mixture of mastic and gravel or crushed stone. The mastic is a mixture of 60–70% (by weight) sand, 15–25% filler and about 20% bitumen. The amount and sizes of gravel or crushed stone in the mix depends on the size of the voids to be filled. Its use increases the stiffness of the mix, preventing viscous bulging and limiting grout penetration; it also reduces the amount of mastic required.

The design of the grout mixture is usually based on experience, since it must have an appropriate viscosity to flow into the voids, but not escape. The viscosities during placing are typically 30–80 Pa.s. Once cooled, the grout must not flow and will have viscosities in the range 10^6–10^9 Pa.s, depending on temperature, slope and stone size.

Practical experience suggests that for a grout penetration depth of $2D_{n50}$, the ratio of the D_{n15} of a quarried rock layer that is to be grouted and the D_{n85} of the stone in the grout should be:

$$D_{n15}/D_{n85} = 5\text{–}10 \text{ (above water)}$$

$$D_{n15}/D_{n85} = 10\text{–}20 \text{ (below water)}$$

The maximum particle size of the stone in the grout also depends on the production equipment, the maximum size normally being 50–60 mm. To achieve sufficient penetration of this grout, the minimum amount of mastic in the mastic/stone mixture needs to be between 50% and 55%.

Production

Grouting mortars are mixed in normal asphalt plants. Sand, gravel and crushed stone are stored in such a way that they cannot be contaminated by other materials. Filler is stored in silos, bitumen in heated tanks.

The mixing time depends on the type of installation and the production method. The temperature of the mix is determined by the transport distance and the required application viscosity and normally lies between 130°C and 190°C. Since the mix is prone to segregation it should be stored in a stirring kettle, and mixes containing larger stones should, preferably, not be stored.

Transport

If possible, mixes should be transported in a stirring kettle. In general, this applies to mixes with up to 50% fine crushed stone. Mix with a higher stone content or with coarser stones needs to be transported in an asphalt container and it should be carefully re-mixed before application.

The material should preferably be applied directly. Stirring kettles should be used for long-term storage. Coarse stone mixes should be stored in containers and carefully re-mixed immediately prior to application.

Application

Hot grouting mortars can be applied using chutes, a crane bucket, or directly from a stirring kettle or asphalt container. The mortar is then spread further by hand with scrapers. Alternatively, the container can be lifted above the slope, so that the mix can flow directly into the rubble layer.
On relatively steep slopes which have to be grouted from the top a careful check is needed on the outflow rate and the quantity of grout used per unit area. If the flow is too high, small stones can be carried down the slope and grout can be lost. In such cases it is better to work from the bottom of the slope upwards. On the other hand, if the grout flows too slowly and cools too quickly there is a risk that voids will remain in the toe of the revetment. Sometimes the grouting needs to be carried out in several layers, and experience and skill are required here. These methods can also be applied to the underwater sections of a revetment, but extra care is necessary because the grouting cannot be observed directly.

Other techniques are available for grouting under water (for example, dosing buckets or, if a steady flow of grout is required, an insulated pipe with a special nozzle). The nozzle should be kept at a height of 0.5–1 m above the rock layer being grouted.

The quantity of grouting mortar required depends on the thickness of the stone layer and the dimensions, form and grading of the stone layer and the extent of grouting needed. For example, about 400 kg/m² of mastic are needed to fully grout a 50 cm thick crushed stone layer, and about 300 kg/m² for a 30 cm thick crushed stone layer, based on a stone voids ratio of about 40%.

The quality of the stone to be grouted should be as good as for the ungrouted layers (see Table 21) and pieces of tabular shape should be avoided. The stone to be grouted must also be clean. In order to prevent fouling with sand and debris, ideally the stones should be grouted immediately following construction. Sometimes it is necessary to hose the revetment clean before grouting. Pitched stones (see Section 5.1.3.9) can also be grouted.

The maximum slope which can be grouted with the more common grouting methods is 1:1,7 above water and 1:1,3 below water. If steeper slopes are required the method of application must be changed.

The application temperature above water should be between 100°C and 190°C, depending on the required viscosity. Below water the temperature must be less than 150°C. If the mix temperature is too high it can damage any underlying geotextile fabric. The maximum safe temperature dpends on the type of fabric, but for polypropylene-based fabrics it is about 140°C.

3.5.3.2 Colloidal cement mortar

Properties

A cement mortar for stone pitching requires a good flow behaviour and an optimal resistance to segregation. This latter factor is very important below water, particularly if there are currents. Colloidal concrete has been specially developed to meet this requirement, and there are two types: colcrete and hydrocrete.

Colcrete was developed in the 1930s by British engineers working with the National Physical Laboratory, Teddington, and is manufactured by forcing the water/cement mixture through a narrow slot. The high frictional forces optimise the wetting of the cement particles, thus maximising the formation of cement hydrate gels. Sand, aggregate and possibly filters and accelerators are added at the next stage in the process, depending on the particular application. Hydrocrete is a modern colloidal concrete, and the colloidal character is obtained by addition of a modified natural polymer. Both colcrete and hydrocrete are designed to optimum compositions to fulfil specific requirements, depending on the grouting depth and the size and grading of the stone layer being grouted or the gap width in the case of pitched stones.

Hydrocrete mortar is relatively stiff. Because of the low workability and the desired density the optimal slump is usually 180–200 mm, although due to the special additive, the water/cement ratio is relatively high. Dense and open-textured varieties of hydrocrete are used. The stiffer open-textured variety uses very little fine aggregate and has a permeability of between 3 and 5×10^{-3} m/s. Grouting depths achievable with the open-textured variety are less than with the dense mortar.

The crushing and tensile strength of dense hydrocrete is about 10–15% lower than normal concrete with the same composition, and the strength of the open mortar is lower still. The Young's Modulus of hydrocrete is also about 18% lower than normal concrete and shrinkage is greater, though frost resistance is similar. The usual tests for normal concrete, such as slump, air content and density, are applicable to hydrocrete, but a special test has been developed for resistance to washing out.

Production and application

The stiffness (low workability) of hydrocrete means that, although it has the advantage of not segregating during transport, it cannot be pumped very well and the open variety not at all. Special methods need to be introduced for below-water grouting to avoid washing out of the fines and for accurate placing.

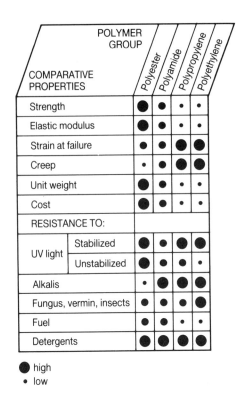

Figure 90 *Comparative properties of general polymer families*

Before application of the mortar the stone to be grouted must be cleaned if there is to be good adherence between the grout and stones. Since applications vary widely, trial mixes and placing may be necessary.

3.5.4 GEOTEXTILES

Geotextiles are permeable textiles made from articifial fibres used in conjunction with soil or rocks as an integral part of a man-made project and were first used in the Netherlands. They are frequently employed as filter membranes and as the interface between differently graded layers. Geotextiles are also used as bottom protection and can be loaded with concrete blocks (block mattress), bituminous bound crushed stone and sand (fixtone mattress) and geotextile tubes filled with gravel (gravel-sausage mattress). Gravel bags have also been used for special filter requirements.

The basic functions of geotextiles may be listed as follows:

1. *Separation*: the geotextile separates layers of different grain size;
2. *Filtration*: the geotextile retains the soil particles while allowing water to pass through;
3. *Reinforcement*: the geotextile increases the stability of the soil body;
4. *Fluid transmission*: the geotextile functions as a drain because it has a higher water-transporting capacity than the surrounding materials.

3.5.4.1 Geotextile manufacture

Geotextiles are manufactured from a variety of artificial polymers:

1. Polyamide (PA);
2. Polyester (PETP);
3. Polyethylene (low-density LDPE and high-density HDPE);

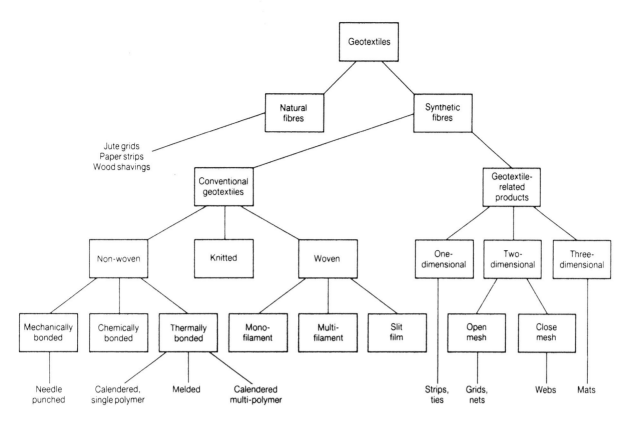

Figure 91 *Geotextile classification groups*

4. Polypropylene (PP);
5. Polyvinylchloride (PVC);
6. Chlorinated polyethylene (CPE).

The first four are the most widely used although many variations are possible. Additives are also employed in geotextile manufacture to minimise ageing, to introduce colour and as anti-oxidants and UV stabilisers.

Comparisons of properties of the four main polymer families are shown in Figure 90. These are very broad, because there are many variants within each group. Some properties (such as strength) are also greatly influenced by the different processes of manufacture. A classification of geotextiles based on the type of production and the form of the basic elements is given in Figure 91.

The basic elements used in geotextiles are monofilaments, multifilaments, tapes, weaving film and stable fibres. Monofilaments are single, thick, generally circular cross-sectioned threads with a diameter ranging from 0.1 mm up to a few millimetres. Multifilaments (yarns) are composed of a bundle of very thin threads. Yarns are also obtained from strips and from wide films. Tapes are flat, very long plastic strips between 1 and 15 mm wide with a thickness of 20–80 μm. A weaving film is sometimes used for the warp 'threads' in a fabric.

The basic fibre is a fibre of length and fineness suitable for conversion into yarns or non-woven geotextiles. For non-woven fabrics the length is usually about 60 mm.

Woven geotextiles

A woven fabric is a flat structure of at least two sets of threads. The sets are woven together, one referred to as the warp, running in a lengthwise direction, and the other, the weft, running across. Woven geotextiles can be categorized by the type of thread from which the fabric is manufactured.

Figure 92 *Tape fabric (courtesy Nicolon BV, Almelo, Holland)*

Monofilament fabrics are used for gauzes of meshes which offer relatively small resistance to through-flow. The mesh size must obviously be adapted to the grain size of the material to be retained. Monofilament fabrics are made principally from HDPE and PP.

Tape fabrics are made from very long strips of usually stretched HDPE or PP film, which are laid untwisted and flat in the fabric. They are laid closely together, and as a result there are only limited openings in the fabric (Figure 92).

Split-film fabrics are made from mostly fibrillated yarns of PP or HDPE (Figure 93). The size of the openings in the fabric depends on the thickness and form of the cross-section of the yarns and on the fabric construction. These split-film fabrics are generally heavy. Tape and split-film fabrics are often called slit-films.

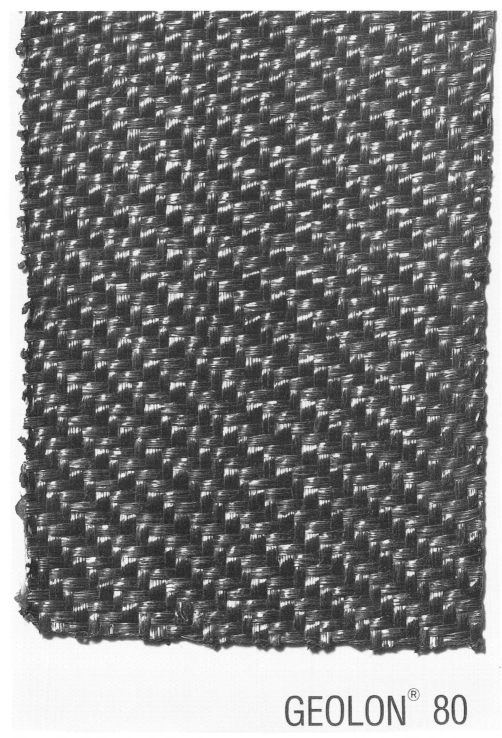

Figure 93 *Fibrillated split-film fabric (scale 1.5:1) (courtesy Nicolon BV, Almelo, Holland)*

Multifilament fabrics are often described as cloth, because they tend to have a textile appearance and are twisted or untwisted multifilament yarns. The fabrics are usually made from PA 6, PA 6.6 or from PETP.

Besides the above-mentioned monofilament fabrics, special *mesh-type constructions* are produced such as those with a monofilament warp and a multifilament weft which have outstanding water-permeability and sand-retention properties. Other examples include open meshes in which the woven or unwoven warp and weft threads are attached at crossing points by chemical or thermal bonding and other meshes constructed by using knitting techniques.

Non-woven geotextiles

A non-woven geotextile is a textile structure produced by bonding or interlocking of staple fibres, either monofilaments or multifilaments arranged at random, accomplished by mechanical, chemical, thermal or solvent means. Non-woven gauzes are structures with large meshes which are formed by placing threads or tapes at predetermined distances on top of one another and bonding them at the intersections by a chemical, thermal or mechanical process.

Geotextile-related products

These products are distinguished in one-dimensional (strips, ties), two-dimensional (gids, nets, webs) and three-dimensional (mats) products. Grids are lattices made from perforated and then stretched polymer sheets. Three-dimensional mats are produced by extruding monofilaments into a rotating profile roller, followed by coating so that the threads adhere to each other at crossings which are spatially arranged. The matting material itself occupies less then 10% of the mat volume. The mats are 5–25 mm thick and about 1–6 m wide.

3.5.4.2 Characteristics and properties

The geotextile properties stem primarily from their functional requirements. Since the geotextile can have a variety of functions, requirements are diverse. For reinforcement the emphasis is on mechanical properties such as E-modulus and strength, for filters it is on properties such as water permeability and soil tightness. The durability required will depend on the specific application and lifetime required. Geotextiles must also fulfil secondary functional requirements related to the execution of the work (e.g. a certain amount of UV resistance is needed) or it must have resistance to mechanical wear and tear if construction equipment is to be driven over the fabric. The suitability of a geotextile should be checked against these functional requirements during the design phase of the project.

Although specification requirement tests need only be carried out once, the following quality control tests may be required during production: tearing strength, grab test strength, tensile strength, strain at breaking load, moduli and mass distribution. These tests should be made in both the length and width directions. The thickness, the mass per unit area and the bursting strength may also need to be checked, and in some applications water permeability and sand-retaining properties will be important. A large number of national and a few international standard test methods are available covering these requirements.

Dimensions

The maximum standard width available for both woven and non-woven fabrics is 5–5.5 m. The length is limited by the available transport facilities and ease of handling on-site. Depending on the mass per unit area, the length generally lies in the range 50–200 m.

Jointing is necessary to obtain greater dimensions. In practice, large areas can be covered by overlapping sheets. Where physical continuity is required without overlap then heat welding (some non-wovens) or stiching may be used. The seam forms the weakest link in the geotextile construction and should therefore be checked thoroughly against the specifications. The thickness of most geotextiles lies between 0.2 and 10 mm when unloaded, although this may sometimes reduce under pressure.

In general, the mass of non-woven geotextiles lies in the range 100–1000 g/m^2, 100–300 g/m^2 being the most commonly used. Woven fabrics can be heavier and masses between 100 and 2000 g/m^2 are possible. The greater demand is for the lighter grades in the range 100–200 g/m^2. Generally, the lighter types of geotextiles are used as separators, the heavier woven fabrics for reinforcement and the heavier non-wovens for fluid transmission.

Mechanical properties

The mechanical properties of geotextiles depend on a number of factors: temperature, atmospheric conditions, the stress–strain history, the mechanical properties of the material and fibre structure, the structures of the yarn and of the geotextile, the direction of anisotropy, and the rate of loading and ageing. Most fabrics exhibit cross-contraction under loading. However, light tape fabrics and fabrics with so-called 'straight' warp construction do not exhibit cross-contraction and construction strain (strain due to fibre straightening). Test methods can be categorised into those that do not prevent cross-contraction (uniaxial) and those that do (biaxial). Methods commonly employed for tensile testing are strip tensile, grab tensile, manchet tensile, plain-strain tensile, wide width tensile and biaxial.

A great variation in both strength and stiffness exists. The strength varies generally between 10 and 250 kN/m. Non-woven and woven PP fabrics are not ideal in situations where high strength is combined with low strain because of the large elasticity of these geotextiles.

All meltspun synthetic polymers, as used in geotextiles, have visco-elastic behaviour, which means that the mechanical behaviour is time-dependent. This becomes manifest in creep and relaxation phenomena. Creep data for polymer materials can be presented in several ways. Often $\log \varepsilon$ (strain) is plotted against $\log t$ (time) for various levels of the ultimate short-term load U, i.e. 50% U, 25% U, etc. The sensitivity to creep of polymers increases considerably in the sequence PETP, PA, PP and PE. For geotextiles that are loaded for prolonged periods of time (10–100 years) the permissible load for polyester is the order of 50% of the tensile strength, for polyamide 40%, and for polypropylene and polyethylene below 25%.

Burst and puncture strength of geotextiles is important in coastal and shoreline rock structures. In these tests a circular piece of geotextile is clamped between two rings and loaded directly by gas or water pressure or by a physical object. Puncture tests can be used for investigating the resistance of a geotextile to puncturing by, for instance, falling stones. The other tests available include the California Bearing Ratio (CBR) plunger tests, the cone drop test and the rammer test (BAW). Test methods are also available for strength parameters such as tear strength, abrasion resistance and friction coefficient.

Hydraulic properties

The water permeability of a geotextile incorporated into a structure depends on the geotextile itself, the subsoil, the load imposed on the face of the geotextile, the hydraulic load, the blocking of the geotextile, the clogging of the geotextile, the water temperature and the composition of the water. The pore size and the number of pores per unit area of a geotextile primarily determine the permeability. The subsoil and imposed load determine how much water has to be discharged and the compression of the geotextile.

Flow through the fabric is normally laminar when the geotextile is embedded in the soil, but becomes turbulent when subjected to wave action (for example, under a coverlayer of rip-rap or blocks). Blocking of flow occurs when soil particles partly wedge into the pores. This normally only arises in situations of uni-directional flow rather than the oscillating flows, which frequently occur under wave action. Clogging of flow occurs when fine particles (for example, iron particles) settle on or in the geotextile or at the interface between the geotextile and the subsoil. Reductional factors in permeability of geotextiles due to clogging of the order of 10 have been found.

Permeability is usually measured in the laboratory using values of hydraulic gradient low enough for laminar flow. For thin, more permeable fabrics, the permeability test may be performed on several separated layers to increase the measureable water head and still maintain laminar flow. The equation for flow through the fabric is

$$k_g = \frac{q \cdot t_g}{a_g \cdot \Delta H}$$

where k_g = permeability of geotextile (m/s),
q = rate of flow (m³/s),
a_g = surface area of geotextile (m²),
ΔH = head loss (m),
t_g = geotextile thickness (m).

The permeability may be expressed as permittivity (k_g/t_g), which is useful in comparing geotextiles because permittivity varies less than peremeability when there is a change of normal stress applied to the geotextile. The permittivity can be found by dividing the permeability by the nominal fabric thickness. The permittivity of geotextiles varies between 10^{-2}s^{-1} and 10s^{-1}.

Geotextiles will only be completely soiltight when the largest opening is smaller than the smallest particle of the subsoil. This is usually not required because of the filter function of the geotextile. When the openings are larger the soil tightness depends basically on the hydraulic loads, the soil grading and the geotextile characteristics. The soil tightness of a geotextile on a particular type of soil can be determined by reconstructing the situation in a model laboratory apparatus and then carrying out measurements using the hydraulic boundary conditions. In addition, the soil tightness can be characterized by a ratio of O/D, where O represents the diameter of holes in a fabric and D the diameter of particles (e.g. O_{90}/D_{50}). A number of methods have been developed in which an opening characteristic (e.g. O_{90} or O_{95}) is determined in terms of a sieve size. These tests are based on either wet or dry sieving. Further information on filter design using geotextiles is given in Section 5.2.5.4.

Chemical properties

One of the characteristic features of synthetic polymers is their relative insensitivity to the action of a great number of chemicals and environmental effects. Nonetheless, each plastic has a number of weaknesses which must be taken into account in the design and application. Specifically, the life of geotextiles can be affected by oxidation and by some types of soil/water/air pollution. Many synthetic polymers are sensitive to oxidation. The end result of oxidation is that mechanical properties such as strength, elasticity and strain absorption capacity deteriorate and the geotextile eventually becomes brittle and cracks.

Investigations have shown that, provided the geotextile is not loaded above a certain percentage of the instantaneous breaking strength, the thermo-oxidative resistance will determine the theoretical life of the material. The allowable load for PETP is, at most, 50%, for polyamides it is somewhat lower, and for polypropylene and polyethylene it is about 10–30%. (This guidance only applies where the geotextile functions as a filter and is not withstanding mechanical loads.)

Specific additives have been developed to counteract these processes. These can be grouped according to their protection function as either anti-oxidants or UV-stabilisers. In fact, the thermo-oxidative resistance of a geotextile is determined by a number of factors: the thermo-oxidative resistance of the polymer itself, the composition of the anti-oxidant packet, the effect of the thermo-oxidative catalytic compounds in the environment, the effects of processing on the long-term thermo-oxidative resistance, the resistance of the anti-oxidant additives to leaching by water and the practical site conditions.

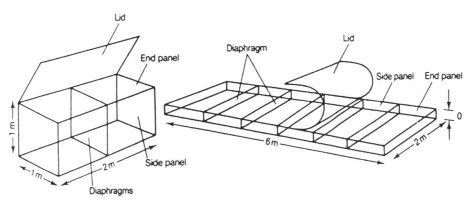

Figure 94 *Gabions*

3.5.5 COMPOSITE MATERIALS

Composite systems are normally adopted where improved stability of the rock materials being used is sought (see Section 5.1.3.9) or where ease of placing the protection system is important.

3.5.5.1 Gabions

A gabion is a box or mattress-shaped container made out of hexagonal (or sometimes square) steel wire mesh strengthened by selvedges of heavier wire, and in some cases by mesh diaphragms which divide it into compartments (Figure 94). Assembled gabions are wired together in position and filled with quarried stone or coarse shingle to form a retaining or anti-erosion structure. The wire diameter varies but is typically 2–3 mm. The wire is usually galvanised or PVC-coated. PVC-coated wire should be used for marine applications and for polluted conditions.

The durability of gabions depends on the chemical quality of the water and the presence of waterborne attrition agents. The influence of the pH on the loss of the galvanic zinc protection is small for pH values in the range 6–12 and there are examples of gabions with negligible loss over 15 years. Grouting of the stone-filled gabions or mattresses can give some protection to the wire mesh against abrasion and corrosion, but this depends on the type of grout and the amount used.

The dimensions of gabions vary, but typically range in length from 2 to 4 m (mattress, 6 m), with widths about 1 m (mattress, 2 m), and height 0.3–1.0 m (mattress, max. 0.3 m). The mesh size is typically 50–100 mm.

The units are flexible and conform to changes in the ground surface due to settlement. Prefabricated gabions can be placed under water. Gabions can thus be used in a wide variety of marine works: groynes, dune and cliff protection, protection of pipelines and cables, and as toe protection. Mattress-shaped gabions are flexible and are therefore able to follow bed profiles both initially and after any scouring which may take place. Gabions can also be piled up to the form retaining walls or revetments (see Section 6.2, Box 80). In order to prevent migration of solid through the structure they may be used in conjunction with geotextile filter layers.

In certain applications the gabion structure needs impermeability or weight to counter uplift. To give these characteristics, the stone is grouted with mastic or a cement-bound grouting. The weight of the structure can also be influenced by the choice of the density of the stone blocks with which the gabion or gabion-mattress is filled.

3.5.5.2 Mattresses

Prefabricated mats for hydraulic constructions have been used extensively as a medium for combating bottom and embankment erosion. The range of mat types have increased considerably during the last decade. Formerly, use was made of handmade mats of natural materials such as reed, twigs/branches (willow), rope, quarry stone and rubble. Today, geotextiles, bituminous and cement-bonded materials, steel cables and wire gauze are used as well as the more traditional materials.

Functions and principles

A distinction must be made between mattresses used for the protection of continually submerged beds and foreshores and mats employed for slope or front face protection of embankments. Bottom protection mats are concerned with:

- The mitigation of groundwater flow in the underlying seabed to prevent horizontal transport and erosion;
- Prevention of piping of soil particles;
- Assisting in the overall geotechnical stability of a larger structure;
- Offering resistance to loads imposed by anchors, trawler boards, etc. and thereby protecting pipes and cable lines;
- Providing a protection system with the flexibility to allow for anticipated settlements.

Mattresses for embankment retention have similar functions, i.e:

- Prevention of soil transport;
- Resistance to wave action;
- Provision of flexibility to allow for settlement.

Additionally, they may be required to:

Not trap refuse; and
- Support vegetative growth.

Mats are manufactured from the following elements:

1. *The carrier*: Fabric, gauze, cables and bundlings by which the necessary strength is obtained for its manufacture, transport and installation.
2. *The filter*: A geotextile which satisfies the conditions with respect to permeability, soil density, strength, etc. In some embankment-mat constructions the geotextile is applied separately.
3. *The ballast*: Concrete blocks, packaged sand, gravel, stone or asphalt necessary for the stability of the protection. Quarry stone can be used to serve as a material to sink the mat.
4. *The connections*: Pegs, wires, etc. by which the ballast material is fixed to the filter and the carrier materials.

In addition to the mat proper, a further deposit of gravel, aggregates, quarry stone, slag, etc. may be necessary when the ballast material stability in itself is insufficient or when extra protection is demanded.

The carrier and the filter can often be combined in the form of a single geotextile with the necessary mechanical and hydraulic properties. Sometimes the filter is combined with the ballast, which is then dimensioned as a granular filter.

Manufacture and placing

Fascine mats are typically up to 100 m long and 16–20 m wide. Historically, these mats were manufactured from osiers which were clamped between the so-called 'raster' of bundles of twigs/branches bound together to a circumference of 0.3 m.

'Plugs' were driven into the crossing-points of the bundles to serve as anchor points for towing and sinking the mats (for bottom protection) or collar-pieces (for foreshore protection). Because of the destruction of the osiers by pile-worm attack, a reed layer was incorporated into the mats. These fascine mats were made on embankments in tidal areas or on special slipways.

The finished mat or collar-piece was then towed to its destination and sunk using quarry stones which were placed in position by hand, with the uppermost rasterwork of bundles serving to maintain the stone in place during sinking. After sinking, more and heavier stones were discharged onto the mat if this was found to be necessary.

Modern mats are made in accordance with the same principles. The twigs/branches and reed are replaced by a geotextile into which loops are woven and to which the rasterwork of bundles are affixed. A reed mat may be affixed onto the geotextile in order to prevent damage to the cloth during the discharge of stones onto the mat. The geotextile extends out beyond the rasterwork with the aid of lath outriggers so that, on sinking, the mats overlap. Sometimes a double rasterwork is applied to give the mats a greater rigidity and more edge support to the quarry stone load. The mat construction can serve as a protection for the bottom, a foreshore and even as an embankment protector. In the last case the mat is hauled up against the embankment itself.

Modern alternatives to fascine mats include mats where the quarried stone is replaced by open-stone asphalt reinforced with steel mesh, or by 'sausages' of cloth or gauze filled with a ballast material such as sand, gravel, bituminous or cement-bound mixtures.

3.6 Quality control

3.6.1 Objective

The objective of quality control is to ensure that the client gets the product required. Thus it forms a vital part of a quality assurance scheme and, with particular reference to quarried rock, entails ensuring that the materials are supplied to at least the minimum requirements specified by the designer and placed in accordance with his criteria. This task is simplified if standard specifications are being used by the producers and the scheme is designed to take advantage of them. Engineers carrying out designs should be aware that armourstone in particular is usually a by-product of normal quarrying, so that what may appear to be an elegant design may only be achieved in practice at very high cost, even though production of appropriate block sizes are maximised. Because of the highly variable characteristics of this naturally occurring material the designer must always be realistic about what is economically available and, whenever possible, tailor the design accordingly (e.g. by introducing the berm breakwater concept).

Post-design quality control divides naturally into three parts:

1. Quarry control—this covers selection of faces/beds, blasting techniques, handling, sorting, selection and loading for delivery (Section 3.6.2). Responsibility and measures for safety during blasting should always be clear.
2. Construction control starts with checking deliveries, and progresses via stockpiles to the detailed placement, control of layer thicknesses and profiles to the finished product (Section 3.6.3).
3. Post-construction monitoring and maintenance (see Chapter 7).

3.6.2 QUALITY CONTROL IN THE QUARRY AND DURING TRANSPORTATION

Even with the use of standard specifications, there will always be a subjective element in the selection of armourstone. This may range from the degree and type of flaws which are acceptable (which in turn will depend on the wave and sediment environment of the finished structure) to the exact colour of the material where it will have a strong visual impact. To this end, it is useful, where suitable local quarries exist, for the client and designer to co-operate with producers in advance of the construction tendering procedure and determine what facilities (including production forecasts and stockpiles, space for secondary stockpiles, and dedicated faces) are available so that tenderers may be informed. Similar exercises may be carried out on more remote sources known to have been suitable and competitive in the past. The advantages of this procedure are that it may enable:

1. Quarries to start stockpiling in advance of the choice of a contractor;
2. The selection of realistic construction times; and
3. The resolution and costing of such matters as haul routes and times through residential areas.

After initial acceptance of material in terms of durability characteristics, quality control is dominated by weight, shape and grading requirements. During the construction phase it is therefore logical to concentrate materials control at the quarry, as this reduces the amount of handling of the material, obviates return loads (a very considerable cost, particularly if shipping is involved) and in many cases makes available to the contractor an accurate calibrated measuring device present in nearly all quarries as part of their basic equipment; the weighbridge. A degree of official control on-site is, however, always necessary to account for possible transport effects.

3.6.2.1 Armourstone stocks

Using such aids as scale bars, metal grids and loaders with weighing devices, the designer should obtain an estimate of the grading and total tonnage of the stockpile and compare this independent estimate with those of the producer and the contractor. Whatever the experience of the individuals concerned, any major differences of opinion should be raised and resolved early. The quarrying (especially stockpiling) should be well organized and well run. Most importantly, the quarry must be capable of producing rock of the type and sizes required.

3.6.2.2 Quarry faces and production

Any significant rock quality variations in the quarry (particularly weathered zones) should be identified by visual inspection. A quick additional method is to sample along quarry faces and test for density. This can be followed by more detailed testing as appropriate. If, as is usual, further production of armourstone will be required to meet the full contract order, the designer should take advice from the producer as to the proposed blasting methods and face locations. Potential problems with generating the specified weight ranges should be identified and the possibility of trial blasts discussed (see Section 3.4). The designer should seek advice on possible future variations of rock quality, block shape and integrity compared with those seen in the stockpile and should agree the production areas within the quarry (or the quarry location in the case of a dedicated quarry).

3.6.2.3 Integrity of blocks

The model specification indicates that blast-induced and natural discontinuities and flaws which could lead to Type 1 breakages (Section 3.2.2.2) during block handling and transportation or in the final works are to be rejected (Appendix 1, Clause A1.6.7). The designer should investigate the stockpile blocks for cracks and flaws, especially where the rock types vary in quality, such as in seams within limestones (see Figure 95). The designer should also note the normal handling procedures during stockpile creations. Rough handling in order to break up blocks

Figure 95(a)

Figure 95(b)

Figure 95 (c)

Figure 95 *Rejected blocks with flaws and cracks. (a) Carboniferous limestone block with three bedding joints, one at each end and one through the centre. The centre join has not quite broken through, leaving one large block of poor integrity which could make two excellent smaller blocks; (b) and (c) weathered joint surfaces parallel to banding in gneiss giving poor integrity— to be rejected (courtesy Queen Mary & Westfield College)*

with incipient failure planes (Figure 96) should be encouraged until the stage immediately prior to final selection and loading onto transport. Care must be taken to prevent further breakages arising at site. The dropping of a number of flawed blocks from a 3 m height onto other blocks using the techniques described in Appendix 2, Section A2.11 may provide the designer with useful observational data. It is advisable to set aside between three and six marked example blocks which visually fail on integrity requirements, the flaws being clearly indicated on each bloc. The percentage of such unsatisfactory blocks in the stockpile can then be roughly estimated for the producer's and contractor's quality control information.

Transit damage can be a major problem, particularly with friable rocks, and the designer should consider which breakage-monitoring procedures are appropriate (Section 3.3.8). These should be instituted in time to check the first consignments. Before leaving the quarry, the designer's representative should inspect and approve

Figure 96(a)

Figure 96(b)

Figure 96 (a) Blocks of poor integrity revealed by rough handling in stockpile; (b) blocks to be rejected (courtesy Queen Mary & Westfield College)

Table 26 Frequency of tests at quarry supplying large total quantities (>25 000 t)

Test	Maximum intervals between tests (tonnages)			
	Heavy gradings		Light gradings	
	0.3–3 tonne	3–10 tonne	5–300 kg	Fine gradings + core materials
Appendix 2, Section A2.2 (Grading and Average Weight)	5000–20 000	5000–20 000	5000–20 000	5000–20 000 plus any load with high percentage of small sizes
Appendix 2, Section A2.1 (size distribution)				
Appendix 2, Section A2.3 (Aspect Ratio l/d)	5000–20 000	5000–20 000	5000–20 000	
Density Appendix 2, Section A2.6 and water absorption Appendix 2, Section A2.7	5000	5000	5000	5000
Complete suit of specific materials tests (e.g. Section A1.6 of Appendix 1)	At start only or depending on variability indicated (e.g. from Density)			
The most appropriate weathering resistance test chosen from Section A1.6.4 of Appendix 1	At start only or depending on variability indicated (e.g. from Density)			

Note
The table can be more precisely detailed for known sources. The above figure of 5000–20 000 t means that a figure between 5000 and 20 000 t is to be agreed for the frequency of production of independently obtained certificates for rock qualities. These figures apply to a large order with constant production personnel and if small variations in quality can be expected. A higher frequency of testing may be required in other cases (e.g. every 1000 to 5000 tonnes). In addition to the above official tests, the producer may wish to present his own private test results during production control. See also Reed (1988) for an example of a testing frequency schedule

all armourstones, and all materials arriving at site should be incorporated within the layers of the structure. Where more than a 5–10% difference in W_{em} (effective mean weight) or W_{50} occurs in transit during the contract the deficiency should be made up at no additional cost (Appendix 3, Clause A3.3.3.4) and steps should be taken to reduce breakage and/or adjust the W_{em} or W_{50} of the quarry output to compensate.

3.6.2.4 Rock quality tests

Most quarries carry out regular tests on the properties of their materials, and the designer may be satisfied about the suitability of the proffered material based on these alone (Sections A1.6.1–A1.6.6, Appendix 1) if they are representative of the material to be supplied to the contract. If appropriate test data are not available, a new collection of representative samples should be made in accordance with Appendix 1, Section A1.8. This will entail identifying the faces/stockpiles which are proposed for the contract and paying particular attention to borderline or potentially weathered materials which may need eliminating. Results of appropriate tests on the new representative samples should be available before any material is dispatched.

The designer may elect to take several samples of material to represent all the distinct rock types or weathering grades in the stockpile. A common mistake is to take insufficient sample materials for a full test suite. In addition to the full test suite at commencement of construction, tests for density and water absorption should be carried out routinely during production. A suggested frequency of various tests for various rock gradings is given in Table 26, and further information on testing frequencies, sampling statistics, and cumulative sum charts for controlling measures variables in a production sequence may be found in BSI handbook No. 24 (*Quality Control*).

The minimum acceptable test values for which the contractor (i.e. producer) is obliged to conform should be indicated in the contract specification. Guidance to this end is provided in Table 21 and Appendix 1.

3.6.2.5 Armourstone shape

With reference to Section 5 of the model specification (Appendix 1) and test method (Appendix 2, Section A2.3), a number of blocks which appear to have an aspect ratio (l/d) which exceeds 3 should be selected for measurement. Up to six of those with ratios greater than 3 are marked with paint indicating the l/d value (see Box 35 for guidance on quality control of shape and Figure 51 and Table 18 for shape classes).

3.6.2.6 Armourstone grading

Poor-quality control over grading of armourstone is the most likely cause of construction difficulties and contractural disagreements. Even quarries working to the standard weight ranges (Table 19) may produce stone with an uneven block weight distribution leading to W_{em} falling outside the required range. Grading control should be done at critical stages:

- During production (most important)
- During barge, truck or rail wagon loading
- On arrival during unloading
- Before placing.

It will therefore often be in the designer's and contractor's interests to prepare and work to a grading plan for the heavy grading classes specified in the contract. An example is given in Box 36 (see also Box 24 and Table 19 for information on the class limit system for gradings), which is similar to the grading plan implemented at Helguvik, Iceland (Read, 1988).

A grading plan may not always be necessary for the standard narrow gradings because they are sufficiently narrowly graded for a correct distribution to be almost automatically generated when working to the LCL and UCL limiting weights. The grading plan principle is perhaps more useful and very effective when applied to wide standard or non-standard gradings (see Table 19 and Box 27).

3.6.2.7 Grading control possibilities for heavy armourstone (<300 kg)

A number of possibilities exist for controlling the grading that may or may not follow the grading plan concept. Those given below are appropriate for heavy gradings with a lower class limit (LCL) in excess of 300 kg:

1. Blocks of approximate weight equal to the LCL and UCL weights (see Box 24) are strategically placed with their weights painted. They then act as visual aids to block selection by the machine operators;
2. Bulk weighing of truck or dumper loads on a weighbridge, counting the number of blocks weighing more than the ELCL, having carefully estimated the weight of blocks less than the ELCL and greater than EUCL with the aid of a measuring tape;
3. For rock types which develop particularly rectanguloid armour block shapes (e.g. bedded sedimentary rocks such as carboniferous limestone and dimension stone rejects), the effective volume of blocks can be assessed from the XYZ enclosing cuboid dimensions. With the known rock density, the effective volume is converted to effective weight for each block. This value will be a overestimate. Knowing the true weight of a load, of, say, four blocks from a truckload weighbridge ticket and dividing this by the effective weights summed together gives a correction factor (usually about 0.55 for blocky equant shapes

Box 35 Shape control

Shape control can ideally be combined with grading control, because the weight or number of stones above the ELCL weight of the sample is also determined during grading control tests. The sample for the shape control test (Appendix 2, Section A2.3) contains at least 50 stones, whereas for the distribution of a light or heavy grading, at least 200 stones are weighed. This larger amount gives a more accurate result for the shape control test, which is desirable when only a small specified percentage is allowed to exceed the critical l/d, value which is usually set at 3.

The sample to be tested should be visually divided into three groups:

1. The flat or long stones certain to exceed the limit;
2. The cubic stones certain not to exceed the limit;
3. The stones possibly exceeding the limit.

The stones of group (3) are classified into groups (1) and (2) by measuring the thickness d and the length l. An accurate measurement is possible when using a manufactured caliper such as that indicated in Figure 97.

The total weight or number of stones of group (1) is determined and divided by the total weight or number of the sample to obtain the percentage exceeding values (see Appendix 2, Section A2.3).

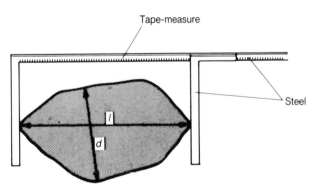

Figure 97 *Measurement of thickness and length (courtesy G. J. Laan)*

Box 36 Grading plan: 3–6 tonne

Sub-class	C1 Weight range (t)	C2 Average weight (t)	C3 Target weight (%)	C4 Target rel. no. of blocks	C5 Target no. and range of blocks per 100 loaded	C6 Target weight per 100 blocks (t)
UCL–EUCL	6–9	7.5	15±15	2.0	9(0–18)	68
W_{50}–UCL	4.5–6	5.25	35±15	6.7	29(17–42)	155
LCL–W_{50}	3–4.5	3.75	45±5	12.0	53(47–58)	204
ELCL–LCL	2–3	2.5	5±5	2.0	9(0–18)	23

EUCL = greater than 97% by wt lighter than this weight
UCL = between 70% and 100% by wt lighter than this weight
LCL = between 0% and 10% by wt lighter than this weight
ELCL = less than 2% by wt lighter than this weight

See Section 3.2.2.3.

Notes

1. W_{50} is the mid-range value from Table 19. Alternative sub-class weight ranges can be set, but in such cases the target weight percentage in each sub-class will be less well defined for calculating the target number of blocks per 100 loaded. The plan may be presented with additional guidance data on typical block dimensions per sub-class for equant ($l/d = 1.5$ to 2) and more irregular-shaped blocks (Figure 51) of a given density. The quarry owner can use the plan to check the grading during routine preparation of standardised stockpiles. The grading plan approach is most useful if the selecting machine has a built-in weighing device with computer software support.

2. $C3_i$ is set by the mid-point of the class limit bands: e.g. LCL is 0–10% therefore target weight for fourth sub-class, $C3_4$, is $5 \pm 5\%$.

3. $C4_i = C3_i/C2_i$.

4. $C5_i = 100 * C4_i / \sum_{i=1}^{4} C4_i$:

   e.g. for fourth sub-class, $C5_4 = 100 * 2/(2 + 12 + 6.7 + 2)$.

5. $C6_i = C5_i * C2_i$.

---

but typically 0.3–0.7). This factor should be reasonably constant from the one rock source and an average factor value can be used to convert the measurements of effective volume, which are easy to carry out, to cumulative weight grading curves;

4. Bulk weighing in accordance with Appendix 2, Section A2.2: Part 1, Section 3.2;

5. Weighing individual stones in accordance with Appendix 2, Part 1, Section 3.1.

Methods (1) to (5) give increasing accuracy and may be applied as follows.

Control during production

Methods (1), (2) and (5) are recommended, but note the following. The accuracy of method (1) (visual selection) does not usually justify the creation of many sub-classes unless the grading being controlled is over a wide tonnage range of if the selection machine has a built-in weighing device. Method (2) must be considered as just an additional unofficial technique for a rough estimate of the grading. Method (5) is an accurate procedure, ideal for feedback. Accumulation of sub-sample results during a certain production time until a total of 200 blocks is obtained gives greatest efficiency. A higher frequency of testing is necessary at the start of production for training purposes.

By establishing the relationship between the grading of material arriving on-site or as-placed in the structure (see below) and that during production, it is possible to compensate the eventual unacceptable difference by (1) secondary selection or (2) readjustment of the production requirements. For the Eastern Scheldt barrier, production control method (4) was required by the client. Production was directed and adjusted using accumulated results of weighed sub-samples for each week's production. Control of production by sampling, weighing and measuring every fifth

> **Box 37** Example of grading control for 9000-tonne capacity barge using the grading plan of Box 36 for 3–6 tonne standard heavy grading
>
> The machine operators selecting blocks must be experienced in classifying blocks into the four sub-classes defined in Box 36. To help inexperienced (and experienced) operators, several marked-up reference blocks of known weight, preferably of the approximate sub-class boundary weights 2, 3, 4.5, 6 and 9 tonnes, are set aside for visual aids to selection. For each block, the operator should monitor and update the number in each sub-class that has been loaded onto the barge (e.g. using four manually operated counters). The operator can then monitor each 100-stone batch and regulate according to the sub-class target values. He should also check-weigh randomly selected blocks (e.g. every 2 hours) to help restore accuracy to eye selection.
>
> At several stages during loading the armourstone tonnage deduced from displaced water volumes (and/or weighbridge tickets), together with the total number of blocks loaded from counter log sheets, gives an estimate of the average weight of blocks. These estimates can also be used to check and regulate the loading to achieve the grading requirement tolerances for average weight, and thereby also to correct for selection bias. For this example, the finally laden barge with 9000 tonnes should have about 2000 blocks of weight greater than the ELCL, with an average of 4.5 tonnes.
>
> This example procedure for grading control on loading does not constitute an official verification of the grading requirement. However, estimated cumulative grading curves can be prepared from such a set of well-kept counter logs. If, following these procedures, loading is also overseen by the designer's representative, the required frequency (Table 26) of accurate grading tests (Appendix 2, Part 1, Section 3.1) on random samples of at least 200 blocks could be significantly reduced. Note that for verification purposes, inspection by bulk weighing of a sample (Appendix 2, Part 1, Section 3.2), which also uses visual estimation, will probably give only marginally greater accuracy in assessing grading than the example procedure outlined.
>
> It is possible that the grading 'as loaded' falls outside the specification when tested by bulk weighing. The results are too late to feedback to production methods but may be useful to estimates the breakage effect of handling. Acceptance control during loading needs the presence of the client/designer's representative.

block and keeping a record has proved satisfactory in many Scandinavian quarries supplying armourstone.

Control from trucks

Methods (2)–(4) may all be used but for official control in the presence of the client/designer, methods (3), (4) or (5) will be required.

Control from ships or barges

Method (2) is often used, although for official control, methods (4) or (5) are recommended if time is available during loading.

Control at site after arrival or during/before placing

Methods (2)–(5) may all be used, depending on the client/designer's wishes, but method (5) can give the W_{50} and W_{em} employed for assessing breakages during delivery.

Barge transport grading control

Barge loading is a critical and difficult moment for quality control because of the intense activity to get the barge loaded as soon as possible. The specialised armourstone quarry should ideally have blocks already sorted into standard grading classes. Careful block stacking and weight distribution (trimming) during barge loading is necessary. Ideally, the correct mix of block sizes should be evenly distributed throughout the barge (or allocated area, if several grading classes are to be transported together) for ease of construction on-site. The grading plan principle may be applied during barge loading (see Box 37).

Road transport grading control

Occasionally, stone can be loaded at the quarry face for direct transport and unloading into the construction at site. This is optimal, but more usually, where the quarry-to-site distances are at least 20 km, a stockpile is created both in the quarry and at site to ensure that work can always proceed. Road haulage is typically in loads of 16–20 tonnes in solid-bodied trucks. Thus, with 3–6 or 6–10 tonne gradings it is not possible for each truckload to contain the target grading needed at site for ease of construction. Grading control is easier than for barge transport because weighbridge tickets combined with block counts gives a continuous record of the total and average weight of blocks delivered to site, and this can be kept to within the specified requirement by referring to a grading plan.

There are many different devices available capable of weighing every block with varying degree of accuracy and applicability to the handling equipment. All need frequent checking (e.g. six times per shift) and are prone to failure with consequent downtime. Selected marked comparison blocks should always be checked over a weighbridge, and for most contracts a simple check on the number and gross weight of blocks in a load passing over the weighbridge will be sufficient. Information on weighing devices is given in Box 38.

3.6.2.8 Underlayer and core materials

Quality control of rock material properties for quarried underlayer and core materials will be identical to that for armourstone (see Section 3.6.2.4). In regard to grading control, such materials in the light and fine grading classes may well have been produced using controlled quarry plant crushing and screening processes. The need for repeated grading tests after the initial evaluation will therefore generally be minimal unless:

1. The upper size material is selected by eye;
2. Wearing of installation and breaking of screen or grid parts occurs;
3. Renewal or correction of crusher jaw openings is required;
4. Variation in intrinsic properties or sizes of blasted rock is present.

Core materials taken from the blastpile using slotted buckets to reduce the fines must be expected to vary in fines content. A moving average based on ten truckloads can be used to check trends in fines content from such a quarry run material (Read, 1988). A grizzly of agreed bar separation may be used to separate that which is designated fines for the purpose of testing. The use of a grizzly is essential for separating unwanted fines from powdery material such as duricrust limestones in the Middle East.

For gravel used as core material or in gravel beaches, quality tests are not usually considered necessary because their durability has been proved during natural processes. Grading and sand content are a function of the borrow area and cannot be controlled other than by correct initial selection of the appropriate borrow area. However, contaminant weak material, such as chalk in flint gravel areas, should be controlled or at least monitored where sources are known to have this problem.

3.6.3 QUALITY CONTROL DURING CONSTRUCTION

The quality control of materials will be concentrated at the quarry while the specification requirement is for material incorporated into the works. Some further tests may therefore be required at the construction site but their frequency will be determined by:

- The importance of fulfilling the requirements of the specification;
- The amount and rate of material arriving;
- The length and continuity of the production period;
- The part played by control in the quarry.

Site testing on a routine basis should rarely be less frequent than recommended in Table 26. The site engineer must be able to call for appropriate testing at any

Box 38 Weighing devices

Some or all of the following types of weighing devices may be useful for quality control and testing of grading requirements:

1. Weighbridge
2. Mechanical scales
3. Electronic scales
4. Spring-type tensile devices ('fisherman' or 'steel-yard')
5. Moment arm
6. Tensile cells
7. Axial load cells (e.g. French AX series (details in *Les Enrochments*, LCPC, 1989))

Principles

(1) Balance-lever or pressure cells; (2) balance-level (counterweights of standard values, weights sliding along a measure beam or transferred to a dial plate); (3) usually pressure load-cells; (4) linear force-elongation springs with elongation transferred to dial-plate or scale; (5) strain gauge distortion measurement of bending in moment arm or hydraulic (oil) pressure in wheel loaders; (6) strain gauge distortion measurement in tensile cell; (7) strain gauge distortion measurement in axis of grab.

Capacity

(1) Large, suitable for single stones, truckloads, dumperloads. For dumper, the width may be too small. In this case separate weighing of left and right side is possible, but this requires a horizontal position of the dumper. (Smaller, transportable ones are extremely useful.) (2) A large variation exists up to several tonnes. (3) A large variation up to maximum block weight. (4) Several types and capacity up to 50 tonnes (e.g. the German PIAB dynamometer which can be read (a) directly off a scale with an enlarged lens, (b) with a dial-plate attachment or (c) with an electrical distance system.) (5) Varies according to lifting capacity of loaders. (6), (7) Large variation, which makes it possible to choose the capacity which is optimum for the grading class being selected.

Application

Stone resting upon the device: types (1), (2), (3)
Stone lifted via a cable or chain: (4), (6), (7)
Stone lifted with a hydraulic grab: (6), (7)
Stone lifted in the bucket of a loader: (5)—not accurate for measuring single blocks because of variation of the position of the centre of gravity of rock relative to bucket axis.

Robustness

(1) Very robust. (2) Varies; in general, good. Damping of shocks due to placing of the stone on the scale may be necessary. A rubber tyre may be used. (3) Varies; in general, good. (4) Every device on a cable is under great risk from damage when used during production and handling. (5) Varies. (6) Very risky in combination with a hydraulic grab. (7) The French report (LCPC, 1989) indicates that the device can be used when handling the stone, can give automated (possibly computerised) results in the operator's cabin, but that weighing results may take a minute or so to settle to accuracies sufficient for testing purposes.

Accuracy

(1) About 20 kg. (2) Varies; in principle, accurate, but the reading accuracy is important, which is determined in part by the scale divisions (dial-plate gives an estimated maximum reading accuracy of 0.5%). (3), (6), (7) Accuracy is relative to the capacity of the device and the lowest capacity, taking into account a safety factor for shocks. Non-linearity and hysteresis of 0.3% with reproducibility of 0.1% have been met for many of these devices. (4) 0.5% of the capacity. (5) Often inaccurate.

Figure 98 *Dynamometer weighing device (courtesy G. J. Laan)*

time that a load or series of loads consistently appears defective. Block breakages during unloading must be monitored and action taken as described in Section 3.6.2.3 above. However, if breakage is less than 2%, grading will not be significantly affected.

3.6.3.1 Bulk placement of core materials

The preferred measurement method for underwater placed core materials is by sounding ball and chain, but other methods are discussed in Appendix 6 under strict monitoring. The accuracy of the grid system of dumping in deep water by barge should be as tightly controlled as practicable. Average errors in dump location as high as 7 m are not uncommon for difficult conditions, while for normal conditions during construction of one breakwater, average errors parallel and normal to the longitudinal breakwater axis were typically 2 and 6 m, respectively. Reference to radio fixing and survey control methods using modern technology are given in Read (1988).

3.6.3.2 Placement of armour

The principal objective during placement must be to achieve good interlocking without creating a 'dry-stone wall' type construction with consequent low porosity. Further information on placing techniques is given in the construction sections of Chapter 6.

Verification test panel

Placement of armour should initially be performed on a verification test panel or section which, upon acceptance, becomes a model for further placement and remains part of the finished works. The length of this section (e.g. 10 or 20 m) should be agreed before construction. With tolerances and survey methods as defined in the specification, the objective of the test panel is verification of:

1. Target tonnage per metre cubed for the bulk density, γ_b, as laid, set out in the preamble to the bill for a design minimum density;
2. The average layer thickness in the drawings.

This exercise is of critical importance to the contractor and designer for the case where a double armour layer of given shape and grading requirement has been specified and there is some doubt as to whether the layer thickness of the drawing and target tonnage per metre cubed are practically achievable within the given tolerances. The possibility of construction difficulties should be reduced if the designer has taken guidance from Table 20.

For the verification test panel, blocks will preferably be weighed individually and placed into a two-layer section using the mix of block weights representative of the specified grading and shape requirements. Alternatively, only their total weights may be determined from the bulk weighings.

Profiles should be established by the survey techniques outlined in Appendix 1, Clause 13, using a survey staff, traveller or dippings as appropriate, to the specified tolerances as appropriate and taking account of any special placing requirements. Additional test panels may be required. Where possible, test panels should be incorporated into the finished works, but in any case approved panels should be preserved to act as a check on subsequent work. Test panels should also serve to give early notice of any difficulties in meeting the specified requirements (e.g. layer thicknesses) and allow any necessary adjustments to be made at a stage where additional costs are minimised. Appropriate calculations on such test panels can be carried out using the method described in Box 39 (overleaf).

Box 39 Verification test panel calculations

Layer thickness coefficients in *SPM* (1984) have created problems in construction and measurement for payment, as they seem to be wrong. The test panel approach is an attempt to solve the problem by on-site verification at the start of construction.

The calculations apply to narrow gradings, where D_{n50} will be approximately equal to the average nominal diameter. The tonnages per metre cubed, γ_b, are calculated from two formulae:

$$\gamma_{b1} = \frac{\text{Measured weight of armour in sample volume of structure}}{\text{Total volume of armour layer sample} (= t_a . A)}$$

where t_a is the average perpendicular thickness calculated from the survey and A is the survey area;

$$\gamma_{b2} = \frac{\text{Calculated weight of armour in sample volume of structure} (= N_a . \gamma_a . D_{n50}^3)}{\text{Total volume of armour layer sample} (= t_a . A)}$$

where N_a is the number of blocks in the sample area

The first and second formula should give the same result. They differ significantly only if the grading of weights or density of rock, γ_a, in the test area are not as specified. The test panel results are to be considered invalid if γ_{b1} and γ_{b2} differ by more than a value to be agreed (say, $0.2 \, t/m^3$) because the material is clearly not to specification.

The test panel should be rebuilt with more carefully selected block weights until γ_{b1} and γ_{b2} differ by less than the agreed value. It is suggested that layer thickness and target tonnage per metre cubed can only be renegotiated (if necessary) once this condition is reached and if the design requirements for γ_b and t_a are demonstrably in need of improvement. The achieved porosity, n_v, and layer thickness coefficient, k_t, can be calculated from

$$\left. \begin{array}{l} n_v = 1 - \gamma_b/\gamma_a \\ k_t = t_a/2D_{n50} \end{array} \right\} \text{ for a double layer}$$

The values obtained should be compared with those in Table 20 and noted for future reference.

4. Physical site conditions and data collection

The needs for data may vary through the life cycle of a structure. The sort of data needed depends on the type of structure, the choice of design approach, structural design, construction methods and on how the structure is managed after construction

Physical site (bathymetric, meteorological, hydraulic, geotechnical) data are required to define the various boundary conditions for design and construction. The planning of maintenance and repair may require, additionally, the setting up of a routine environmental data-acquisition system. A feasibility study may include definition of the social and economic boundary conditions for a project, which also needs specific data to be gathered (see Section 2.1). Figure 99 presents an overview of all types of data which may be needed, the method of obtaining and processing the data, and for which boundary conditions and design formulae each type of data is required.

Given these possible needs, this chapter provides methods of measuring and assessing the physical site conditions—bathymetric, hydraulic and geotechnical—which define the boundary conditions at a rock structure. The kinds of instruments available for data collection are described in Appendices 4 and 5, but these descriptions are in general terms, and for technical specifications of specific instruments the reader should consult the manufacturers' brochures and manuals. The designer should adjust his data requirements to the various design stages described in Section 2.1 and the envisaged use of the data.

The data collection and analysis methods discussed enable the conditions at the structure to be described. Chapter 5 then shows how these conditions (e.g. wave climate) interact with the structure (e.g. in terms of reflection, transmission, run-up and overtopping) to determine the resulting final boundary conditions that are used as input for the structural design.

Data for feasibility study and design

A wide variety of approximate data will usually be needed for a first feasibility study. Such data may be obtained relatively easily (for instance, by a desk study or by consulting a manual). Usually, immediately available data will be insufficient for the final design and the need to gather more will be an uregnt requirement. The cost of data acquisition will be high if routine methods cannot be used or a high degree of detail or special equipment is needed.

Design approach, data specification and use

The quantity and types of data required are also a function of the design approach. For example, a probabilistic design will require more statistical information (e.g. short-term distributions and standard deviations) than a deterministic one (where long-term distributions are often sufficient). Extreme value analyses require data covering a relatively wide range of conditions, which usually dictates a longer period of measurement. For a joint probability analysis to be contemplated, the data must enable statistical correlations to be determined. If modelling is planned, the nature of the data-collection programme will be influenced by the decision as to whether numerical or physical modelling is to be used. If a numerical model is to be used, the need to tune the model coefficients will require measurement of at least one set of all variables. If a physical model is to be employed, a simpler data set (e.g., H_s, T_m, bathymetry) will suffice.

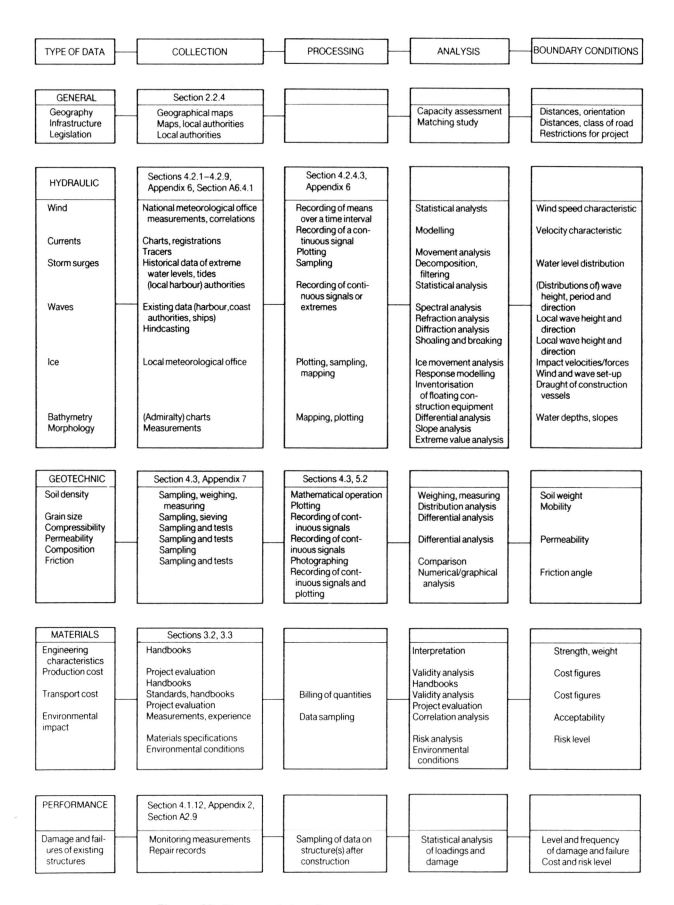

Figure 99 *Diagram of data flow*

Data for maintenance

Maintenance of a structure (Section 2.1.7 and Chapter 7) will also involve data gathering by continuous monitoring and surveys on a regular basis, both during and after extreme storm events. The information required include the environmental data discussed in this chapter, data on the hydraulic and geotechnic response to the presence of the structure and data on the structural response itself. The monitoring of structural response is covered in Chapter 7 and Appendix 6. The data obtained facilitates evaluation of:

- Damage due to actual loadings (known or estimated);
- Decisions on (emergency) repair;
- Planning and scheduling necessary maintenance work;
- Evaluation of the functioning of the structure;
- Initiation of rehabilitation or replacement;
- The setting up of a data bank for the structure and its performance.

Physical site data (e.g. water levels, waves) thus collected are often of general future interest both in terms of future replacement of the structure and of the ability of future functioning of the structure (for example, to limit hinterland flooding or to affect beach forms).

Options for data collection

Before examining the methods used to obtain the necessary environmental data in more detail, it is worth noting in a general way that the principal options to obtain data for a conceptual or structural design are to:

- Search for existing data that apply to the site (from authorities, companies, local sources);
- Set up a measurement programme;
- Make use of data from neighbouring locations (find correlations, make inter/extrapolation);
- Use physical modelling (scale models);
- Use numerical models (hindcasting);
- Choose an alternative design concept that is insensitive to the missing data;
- Use unreliable data and account for that in the design (probabilistic approach).

The requirements for equipment, instrumentation, energy supply, transport, labour, organisation and hard- and software requirements must be considered for each of the options mentioned (Box 40).

Box 40 Supporting equipment for physical site data collection

Depending on the type of measurement operation (on–off or continuous measurements), the sensitivity of the instruments being used, safety considerations and the environment in which the data are being collected, environmental measurements may be made from:

1. *Fixed frames* (*e.g. piles or pile-supported platforms*). A pile or platform is advantageous, particularly for continuous measurements on a very limited area and in shallow water. Ignoring the question of the design of the measurement system itself, the loss of accuracy due to wave- and current-induced motions of the fixed frame is relatively low.
2. *Floating equipment* (*ship, barge, semi-submersible*). Deeper water and mobility are the most evident reasons to choose floating equipment. Wave and current action (anchoring) are limiting factors to the measurement operation. An additional consideration is the limited accuracy of the measurements due to wave- and current-induced movements.
3. *Submarines*. Large water depths, eventually in combination with large wave height or currents, can make submarines competitive. This holds particularly when visual inspection of (objects on) the seabottom is also needed. Rapid developments of remote-operated vehicles, equipped with mechanical arms, have reduced the advantages of manned submarines.

Some equipment (e.g. wave rider buoy) will not require any of the above during the time of measurement, but may need it before and afterwards in order to be deployed at the measurement site.

4.1 Bathymetry and morphology

The bathymetry and morphology of the seabed at the structure site is fundamental to the design of coastal and shoreline rock structures. Bathymetry is the description of the variation in the bed level across an area relative to a datum, and is the major boundary condition for geometric and structural design having a significant influence on the volume of rock needed and the hydraulic loadings (e.g. waves), the latter often being largely dictated at coastal sites by the water depth. On charts for navigation purpose the seabed level is defined relative to the chart datum, commonly taken to be the lowest astronomic tide level. Morphology, the sedimentation and erosion processes occurring at the bed combined with the presence of very soft sediments and/or layers of high concentrations of suspended sediments (mud, clay) sometimes cause problems in measurement of bathymetry.

In many cases knowledge of morphological changes of the bed with time are as critical to design as the mean bed level, enabling definition, for example, of the lowest bottom level in front of a structure. Therefore where a seabed is either in dynamic equilibrium or is morphologically unstable, the range of bed level changes that should be taken into account in the design must be determined. Morphologiocal changes may be due to (tidal) currents and/or wave action. If no data on bed levels are available, a first indication of their variability can be obtained from comparing known or predicted current velocities and/or wave characteristics (height and period) with threshold values for the initiation of sediment transport for currents and waves (Shields, 1936; Komar and Miller, 1974; see Section 6.4). The prevailing local current, wave and sediment conditions may lead to bed variations exhibiting typical mean dimensions and timescales. For example, sand waves observed in the southern part of the North Sea show lengths and heights of the order 100 m and 10 m, respectively, with related 'wave' periods ranging from one to 10 or even 100 years. To ensure representative data on bottom changes, measurements must cover at least one wave length per section and, in repeated measurements at one single position, a minimum sampling frequency of twice the typical frequency of the natural changes of the seabottom should be chosen.

4.2 Hydraulic boundary conditions and data collection

The principal physical relations between the relevant hydraulic boundary conditions are shown in Figure 100. The figure also indicates for a specific case which relevant design parameters should be determined.

Possible combinations of two or more parameters together often determine the design loading of a structure. In these cases not only do the separate design values have to be known but a joint probability analysis is also needed to find the probability that any combination of both will exceed a predefined design level. Examples of combined loading include:

1. Water level and wave height, which will determine the required crest level of a seawall or current; and
2. Orbital velocities, which will determine the stone weight of a scour protection.

Joint probabilities of such combinations are treated further in Section 4.2.8.

4.2.1 WATER LEVELS

Water levels are a principal boundary condition for the design of coastal structures, the measurement of which is discussed in Appendix 4. This is because other hydraulic boundary conditions used in the design of rock structures are

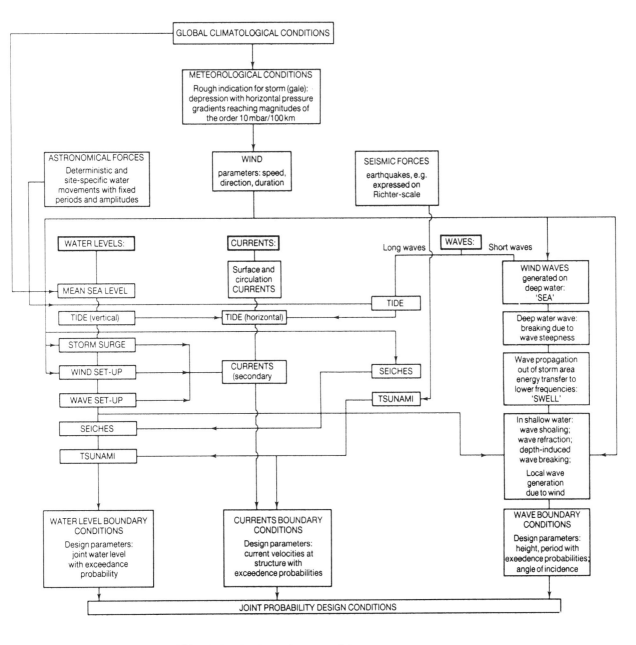

Figure 100 *Hydraulic boundary conditions*

closely related to the actual water level (e.g. limitation of wave height due to wave breaking—see Section 4.2.3.4). A number of water level components are, in fact, long wave phenomena (e.g. tides) but may, for practical purposes, be treated as quasi-static water levels.

Variations in water level are due to meteorological, astronomical tide and seismic influences (Figure 101) and also depend on the local bathymetry. The various

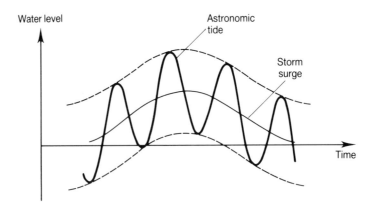

Figure 101 *Variation in water level due to storm surge and astronomic tide*

water level components arising from these influences are treated below and the joint probability of their occurrence is discussed in Sections 4.2.1.9 and 4.2.8.

Of the components, only those due to water level fluctuations depending on meteorological and seismic conditions are highly indeterministic, typical time scales rangeing from seconds (short waves) to days (storm surges). As it is difficult to separate storm surge level, wind set-up and wave set-up in a water level recording, these are generally taken together as an overall addition to the tidal component which dictates the final 'storm surge level'. The variation with time of water level due to astronomic tide and storm surge is illustrated in Figure 101.

4.2.1.1 Mean water level

For coastal waters in open communication with the sea, the mean water level (MWL) can often effectively be taken as a site-specific constant, being related to the mean sea level (MSL) of the oceans. However, over longer time scales of the order of 100 years the MSL is not constant. Due to long-term climatic variations, the MSL and hence most MWLs are presently showing a rising trend of $0.10 \simeq 0.15$ m per 100 years since 1879 when data began to become available. The 'greenhouse effect' caused by human impact on the global environment (which is projected to result in a 2–4°C rise in global air temperatures over the next 100 years compared with a 0.5°C increase over the last 100 years) suggests that this rate will increase. Present estimates of the future rate of MSL rise vary widely, but figures between 0.6 m and 1.0 m are typical. Pugh (1990) gives a useful overview of present estimates and implications. However, site-specific information which takes account of local tectonic movements and/or subsidence/heave arising from groundwater level changes must be obtained for specific designs.

4.2.1.2 Tides

The basic driving forces of tidal movements are astronomical and therefore entirely deterministic, which enables accurate prediction of tidal levels (and currents). Since tides are a long wave phenomenon, resonance and shoaling effects caused by geography and bathymetry can lead to considerable amplification of tidal levels in shallow seas and estuaries. Tidal range, approximately equal to twice the tidal amplitude, is generally below 1 m in open oceans and increases slightly towards the continents and may increase considerably in shelf seas. Large amplifications are found, for example, in bays along the coasts of England, France and Wales (7 m) and in the Bay of Fundy, Canada (10 m), while 2–3 m are common ranges for the southern North Sea. The predictive character of tides can be useful when timing of critical operations and manouvering during construction is needed.

The dominant tidal components have periods of approximately half a day (semi-diurnal tides) and a day (diurnal tides). Dominant semi-diurnal tidal components include principal lunar (M_2, period 12.42 hours) and principal solar (S_2, period 12.00 hours). Dominant diurnal components include principal lunar diurnal (O_1, period 25.82 hours) and luni-solar diurnal (K_1, period 23.93 hours). Specific coastline geometry (e.g. of channels, bays, estuaries) and bottom friction can generate frequencies equal to the sum or difference of basic frequencies. The contributions of these 'secondary' frequencies may be significant.

When sufficient tidal data or details are lacking, an analysis of tidal movements may be necessary to determine the local astronomic tidal constants. This involves analysis of a limited series of water level recordings that are known not to have been affected by other stochastic components, using a Fourier transformation technique, the output of which is the local amplitudes of the known local principal tidal frequencies. From such measurements and analyses, tide tables are composed on a routine basis and issued yearly by port or coast authorities and by national admiralities. (The US and British Admiralties, in particular, have extensive data files.) These provide high water (HW) and low water (LW) levels and times for major ports, usually one year ahead, but they can also be used to derive site-specific tidal constants needed for further prediction at intermediate sites.

4.2.1.3 Storm surges

Local minima of atmospheric pressures (depressions) cause a corresponding rise of mean water level (MWL) (similarly, high pressures cause low water levels). Mean air pressures at sea level are 1013 mbar (=hPa) approximately. In the storm zones of higher latitudes ($\geq 40°$) variations from 970 to 1040 mbar are common, while in tropical storms pressures may drop to 900 mbar.

The height of the corresponding static rise of MWL (z_a, in metres) is:

$$z_a = 0.01\ (1013 - p_a) \qquad (4.1)$$

where p_a is the atmospheric pressure at sea level, in mbar or hPa. Due to dynamic effects, however, the rise in water level can be amplified significantly. When the depression moves quickly, the elevation of the water level moves correspondingly as a storm surge. A storm surge behaves as a long wave with a wave length approximately equal to the width of the centre of the depression. The height of these long waves may increase considerably due to shoaling. For example, along the coasts of the southern North Sea, storm surges with a height of 3 m have been recorded. As stated above, in practice the term 'storm surge level' has a wider interpretation, generally including the astronomic tidal component and other meteorological effects.

4.2.1.4 Wind set-up

Shear stress exerted by wind on the water surface causes a slope in the water surface (Figure 102) as a result of which wind set-up and set-down occur at

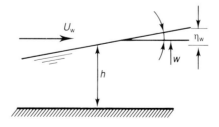

Figure 102 *Wind set-up*

down- and upwind shorelines respectively. Wind set-up is most pronounced along relatively shallow waters such as estuaries and shelf seas (for example, along the southern coasts of the North Sea, Belgium, The Netherlands and Germany, storm-induced wind set-up may amount to 3 m). If the water depth and wind field are constant as shown in Figure 102, the wind-induced gradient (i_w) of the still-water surface can be found from

$$i_w = c_w(\rho_{air}/\rho)U_w^2/(gh) \qquad (4.2)$$

where U_w = wind speed,
h = water depth,
ρ, ρ_{air} = mass density of sea water and air (1030, 1.21 kg/m^3),
c_w = air/water friction coefficient (0.8 10^{-3}–3.0 10^{-3})—
values increase with wind speed (Abrahametal, 1979).

The resulting maximum set-up (η_w) at the downwind coast or shoreline is

$$\eta_w = i_w F/2 \qquad (4.3)$$

where

$$F = \text{fetch length}$$

The actual wind set-up is strongly affected by local alignment of coastlines and bathymetry. It is therefore recommended that the above formulae should not be used unless comparison is being made with existing local records of wind and water levels, where a useful indication of the distribution of the wind set-up component can be made. Reference may also be made to Ippen (1966), who presents methods to derive wind set-up for a number of schematisations of local geography and bathymetry. Wind set-up is generally included in the present generation of numerical models by taking account of surface shear.

4.2.1.5 Wave set-up

Wave set-up is caused by energy dissipation due to shoaling of the incoming waves (see Figure 103). For small-amplitude sinusoidal waves the wave set-up can be roughly approximated by linear wave theory.

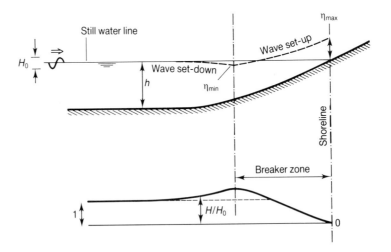

Figure 103 *Wave set-up*

Using linear wave theory (see the introduction to Section 4.2.2) for normally incident small-amplitude regular waves, a first approximation of the maximum set-up at the water line is derived by Battjes (1974). Referring to the wave conditions at the breaker line, the theory enables the derivation of the following approximate relationship, which can be used as a first estimate of wave set-up:

$$\eta_{max} = 0.3 \gamma_b H_b \qquad (4.4)$$

where γ_b = breaker index or maximum H/h ratio (Section 4.2.3.4),
H_b = wave height at the breaker line (regular waves).

H_b can be found by applying a wave model to the local bathymetry, using deep-water waves as a boundary condition (H_0, T_p, β_0, see Section 4.2.3).

Distributions of the wave set-up component can be obtained from wave data or (if these data are lacking) through hindcasting with existing wind data. However, as correct wave calculations require actual water depths, other water level components must be known. Present numerical wave models therefore determine wave set-up as an integral part of their calculations.

4.2.1.6 Seiches

Seiches are long waves caused by local meteorological phenomena such as squalls (small depressions) and gusts. Typical time scales are of the order of minutes (frequencies are usually less than 0.01 Hz). Only analysis of local water level recordings will enable proper account of seiches to be taken for design purposes. Where measurements have to be carried out, the minimum sampling frequency should be at least twice the expected maximum frequency of the phenomenon of interest.

4.2.1.7 Tsunami

Tsunami are seismically induced gravity waves, characterised by wave lengths that are in the order of minutes rather than seconds. They often originate from earthquakes below the ocean, where water depths can be more than 1000 m, and may travel long distances without reaching any noticeable wave height. However, when approaching coastlines the height may increase considerably. Due to the large wave length, increases that result from shoaling and refraction coming inshore from quite large water depths can be calculated using shallow-water wave theory. Wave reflection from relatively deep slopes of continental shelves may also be an important consideration.

Although theoretical work is available (e.g. Wilson *et al.*, 1962; Wilson, 1963) the only published observations for height and period of actual tsunami are those found in Japanese sources, and concern tsunami observed at coasts within a range of about 750 km from the epicentre of sub-ocean earthquakes (Figure 104).

Tsunami are as unpredictable as earthquakes. In principle, they can be accounted for in a design, although use of a 'no-damage' criterion may not lead to an economically optimal solution.

4.2.1.8 Flood waves

In river mouths, flood waves may contribute to the instantaneous water level. However, this component is generally of minor importance in comparison with tidal influences and is limited to the direct vicinity of the river mouth. Moving further upstream into the river, the flood wave component eventually becomes dominant, but here bank structure design is generally governed mainly by currents and not by waves, and hence is outside the scope of this manual.

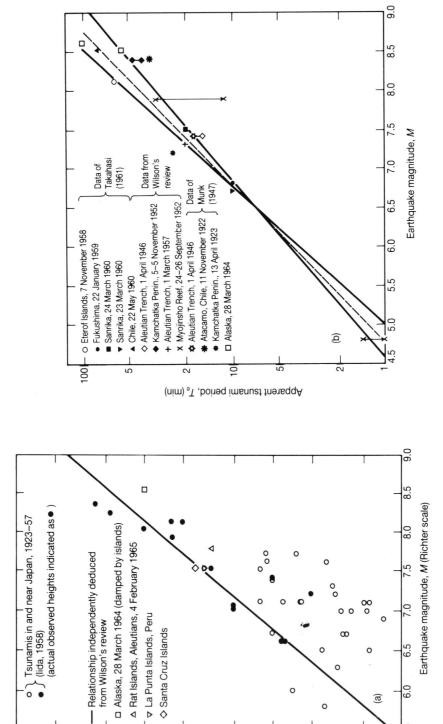

Figure 104 (a) Heights of tsunamis to be expected at a coast from submarine earthquakes of various magnitudes within a range of about 500 miles (mainly based on Japanese data); (b) relationship between the dominant period of a tsunami and the magnitude of the earthquake giving rise to it

4.2.1.9 Design extreme water level

To determine the extreme design water level, all components must be determined as a function of the (average) probability of exceedance (alternatively expressed as *average* exceedance frequency or return period). These exceedance curves, such as that shown in Figure 105, are based upon a *long-term distribution* curve, obtained by fitting water level data to a standard statistical distribution (see Box 41). Unfortunately, the lack of data for low frequencies (long return periods) means that extrapolation is usually necessary. The extrapolation can be checked or supported by numerical modelling of the underlying physical processes to give a better understanding of the results.

Once exceedance curves for all components are found, the next step to be made is

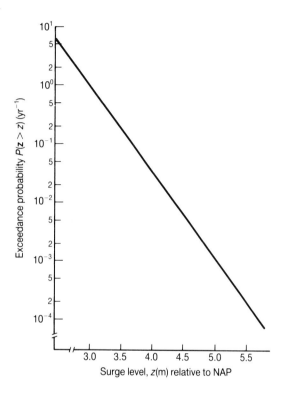

Figure 105 *Example of storm surge levels for a location on the Dutch coast*

Box 41 Theoretical extreme value probability distributions

The following theoretical extreme value probability distributions are commonly used for fitting to long-term distributions of wave heigher, water levels, etc.:

Gumbel $P(X \leq x) = \exp[-\exp(-Ax + B)]$

Weibull $P(X \leq x) = 1 - \exp\left[-\left(\frac{x-A}{B}\right)^c\right]$

Log-normal $p(x) = 1/(Ax\sqrt{\pi}) \exp\left[-\left(\frac{\ln x - B}{A}\right)^2\right]$

Exponential $P(X \leq x) = \exp[-(x-A)/B]$

where $P(X \leq x)$ is the probability distribution function, i.e. the probability that X will not exceed x, $p(x)$ is the probability density function of x and $p(x) = dP/dx$.

to compose the combined water level so that the frequencies of the components combine into the desired exceedance frequency. In the worst case, when all components are fully dependent, they occur simultaneously and the design level is simply obtained from the sum of all components corresponding to this frequency. Possible reductions in the severity of the design level with respect to this worst case will arise from the degree of mutual correlation between the water level components. Analysis of the underlying physical processes (Figure 99) will permit some assessment of possible dependencies, and eventually enable a joint probability analysis of the water level effects (Section 4.2.8) to be carried out to produce an exceedance curve for the resultant water level.

Very limited data are often available on the separate component effects leading to extreme water levels. Here a simple empirical approach that may be usefully adopted is the method of analysis of annual extreme water levels, data for which are often readily available from national and local ports and published sources (e.g. Graff, 1981; Blackman, 1985). The method simply involves fitting the annual extreme water level data to an appropriate long-term distribution (see Box 42. Its limitation is its sensitivity to outliers, but the overall approach is justified by the rarity of any correlation between astronomic tide and meteorologic surge. Having determined the probability distribution of water levels, economic optimisation (Section 2.1.7) may then be used to select the appropriate extreme design level and the corresponding risk.

Sometimes there are very few data on water levels for the site of interest but they may be available for neighbouring or comparable locations. Correlation techniques (interpolation and extrapolation), are a useful way of translating existing data from neighbouring locations to the site of interest. Although these methods may save time and cost, these should be applied with care. Wind effects, (strong) currents and differences in wave set-up will invalidate use of these methods. Correlation factors between locations can be derived or verified from a limited number of simultaneous measurements, both at the site for which the main data source exists and at those to be correlated with it. Erroneous design data may result from the use of such correlations to water level that are beyond the range of the verification.

Box 42 Simple approach to correlating extreme water levels

An approach to deriving a first estimate of a probability distribution of extreme water levels for a site for which there are only basic astronomic tidal information is to correlate this site with one nearby for which both tidal data and extreme water level predictions are available. Correlation is then achieved by assuming (Graff, 1981) that the following ratio is the same for the two sites:

$$\frac{\text{Extreme level} - \text{mean high water spring (MHWS) level}}{\text{Spring tide range (MHWS} - \text{MLWS)}}$$

Where available and appropriate, a slightly more accurate estimate can be achieved by replacing spring tidal range in the above ratio by the sum of the principal semi-diurnal tidal components, $M_2 + S_2$.

4.2.2 DEEP-WATER WAVE CONDITIONS

This section describes how waves can be described and estimated for design purposes in the 'offshore' situation. Here waves are not limited by water depths and are therefore generally described as 'deep-water' waves. Section 4.2.3 then describes the various shallow-water transformation processes occurring to these waves, which often ultimately determine the loading they impose on coastal rock structures. Such transformations are, however, often not required in the case of rock protection to reservoir dams where the water is sufficiently deep to preclude them, and here the wave conditions calculated as described in this section can be applied directly. Long-wave phenomena such as *tides*, *tsunami* and *storm surges*

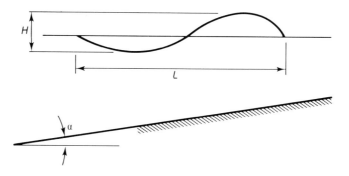

Figure 106 *Definition sketch for regular wave parameters*

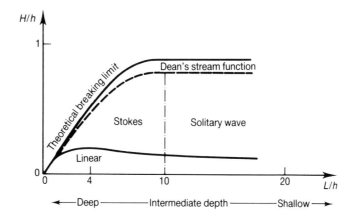

Figure 107 *Applicability range for wave theories*

are, for engineering purposes, treated as quasi-static water level components rather than as waves (Sections 4.2.1.2, 4.2.1.7 and 4.2.1.3, respectively).

The principal loadings exerted by waves upon marine structures can be subdivided into wave impact loads (short-duration pressures), drag loads (pressure plus shear) and inertia loads. In the design process these loadings can be schematised by describing them as functions of the wave parameters. The basic wave parameters are wave height (H), wave period T), wave length (L), propagation direction with respect to true North (θ) or to the direction normal to a structure (β), and velocity of propagation ($c = L/T$). A definition sketch for regular waves is given in Figure 106. For values in deep water subscript 'o' is used conventionally (e.g. H_o).

Many wave theories are now available to derive other wave parameters from the basic parameters described aboved. The majority of design methods are based on Stokes' linear wave theory. A major advantage of linear theory in design procedures is that the principle of superposition can be applied to wave-related data, obtained from a composite wave field. Using linear wave theory, practical engineering approximations have been derived for regular waves propagating in deep and shallow water, respectively. The validity range of these approximations can be given in terms of the relative water depth (h/L_0). Deep- and shallow-water approximations may be used for $h/L_0 > 1/4$ and for $h/L_0 < 1/20$, respectively.

Mathematically, linear wave theory becomes less applicable when the shape of the waves deviates from pure sinusoidal and when wave steepness ($s = H/L$) increases. Other wave theories that may be used in these situations include Stokes' higher-order wave theories, cnoidal wave theory, solitary wave theory and Dean's Stream Function theory. The description of these theories is beyond the scope of this manual but can be found in relevant textbooks and in the literature (e.g. *SPM*, 1984). An overview of the applicability of wave theories is given in Figure 107.

4.2.2.1 Statistical properties of waves

Wave statistics play the major role in determination of design loads and risk assessment and thus in the overall design of rock structures in coastal and shoreline engineering. Some explanation of basic statistical wave properties is therefore necessary.

Wind-generated waves are irregular (non-periodic), and a typical record of sea surface elevation is shown in Figure 108. This figure also illustrates the definition of the 'zero-crossing' individual wave period T_z which identifies individual waves, and the corresponding 'lowest trough' to 'highest crest' height which defines the height, H, of that individual wave.

The standard recording period is 20 minutes and is generally taken to represent wave conditions over a 3-hour period, during which the conditions are assumed to be statistically 'stationary'. From the records, a number of commonly used characterisations of the wave climate (IAHR/PIANC, 1986) may be determined, including the significant wave height H_s (defined as the mean of the highest one-third of wave heights in a record or as $4\sqrt{m_o}$ from the spectrum—see Table 28 in Section 4.2.2.5) and the mean wave period, T_m (defined as the average of all zero-crossing wave periods, T_z in the record) or the spectral peak wave period, T_p (see Section 4.2.2.5). For deep-water conditions a storm can be characterised by H_s and T_p, assuming a statistically stationary sea state during the storm. For shallow-water conditions, this assumptions can be unrealistic due to variations in water level caused by tidal and/or set-up effects.

During each stationary (3-hour) wave event or sea state a short-term wave height distribution applies. In deep water the water surface elevation follows a Gaussian process, and thus the wave heights closely follow the Rayleigh distribution. In shallow water, by contrast, the wave height distribution is affected by wave breaking (Section 4.2.3.4).

Figure 109 shows (on a linear-log scale) the Rayleigh distribution

$$P(\mathbf{H} < H) = 1 - \exp-(H^2/8m_o)$$

where m_o is obtained from the wave spectrum (see Section 4.2.2.5). An example is also given in Figure 109 of a shallow-water wave height distribution, which shows a typical shallow-water deviation from the Rayleigh line and the limitation of wave heights due to breaking is clearly seen. For design purposes, wave heights in an individual sea state can be related to a certain probability of exceedance (e.g. $H_{2\%}$, the wave height exceeded by 2% of the waves in record). From Figure 109 it can be seen that, for example, H_s is related to the mean wave height ($\bar{H}$) by $H_s = 1.60\bar{H}$ (other principal relations based on the Rayleigh distribution are summarised in Section 4.2.2.5, Table 28).

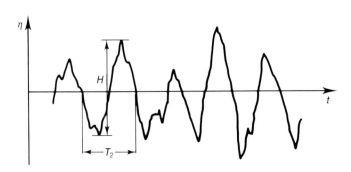

Figure 108 *Typical record of irregular sea*

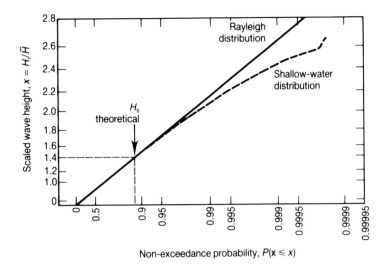

Figure 109 *Rayleigh distribution of wave heights compared with a shallow-water distribution*

The Rayleigh distribution requires determination of only one parameter or reference wave height, which can be found by applying a spectral analysis (Section 4.2.2.4) to a digital surface elevation record (Priestley, 1981). If only analogue traces are available, the Draper/Tucker method may be used to estimate H_s and T_m (Draper, 1963).

Where wave grouping effects occur such that high waves appear in groups rather than individually the Rayleigh distribution will not apply, the grouping of the waves implying that individual wave heights are *not* uncorrelated. Wave grouping is often associated with long period waves or swell, but the phenomena are different. (The presence of swell can be identified by the existence of a double-peaked wave spectrum.) However in practice, swell (and also other long period waves, caused by secondary effects within the sea itself may cause an apparent wave grouping effect as may the combination of waves from two directions. Kimmra (1980) and Goda (1976) provide a statistical approach to assessing the phenomena of wave grouping. For design purposes, wave grouping may be important when considering such phenomena as wave run-up and overtopping where sequences of individual large waves can create a cumulative severe effect that would not otherwise arise.

4.2.2.2 Long-term distribution of wave height and period

Individual waves (see Figure 108) show no significant relation between height and period, but when characteristic values of individual sea states are compared (for instance, H_s and T_m or T_p) such relations become significant, and a so-called long-term distribution can be established in the form of a scatter diagram (Figure 110). As wave height and period determine the major design loadings on rock structures in coastal and shoreline engineering, joint probability long-term distributions of H_s and T_m are required for subsequent extreme value analyses. A theoretical description of a joint distribution function, $P(H_s, T_m)$, for individual waves has been given by Longuet-Higgins (1975). Coefficients used in this function must be obtained from the wave frequency spectrum (Section 4.2.2.5). In practice, a joint distribution can also be derived directly from wave measurements. Wave heights and periods determined from a series of 3-hourly recordings can be presented in a scatter diagram which gives the fraction of waves found within each of a number of predefined classes of H and T_m (see Figure 110). The scatter diagram is created by counting over a long period of time, covering a range of storm conditions, the total number of individual sea states falling within classes ΔH_s

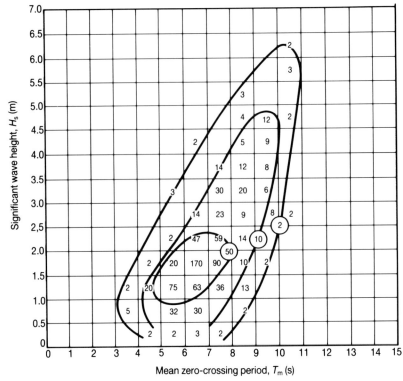

Note: Figures in circles on heavy contour lines indicate percentage of observations falling outside a particular contour line.

Figure 110 *Scatter diagram indicating joint distribution of H_s and T_m*

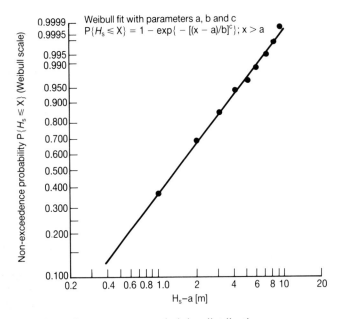

Figure 111 *Weibull fit to long-term wave height distribution*

and ΔT_m. Division by the total number of sea states gives an estimate of the two-dimensional (H_s, T_m) distribution function.

Extrapolation of the validity of a distribution obtained from a scatter diagram beyond the range covered by the measurement (H_s, T_m and β_0) must be done with care. However, this is generally the only way of predicting low-frequency (long term period) sea state parameters. The procedure adopted is to fit to a theoretical extreme value distribution (see Box 41 and Figure 111). It should be noted that the distribution may deviate from the ideal where wave breaking plays a role or when the data are mixed (e.g. from ordinary and tropical storms).

4.2.2.3 Visual wave observations

Visual observations of waves have been made for many years from ships, significant data banks of which are held by meteorological offices, and provide an estimate for the mean height of the highest one-third of the wave heights on each occasion when an observation was made. Observations made from ships may be the only available source of data, and are limited to height and period observations of locally generated wind waves, known by mariners as 'sea' (H_{v1}, T_{v1}), and of background swell (H_{v2}, T_{v2}). Comparisons of such information with instrumental data (Hogben *et al.*, 1986), providing a corresponding 'true' estimate of H_s and T_s, have shown that:

$$H_s \approx \sqrt{(H_{v1}^2 + H_{v2}^2)} \tag{4.5}$$

with good correlation, but that correlations between visually observed and measured wave periods are unreliable.

Scatter diagrams of such data can be used directly to determine long-term wave climate distributions and extreme wave events, but clearly the reliability of the wave period estimates and predictions is limited. Improved distributions can be obtained using computer smoothing programs such as British Maritime Technology's NMIMET (Hogben *et al.*, 1986) or the British Meteorological Office's METWAVE, which are based on a technique of correlating the simultaneous wave height and wind speed observations.

4.2.2.4 Wind statistics

Where wave data are not available but can be obtained, hindcasting from wind records may be the only way to achieve an estimate of the design wave conditions. Wind speeds (usually hourly mean values of speed and direction) may be obtained from local permanent meteorological stations, which may be situated in harbours and near airports. It should be noted that on land, records may easily show 10–20% reduction of wind speeds (due to increased surface roughness) compared to values measured over water.

Records may also be available from ship observations, and these may form a useful starting data set for design, while better records are obtained or a measurement programme is carried out as part of a project-related survey. Depending on the extent of the record, wind speed distribution curves can be derived for directional sectors of (say) 30° or for a predefined dominant direction. Corresponding wave parameters can be found by using empirical wave growth formulae (Section 4.2.2.6) or numerical simulation models (Section 4.2.9). For wave prediction the wind speed (U_w) at 10 m above the sea surface (U_{10}) is commonly used as an input parameter. Measurements at other levels might be correlated to U_{10} by using a vertical wind speed distribution function with respect to the height above the sea surface:

$$U_w(z)/U_{10} = (z/10)^{1/7} \tag{4.6}$$

When no wind records are available, synoptic weather charts filed on a daily basis by national meteorological offices may be used with specialist advice to reconstruct historical (day mean) wind speeds.

Duration of wind speeds is also important when deriving wave parameters (Section 4.2.2.6) or carrying out joint probability analyses (Section 4.2.8). For this purpose, available data on mean wind speeds for a certain duration may have to

Table 27 Wind speed (U_w) conversion factors related to duration of wind speed

Time base (hours)	Factor (−)
$\frac{1}{4}$	1.05
$\frac{1}{2}$	1.03
1	1.00
3	0.96
6	0.93
12	0.87
24	0.80

be transformed to mean wind speeds on a different time base. This transformation can be achieved using ratios of the factors in Table 27.

4.2.2.5 Spectral description of waves

In general, an observed wave field can be decomposed into a number of individual wave components, each with their own frequency ($f = 1/T$) and direction (θ). The distribution of wave energy as a function of the wave frequency is commonly presented by means of the one-dimensional wave energy density spectrum, denoted as $E_{\eta\eta}(f, \theta)$.

Examples of a spectrum are given in Figure 108, where the parameters are also shown, which are frequently used to describe random wave fields. These parameters will be described below. A variety of semi-empirical wave spectra have been presented, each having their specific range of applicability. Two of the most widely used are the spectrum described by Pierson and Moskowitz (1964) and JONSWAP (Hasselmann, 1973), both of which are shown in Figure 112. In the figure, the peak frequency (f_p) is a characteristic parameter which is defined as the frequency of peak energy in a wave spectrum. The peak period T_p is the inverse of f_p.

The *Pierson–Moskowitz* (P–M) spectrum represents a *fully developed sea* in deep water. The *JONSWAP* (J) spectrum represents *fetch-limited sea* states, i.e. growing

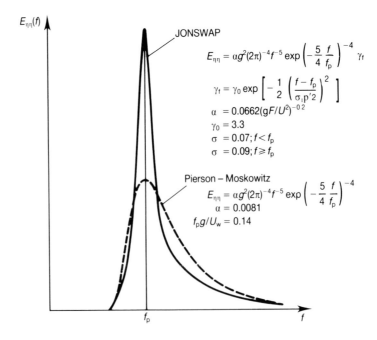

Figure 112 *Pierson–Moskowitz and JONSWAP spectra*

Table 28 Estimates and relation of wave parameters in a stationary (Gaussian) sea state

Characteristic height		H/H_{rms}	$H/\sqrt{m_o}$	H/H_s
Standard deviation of free surface	$\sigma_\eta = \sqrt{m_o}$	$1/\sqrt{2}$	1.0	0.250
Root-mean-square height	H_{rms}	1.0	$2\sqrt{2}$	0.706
Mode	$\mu(H)$	$1/\sqrt{2}$	2.0	0.499
Median height	H such that $P(\underline{H} < H) = 0.5$			
Mean height	$\bar{H} = H_1$	$\sqrt{\ln 2}$	$\sqrt{8 \ln 2}$	0.588
Significant height	$H_s = H_{1/2}$	1.416	4.005	1.000
Average of tenth highest waves	$H_{1/10}$	1.80	5.091	1.271
Average of hundredth highest waves	$H_{1/100}$	2.359	6.672	1.666

Characteristic period		Estimate related to spectral moments
Mean period	$\mu_T = T_m$	$\sqrt{(m_o/m_2)}$
Peak period of spectrum	T_p	$T_p = 1/f_p$

sea. The spectra are formulated using a power function with respect to the frequency f ($=1/T$), or angular frequency $\omega = 2\pi/T$) containing several scaling parameters and constants. The *TMA* spectrum has been developed more recently (Hughes, 1984) to cover both *fetch-limited sea* and *shallow water* effects. This spectrum basically consists of factors originating from the Pierson–Moskowitz and JONSWAP spectra. Additionally, a factor (Φ) has been introduced to describe the effect of water depth. This results in the following spectral description, considered to be the most complete presently available:

$$E_{TMA} = \underbrace{\alpha g^2 (2\pi)^{-4} f^{-5}}_{\text{P-M and J}} \quad \underbrace{\Phi(2\pi f, h)}_{\text{TMA}} \quad \underbrace{\exp[-5/4(f/f_m)^{-4}]}_{\text{P-M and J}} \quad \underbrace{\gamma_o^{\exp[-(f/f_m - 1)^2/2\sigma^2]}}_{\text{J}} \tag{4.7}$$

where

$\alpha = 0.076 \, (gF/U_w^2)^{-0.22}$ in TMA; $(0.0662(gF/U_w^2)^{-0.2}$ in J; 0.0081 in P–M),
$\gamma_o \approx 3.3$,
$\sigma \approx 0.07$ for $f < f_m$ and $\sigma \approx 0.09$ for $f \geq f_m$,
$f_m \approx 0.14 \, g/U_{19.5}$ (practically this is a wind-dependent f_p),
U_w = wind speed at a height z (Section 4.2.2.6),
  Φ describes the influence of the water depth and can be

approximated to within 4% by:

$$\Phi = \begin{cases} 2\pi^2 f^2 h/g & \text{for } f \leq 1/[2\pi\sqrt{(h/g)}] \\ 1 - 1/2[2 - 2\pi f \sqrt{(h/g)}]^2 & \text{for } f > 1/[2\pi\sqrt{(h/g)}] \end{cases} \tag{4.8}$$

Estimation of wave parameters from a spectrum

For convenience the *n*th moment of a one-dimensional spectrum $E_{\eta\eta}(f)$ is defined as:

$$m_n = \int_{f=0}^{\infty} f^n E_{\eta\eta}(f) \, df; \qquad n = 0, 1, 2, \ldots. \tag{4.9}$$

For example, m_0 represents the area enclosed by the spectrum, which is obtained by integration of the spectrum $E_{\eta\eta}(f)$ with respect to f (selecting $n=0$ in equation (4.9)). Assuming the water surface elevation follows a stationary Gaussian process, and hence the heights are Rayleigh distributed, most of the principal wave field or sea state parameters can be expressed in terms of these moments, as seen in Table 28. This table also gives some practical conversion factors for height parameters of

Rayleigh-distributed waves. The factor '4' in the relationship between H_s and $\sqrt{m_o}$ is a theoretical value and, practically, analysis of real records show values down to 3.6.

Wave periods can be characterised by T_p, T_m and T_s (T_s is defined similarly to H_s). Wave periods do not allow conversion with factors related to a distribution function. Instead, such factors depend on the spectral shape. Wave data analysis by Goda (1979) has revealed a range for various conversion factors. For example, $T_m/T_p = 0.71$–0.82 for a P–M spectrum and 0.79–0.87 for a JONSWAP spectrum. Generally, $T_s/T_p = 0.90$–0.96 and $T_s/T_m = 1.13$–1.33 was found to apply. From the analysis it was concluded that T_s is a more reliable characterization than T_m for wind-generated waves. On the other hand T_m is more reliable for simulation of waves in physical models.

4.2.2.6 Sea development during storm and wave forecasting

The relations given in the previous section can only be used when a stationary sea state can be assumed. However, during storm conditions this does not apply, as the wave field is developing under the influence of wind shear upon the water. A wave field under the influence of wind is called 'sea', as opposed to a wave field that is not exposed to wind, which is known as 'swell'. In general, swell can be related to distant storms and is characterised by a narrow spectrum of relatively low frequencies.

Like individual waves under the influence of wind, the wave spectrum also shows an evolution with time. Wind-induced wave growth has traditionally been described by using empirical formulae. The approach of determining sea growth through analyses of the wave spectrum has been developed, but empirical methods still play an important role. A great effort has been made on *numerical wave forecasting* by meteorological and hydraulic institutes. With these numerical models the instantaneous wave parameters are derived from the energy content of the wave field, which is computed as being distributed over the frequencies within the wave spectrum. The necessary input for these models are wind fields, obtained from synoptic weather charts. The models take account of energy gain due to transfer from wind energy and energy loss due to dissipation (breaking, bottom friction). More advanced models have also included the interaction and energy transfer among individual wave components of the frequency spectrum.

Empirical wave growth formulae are based upon the relations between the 'characteristic' wave in the 'standard wind field'. This wind field is given through the average wind speed (U_w, see Section 4.2.2.4): U_{10}), the fetch (F, the distance to the coast, measured in the upwind direction) and the duration (t) of the wind field. An additional characteristic parameter is the water depth (h), which is usually assumed constant.

A fully developed sea has, for a given U_w, reached its maximum wave height and period. In a fetch-limited or growing sea at least one of the parameters F, t or h poses a limiting condition to the actual sea as long as they have not reached a certain minimum value beyond which the limiting condition vanishes. Using g/U_w^2 and g/U_w as scale factors for H, F and h and for t and T, respectively, the empirical wave growth formulae are usually written in a dimensionless form as $\tilde{H} = F_n\{\tilde{h}, \tilde{F}, \tilde{t}\}$ and $\tilde{T} = F_n\{\tilde{h}, \tilde{F}, \tilde{t}\}$. These relations are given in the *SPM* (1978), which also addresses determination of an effective fetch (fetch reduction due to fetch width). For design purpose, diagrams have been prepared to estimate H_s and T_s of the expected wave field, at a given fetch (F) and wind speed (U_w). The diagrams are given in Figures 113 (deep water) Green and Dorrestein (1976), and 114 (shallow water) from Bretschneider (1954), respectively. The reader should note that the latest (4th) edition of *SPM* (1984) contains new wave-prediction curves,

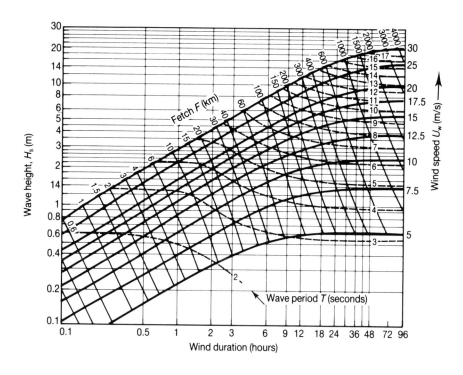

Figure 113 *Deep water wave forecasting diagram for a standard wind field*

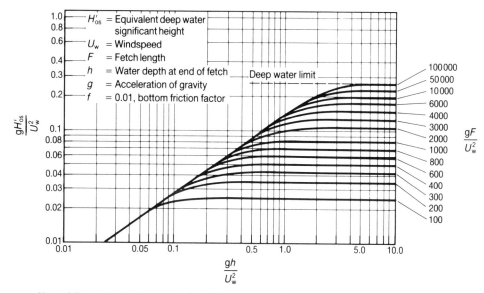

Note: A final estimate of wave period in shallow water maybe made from Figures 113 or 116

Figure 114 *Shallow-water wave forecasting diagram in a standard wind field. Note: A first estimate of wave period in shallow water may be made from Figures 113 and 116*

based on an intermediate calculation of wind stress, whose reliability for all situations has recently been questioned, particularly for extreme events.

Spectral methods (Section 4.2.2.5) also offer means to estimate sea growth. Suitable prediction curves for wave height and period, based on the JONSWAP spectrum, are given in Figures 115 and 116. It should be noted that waves thus obtained must be modified, if appropriate, to take account of possible shallow-water effects (Section 4.2.3).

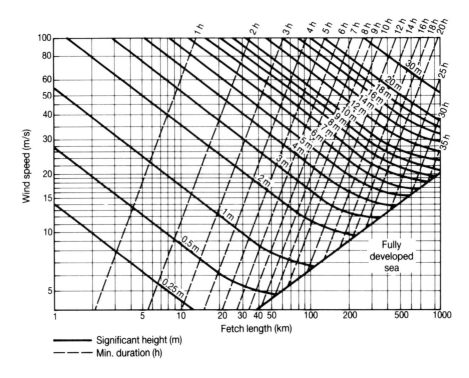

Figure 115 *Prediction of significant wave height from wind speed for JONSWAP spectrum*

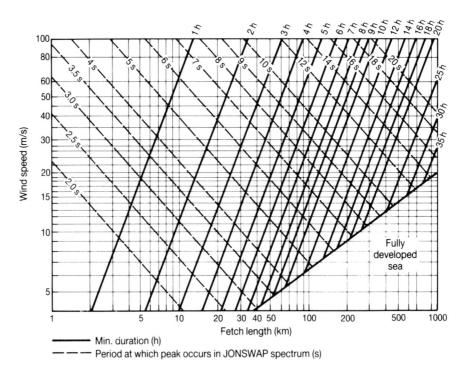

Figure 116 *Prediction of peak wave period from wind speed for JONSWAP spectrum*

4.2.2.7 Wave forecasting in inland waters and reservoirs

Methods of prediction of waves on inland waters have not been extensively studied or documented. The majority of all wave measurements (and from these the prediction methods) have been carried out in open-sea conditions. However, it is a particularly important consideration where reservoirs are concerned, as construction of the appropriate wave protection must be complete in advance of the filling of the reservoir, and hence direct measurement of wave climate is excluded from the design process.

Measurements obtained recently on UK reservoirs (Owen, 1988) showed that none of these methods gave particularly good agreement for all conditions. The traditional Saville or the Donelan–JONSWAP methods both gave reasonable results, and it is suggested that these should be adopted for small and medium lakes and reservoirs. For very large fetches, open-sea methods are probably the best to apply. This section describes four methods commonly adopted for wave prediction on inland waters.

Saville method

This uses SMB wave-prediction curves for open waters, and adapts them to reservoirs using the concept of effective fetch (Saville *et al.*, 1962). The SMB curves can be represented by the formulae:

$$\tilde{H}_s = 0.283 \tanh(0.0125 \tilde{F}^{0.42}) \tag{4.10}$$

and

$$\tilde{T}_m = 7.54 \tanh(0.077 \tilde{F}^{0.25}) \tag{4.11}$$

where the dimensionless parameters referred to in Section 4.2.2.5 above in this case are:

$$\tilde{H}_s = gH_s/U_w^2 \tag{4.12}$$

$$\tilde{F} = gF/U_w^2 \tag{4.13}$$

$$\tilde{T}_m = gT_m/U_w \tag{4.14}$$

and the definition of the effective fetch is illustrated in Figure 117. A noticeable feature is that the effective fetch is independent of wind speed. Saville's effective fetch should *not* be used with an wave prediction formulae other than SMB: serious underestimates of wave height will result otherwise.

JONSWAP method

This uses the JONSWAP curves (Figures 115 and 116) in exactly the same way as for open water, which can be represented by the formulae:

$$\tilde{H}_s = 0.00178 \tilde{F}^{0.5} \tag{4.15}$$

$$\tilde{T}_p = 0.352 \tilde{F}^{0.3} \tag{4.16}$$

where

$$\tilde{T}_p = gT_p/U_w \tag{4.17}$$

With a typical JONSWAP spectrum:

$$\tilde{T}_m = 0.87 \tilde{T}_p \tag{4.18}$$

and therefore

$$\tilde{T}_m = 0.306 \tilde{F}^{0.3} \tag{4.19}$$

Seymour's method

This uses the JONSWAP curves but adapts them for fetches of restricted width using a different concept of effective fetch (Seymour, 1979). The method is

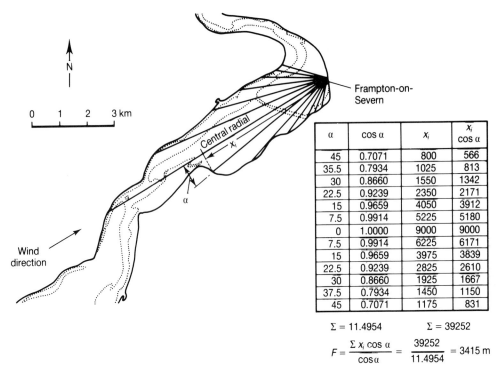

Figure 117 *Example calculation of effective fetch length by Saville's method*

sometimes referred to as the JONSEY method (JON(SWAP)SEY(MOUR)). Seymour's expression for wave height becomes:

$$H_s = \sqrt{E} \qquad (4.20)$$

where

$$E = (2/\pi) \sum_i E_i \cos^2(\phi_i - \phi_w) \Delta\phi \qquad (4.21)$$

In this expression E_i is the wave energy ($H_s^2/4$) generated along a fetch direction i_i by the selected wind speed U_w acting over a fetch length F_i, and $\Delta\phi$ is the increment of fetch directions (typically, 5–10°). The term $\cos^2(\phi_i - \phi_w)$ is the wave energy directional spreading function, and the summation is carried out over the range $\phi_i - \phi_w = 90°$.

Donelan–JONSWAP method

This also uses the JONSWAP wave-prediction curves, but combines them with Donelan's concept that the fetch length should be measured along the wave direction (θ) rather than the wind direction (ϕ), and that the wind speed used for wave prediction should therefore be the component along the wave direction (Owen, 1988). This presupposes that the wave direction is known or can be calculated. For long, narrow water bodies the wave direction will probably be along the water body axis for a wide range of wind directions. For fetches of general shape, Donelan assumed that the predominant wave direction was that which produces the maximum value of wave period (for a given wind speed). From the JONSWAP formula, the maximum wave period is obtained when the product:

$$\cos(\phi_w - \theta)^{0.4} F_\theta^{0.3}$$

reaches its peak within the range $/\phi_w - \theta/ = 90°$. Here θ is the predominant wave direction and F_θ is the fetch along that direction. For any irregular shoreline, and a given wind direction, the value of θ satisfying this condition can only be determined by trial and error. However, θ is independent of wind speed, so only one set of calculations is needed for a particular water body. Once θ has been determined, the wave height and period are derived from the modified JONSWAP formulae:

$$\tilde{H}_{s\theta} = 0.00178 \tilde{F}^{0.5} \qquad (4.22)$$

$$\tilde{T}_{p\theta} = 0.352 \tilde{F}_{\theta}^{0.3} \qquad (4.23)$$

where

$$\tilde{H}_{s\theta} = gH_s/[U^2 \cos^2(\phi_w - \theta)] \qquad (4.24)$$

$$\tilde{T}_{p\theta} = gT_p/[U \cos(\phi_w - \theta)] \qquad (4.25)$$

$$\tilde{F}_{\theta} = gF/[U^2 \cos^2(\phi_w - \theta)] \qquad (4.26)$$

4.2.2.8 Deep-water wave breaking and diffraction

Besides wave generation by wind, very few processes need to be taken into account when waves are in deep water. When designing for waves in shallow water, more influences have to be accounted for (Section 4.2.3). In deep water only three principal factors may need to be considered:

1. Deep-water wave breaking. Waves in deep water may break when a certain limiting wave steepness ($s = H/L$) is exceeded. For *regular waves* in deep water the wave height is limited by steepness according to the breaking criterion of Miche (1944) to about one-seventh of the wavelength. In practice, only individual waves in a random sea approach this value, and calculated steepnesses using significant wave height and peak wave period rarely exceed one-twentieth. This factor is, of course, implicit in the empirical wave growth formulae.
2. Wave reflection and diffraction by islands, rocks, etc., information on which is provided in Sections 4.2.3.3 and 4.2.3.6.
3. Energy transfer between frequencies. Energy transfer from higher to lower frequencies is the cause of swell, which may be observed at large distances from the area of wave generation. In deep water, mutual interactions between wave frequencies may cause energy transfer towards higher frequencies. Both processes can only be predicted by the use of advanced wave models, although swell can, of course, be measured directly.

4.2.3 TRANSFORMATION OF WAVES IN SHALLOW WATER

In coastal structures the effects of reducing water depths and coastal forms on the incoming (deep-water) waves (Section 4.2.2) have to be considered. These factors lead to transformation of the incoming waves (refraction—Section 4.2.3.1; shoaling—Section 4.2.3.2; diffraction—Section 4.2.3.3) and eventually to breaking (Section 4.2.3.4), as illustrated in Figure 118. Wave breaking results in significant dissipation of energy and may be a major factor limiting the design wave height and, consequently, the loading on the structure. All these phenomena are a function of water depth, and hence a proper description of bathymetry is required. Graphs and formulae are presented here and elsewhere (e.g. *SPM*, 1984) to enable the designer to make a first assessment of the influence of these phenomena. However, for a proper spatial description of wave parameters, appropriate numerical models of wave propagation should be used.

For the identification of many phenomena in shallow water, such as wave breaking, a useful parameter is the surf similarity parameter ξ (also known as the Iribarren number, Ir). ξ (or Ir) reflects the ratio of bed slope, m, and wave steepness, s ($= H/L$), and is defined as:

$$\xi = \tan(\alpha)/\sqrt{(H/L)} \qquad (4.27)$$

Figure 118 *Transformation of waves in shallow water (refraction, diffraction, shoaling and breaking) (courtesy HR Wallingford)*

4.2.3.1. Refraction

Refraction is the change in the wave-propagation velocity and, consequently, also in the direction of wave propagation, when waves propagate in varying water depth. As a result, the direction of wave incidence (β) relative to the structure inclines towards the direction normal to the local depth contours. This usually implies that the wave crests tend to become more parallel to the coastline when approaching more shalow water (Figure 119(a)). The corresponding change in wave height (relative to H_0), caused by redistribution of energy along the wave crests, is usually expressed in the refraction coefficient, K_R.

Applying linear wave theory to a regular wave with wave number k ($=2\pi/L$) and direction β_0 (or θ_0) in deep water, the local wave direction β (or θ_0) at a water depth h is found from:

$$\beta = \arcsin\{\sin(\beta_0)\tanh(k_h)\} \qquad (4.28)$$

and the corresponding refraction coefficient, K_R, from:

$$K_R = \sqrt{\{\cos(\beta_o)/\cos(\beta)\}} \qquad (4.29)$$

For irregular sea, a representative effective K_{Reff} value can be obtained by applying an averaging procedure to a range of relevant frequencies (f) (or corresponding wave number (k)) and directions (β_o). Thereby weight factors should be determined from the relative contributions $(\Delta E)_{ij}$ from intervals $(\Delta f, \Delta \beta)$ to the total energy content (E) of the spectrum.

Neglecting shoaling effects, this leads to the following procedure:

$$(\Delta E)_{ij} = (1/m_o)\iint_{\Delta f \, \Delta \beta} E_{\eta\eta}(f,\beta)\,d\beta\,df \qquad (4.30)$$

and

$$K_{Reff} = [\Sigma_i \Sigma_j (\Delta E)_{ij}(K_R)_{ij}^2]^{1/2} \qquad (4.31)$$

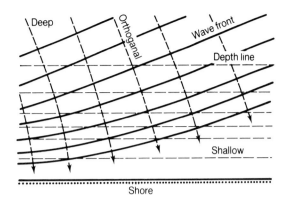

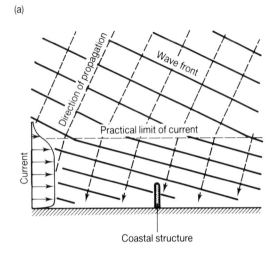

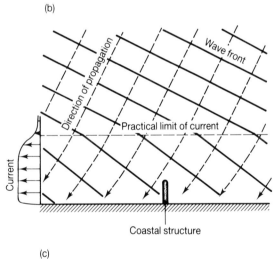

Figure 119 *Examples of wave refraction. (a) Depth refraction on a foreshore; (b) current refraction (opposing waves); (c) current refraction (following waves)*

For practical reasons, intervals Δf and $\Delta \beta$ of varying width are usually chosen, centred around a number of representative values for f and β, respectively (e.g. requiring equal energy contributions for each interval). Applying an extended form of this procedure, Goda (1985) has given diagrams (Figure 120) for K_{Reff} in a directional wave field on a coast with straight, parallel depth contours.

In addition to variations in water depth, the interaction of waves with current also causes refraction. This is due to the change of wave propagation velocity when waves are running into a current discussed in Section 4.2.3.7. Figures 119(b)

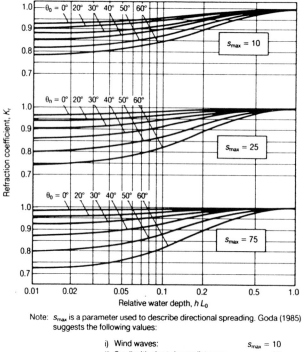

Note: s_{max} is a parameter used to describe directional spreading. Goda (1985) suggests the following values:

i) Wind waves: $s_{max} = 10$
ii) Swell with short decay distance: $s_{max} = 25$
(with relatively large wave steepness)
iii) Swell with long decay distance: $s_{max} = 75$
(with relatively small wave steepness)

Figure 120 K_R *for an irregular directional wave field on a coast with straight, parallel depth contours (Goda, 1985). Note: s_{max} is a parameter used to describe directional spreading. Goda (1985) suggests the following values:*

1. *Wind waves:* $s_{max} = 10$
2. *Swell with short decay distance:* $s_{max} = 25$
   (with relatively large wave steepness)
3. *Swell with long decay distance:* $s_{max} = 75$
   (with relatively small wave steepness)

and (c) show the effect of current refraction for waves opposing and following a local current with a certain inclination. Examples of this phenomenon may be found in tidal and longshore currents and near the mouths of rivers.

4.2.3.2 Shoaling

Shoaling is a change in wave height when waves propagate in varying water depths. The effect is normally expressed in terms of the shoaling coefficient, K_s, which is defined as the local wave height relative to H_o. Using linear wave theory, K_s can, for a given wave period, be written as a function of water depth:

$$K_s = \frac{1}{\{[1 + (2kh/\sinh(2kh))]\tanh(kh)\}^{0.5}} \tag{4.32}$$

Equation (4.32) gives, under the usual limitations related to the linear wave theory, appropriate estimates for design purpose. Non-linear effects may cause deviations from equation (4.32), values of approximately 10% being reported by Goda (1985). In shallow water depths wave breaking becomes increasingly important and wave models, accounting for breaking, should be used instead. Numerical wave models are treated in Section 4.2.8.3. A numerical model (ENDEC) has been used to generate the design graphs given in this section, in which the influence of both shoaling and wave breaking is included (van der Meer, 1990c, e). Tests have

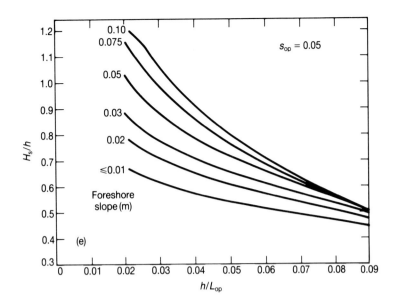
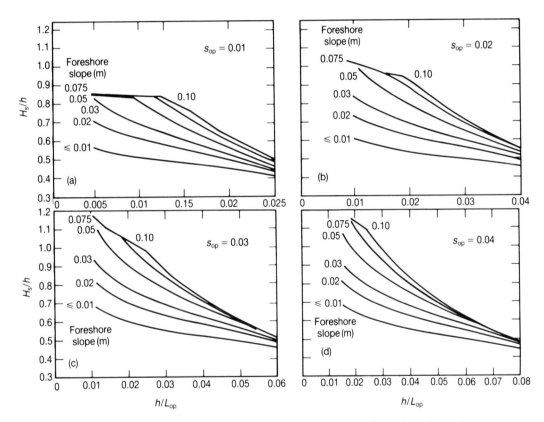

Figure 121 *Shallow-water significant wave heights for uniform foreshore slopes*

shown that wave height predictions using the design graphs from this model are accurate for slopes ranging from 1:10 to 1:100. For slopes flatter than 1:100, the predictions for the 1:100 slopes should be used. Wave breaking is discussed further in Section 4.2.3.4.

Shoaling on uniform slopes (with breaking)

In Figure 121 a series of example design graphs are given for uniform foreshore slopes. In order to use these graphs the following input values are required:

1. Local relative water depth, h/L_{op}, where $L_{op} = gT_p^2/2\pi$;
2. Slope of foreshore, m, with $m =$ tangent of slope angle, curves being given for the range $m = 0.075$–0.01 (1:13–1:100);

3. Deep-water wave steepness, $s_{op} = H_{os}/L_{op}$, curves are given for $s_{op} = 0.01$, 0.02, 0.03, 0.04, 0.05.

The output from the graphs is the local relative wave height: H_s/h.

Additionally, the effect (through refraction) of wave incidence (β_o) may be computed from Figure 122 for $\beta_o = 30°$ and $50°$ and compared with normal incidence ($\beta_o = 0°$). This can be done for combinations of $s_{op} = 0.01$ or 0.05 and $m = 1:13$ or $1:50$. Refraction is treated in Section 4.2.3.1.

Alternative graphs and formulae to determine the effect of both shoaling and wave breaking are presented by Goda (1985) for uniform foreshore slopes which have been adopted widely (see e.g. BS 6349: Part 1). The formulae are summarised in Box 43.

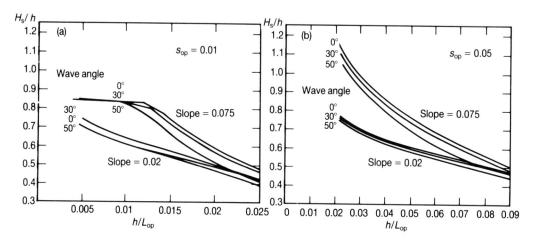

Figure 122 *Effect of angle of wave incidence on shallow water significant wave height for uniform foreshore slopes*

Box 43 Formulae due to Goda (1985) for wave height estimation in the surf zone

$$H_s = \begin{cases} K_s H'_{os} & h/L_o \geq 0.2 \\ \min\{(\beta_0 H'_{os} + \beta_1 h), \beta_{max} H'_{os}, K_s H'_{os}\} & h/L_o < 0.2 \end{cases}$$

$$H_{max} \equiv H_{1/250} = \begin{cases} 1.8 K_s H'_{os} & h/L_o \geq 0.2 \\ \min\{(\beta_0^* H'_{os} + \beta_1^* h), \beta_{max}^* H'_{os}, 1.8 K_s H'_{os}\} & h/L_o < 0.2 \end{cases}$$

The symbol $\min\{a, b, c\}$ represents the minimum value among a, b and c. The coefficients $\beta_0, \beta_1, \ldots$ have been formulated as listed below.

Coefficients for $H_{1/3}$	Coefficients for H_{max}
$\beta_0 = 0.028(H'_o/L_o)^{-0.38} \exp[20\tan^{1.5}\Theta]$	$\beta_0^* = 0.052(H'_o/L_o)^{-0.38} \exp[20\tan^{1.5}\Theta]$
$\beta_1 = 0.52 \exp[4.2\tan\Theta]$	$\beta_1^* = 0.63 \exp[3.8\tan\Theta]$
$\beta_{max} = \max\{0.92, 0.32(H'_o/L_o)^{-0.29} \exp[2.4\tan\Theta]\}$	$\beta_{max}^* = \max[1.65, 0.53(H'_o/L_o)^{-0.29} \exp[2.4\tan\Theta]]$

Note: $\max\{a, b\}$ gives the larger of a or b.

The shoaling coefficient, K_s, is that obtained using linear wave theory (equation (4.30)).

Goda (1985) advises that his numerical formulae may overestimate wave heights by several per cent. In particular, for waves of steepness greater than 0.04, the formulae overestimate significant wave heights by at least 10% around the water depth, at which the value of $H_{1/3} = \beta_0 H'_o + \beta_1 h$ becomes equal to the value of $H_{1/3} = \beta_{max} H'_o$. A similar difference also appears for the case of H_{max}. Waves of large steepness may have a discontinuity in the estimated height of H_{max}, at the boundary $h/L_o = 0.2$. Caution should be exercised in applying Goda's formulae in regard to such differences and discontinuities.

Shoaling on composite slopes (with breaking)

There may be cases in which the foreshore slope cannot be regarded as uniform, because different slopes can be identified. An example is the type of profile characterised by a relatively wide and shallow foreshore with a gentle slope (m), with a steep sloping edge (slope m_o) towards deep water. This situation can be found when structures have to be designed on shoals or beaches (e.g. groynes). For this type of 'double-slope' profile a set of graphs has been generated using the ENDEC model to show the influence of the transition of slope angle. This transition depth (h_k, see Figure 123(a)) determines the edge of the shallow foreshore. Figures 123(b)–(e) show the relative wave heights H_s/H_o for a number of double-slope profiles and require the following input values:

1. Local relative water depth, h/L_{op}, along the X-axis;
2. Slopes of foreshore and edge, curves being given for foreshore slopes $m = 1:300$ and 1:50; and for edge slopes of $m_o = 1:30$ and 1:10.
3. Deep water wave steepness $S_{op} = H_{os}/L_{op}$, curves are given for $S_{op} = 0.01$ and 0.05.

For detailed wave prediction on composite profiles a numerical simulation is

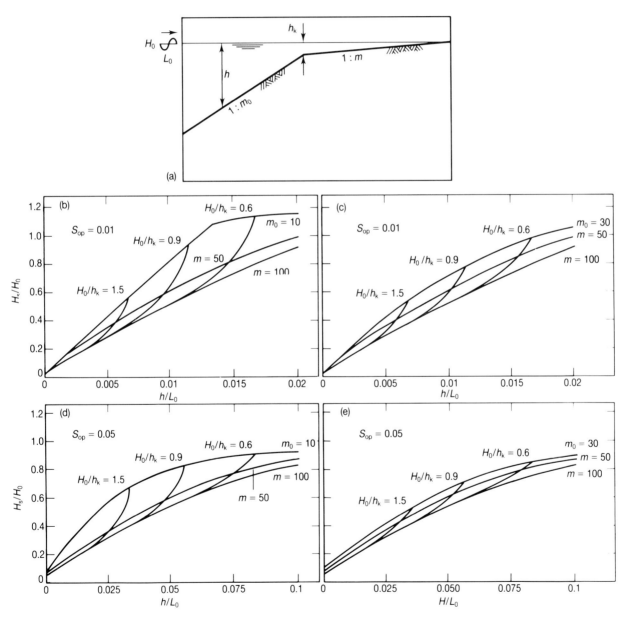

Figure 123 *Shallow-water significant wave heights for a two-slope foreshore profile*

preferable. This holds for all cases in which the required accuracy of the predicted wave heights does not allow for any of the profile schematisations given in this manual. If the edge slope commences further than two (local) wave lengths from the structure, the effect of this slope is much reduced and the uniform foreshore slope curves may provide an adequate prediction.

4.2.3.3 Diffraction

Propagating waves, impinging on obstacles such as piles, breakwaters, headlands and islands, interfere with these structures. The resulting wave field around the structures generally shows a marked change relative to the undisturbed wave field. The resulting wave field is a recomposition due to:

1. Incoming waves (partly attenuated by the structure);
2. Waves reflected by the structure;
3. Wave energy radiated from limiting points of the exposed part of the structure (according to the Huygens' principle).

The resulting change of wave height, relative to the original undisturbed wave, is expressed by the diffraction coefficient (K_d), and except for factor (1) above the amplitude of the components is, to a greate extent, determined by the reflection characteristics of the structure.

A diffraction analysis is often performed by using numerical models, as alternatives are only available for very simple geometries, i.e. alignment and cross-section of seawalls, breakwaters, groynes, etc. The principle of the numerical methods is the solution of the stationary Laplace equation for the wave velocity potential (Sommerfeld, 1896). Common boundary conditions applied are impermeable structures with a vertical wall. The assumptions made are usually constant water depth and small-amplitude waves (linear wave theory).

This principle can be used to construct normalised diagrams of the spatial distribution of diffracted curve conditions. These should, however, be based on random rather than regular waves, as the classic regular wave diagrams can lead to underestimation of wave heights. Using a procedure similar to that described in Section 4.2.3.1 for refraction, effective diffraction coefficients (K_d) can be calculated for a directional random wave field. Results (Goda, 1985) are shown in Figures 124–126 for normally incident waves on breakwaters. Each figure comprises four diagrams covering both locally generated wind waves ($S_{max}=10$) and almost unidirectional swell waves ($S_{max}=75$) in areas both near to and distant from the breakwater(s). Figure 124 gives the results for a semi-infinite breakwater and Figures 125 and 126 show those for a gap in an infinite breakwater. In the gap case, the leading parameter is the relative gap width (B/L, $L=$ local wave length at the gap). Two cases are shown: in Figure 125 $B/L=1$ and in Figure 126 $B/L=4$. For B/L greater than about 5, both breakwater parts interfere with the waves independently and superposition of results for each breakwater considered separately (using Figure 124) gives a better approximation for K_d. For oblique angles of incidence (β relative to normal) the diagrams can still be used when the breakwater is rotated, maintaining the original co-ordinates, but with B reduced to an imaginary width of $B\cos\beta$.

In conclusion, it should be appreciated that Figures 124–126 can only be used for a rough estimate of K_d when the conditions do not differ too much from those applying to the cases given. It should also be stressed that the regular wave diffraction diagrams in *SPM* (1984) should not generally be used, as these tend to underestimate K_d and hence also wave heights.

4.2.3.4 Wave breaking

Wave breaking occurs mainly due to the two criteria—depth or steepness, each

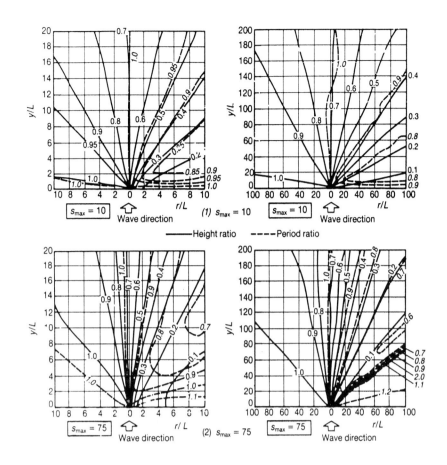

Figure 124 *Diffraction diagrams for a semi-infinite breakwater for random waves of normal incidence. Solid lines = wave-height ratio; dashed lines = wave-period ratio*

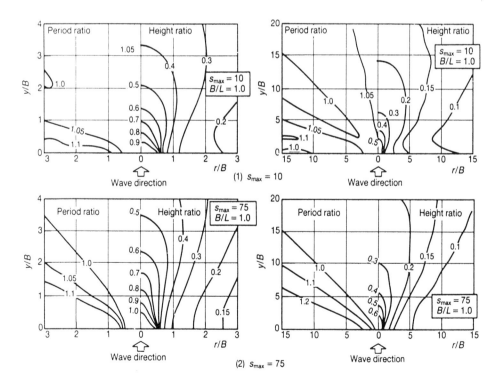

Figure 125 *Diffraction diagrams for a breakwater opening with $B/L = 1.0$ for random waves of normal incidence*

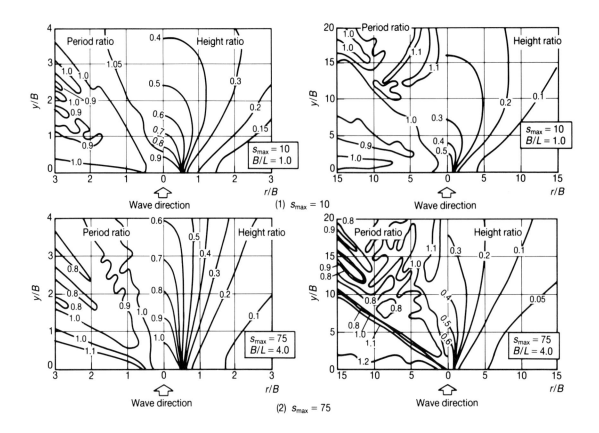

Figure 126 *Diffraction diagrams for a breakwater opening with B/L = 4.0 for random waves of normal incidence*

limiting the maximum wave height. While depth-induced breaking is usually the determining factor in shallow water, the limit of steepness should still be considered.

Breaking due to exceedance of the steepness criterion has been mentioned in Section 4.2.2.8 as the main limiting factor in deep water. The steepness criterion is influenced by water depth, and where the water is not deep is given (Miche, 1944) for regular waves as:

$$H/L \leq [H/L]_{max} = 0.14 \tanh(2\pi h/L) \qquad (4.33)$$

The breaking criterion due to water depth is normally given by a useful non-dimensional parameter called the breaker index (γ_{br}), defined as the maximum wave height-to-depth ratio (H/h):

H limited by *depth*: $\qquad H/h \leq [H/h]_{max} = \gamma_{br} \qquad (4.34)$

γ_{br} has a theoretical value of 0.78 for regular waves. However, it should be noted that γ_{br} is not constant, but ranges roughly between 0.5 and 1.5, while for irregular waves (represented by H_s) typical values are found to be $\gamma_{br} = 0.5$–0.6. The actual limiting wave height ratio γ_{br} depends mainly on such parameters as ξ ($m/\sqrt{s}$ (see equation (4.27)) and may reach values as large as 1.5 for individual waves. Figure 127 gives a good impression of the relationship between γ_{br} and ξ_o and of the related scatter of the data. The form of the breaking waves is an important factor in the loading they impose (see Chapter 5) and also appears to be a function of ξ (see 141 in Chapter 5). The breaking wave form due to depth limitation ranges from spilling (see Figure 128), through plunging and collapsing to surging.

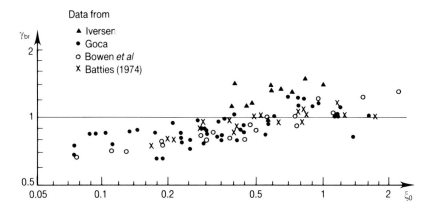

Figure 127 γ_{br} as a function of ξ

Design curves for the combined effect of wave shoaling and wave breaking, obtained from an energy-based wave model (ENDEC) developed by Delft Hydraulics, have been presented in Figures 118–120. This model calibrates optimally when $\gamma_{br} = 0.5 + 0.4 \tanh(33s_o)$, where s_o is the deep-water wave steepness. For more information on the principles of such models reference should be made to Battjes and Janssen (1978).

4.2.3.5 Extreme wave heights in shallow water

One effect of shoaling, refraction and breaking is to change the shape of the wave height distribution from that of the Rayleigh distribution (see Figure 109). Practical relations for extreme wave heights in shallow water are given by Stive (1986). The relations are based upon prototype and laboratory data, and give $H_{1\%}$ and $H_{0.1\%}$ as a function of H_s and local depth h:

$$H_{1\%} = 1.517 H_s \, 1/(1 + H_s/h)^{1/3} \qquad (4.35)$$

$$H_{0.1\%} = 1.859 H_s \, 1/(1 + H_s/h)^{1/2} \qquad (4.36)$$

In equations (4.35) and (4.36) the coefficients 1.517 and 1.859 represent conversion coefficients to the significant wave height which follow the Rayleigh distribution and the remaining part of the right-hand side of the equation reflects the depth limitation.

4.2.3.6 Reflection

Wave reflection generally occurs when waves are propagating onto breakwaters, seawalls and breaches (Figure 129). It may need to be determined when it forms a component of the design boundary conditions (for example, when assessing toe scour or the design of bottom protection in front of the wave-reflecting structure).

Where reflection occurs, orbital (bed) velocities should be calculated based on the combined effect of incoming and reflected waves. Reflected waves can be added to the incoming waves using the principle of superposition, so long as linear wave theory is being used. Superposition can also be used for random waves, since the theory for these also relies on linear wave theory. Reflected wave heights are

Figure 128 *Spilling breakers on Welsh coast, UK (courtesy J. D. Simm, Robert West & Partners)*

determined by the relevant reflection coefficient for the structure, C_r (see Section 5.1.1.1 for a definition), and depend largely on the geometry of the structure.

The spatial distribution of the height of waves reflected by a structure of limited length is strongly affected by diffraction. Because of this, a breakwater gap analogy can be used to determine reflected wave heights for such structures according to the principle indicated in Figure 130 (Goda, 1985), where if H is the incident wave height, the 'imaginary' undisturbed incident wave height arising from reflection is $C_r H$.

4.2.3.7 Wave–current interaction

When waves interfere with a (tidal) current the wave propagation and the wave spectrum are affected, and thus to obtain appropriate design conditions the problem of joint probabilities arises (see Section 4.2.7). The effect of a current with velocity U and direction θ_u is a resulting angular wave frequency of:

$$\omega' = \omega + kU \cos(\theta_0 - \theta_u) \qquad (4.37)$$

where $k = 2\pi/L$ (wave number)
  $\omega = 2\pi/T$ (angular frequency)
  $\theta_0 =$ direction of wave propagation.

In equation (4.37) ω refers to the value of the angular frequency without current, which may be solved from the dispersion relation:

$$\omega^2 = gk \tanh(kh) \qquad (4.38)$$

Another design aspect of the combination of currents and waves is the influence on design flow velocities near the seabed (e.g. for a bottom or scour protection). A method of assessing and designing for this combined flow loading is described in Section 5.1.3.11.

Figure 129 *Wave reflection from a breakwater*

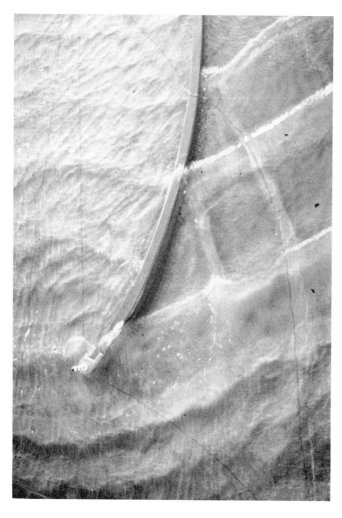

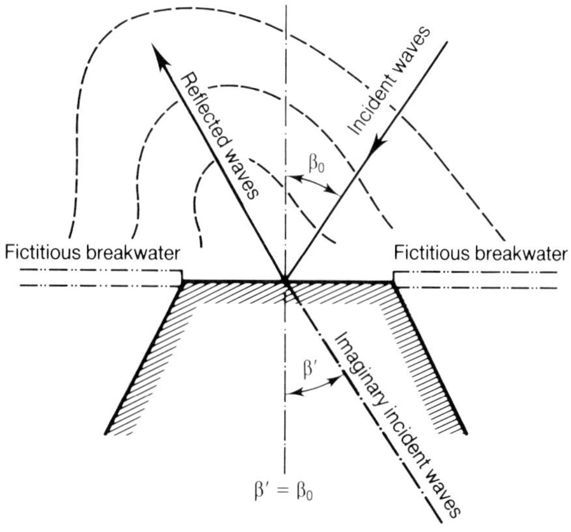

Figure 130 *Diffraction approach to determine the height of waves reflecting from a limited length structure*

4.2.4 SEABOTTOM ORBITAL VELOCITIES

These are required for design of rockfill used in offshore protection of pipelines under combined current and wave attack (see Section 5.1.3.11). Kirkgöz (1986) has shown that linear wave theory gives reasonable agreement with observed orbital velocities, even at the transformation point of plunging breakers. Soulsby (1987) has thus provided design curves based on linear wave theory (see Figure 131) which enable calculation of bottom orbital velocities for both monochromatic and random waves, where the maximum horizontal orbital bed velocity U_o was obtained from the relation:

$$U_o = \tfrac{1}{2}\omega H/\sinh(kh) \qquad (4.39)$$

ω being obtained from the dispersion relation (equation (4.38)).

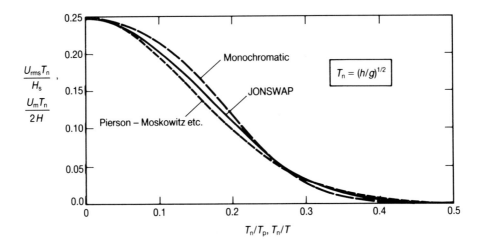

Figure 131 *Bottom orbital velocity for monochromatic waves ($U_m T_n/2H$ versus T_n/T) and a spectrum of waves ($U_{rms} T_n/H_s$ versus T_n/T_p). The curve labelled 'Pierson–Moskowitz etc.' also applies to the Bretschneider, ISSC and ITTC forms*

4.2.5 CURRENTS

Although in a marine environment waves are usually the dominant loading, currents should also be considered. Structural design (rock stability), construction and transport (required anchoring, possible speed of vessels) are examples of situations where currents may need to be known. In a more indirect sense, currents may affect a structure through erosion of the seabed. The principal sources and types of currents are:

- Tidal currents;
- Wind-induced currents;
- Density currents;
- Wave-induced currents (e.g. longshore currents);
- Circulation currents (due to the Coriolis effect induced by rotation of the earth);
- River discharge.

Wind-induced currents may result from local differences in wind and/or wave set-up (Section 4.2.1, Figures 102 and 103). Currents can be described over a given time scale by a time-mean magnitude, direction, and spatial and turbulent variability. In addition, current velocities may vary in a vertical sense over the water depth and, except for density- and wind-induced currents, the vertical distribution of velocities can often be described by a logarithmic function.

Data on current velocities can be obtained by direct measurement, methods for which are discussed in Appendix 4, or by use of numerical and physical models, although in the latter case boundary condition current measurements are often required. Other than these approaches, information on currents for preliminary design purposes can also be obtained from the following sources.

Charts and tables

In many countries, coast or port authorities or the Admiralty can provide tables and charts of surface current velocities, observed in the vicinity of main shipping routes, river mouths and estuaries. These are useful, so long as it is recognised that surface velocities may, for wind- and density-induced currents, differ significantly from the velocities closer to the bottom.

Sources of data include the British and US Admiralties, who hold data on surface currents in many strategic marine areas all over the world. Marine and offshore activities have also often necessitated current (and other) measurements, but such data may be in the private domain and difficult to obtain.

Correlation and transformation

When current data are available from one or more nearby locations, the currents for the site concerned might be estimated from correlation, interpolation or extrapolation. However, only for tidal or wind-induced currents can a more or less reliable relation be assumed between neighbouring locations. The sites to be correlated must show good similarity with regard to geography (alignment or coastline, exposure to wind and waves, location relative to river mouths, bays, breakwaters) and bathymetry (depth contours). Correlation factors for one or more other locations can be derived from a limited number of simultaneous measurements both at the site and at correlated sites.

Analytical models

Few types of currents permit derivation of useful analytical expressions. Where such derivations are possible, solutions are found for the governing equations of momentum and continuity by making geometric simplifications and/or by neglecting terms in the equations. Even when such solutions are available, empirical input is often needed. Examples of possibilities for analytical solution include:

1. The tidal current in the entrance of a harbour basin or estuary, where if the geometry allows for schematisation by a simple rectangular shape, the storage equation (based on continuity only) can be used to relate the current velocity to the (known) water levels and width and length of the basin;
2. Longshore and density currents which may allow for an analytical approach, but rarely without any empirical support (see Bowen, 1969; Brocard and Harleman, 1980).

4.2.6 SHIP-INDUCED WATER MOVEMENTS

For the design of the majority of marine structures ship-induced flows and waves are not the dominating loadings. The ship-induced water movements can be decomposed into five main components, which originate from river and channel engineering (Figure 132). Front and stern waves might have to be considered with regard to rock slopes of harbour entrance channels or jetties. Propeller thrust may be a relevant boundary condition for bottom protection in harbour entrance channels. A comprehensive overview of ship-induced water movements is given by PIANC (1987a). Computer techniques are also available for predicting ship-induced water movements (Bouwmeester, 1977). This section provides only an introduction to the subject.

The relevant ship speed parameters are the actual speed (v_s) the limit speed (v_L) in a channel, according to Schijf (1953). Other ship-related parameters are length (L_s), beam width (B_s), draught (D_s), installed engine power (P_s) and area of the

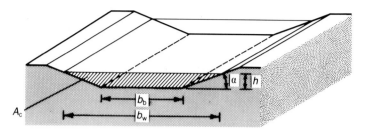

(a) Waterway geometry

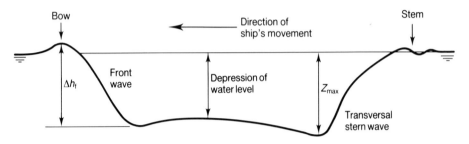

(b) Profile of water surface adjacent to a moving ship

Figure 132 *Definition sketch of waterway characteristics and ship-induced water movements. (a) Waterway geometry; (b) profile of water surface adjacent to a moving ship*

submerged cross-section at midship (A_m). The parameters of the channel are width (b_w at the water surface and b_b at the bottom), side slope angle (α) and area of the submerged cross-section (A_c). Parameters of waterway and water level are also defined in Figure 132.

According to Schijfs' (Schijf and Jansen, 1953) diagram (Figure 133 taken from Schijf and Jansen (1953)) the average water level depression (Δh) and the average return flow velocity ($\bar{u}_r$) can be determined by using the imaginary water depth ($h' = A_m/b_w$), the blockage ratio (A_m/A_c). For a conventional ship, sailing along the axis of the channel, the height of the stern wave is equal to Δh. The front wave (Δh_f) for this case can be estimated as $\Delta h_f = 1.10\ \Delta h$ and its gradient can be approximated by $0.03\Delta h_f$ (Verhey and Van der Wal, 1984).

A rough indication of the bed velocities can be obtained from Verhey (1983):

$$\bar{u}_r = 1.15\alpha_s(P_s/D_{pr}^2)^{0.33}D_{pr}/z_{pr} \qquad (4.40)$$

where α_s = coefficient (0.25–0.75), depending on ship type,
D_{pr} = diameter of propeller (without a nozzle) (m),
z_{pr} = height of propeller axis above the bottom) (m),
P_s = power of propeller (kW).

Direct measurements are, of course, possible, and measurement techniques on the relevant parameters are discussed in Appendix 4. It should, however, be remembered that ships cause waves of relatively high frequencies, water level changes which are rapid and water flows, particularly screw races, which are highly turbulent.

4.2.7 JOINT PROBABILITIES

Design loading conditions are usually determined by more than one parameter, and these parameters are often statistically dependent. Two practical examples are:

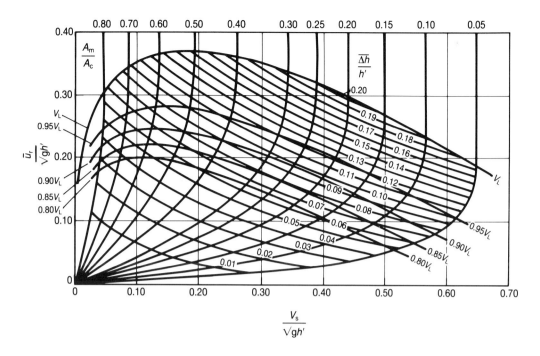

Figure 133 *Schijf's chart for estimating return current and water level depression. (Factor of safety $\alpha_1 (=1.4-0.4 v_s/v_L)=1.1$)*

1. Water level (tide and/or storm surge) and wave height; and
2. Orbital velocities and (tidal) current velocities.

4.2.7.1 Design with joint waves and water levels

If a functional requirement has been specified that a structure should have a certain probability of not failing (P_F) against a loading comprising a combination of wave heights and water levels, the relevant combination or combinations must be found which give P_F as the joint exceedance probability. A joint probability analysis must therefore be carried out to assess the relevant wave height and water level conditions. Otherwise the designer will be forced to design the structure conservatively for wave heights and water levels which independently both have a non-exceedance probability of P_F. (Determination of the independent long-term distributions of water level and wave height have been discussed in Sections 4.2.1.9 and 4.2.2.2.) This latter procedure may lead to a simple summing up of extreme loadings corresponding to a non-exceedance probability of P_F and thus to an uneconomic design. (For example, a crest level may be determined based on an extremely rare combination of water level and wave-induced run-up level.) Joint probability analyses, which overcome this conservatism, may therefore have significant economic advantages.

An example of a joint probability distribution function was shown in Figure 32 in Section 2.3.2. Here, instead of R and S, the two variables on which the probability density function depends are the (total) storm surge level (z) and significant wave height (H_s). The joint probability distribution can be described by a series of conditional probabilities, as shown in Figure 134, which gives an example of data for the entrance of the eastern Scheldt estuary in the Netherlands. The figure illustrates a number of conditional probability distributions for H_s at fixed values for z.

An approximate assessment of joint probabilities can be made assuming that the individual density functions can be described with means (μ_i) and standard deviations (σ_i). In cases of uncorrelated variables these combine into a joint mean

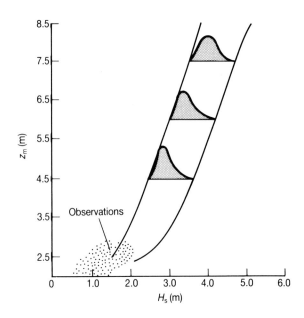

Figure 134 *Conditional probability densities $p(H_s, z)$ for entrance to eastern Scheldt*

and standard deviation, which are obtained by $\Sigma \mu_i$ and $\Sigma \sigma_i^2$, respectively. Benjamin and Cornell (1969) describe procedures applicable to correlated variables.

If both surge level (z) and significant wave height (H_s) have distributions that are in turn dependent on (or correlated to) wind speed by simple analytical relationships or via model simulation techniques, the following joint probability approach may be adopted:

1. Use wind data to find the probability density function $p(U_w)$ for the wind speed U_w;
2. Determine (for example, with equations (4.2) and (4.3)) the wind stress surge level. z as a function of the wind speed U_w and corresponding probability density $p(U_w)$. This gives the conditional probability density for z, given a wind speed U_w, known as $p(z; U_w) = p(U_w)$ (see Figure 135);
3. For a fixed surge level z and wind speed U_w, the wave height H_s is calculated. If necessary, shallow-water effects (Section 4.2.3) are included
   (a) Either directly in the spectrum (e.g. using the TMA spectrum (Section 4.2.2.5)) or
   (b) By refraction, shoaling and diffraction coefficients (Section 4.2.3);
4. The conditional probability $p(H_s; z)$ is found as $p(U_w; z)$, which might appear as shown in Figure 134.

Generally, H_s and z are dependent, and a joint probability density function should be used. In this function the dependence of H_s and z is reflected by their correlation coefficient ($P_{H,z}$. In the particular case that H_s and z are independent, $\rho_{H,z}=0$. Where they are more dependent, the Gaussian joint distribution function (with means μ_H, μ_z and standard deviations σ_H, σ_z), can be written as:

$$p(H,z) = [\phi_H \phi_z]^{1/(1-\rho_{H,z}^2)} \frac{\exp\left\{\frac{(\rho_{H,z})}{(1-\rho_{H,z}^2)} \cdot \frac{[(H-\mu_H)(z-\mu_z)]}{(\sigma_H \sigma_z)}\right\}}{\sqrt{(1-\rho_{H,z}^2)}} \tag{4.41}$$

where ϕ_H, ϕ_z = Gaussian (single) probability density distribution functions for H and z, respectively and ρ_{Hz} can be expressed in terms of the covariance ($\sigma_{H,z}^2$) of H and z by $\rho_{H,z}^2 = \sigma_{H,z}^2/(\sigma_H \sigma_z)$.

Equation (4.41) shows the influence of the mutual dependence between (in this example) H_s and z and can be used for $\rho_{H,z} < 1$. Figure 136 gives an example of joint distributions of wave height (H) and water level (z) for $\rho_{H,z}=0$ (fully

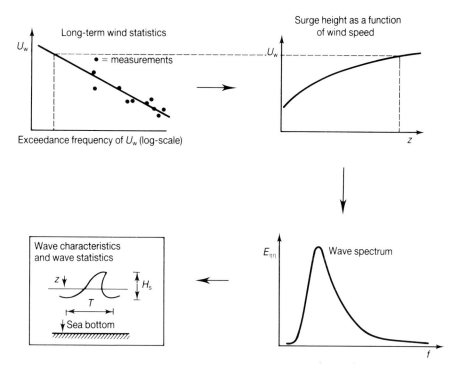

Figure 135 *Design process using joint probability distribution of z and H_s*

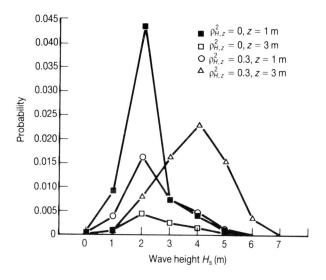

Figure 136 *Example of conditional distributions of H_s obtained from joint Gaussian distributions of correlated parameters H_s and z*

independent) and for $\rho_{H,z} = 0.3$. The figure shows conditional distributions of H_s for $z = 1$ and $z = 3$ m, representing cross-sections through the joint distribution function for z and H_s.

The above example shows the importance of a correct estimate of the actual correlations (instead of assuming *a priori* that $\rho_{H,z} = 0$). Such an estimate often demonstrates that, at a given z, exceedance probabilities are less than estimates made on the basis of no correlation between H_s and z. Expressed another way, this means that at a predefined exceedance probability, a larger H_s can be resisted. The value of the risk/cost consequences in terms of improved safety/reduced cost arising from such improved estimates cannot be overstated.

4.2.7.2 Design with joint waves and currents

Here again it may well be worth carrying out a joint probability analysis, as waves usually demonstrate a clear physical interaction with currents, one example being current refraction (Section 4.2.3.1). In addition, waves and currents often have a common driving force in wind and the waves themselves may induce further currents. In such cases the current velocities will not be independent of wave height and other wave-related parameters such as orbital velocities.

4.2.8 MODELLING

4.2.8.1 Numerical modelling of water levels and currents

Numerical models are essentially tools to solve a set of mathematical equations for the variable(s) of interest (e.g. the depth mean current velocity), the equations in turn being a schematic description of the real underlying processes. One-dimensional flow models usually perform well for average currents in well-defined current systems with pronounced flow concentrations (e.g. channels and rivers) but in most situations two- and three-dimensional flow models will be required, and here the choices of schematisation and computation schemes are important for realistic results.

Perhaps the most important factor to bear in mind with numerical models is to ensure that the implied boundary conditions (water levels, including tidal and wind/wave-induced velocities, and treatment of bottom boundaries) are correct. When models are used to solve questions related to sediment transport, spatial variations may be particularly decisive. In general, boundary conditions involving water levels and/or current velocities should be imposed sufficiently far from the structure to allow possible numerical disturbances to damp out before reaching the area of interest. When this implies large models, efficient use can be made of nested models, the principle of such models being that boundary conditions for a small-area model with a fine grid are provided by a larger (surrounding) model with a coarser one.

4.2.8.2 Physical modelling of water levels and currents

Physical modelling is an important option for complicated (usually three-dimensional) current patterns for which, despite their complexity, the boundary conditions can be reproduced well in the laboratory. Typical cases are structures exposed to combined current and wave action, complex bathymetry and more or less unconventional structure geometry. Physical modelling can be particularly useful in a number of situations:

1. Where interference of currents and waves is concerned, although numerical models are now being developed to cover this situation (Yoo *et al.*, 1989);
2. Where verification of (or comparison with) a numerical model is required;
3. If a physical model can be built and operated at a competitive cost in relation to other options.

Where problems are experienced with physical modelling they often relate to scale effects which arise when the principal forces within the process are not properly scaled. Most flow and wave models are arranged so that the dominant gravity forces, represented by the Froude number, $U_c/\sqrt{(gL_c)}$ (where U_c and L_c are characteristic flow velocities and dimensions, respectively), are scaled correctly. Where there are conflicts with other forces (e.g. friction forces represented by the Reynolds number, $U_c L_c/\nu$), adjustments are made accordingly (e.g. by providing additional corrective friction elements in the case of Reynolds number problems).

4.2.8.3 Numerical wave modelling

Most reliable numerical wave models for prediction of the wave boundary conditions at the structure are based on linear wave theory and incorporate, as a minimum, representation of refraction and shoaling effects together with diffraction effects when appropriate. Many rock structures under design conditions will also be exposed to breaking or broken waves, and here it is necessary that energy dissipation due to breaking be also properly estimated with the model. In the use of linear wave models, the transformation of a random wave field is found from the transformations of its constituent components. For example, the ENDEC model (ENergy DECay), referred to in Sections 4.2.3.2 and 4.2.3.4, can be used to find the local wave characteristics for any arbitrarily specified beach profile with slopes between 1:10 and 1:100, taking into account the energy dissipation due to breaking. When using detailed bathymetrical data and an incoming deep-water wave, characterised by H_{0s}, T_p and β_0, the model produces, for the chosen frequency (e.g. $f_p = 1/T_p$), the corresponding local values of H and β. For design purposes, the statistical distribution of H is also given (H_s, $H_{10\%}$, $H_{1\%}$, etc.). Energy-based random wave models, such as the one referred to, seem most suited to provide for the general design boundary conditions. A model developed at Liverpool University (Yoo *et al.*, 1989) is based on the energy principle and also includes currents within it, but as yet has only been developed for regular waves.

It should be noted that in most cases the proper *water level* has to be supplied *a priori* to ensure correct results from any wave model, and the procedures discussed in Section 4.2.1 should be followed to ensure this.

4.2.8.4 Physical wave modelling

Physical wave models can be used as a predictive scale model for the prototype or as a verification model for a numerical one. As the present state of the art in numerical wave modelling is often considered sufficient for standard design purposes, physical modelling is mainly applied when a complicated bathymetry in front of a structure causes significant variations in the near-structure sea state and/or interrelated specific wave-related structural design aspects (e.g. run-up, overtopping, toe scour, rock movements) also have to be clarified. It is mainly due to this capacity to deal with complex interactions that leads to physical models being selected to obtain the necessary design data. However, for many of the more standard design problems numerical models may be the more economical option. Thus expected accuracy must be balanced against cost of both numerical and physical modelling. Scale factors and scaling effects (already discussed in Section 4.2.9.2) must also be considered in this balancing of options and in the final design of the physical model, if selected.

4.3 Geotechnical boundary conditions and data collection

Depending on the functional requirements, coastal and shoreline structures have to withstand a combination of actions induced by waves, currents, differences in water levels, seismicity and other specific loadings (such as ship collisions or ice). These actions, including the self-weight of the structure, have to be transferred to the subsoil in such a way that:

1. The deformations of the structure are acceptable; and
2. The probability of loss of stability is sufficiently low.

The transfer of actions to the subsoil involves (time-dependent) changes in soil stresses, both within the structure and in the soil layers. Due to these stress increments, the soil layers and, as a consequence, the coastal or shoreline structure may deform vertically and horizontally or may even lose its stability. This deformation behaviour depends not only on the external loadings but also on the geometry (slope steepness), the structure's weight and the permeability, stiffness and shear strength of the subsequent structure and soil layers (see Figure 137). In

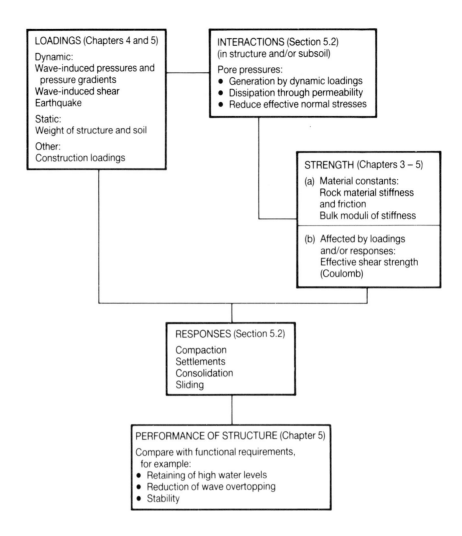

Figure 137 *Relational scheme for geotechnical structural performance*

fact, the performance of a coastal or shoreline structure depends to a large extent on the interaction between that structure and the subsoil. For this reason, a good knowledge of the main geotechnical properties of the soil layers and construction materials is needed.

The set-up of a soil-investigation programme is discussed below, and is focused on the collection of the relevant soil data. The most commonly used field and laboratory tests are outlined in Appendix 5. Special attention in this Appendix is also paid to the equipment applicable for overwater site investigations (see, for example, Appendix 5, Section A5.2).

4.3.1 GENERAL ASPECTS OF PLANNING A SOIL-INVESTIGATION PROGRAMME

A proper specification of a soil-investigation programme should provide the following information:

- Which soil data have to be collected;
- How reliably they should be obtained;
- At what locations (number, spatial distribution and depths);
- Which site investigation and laboratory tests should be performed and what type of equipment is needed;
- When the programm is to be carried out; and:

- Who is responsible for the contracting work in the field, the laboratory tests and the interpretation of the tests results.

Prescribing a standard form of a soil-investigation programme is hardly possible, because the above specifications depend to a large extent on the project to be considered. The following conditions play a major role:

- The global structural requirements of the project, including location, layout, and nature and extent of structure;
- The availability of existing geotechnical data;
- The knowledge and experience of geotechnical experts;
- The detailed structural requirements expressed in terms of geotechnical limit states (allowable stability and deformations), including the type of loading schemes and failure mechanisms;
- The applicability (including the terrain accessibility) of the available investigation tools and qualified personnel;
- Last (but not least) the boundary conditions stipulated by the client or owner in terms of time and money.

A high-quality investigation must be economically efficient in the sense that the cost of the soil investigation must be justified by building up additional knowledge of the geotechnical behaviour of the structure. It is emphasised that in the case of over-water site investigations the costs of operating equipment (vessels or platforms) forms the major part of the total costs of geotechnical investigations.

For important structural projects, programming and execution of soil investigations is a growing process, in which the need for additional soil data depends on the previously collected data and on the phase of the design stage. Related to the design stage, three steps of a soil-investigation programme can generally be distinguished and are related to the overall design process (Figure 1):

1. *Desk study*: collecting available topographical, geological, hydrological and geotechnical data for an initial appraisal of the boundary conditions;
2. *Preliminary soil investigations*: obtaining global information on soil data over a wide area aiming at recognising the main geotechnical problems, making a proper choice of the location of the structure and allowing alternative designs to be considered and a preferred solution to be selected.
3. *Final soil investigations*: obtaining detailed soil data defining specified geotechnical parameters in order to prepare the final design and construction method of the structure.

4.3.2 DESK STUDIES

For major projects a good geological analysis, based on the overall geological structure of the country in which they are to be built, is of the greatest importance for an understanding of the geophysical and hydrogeological conditions. The most important geological aspects are:

- Geological stratification, formation and history;
- Groundwater regime;
- Seismicity.

In most countries topographical and geological data are available in the scientific libraries of universities. In developed countries the geological structure is known to a large extent and in much detail. Elsewhere geological data may be very limited and restricted to the shallow strata.

In addition to geological maps or descriptions the following data should be collected:

- Topographical and bathymetrical maps;
- Groundwater maps (on land site);
- Climatological maps;
- Data on sea water levels, waves and currents;
- Local maps from public organisations;

- Historical data on observed geotechnical failure mechanisms;
- Published geotechnical data relating to the area or geological formations.

It is emphasised that in most cases the available soil data on the part of a coastal area permanently covered by the sea will be much less than on the landward and intertidal part. Where little can be determined of the subsoil conditions from the desk study a preliminary soil investigation must be performed.

4.3.3 PRELIMINARY SOIL INVESTIGATIONS

These should be designed to provide a proper insight into the subsoil conditions in order to identify the main geotechnical problems that may influence the location and the global design alternatives for the coastal or shoreline structure. The investigation should therefore permit the type and thickness of all soil strata to be established, both soft and compressible layers and harder layers. In addition, during the preliminary investigations, information will be gained on the possible sources of construction materials such as sands, gravel and rockfill.

Depending on the data available from the desk study and the global functional requirements of the coastal structure, a preliminary soil-investigation programme will generally consist of:

- Stratigraphical surveys that may comprise *geophysical measurements* over an extended area supplemented by a small number of *borings* (with *disturbed sampling*) and *penetration tests* (see Appendix 5, Section A5.1).
- For land investigations both *electro-resistivity*, *electro-magnetic* and *seismic* (*reflection* and *refraction*) survey techniques may be used (Section A5.1.1). For marine conditions continuous seismic reflection surveys are most commonly applied.
- In addition, extra *shallow borings* may be performed to assess the applicability of the upper soil layers as construction materials. For soft soil conditions the *vibro-coring* system may be very suitable for marine sites.

After preliminary design alternatives have been selected, detailed soil investigations have to be carried out to prepare the final design.

4.3.4 DETAILED SOIL INVESTIGATIONS

The detailed or final soil-investigation programme should provide sufficiently additional information on those soil parameters which are decisive for the selection of the (ecnomic) optimum design, construction and maintenance alternative. Although it is not possible to give a standard description of the detailed programme, some general comments should assist in setting up a suitable soil-investigation programme.

The type of soil properties to be collected and the reliability and the locations must be directly related to the functional requirements of the proposed design alternatives. These specifications may be derived from the evaluation of the associated geotechnical limit states as described in Section 5.2.

In Table 29 a summary is given of the main soil parameters to be determined for five types of geotechnical failure mechanisms, including shallow and deep-seated slip failure, liquefaction, dynamic failure, excessive settlements and filter erosion. The symbols of the soil parameters correspond with the parameters specified in Section 5.2.2.

The reliability of the data to be collected is mainly a matter of economics. More accurate description of the geotechnical failure mechanisms (at higher costs of soil investigations) may lead to a more optimised design at lower construction costs. Once the relevant soil parameters have been identified (according to Table 29), the suitable soil tests and associated *in-situ* equipment can be selected.

Table 29 Required soil data for the evaluation of the geotechnical limit states

Geotechnical limit states					Geotechnical information	
Macro-instability			Macro-failure	Micro-instability		
Slip failure	Lique-faction	Dynamic failure	Settle-ments	Filter erosion	Name	Symbol
A	A	A	A	A	Soil profile	—
A	A	A	A	A	Classification/grain size	D_i
A	A	A	B	A	Piezometric pressure	P
B	B	B	A	A	Permeability	k
A	B	B	A	B	Dry/wet density	$\rho_d, \rho_{sat}, \rho_{sub}, \gamma = \rho g$
—	A	B	—	—	Relative density, porosity	n, n_c
A	B	B	—	C	Drained shear strength	c', ϕ'
A	—	—	—	C	Undrained shear strength	s_u
B	—	—	A	—	Compressibility	C, C_s
A	—	—	A	—	Rate of consolidation	c_v
B	B	A	A	—	Moduli of elasticity	G, E
B	A	A	A	—	In situ stress	σ'
—	A	B	A	—	Stress history	OCR
B	A	A	B	—	Stress/strain curve	G, E

A: Very important
B: Important
C: Less important

Although there is a great variety of testing methods, each with its own merits and restrictions, Table 30 may give a first guide to selecting appropriate site-investigation techniques. This table provides a summary of the applicability of the most commonly used *in-situ* tests for the determination of the soil parameters as mentioned in Table 29. *In-situ* test methods may not be able to provide all the desired information, and therefore will need to be supplemented by laboratory tests (Appendix 5, Section A5.3) on borehole and other samples.

In many cases *Cone Penetration Tests* (CPTs) with and without sleeve friction measurements are very suitable (Section A5.1.2) and are relatively cheap for detailed surveys, except for circumstances where (bed)rock exists at shallow depths. They should, however, be supplemented by a small number of *borings with undisturbed sampling* (Section A5.1.3) at locations indicated from the CPTs. The borings should always be correlated to a CPT in the direct vicinity.

CPTs with simultaneous measurement of the pore water pressure during penetration (*Piezo cone, CPTU*) may provide more detailed information on the presence of alternating thin sand and clay layers. This may be of particular interest for cyclically loaded structures, where the drainage capacity of the soil is important. Where possible, these measurements should be supplemented by those obtained from piezometers installed in boreholes.

If the preliminary soil surveys show the existence of compressible clay and peat layers, which may directly affect the settlements and the stability of the marine structure, consolidation tests using an *oedometer* or *triaxial cell* (Appendix 5, Section A5.3.3) should be performed on undisturbed samples.

If, on the other hand, loose-packed sand layers may be expected, which are of major importance when considering the liquefaction potential, *critical density tests* or *cyclic triaxial tests* on sand samples (built up at different porosities) are recommended. To correlate the laboratory results to the *in-situ* density, empirical correlations relating SPT and other penetration tests to the relative density may be used. A good direct measurement of the *in-situ* density may be obtained using a *nuclear probe* down a borehole.

Worldwide, the *standard penetration test* (SPT), which provides (Section A5.1.2) both dynamic penetration resistances (number of blow counts) and soil samples, is very frequently used. A borehole is required to facilitate SPT. Many countries

Table 30 *In-situ* test methods and their perceived applicability (modified after Robertson and Campanella, 1983)

	Site-investigation methods										
	Geophysical methods (Section A5.1.1)			Penetration methods (Section A5.1.2)						Borings (Section A5.1.3)		
	Seismic	Electr. resist.	Electro-magnetic	nuclear	Cone penetr. test (CPT)	Piezo cone test (CPTU)	Stand. penetr. test (SPT)	Field vane test (VST)	Press. meter test (PMT)	Dilato meter test (DMT)	Dist. samples	Undist. samples + Lab. tests	Moni-toring wells
Soil profile	C	C	C	—	A	A	A	B	B	A	A	A	—
Classification	—	—	—	—	B	B	B	B	B	B	A	A	—
Piezometric pressure	—	—	—	—	—	A	—	—	B	—	—	—	A
Permeability	—	—	—	—	C	C	C	—	B	—	C	A	C
Dry/wet density	—	—	—	A	C	C	C	—	C	C	C	A	—
Relative/density	—	—	—	—	B	B	B	—	C	C	—	A	—
Friction angle	—	—	—	—	B	B	B	C	C	B	—	A	—
Undr. shear strength	—	—	—	—	B	B	C	A	B	C	—	A	—
Compressibility	—	—	—	—	C	C	—	—	C	C	—	A	—
Rate of consolidation	—	—	—	—	—	A	—	—	A	—	—	A	C
Moduli of elasticity	A	—	—	—	B	B	B	B	B	B	—	A	—
In-situ stress	—	—	—	—	C	C	C	C	B	B	—	A	—
Stress history OCR	—	—	—	—	C	C	—	B	B	B	—	A	—
Stress/strain curve	—	—	—	—	—	C	—	B	B	C	—	A	—

Ground conditions													
Hard rock	A	—	A	A	—	—	—	—	A	—	A	A	C
Soft rock-till, etc.	A	—	A	A	C	C	C	—	A	C	A	A	A
Gravel	A	B	A	A	C	C	B	—	B	—	A	C	A
Sand	A	A	A	A	A	A	A	—	B	A	A	A	A
Silt	A	A	A	A	A	A	B	B	B	A	A	A	A
Clay	A	A	A	A	A	A	C	A	A	A	A	A	A
Peat-organics	C	A	A	A	A	A	C	B	B	A	A	A	A

A: High applicability
B: moderate applicability
C: Limited applicability

have their own preference and experience for using SPT with borings rather than CPT, except where soft soil conditions prevail, where CPT has significant advantages.

In Scandinavia the *field vane shear test* (VST) is a very popular penetration *in-situ* test for silt, clay and peat soils (Section A5.1.2). This test is generally carried out from the bottom of a borehole, but can also be undertaken (at specified depths) with a CPT hydraulic penetration system.

Pressure meter tests (PMT) are used for the determination of the modulus of elasticity of the soil (Section A5.1.2). Information on the preloading and the *in-situ* stress of the soil can also be obtained from this test, which can be applied to a great variety of soil types. The standard (originally French) PMT is performed from boreholes. Several developments took place to carry out the PMT as a penetration test from ground level similar to the CPT (e.g. the self-boring pressure meter and the full displacement pressure meter). These systems are restricted to soft soil conditions.

The use of the *dilatometer test* (DMT), developed in Italy, is increasing rapidly in Western countries. Due to its relatively simple performance, it is a penetration tests that can be carried out from ground level (Section A5.1.2). In a similar way to the pressure meter, the DMT provides information of the stress–strain behaviour of the soil in a limited strain range. Site-investigation techniques are discussed further in Appendix 5 together with a brief description of the most commonly used laboratory tests.

5. Physical processes and design tools

The processes involved with stability of coastal and shoreline structures under wave (possibly combined with current) attack are given in a basic scheme in Figure 138. The environmental conditions (wave, current and geotechnical characteristics) lead to a number of parameters which describe *the boundary conditions at or in front of the structure* (A in Figure 138). These parameters are not affected by the structure itself, and generally, the designer of a structure has no influence on them. Wave height, wave height distribution, wave breaking, wave period, spectral shape, wave angle, currents, foreshore geometry, water depth, set-up and water levels are the main hydraulic environmental parameters and were described in Section 4.1. A specific geotechnical environmental condition is an earthquake, which is covered in Section 5.2.

Governing parameters can be divided into parameters related to *hydraulics* (B in Figure 138), *geotechnics* (C) and to the *structure* (D). *Hydraulic* parameters are related to the description of the wave action on the structure (hydraulic response). These parameters are described in Section 5.1.1.1. The main hydraulic responses are wave run-up and run-down, wave overtopping, wave transmission and reflection. These are covered in Section 5.1.2. *Geotechnical* parameters are related to, for instance, liquefaction, dynamic gradients and excessive pore pressures, and are treated in Section 5.2.

The structure can be described by a large number of *structural* parameters (D) and some important ones are the slope of the structure, the mass and mass density of the rock, rock or grain shape, surface smoothness, cohesion, porosity, permeability, shear and bulk moduli and the dimensions and cross-section of the structure. The structural parameters related to hydraulic stability are described in Section 5.1.1.2 and those related to geotechnical stability in Section 5.2.

The *loads on the structure* or on structural elements are given by the environmental, hydraulic, geotechnical and structural parameters together (E in

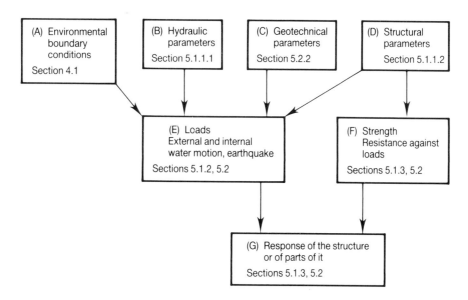

Figure 138 *Basic scheme for assessment of coastal structure response*

Figure 138). These loads can be divided into those due to external water motion on the slope and loads generated by internal water motion in the structure and earthquakes. The external water motion is affected by, among others, the deformation of the wave (breaking or not breaking), run-up and run-down, transmission, overtopping and reflection. These topics are described in Section 5.1.2. The internal water motion describes the penetration or dissipation of water into the structure and the variation of pore pressures and of the phreatic surface. These topics are treated in Section 5.2. Loads by ship waves are only briefly mentioned in this manual (see Section 4.2.6) and readers are referred to Verhey and Bogaerts (1989) for more information on this aspect.

Almost all structural parameters may have some or even a large influence on the loads. Size, shape and grading of armourstones affect the roughness of the slope, and therefore run-up and run-down. Filter size and grading, together with the above-mentioned characteristics of the armourstones, influence the permeability of the structure, and hence the internal water motion.

The resistance against the loads (waves, earthquakes) can be called the *strength of the structure* (F in Figure 138). Structural parameters are essential in the formulation of the strength of the structure. Most of them also influence the loads, as described above.

Finally, the comparison of strength with the loads leads to a description of the response of the structure or elements of the structure (G in Figure 138), the description of the so-called 'failure mechanisms'. This mechanism may be treated in a deterministic or probabilistic way.

Hydraulic structural responses are stability of armour layers, filter layers, crest and rear, toe berms and of crest walls and dynamically stable slopes, and these responses are described in Section 5.1.3. Geotechnical responses or interactions are slip failure, settlement, liquefaction, dynamic response internal erosion and impacts, and are covered in Section 5.2.

Figure 138 can also be used to describe the various ways of physical and numerical modelling of the stability of coastal and shoreline structures. A black-box method is used if the environmental parameters (A in Figure 138) and the hydraulic (B) and structural (D) parameters are modelled physically, and the responses (G) are given in graphs or formulae. A description of water motion (E) and strength (F) is not considered here.

A grey-box method is used if parts of the loads (E) are described by theoretical formulations or numerical models which are related to the strength (F in Figure 138) of the structure by means of a failure criterion or reliability function. The theoretical derivation of a stability formula may be the simplest example of this.

Finally, a white box is used if all relevant loads and failure criteria can be described by theoretical/physical formulations or numerical models, without empirical constants. It is obvious that it will take a long time and a lengthy research effort before coastal and shoreline structures can be designed by means of a white box. The colours black, grey and white, used for the methods described above, do not suggest a preference. Each can be useful in a design procedure.

This chapter will deal with physical processes and design tools, which means that design tools should be described so that:

- They are easily applicable;
- The range of application should be as wide as possible;
- Research data from various investigations should, wherever possible, be combined and compared, rather than giving the data of different investigations separately.

A few studies were performed for this manual in order to meet the above statements, i.e.:

- Testing of very wide gradings (Allsop, 1990);
- A desk study on the stability of low-crested structures (van der Meer, 1990a);
- A desk study on wave transmission (van der Meer, 1990b).

The final results of these studies are described in this chapter.

5.1 Hydraulic interactions

5.1.1 GOVERNING PARAMETERS

5.1.1.1 Hydraulic parameters

The main hydraulic responses of coastal and shoreline structures are wave run-up and run-down, overtopping, transmission and reflections. The governing parameters related to these hydraulic responses are illustrated in Figure 139, and are discussed in this section. The hydraulic responses themseves are described in Section 5.1.2.

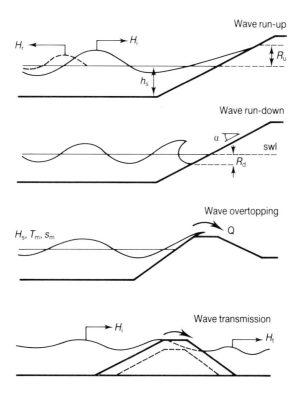

Figure 139 *Governing hydraulic parameters*

Wave steepness and surf similarity or breaker parameter

Before run-up, run-down, overtopping, transmission and reflection are described, the wave boundary conditions will be defined. Wave conditions are given principally by the incident wave height, H_i, usually as the significant wave height, H_s; the mean or peak wave periods, T_m or T_p; the angle of wave attack, β; and the local water depth, h.

The influence of the wave period is often described using the (dimensionless) deep-water wave steepness, defined by:

$$s = 2\pi H/gT^2 \tag{5.1}$$

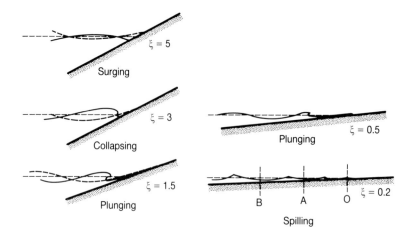

Figure 140 Breaker types as a function of ψ

and calculated using the local wave height at the structure. Use of H_s and T_m or T_p in equation (5.1) gives a subscript to s, respectively s_{om} and s_{op}.

The most useful parameter describing wave action on a slope, and some of its effects, is the surf similarity or breaker parameter, ξ, also termed the Iribarren number Ir:

$$\xi = \tan \alpha / \sqrt{s} \tag{5.2}$$

The surf similarity parameter has often been used to describe the form of wave breaking on a beach or structure (Figure 140, Battjes, 1974). It should be noted that different versions of this parameter are defined in this manual, reflecting the approaches of different researchers. In this section ξ_m and ξ_p are used when s is described by s_{om} or s_{op}.

Run-up and run-down

Wave action on a shoreline structure will cause the water surface to oscillate over a vertical range generally greater than the incident wave height. The extreme levels reached in each wave, termed run-up and run-down, R_u and R_d, respectively, and defined relative to the static water level, constitute important design parameters (see Figure 139). The design run-up level will be used to determine the level of the structure crest, the upper limit of protection or other structural elements, or as an indicator of possible overtopping or wave transmission. The run-down level is often taken to determine the lowest extent of main armour protection and/or a possible level for a toe berm. Run-up and run-down are often given in a dimensionless form R_{ux}/H_s and R_{dx}/H_s, where the subscript x describes the level considered (for instance, 2%) or significant (s).

Overtopping

If extreme run-up levels exceed the crest level the structure will overtop. This may occur for relatively few waves under the design event, and a low overtopping rate may often be accepted without severe consequences for the structure or the area protected by it. Seawalls and breakwaters are often designed on the basis that some (small) overtopping discharge is to be expected under extremne wave conditions. The main design problem therefore reduces to one of dimensioning the cross-section geometry such that the mean overtopping discharge, $\bar{Q}$, under design conditions remains below acceptable limits. A dimensionless parameter for $\bar{Q}$ was defined by Owen (1980), and will be used here:

$$Q^* = \frac{\bar{Q}}{\sqrt{(gH_s^3)}} \sqrt{(s/2\pi)} \tag{5.3}$$

Here $\bar{Q}_m$ and $\bar{Q}_p$ will also be employed when s_{om} and s_{op} are used in equation (5.3).

Wave transmission

Breakwaters with relatively low crest levels may be overtopped with sufficient severity to excite wave action behind. Where a breakwater is constructed of relatively permeable construction, long wave periods may lead to transmission of wave energy through the structure. In some cases the two different responses will be combined. The quantification of wave transmission is important in the design of low-created breakwaters intended to protect beaches or shorelines, and in the design of harbour breakwaters where long wave periods transmitted through the breakwater could cause movement of ships or other floating bodies.

The severity of wave transmission is described by the coefficient of transmission, C_t, defined in terms of the incident and transmitted wave heights, H_i and H_t, respectively, or the total incident and transmitted wave energies, E and E_t:

$$C_t = H_t/H_i = \sqrt{(E_t/E_i)} \qquad (5.4)$$

Wave reflections

Wave reflections are of importance on the open coast and at commercial and small harbours. The interaction of incident and reflected waves often lead to a confused sea in front of the structure, with occasional steep and unstable waves of considerable hazard to small boats. Reflected waves can also propagate into areas of a harbour previously sheltered from wave action. They will lead to greater peak orbital velocities, increasing the likelihood of movement of beach material. Under oblique waves, reflection will increase littoral currents and hence local sediment transport. All coastal structures reflect some proportion of the incident wave energy. This is often described by a reflection coefficient, C_r, defined in terms of the incident and reflected wave heights, H_i and H_r, respectively, or the total incident and reflected wave energies, E_i and E_r:

$$C_r = H_r/H_i = \sqrt{(E_r/E_i)} \qquad (5.5)$$

When considering random waves, values of C_r may be defined using the significant incident and reflected wave heights as representative of the incident and reflected energies.

5.1.1.2 Structural parameters

Coastal structures can be classified by use of the $H/\Delta D$ parameter, where $H =$ wave height, $\Delta =$ relative mass density and $D =$ characteristic diameter of structure, armour unit (rock or concrete), stone, gravel or sand. Small values of $H/\Delta D$ give structures as caissons or structures with large armour units. Large ones imply gravel beaches and sand beaches.

Only two types of structures have to be distinguished if the response of the various structures is concerned. These can be classified into *statically stable* and *dynamically stable*.

Statically stable structures are those where no or minor damage is allowed under design conditions. Damage is defined as displacement of armour units. The mass of individual units must be large enough to withstand the wave forces during design conditions. Caissons and traditionally designed breakwaters belong to the group of statically stable structures. The design is based on an optimum solution between design conditions, allowable damage and costs for construction and maintenance. Static stability is characterised by the design parameter damage, and can roughly be classified by $H/\Delta D = 1-4$.

Dynamically stable structures are those where profile development is concerned.

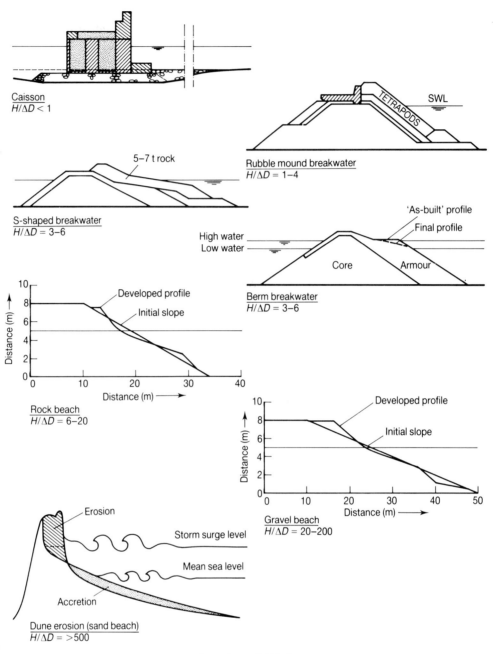

Figure 141 *Type of structure as a function of H/ΔD*

Units (stones, gravel or sand) are displaced by wave action until a profile is reached where the transport capacity along the profile is reduced to a very low level. Material around the still-water level is continuously moving during each run-up and run-down of the waves, but when the net transport capacity has become zero the profile has reached an equilibrium. Dynamic stability is characterised by the design parameter profile, and can roughly be classified by $H/\Delta D > 6$.

The structures concerned in this manual are rock-armoured breakwaters and slopes, rock beraches, berm-type breakwaters and gravel beaches, and are roughly classified by $H/\Delta D = 1 - 500$. An overview of the the types of structures with different $H/\Delta D$ values is shown in Figure 141, which gives the following rough classification:

- $H/\Delta D < 1$ *Caisson or seawalls*

No damage is allowed for these fixed structures. The diameter, D, can be the height or width of the structure.

- $H/\Delta D = 1-4$ *Stable breakwaters*

Generally, uniform slopes are applied with heavy artificial armour units or natural rock. Only little damage (displacement) is allowed under severe design conditions. The diameter is a characteristic diameter of the unit, such as the nominal diameter.

- $H/\Delta D = 3-6$ *S-shaped and berm breakwaters*

These structures are characterised by more or less steep slopes above and below the still-water level with a more gentle slope in between which reduces the wave forces on the armour units. Berm breakwaters are designed with a very steep seaward slope and a horizontal berm just above the still-water level. The first storms develop a more gentle profile which is stable further on. The profile changes to be expected are important.

- $H/\Delta D = 6-20$ *Rock slopes/beaches*

The diameter of the rock is relatively small and cannot withstand severe wave attack without displacement of material. The profile which is being developed under different wave boundary conditions is the design parameter.

- $H/\Delta D = 15-500$ *Gravel beaches*

Grain sizes, roughly between 10 cm and 4 mm, can be classified as gravel. Gravel beaches will change continuously under varying wave conditions and water levels (tide). Again, the development of the profile is one of the design parameters.

- $H/\Delta D > 500$ *Sand beaches* (*during storm surges*)

Material with very small diameters can withstand (dynamically) severe wave attack. The Dutch coast is partly protected by sand dunes. The dune erosion and profile development during storm surges is one of the main design parameters. Extensive basic research has been performed on this topic (Vellinga, 1986).

Structural parameters can be divided into four categories, which will be treated in this section, and are those related to:

1. Waves;
2. Rock;
3. The cross-section;
4. The response of the structure.

Structural parameters related to waves

The most important parameter which gives a relationship between the structure and the wave conditions has been used above. In general, the $H/\Delta D$ gives a good classification. For the design of coastal and shoreline structures this parameter should be defined in more detail.

The wave height is usually the significant wave height, H_s, defined either by the average of the highest one-third of the waves or by $4\sqrt{m_0}$. For deep water both definitions give more or less the same wave height. For shallow-water conditions substantial differences may be present. The relative buoyant density is described by:

$$\Delta = (\rho_{ssd}/\rho_w) - 1 \qquad (5.6)$$

where ρ_{ssd} = saturated surface dry mass density of the rock (see Section 3.2.1) and ρ_w = mass density of water. The diameter used is related to the average mass of the rock and is called the nominal diameter:

$$D_{n50} = (W_{50}/\rho_r g)^{1/3} \qquad (5.7)$$

where D_{n50} = nominal diameter and W_{50} = median weight of unit given by 50% on the weight-distribution curve. The parameter $H/\Delta D$ changes to $H_s/\Delta D_{n50}$.

Another important structural parameter is the surf similarity parameter, which relates the slope angle to the wave steepness, and which gives a classification of breaker types. The surf similarity parameter ξ (ξ_m, ξ_p with T_m, T_p) is defined in Section 5.1.1.1.

For dynamically stable structures with profile development a surf similarity parameter cannot be defined, as the slope is not straight. Furthermore, dynamically stable structures are described by a large range of $H_s/\Delta D_{n50}$ values. In that case it is also possible to relate the wave period to the nominal diameter and to make a combined wave height–period parameter, defined by:

$$H_0 T_0 = H_s/\Delta D_{n50} * T_m \sqrt{(g/D_{n50})} \qquad (5.8)$$

The relationship between $H_s/\Delta D_{n50}$ and $H_o T_o$ is listed below:

Structure	$H_s/\Delta D_{n50}$	$H_o T_o$
Statically stable breakwaters	1–4	<100
Rock slopes and beaches	6–20	200–1500
Gravel beaches	15–500	1000–200 000
Sand beaches	>500	>200 000

Another parameter which relates both wave height and period (or wave steepness) to the nominal diameter was introduced by Ahrens (1987). In the *SPM* (1984), $H_s/\Delta D_{n50}$ is often called N_s. Ahrens included the wave steepness in a modified stability number N_s^*, defined by:

$$N_s^* = N_s s_p^{-1/3} = (H_s/\Delta D_{n50}) s_p^{-1/3} \qquad (5.9)$$

In this equation s_p is the local wave steepness and not the deep-water wave steepness. The local wave steepness is calculated using the local wave length from the Airy theory, where the deep-water wave steepness is calculated by equation (5.1). This modified N_s^* number has a close relationship with $H_o T_o$ defined by equation (5.8). Wave height–period parameters used are summarised in Box. 44.

Box 44 Wave height–period parameters

$H_s/\Delta D_{n50} = N_s$
$H_s/\Delta D_{n50} s_p^{-1/3} = N_s$
$H_s/\Delta D_{n50} T_m \sqrt{(g/D_{n50})} = H_0 T_o$
$H_s/\Delta D_{n50} s_{om}^{-0.5} \sqrt{(2\pi H_s/D_{n50})} = H_o T_o$
$\xi_m = \tan \alpha / \sqrt{s} = \tan \alpha / \sqrt{(2\pi H_s/g T_m^2)}$

The most important parameter which is related to the rock is the nominal diameter defined by equation (5.7). Related to this is, of course, W_{50}, the 50% value on the weight-distribution curve. The grading of the rock can be given by the D_{85}/D_{15}, where D_{85} and D_{15} are the 85% and 15% values of the sieve curves, respectively. These are the most important parameters as far as stability of armour layers is concerned. Rock shape and rock quality are described in Chapter 3, and Appendix 1 and will not be repeated here. Examples of gradings in Box 45 show the relationship between class of stone (here simply taken as W_{85}/W_{15}) and D_{85}/D_{15}.

Box 45 Examples of gradings

Narrow grading $D_{85}/D_{15}<1.5$		Wide grading $1.5<D_{85}/D_{15}<2.5$		Very wide grading $D_{85}/D_{15}>2.5$	
Class	D_{85}/D_{15}	Class	D_{85}/D_{15}	Class	D_{85}/D_{15}
15–20 t	1.10	1–9 t	2.08	50–1000 kg	2.71
10–15 t	1.14	1–6 t	1.82	20–1000 kg	3.68
5–10 t	1.26	100–1000 kg	2.15	10–1000 kg	4.64
3–7 t	1.33	100–500 kg	1.71	10–500 kg	3.68
1–3 t	1.44	10–80 kg	2.00	10–300 kg	3.10
300–1000 kg	1.49	10–60 kg	1.82	20–300 kg	2.46

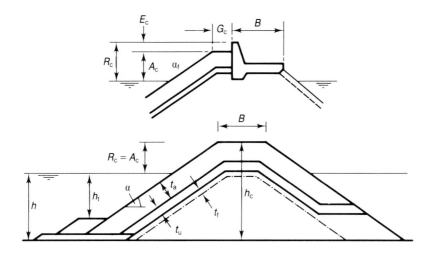

Figure 142 *Governing parameters related to the structure cross-section*

Further details of the recommended method of specifying gradings and of suggested standard gradings are given in Section 3.2.2.4.

Structural parameters related to the cross-section

There are many parameters related to the cross-section, and most of them are obvious. Figure 142 gives an overview. The parameters are:

Crest freeboard, relative to still-water level	R_c
Armour crest freeboard relative to still-water level	A_c
Difference between crown wall and armour crest	F_c
Armour crest level relative to the seabed	h_c
Structure width	B
Width of armour berm at crest	G_c
Thickness of armour, underlayer, filter	t_a, t_u, t_f
Area porosity	n_a
Angle of structure slope	α
Depth of the toe below still-water level	h_t

The permeability of the structure has an influence on the stability of the armour layer. This depends on the size of filter layers and core and can be given by a notional permeability factor, P. Examples of P are shown in Figure 143, based on the work of van der Meer (1988a). The lower limit of P is an armour layer with a thickness of two diameters on an impermeable core (sand or clay) and with only a thin filterlayer. This lower boundary is given by $P=0.1$. The upper limit of P is given by a homogeneous structure which consists only of armourstones. In that case, $P=0.6$. Two other values are shown in Figure 143, and each particular structure should be compared with the given structures in order to make an estimation of the P factor.

Latham *et al.* (1988) tested rock slopes with an impermeable core and with only a 1.5 diameter thick layer. From the results it was concluded that in that case a value of $P=0.07$ might be used.

Structural parameters related to the response of the structure

The behaviour of the structure can be described by a few parameters and statically stable structures by the development of damage. This can be the amount of rock that is displaced or the displaced distance of a crown wall. Dynamically stable structures are described by a developed profile.

The damage to the armour layer can be given as a percentage of displaced stones related to a certain area (the whole or a part of the layer). In this case, however,

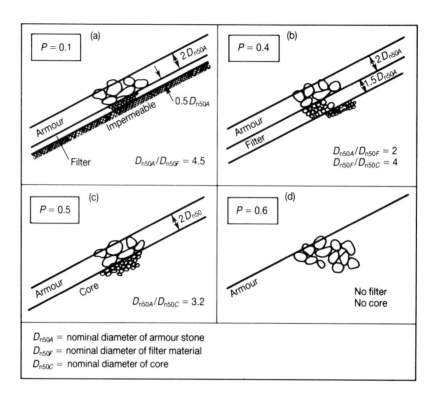

Figure 143 *Notional permeability factor P for various structures*

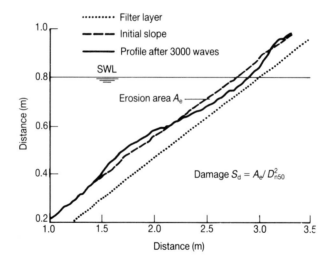

Figure 144 *Damage S_d based on erosion area A_e*

it is difficult to compare various structures as the damage figures are related to different totals for each structure. Another possibility is to describe the damage by the erosion area around still-water level. When this erosion area is related to the size of the stones, a dimensionless damage level is presented which is independent of the size (slope angle and height) of the structure. This damage level is defined by:

$$S_d = A_e/D_{n50}^2 \qquad (5.10)$$

where S_d = damage level and A_e = erosion area around still-water level.

Table 31 Design values of S_d for a $2D_{n50}$ thick armour layer

Slope	Initial damage	Intermediate damage	Failure
1:1.5	2	3–5	8
1:2	2	4–6	8
1:3	2	6–9	12
1:4	3	8–12	17
1:6	3	8–12	17

A plot of a structure with damage is shown in Figure 144, the damage level, S_d, taking both settlement and displacement into account. A physical description of the damage, S_d, is the number of squares with a side D_{n50} which fit into the erosion area. Another description S_d is the number of cubic stones with a side of D_{n50} eroded within a D_{n50} wide strip of the structure. The actual number of stones eroded within this strip can be more or less than S_d, depending of the porosity, the grading of the armourstones and the shape of the stones. Generally, the actual number of stones eroded in a D_{n50} wide strip is equal to 0.7–1 times the damage S_d.

The limits of S_d depend mainly on the slope angle of the structure. For a two-diameter thick armour layer the values in Table 31 can be used. The initial damage of $S_d = 2-3$ is according to the criterion of the Hudson formula, which gives 0–0.5% damage. Failure is defined as exposure of the filter layer. For S_d values higher than 15–20 the deformation of the structure results in an S-shaped profile and should be called dynamically stable.

Dynamically stable structures are those where profile development is accepted. Units (stones, gravel or sand) are displaced by wave action until a profile is reached where the transport capacity along the profile is reduced to a minimum. Dynamic stability is characterised by the design parameter *profile*.

An example of a schematised profile is shown in Figure 145. The initial slope was 1:5, which is relatively gentle and one should note that Figure 145 is shown on a distorted scale. The profile consists of a beach crest (the highest point of the profile), a curved slope around still-water level (above still water steep, below,

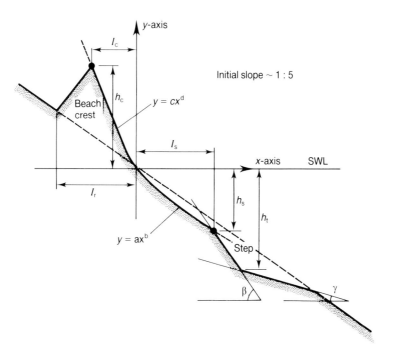

Figure 145 *Schematised shingle beach/rock slope profile on a 1:5 initial slope*

gentle) and a steeper part relatively deep below that level. For gentle slopes (shingle slope >1:4) a step is found at this deep part. The profile is characterised by a number of lengths, heights and angles, and these were related to the wave boundary conditions and structural parameters (van der Meer, 1988a). A computer program was developed at Delft Hydraulics (BREAKWAT) which can calculate the profile. Examples of profiles on steeper slopes than 1:5 and the use of the program are given in Section 5.1.3.7.

5.1.2 HYDRAULIC RESPONSE

This section presents methods that may be used for the calculation of the hydraulic response parameters which were also given in Figure 139:

Run-up and run-down levels;
Overtopping discharges;
Wave transmission;
Wave reflections.

Where possible, the prediction methods are identified with the limits of their application. These methods are generally only available to describe the hydraulic response for a few simplified cases, either because tests have been conducted for a limited range of wave conditions or because the structure geometry tested represents a simplification in relation to many practical structures. It will therefore be necessary to estimate the performance in a real situation from predictions for related (but dissimilar) structure configurations. Where this is not possible, or the predictions are less reliable than are needed, physical model tests should be conducted.

5.1.2.1 Wave run-up and run-down

Prediction of R_u and R_d may be based on simple empirical equations, themselves supported by model test results, or on numerical models of wave/structure interaction. A few simple numerical models of wave run-up have been developed recently, but have only been tested for a few cases and will not be discussed further here.

All calculation methods require parameters to be defined precisely. Run-up and run-down levels are defined relative to still-water level (see Figure 139). On some bermed and shallow slopes run-down levels may not fall below still water. All run-down levels in this manual are given as positive if below still water and all run-up levels will also be given as positive if above it.

The upward excursion is generally greater than the downward, and the mean water level on the slope is often above still-water level. Again, this may be most marked on bermed and shallow slopes. These effects often complicate the definition, calculation or measurement of run-down parameters.

Run-up levels, R_{ux}, will vary as do wave heights and lengths in a random sea. Prediction methods are available for the significant run-up level, R_{us}, or the 2% exceedance level, $R_{u2\%}$. The form of the probability distribution of run-up levels is generally not well established. Results of some tests suggest that, for simple configurations with slopes between 1:1.33 and 1:2.5, a Rayleigh distribution for run-up levels may be assumed where other data are not available. This distribution may be used to relate the 2% run-up and significant run-up levels, giving:

$$R_{u2\%} = 1.4 R_{us} \tag{5.11}$$

Much of the field data available on wave run-up and run-down apply to shallow and smooth slopes. Some laboratory measurements have been made on steeper smooth slopes and on porous armoured ones. Prediction methods for smooth

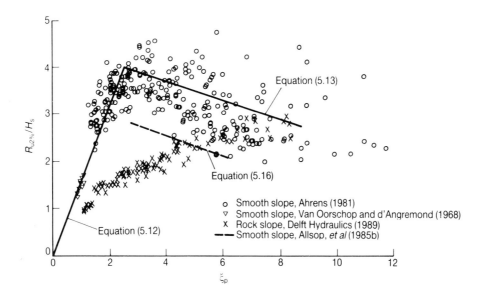

Figure 146 *Comparison of relative 2% run-up for smooth and rubble slopes*

slopes may be used directly for armoured slopes that are filled or fully grouted with concrete or bitumen. These methods can also be used for rough non-porous slopes with an appropriate reduction factor.

The behaviour of waves on rough porous slopes is very different from that on non-porous ones, and the run-up performance is not well predicted by adapting equations for smooth slopes. Different data must be used. This difference is illustrated in Figure 146, where 2% relative run-up, $R_{u2\%}/H_s$, is plotted for both smooth and rock slopes. The greatest divergence between the performance of the different slope types is seen for $1 < \xi_p < 5$. For ξ_p above about 6 or 7 the run-up performance of smooth and porous slopes tend to very similar values.

Run-up and run-down will now be treated for the following structures:

- Smooth slopes
- Rough non-porous slopes
- Armoured rubble slopes
- Gravel beaches
- Bermed and composite slopes

Smooth slopes

Measurements of wave run-up on smooth slopes have been analysed by Ahrens (1981), Delft Hydraulics (M1983, 1989), and Allsop *et al.* (SR2, 1985). In each instance the test results are scattered (Figures 146 and 147) but simple prediction lines have been fitted to the data.

Figure 146 shows the data of Ahrens (1981) for slopes between 1:1 and 1:4 of Van Oorschot and d'Angremond (1968) for slopes 1:4 and 1:6 and of Allsop *et al.* (1985b) for slopes between 1:1.33 and 1:2. All mentioned data points are for smooth slopes. The other points in Figure 146 are for rock slopes (Delft Hydraulics, M1983, 1989). The scatter for Ahrens data is large. He measured only 100–200 waves and the 2% value is not very reliable in that case.

Figure 147 shows the same data, but now for the significant levels. The scatter around Ahrens data is now much less. In both figures the data of Allsop *et al.* is about 20–30% lower than those of Ahrens. Reasons for the differences are hard to give, but possibly different definitions of run-up level and different test methods

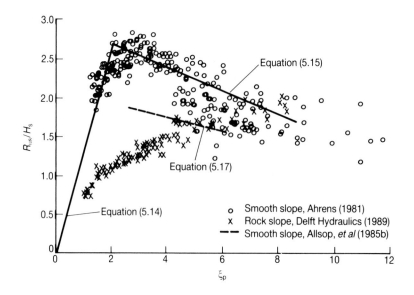

Figure 147 *Comparison of relative significant run-up for smooth and rubble slopes*

could have caused it. Based on these figures the data of Ahrens probably give a conservative estimate.

Relative run-up levels on smooth slopes may be calculated from empirical equations given below ((5.12)–(5.17)) using the surf parameter based on the peak wave period, ξ_p. These equations were based on Figures 146 and 147 and were derived for this manual. For the original equations based on the data of Ahrens reference should be made only to Ahrens (1981). The run-up equations are cross-referenced to the appropriate lines in Figures 146 and 147.

Run-up

$$0 < \xi_p < 2.5 \quad R_{u2\%}/H_s = 1.6\xi_p \quad \text{data of Ahrens} \quad (5.12)$$

$$\xi_p > 2.5 \quad R_{u2\%}/H_s = 4.5 - 0.20\xi_p \quad (5.13)$$

$$0 < \xi_p < 2.0 \quad R_{us}/H_s = 1.35\xi_p \quad \text{data of Ahrens} \quad (5.14)$$

$$\xi_p > 2.0 \quad R_{us}/H_s = 3.0 - 0.25\xi \quad (5.15)$$

$$2.8 < \xi_p < 6 \quad R_{u2\%}/H_s = 3.39 - 0.21\xi_p \,\text{data of Allsop } et\ al. \quad (5.16)$$

$$2.8 < \xi_p < 6 \quad R_{us}/H_s = 2.11 - 0.09\xi_p \quad (5.17)$$

Run-down
Run-down on smooth slopes can be calculated by:

$$0 < \xi_p < 4 \quad R_{d2\%}/H_s = 0.33\xi_p \quad (5.18)$$

$$\xi_p > 4 \quad R_{d2\%/Hs} = 1.5 \quad (5.19)$$

Rough non-porous slopes

The calculation of run-up levels on non-porous rough slopes may be based upon the methods for smooth slopes given above, and the use of a run-up reduction factor, r, which should be multiplied by the run-up on a smooth slope. As can be seen in Figures 146 and 147, the reduction coefficients are only applicable for $\xi_p < 3$–4. The reduction factors are taken from TAW (1972).

Roughness reduction factors for run-up for various types of construction ($\xi_p < 3$–4)

Construction	Run-up reduction factor, r
Smooth, impermeable (including smooth concrete and asphalt)	1.0
Stone blocks, pitched or mortared, and open-stone asphalt	0.95
Concrete blocks	0.9
Stone blocks, granite sets	0.85–0.9
Turf	0.85–0.9
Rough concrete	0.85
One layer of stone rubble on impermeable base	0.8
Stones set in cement	0.75–0.8
Gravel, gabion mattresses	0.7
Dumped round stones	0.6–0.65
Rock/rip-rap with total layer thickness greater than $2D_{n50}$	0.5–0.6

No data on wave run-down levels are available for rough non-porous slopes. It is recommended that predictions be based on the methods given above for smooth slopes.

Armoured rubble slopes

Over most wave conditions and structure slope angles a rubble slope will dissipate significantly more wave energy than the equivalent smooth or non-porous slope. Run-up levels will therefore generally be reduced. This reduction is influenced by the permeability of the armour, filter- and underlayers, and by the steepness and period of the waves. Few methods are available to quantify this effect reliably. Run-up levels on rubble slopes armoured with rock armour or rip-rap have been measured in laboratory tests using either regular or random waves. In many instances the rubble core has been reproduced as fairly permeable, except for those particular cases where an impermeable core has been used. Test results often therefore span a range within which the designer must interpolate.

Analysis of test data from measurements by van der Meer (1988a) has given prediction formulae for rock slopes with an impermeable core, described by a notional permeability factor $P = 0.1$, and porous mounds of relatively high permeability given by $P = 0.5$ and 0.6 (Delft Hydraulics M1983, Part 3, 1988). The notional permeability factor P was described in Section 5.1.1.2. The prediction formulae are:

$$R_{ui}/H_s = a\xi_m \quad \text{for} \quad \xi_m < 1.5 \tag{5.20}$$

$$R_{ui}/H_s = b\xi_m^c \quad \text{for} \quad \xi_m > 1.5 \tag{5.21}$$

The run-up for permeable structures ($P > 0.4$) is limited to a maximum:

$$R_{ui}/H_s = d \tag{5.22}$$

In the above equations coefficients a–d are as given below:

Coefficients in equations (5.20)–(5.22)

Run-up level (i)	a	b	c	d
0.1%	1.12	1.34	0.55	2.58
1%	1.01	1.24	0.48	2.15
2%	0.96	1.17	0.46	1.97
5%	0.86	1.05	0.44	1.68
10%	0.77	0.94	0.42	1.45
Sign.	0.72	0.88	0.41	1.35
Mean	0.47	0.60	0.34	0.82

Values for the coefficients a–d have been determined for exceedance levels of $i = 0.1\%$, 1%, 2%, 5%, 10%, significant, and mean run-up levels.

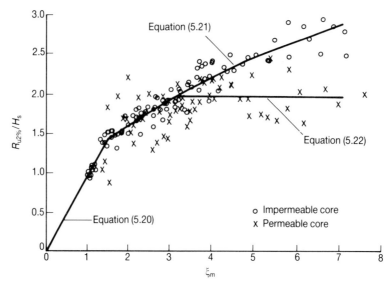

Figure 148 *Relative 2% run-up on rock slopes*

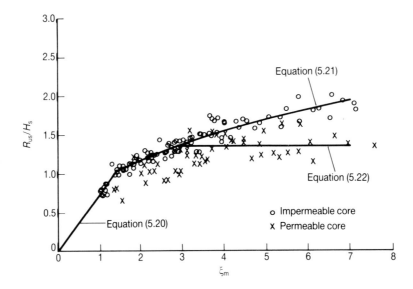

Figure 149 *Relative significant run-up on rock slopes*

Results of the tests and the equations are shown for values of $i = 2\%$, and significant, for each of $P = 0.1$ and $P > 0.4$, in Figures 148 and 149. Run-down levels on porous rubble slopes are also influenced by the permeability of the structure, and on the surf similarity parameter. For wide-graded rock armour, or rip-rap, on an impermeable slope a simple expression for a maximum run-down, taken to be around the 1% level, is derived from test results by Thompson and Shuttler (1977):

$$R_{d1\%}/H_s = 0.34\xi_p - 0.17 \tag{5.23}$$

Analysis of the run-down on the sections tested by van der Meer (1988a) has given an equation which includes the effects of structure permeability and wave steepness:

$$R_{d2\%}/H_s = 2.1\sqrt{(\tan \alpha)} - 1.2P^{0.15} + 1.5\exp(-60s_m) \tag{5.24}$$

Test results are shown in Figure 150 for an impermeable and a permeable core. The presentation with ξ_m only gives a large scatter. Including the slope angle and

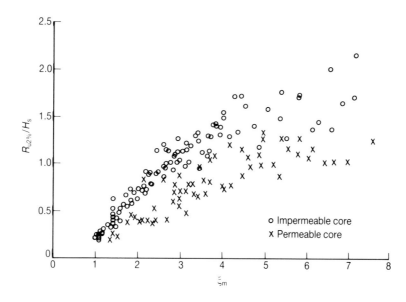

Figure 150 *Relative 2% run-down on impermeable and permeable rock slopes*

the wave steepness separately as well as the permeability, as in equation (5.24), reduces the scatter considerably.

Gravel beaches

Gravel or shingle beaches differ from the armoured slopes considered above principally in the size of the beach material, and hence its mobility. The typical stone size is sufficiently small to permit significant changes of beach profile, even under relatively low levels of wave attack. A shingle beach may be expected to adjust its profile to the incident wave conditions, provided that sufficient beach material is available (see Sections 4.1.3.1 and 5.1.3.7). Run-up levels on a shingle beach are therefore calculated without reference to any initial slope angle.

The equilibrium profile of shingle beaches under (temporary constant) wave conditions is described by van der Meer (1988a) and the schematised profile was given in Section 5.1.1.2 and Figure 145. The crest height above still-water level is described by the parameter h_c. For small diameters (shingle, $D_{n50} < 0.10$ m) the crest height is only a function of the wave height and wave steepness. The relationship is given by:

$$h_c/H_s = 0.3/\sqrt{s_m} \qquad (5.25)$$

Only the highest waves will overtop the beach crest. Therefore equation (5.25) gives a crest height which is more or less close to $R_{u2\%}$. Powell (1990) endorses these conclusions based on a recent comprehensive series of 131 tests on shingle beaches.

Oblique wave attack

The effect of oblique wave attack on run-up levels on a 1:6 slope with regular waves has been studied by Tautenhaim et al. (1982). Their results suggest that run-up levels for normal wave attack, $\beta = 0°$, can be exceeded for small angles (say, $\beta = 10$–$30°$). The general trend of increasing run-up levels at small angles of obliquity, $\beta \leq 30°$, for smooth slopes is supported by Owen's (1980) work on overtopping (see Section 5.1.2.2).

Recently (1989), Delft Hydraulics has performed an extensive model investigation on the influence of oblique and multi-directional seas on run-up and overtopping. The data have not yet fully been analysed, but some preliminary conclusions on important aspects can be given. One of these is the above-mentioned increase in run-up for small angles.

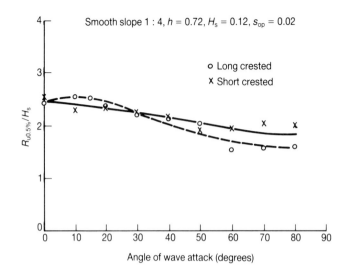

Figure 151 *Influence of oblique and long- and short-crested seas on run-up on a 1:4 smooth slope*

Figure 151 shows the results on a 1:4 smooth slope for a wave steepness of $s_m = 0.02$. Results are shown for both long-crested (random) and short-crested or multi-directional seas. The tests with long-crested waves show the same tendency as described above. The maximum run-up level is found for $\beta = 10-15°$. For a smooth 1:2.5 slope this was $\beta = 20-30°$. However, for short-crested waves this phenomenon is not present! The maximum run-up was found for perpendicular wave attack, both for the 1:4 and the 1:2.5 slope. This means that in nature and under storm conditions the phenomenon is not present either. Only for long swell (no storm conditions) under a small wave angle ($\beta = 20-30°$) might it be possible for the run-up to be higher than for normal wave attack. In that case the swell can be considered as long-crested waves.

Figure 151 shows further that the decrease in run-up for large angles, $\beta > 50°$, is smaller for short-crested than for long-crested seas. For both slopes, 1:4 and 1:2.5, and for wave steepnesses of $s_{op} = 0.02$ and 0.04, the run-up for long-crested waves and large angles ($\beta > 70°$) reduced to about 60% in comparison with normal wave attack. This is also shown in Figure 151. For the same slopes and wave steepnesses, but for short-crested waves, the reduction was a factor of only 0.8 instead of 0.6. This is also clear in Figure 151. The tests on overtopping showed the same tendencies as for run-up described here.

The overall conclusions are:

- Wave run-up and overtopping in short-crested seas is maximum for normal wave attack.
- The increase in run-up at $\beta = 10-30°$ is only present for long-crested waves.
- Reduction of run-up for large wave angles in short-crested waves is not more than a factor of 0.8 compared with normal wave attack.

Bermed and composite slopes

The methods discussed so far have all been developed for simple slopes. In some cases local geometric restrictions require a bermed or composite slope. Such a cross-section may also arise where a roadway is incorporated for access or where the structure is extended or upgraded. A berm at or close to design still-water level will often reduce both run-up and reflections. No data are available to predict the reduction of run-up under random waves on a bermed slope.

Methods to estimate run-up on composite or bermed slopes under regular waves are discussed in *SPM* (1984) and by Thompson (1973). Both are based on the estimate of an equivalent simple slope. Neither method has been tested or validated for random waves. The effect of a berm on overtopping discharges under random waves is discussed later in Section 5.1.2.2, and the effect on wave reflections in Section 5.1.2.4.

5.1.2.2 Overtopping

In the design of many seawalls and breakwaters the controlling hydraulic response is often the wave overtopping discharge. Under random waves this varies greatly from one wave to another. There are very few data available to quantify this variation. For many cases it is sufficient to use the mean discharge, $\bar{Q}$, usually expressed as a discharge per metre run (m³/s.m). Suggested critical values of $\bar{Q}$ for various design situations are summarised in Box 46. The dimensionless discharge, Q_m^* or Q_p^*, was given in Section 5.1.1.1 and equation (5.3).

Box 46 Critical overtopping discharges

Limiting values of $\bar{Q}$ for different design cases have been suggested, and are summarised in Figure 152. This incorporates recommended limiting values of the mean discharge for the stability of crest and rear armour to types of seawalls and or the safety of vehicles and people.

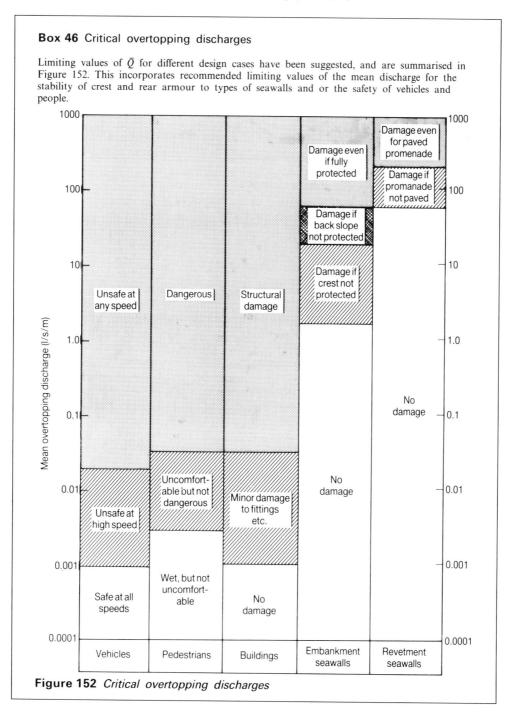

Figure 152 *Critical overtopping discharges*

The calculation of overtopping discharge for a particular structure geometry, water level and wave condition is based on empirical equations fitted to hydraulic model test results. The data available on overtopping performance are restricted to a few structural geometries. The widest data set applies to plain and bermed smooth slopes without crown walls (Owen, 1980). More restricted studies have been reported by Bradbury *et al.* (1988) and Aminti and Franco (1988).

Each of these studies have developed dimensionless parameters of the crest freeboard for use in prediction formulae. Different dimensionless groups have been used by each author, and no direct comparisons have yet been made. The simplest such parameter is the relative freeboard, R_c/H_s. This simple parameter, however, omits the important effects of wave period, and the introduction of other dimensionless parameters has been necessary to include the wave length or steepness.

For *plain and bermed smooth slopes* Owen (1980) relates a dimensionless discharge parameter, Q_m^*, to a dimensionless freeboard parameter, R_m^*, by an exponential equation of the form:

$$Q_m^* = a\exp(-bR_m^*/r) \tag{5.26}$$

where Q_m is defined in equation (5.3) and the dimensionless freeboard is

$$R_m^* = R_c/H_s^* * \sqrt{(s_m/2\pi)} \tag{5.27}$$

Values for the coefficients a and b were derived from the test results and are given in Table 32. Those of a and b were also derived for smooth-bermed seawalls of the generalised profile shown in Figure 153, and these are given in Table 33 for use with equation (5.26). Figure 154 (Owen, 1980) presents $\log Q_m^*$ versus R_m^* for smooth, straight slopes. For low crest heights and large discharges the curves come together in one point, indicating that in that case the slope angle is no longer important. Moreover, the discharges for slopes 1:1, 1:1.5 and 2 are almost equal.

The above-mentioned investigation on the influence of oblique and short-crested waves on run-up and overtopping performed by Delft Hydraulics also gave discharges for a smooth 1:4 slope, which were compared with other investigations. It was concluded that the effect of spectral shape was small if the peak period was used instead of the mean period. Figure 155 gives the results, including a suggested equation for the discharges.

Few data are available to quantify the overtopping performance of *rough non-porous* slopes directly. Owen has suggested that values of the run-up reduction or roughness coefficient, r, tabulated in Section 5.1.2.1, could be used with equation (5.26) to give estimates of overtopping for simple armoured slopes. This is expected to give conservative results, particularly for berm slopes.

Surprisingly, there are very few data available describing the overtopping

Table 32 Values of the coefficients a and b in equation (5.26) for straight, smooth slopes

Slope	a	b
1:1	0.00794	20.12
1:1.5	0.0102	20.12
1:2	0.0125	22.06
1:3	0.0163	31.9
1:4	0.0192	46.96
1:5	0.025	65.2

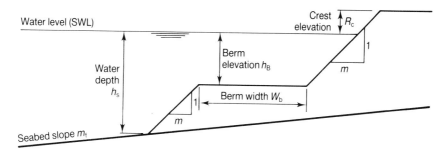

Figure 153 *Generalised profile for smooth, bermed seawalls*

Table 33 Values of the coefficients *a* and *b* in equation (5.26) for bermed smooth slopes

Seawall slope	Berm elevation, h_B (m SWL)	Berm width, w_B (m)	A	B
1:1	−4.0	10	6.40×10^{-3}	19.50
1:2			9.11×10^{-3}	21.50
1:4			1.45×10^{-2}	41.10
1:1	−2.0	5	3.40×10^{-3}	16.52
1:2			9.80×10^{-3}	23.98
1:4			1.59×10^{-2}	46.83
1:1	−2.0	10	4.79×10^{-3}	18.92
1:2			6.78×10^{-3}	24.20
1:4			8.57×10^{-3}	45.80
1:1	−2.0	20	8.80×10^{-4}	14.76
1:2			2.00×10^{-3}	24.81
1:4			8.50×10^{-3}	50.40
1:1	−2.0	40	3.80×10^{-4}	22.65
1:2			5.00×10^{-4}	25.93
1:4			4.70×10^{-3}	51.23
1:1	−2.0	80	2.40×10^{-4}	25.90
1:2			3.80×10^{-4}	25.76
1:4			8.80×10^{-4}	58.24
1:1	−1.0	5	1.55×10^{-2}	32.68
1:2			1.90×10^{-2}	37.27
1:4			5.00×10^{-2}	70.32
1:1	−1.0	10	9.25×10^{-3}	38.90
1:2			3.39×10^{-2}	53.30
1:4			3.03×10^{-2}	79.60
1:1	−1.0	20	7.50×10^{-3}	45.61
1:2			3.40×10^{-3}	49.97
1:4			3.90×10^{-3}	61.57
1:1	−1.0	40	1.20×10^{-3}	49.30
1:2			2.35×10^{-3}	56.18
1:4			1.45×10^{-4}	63.43
1:1	−1.0	80	4.10×10^{-5}	51.41
1:2			6.60×10^{-5}	66.54
1:4			5.40×10^{-5}	71.59
1:1	0.0	10	9.67×10^{-3}	41.90
1:2			2.90×10^{-2}	56.70
1:4			3.03×10^{-2}	79.60

performance of rock-armoured seawalls without crown walls. However, the results from two tests by Bradbury *et al.* (1988) may be used to give estimates of the influence of wave conditions and relative freeboard. Again, the test results have been used to given values of coefficients in an empirical equation. To obtain the best fit to the prediction equation, Bradbury *et al.* have revised Owen's parameter R_m to give F^*:

$$F^* = R_c/H_s * R_m^* = [R_c/H_s]^2 \sqrt{(s_m/2\pi)} \qquad (5.28)$$

Predictions of overtopping discharge can then be made using

$$Q_m^* = a(F^*)^{-b} \qquad (5.29)$$

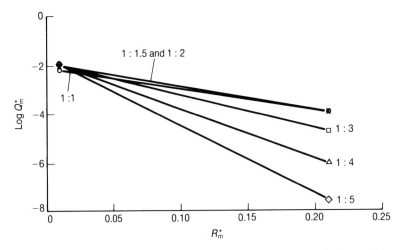

Figure 154 Overtopping discharges for smooth, straight slopes. T_m is used in R_m^* and Q_m^*

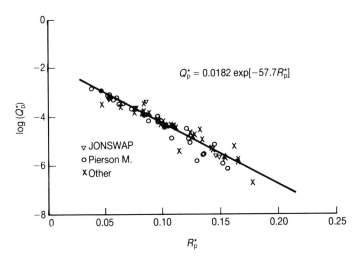

Figure 155 Overtopping of a smooth 1:4 slope described using T_p

Values of a and b have been calculated from the results of tests with a rock-armoured slope at 1:2 with the crest details shown in Figure 156. For section A, $a = 3.7*10^{-10}$ and $b = 2.92$. For section B, $a = 1.3*10^{-9}$ and $b = 3.82$.

More data are available to describe the overtopping performance of rock-armoured structures with crown walls. Tests have been conducted by Bradbury and Allsop (1988) and Aminti and Franco (1988), and the results have been used to determine values for coefficients a and b in equation (5.29). The cross-sections studied are illustrated in Figure 157 and values for the coefficients are given in Table 34. Each of these studies have used different dimensionless freeboard and discharge parameters in empirical equations that are valid for a range of different structural configurations. The data have not been analysed as a single set, and the user may therefore need to compare the results given by more than one prediction method.

5.1.2.3 Transmission

Structures such as breakwaters constructed with low crest levels will transmit wave energy into the area behind the breakwater. The transmission performance of low-crested breakwaters is dependent on the structure geometry, principally the crest

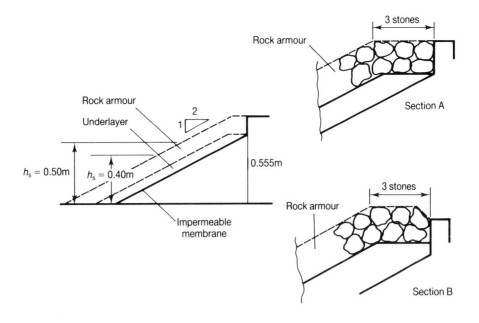

Figure 156 *Overtopped rock structures with low crown walls*

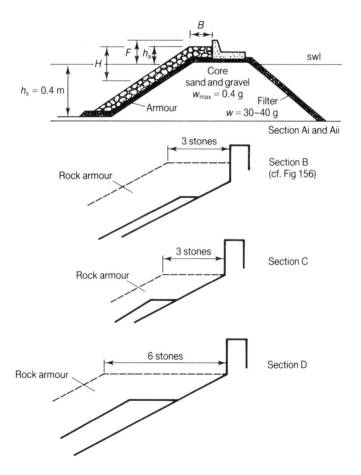

Figure 157 *Studies of tested cross-sections*

Table 34 Coefficients *a* and *b* in equation (5.29) for overtopping discharges over cross-sections in Figure 157

Section	Slope angle	B/H_s	a	b
Ai	1:2.0	1.10	1.7×10^{-8}	2.41
		1.85	1.8×10^{-7}	2.30
		2.60	2.3×10^{-8}	2.68
Aii	1:1.33	1.10	5.0×10^{-8}	3.10
		1.85	6.8×10^{-8}	2.65
		2.60	3.1×10^{-8}	2.69
B	1:2	0.79–1.7	1.6×10^{-9}	3.18
C	1:2	0.79–1.7	5.3×10^{-9}	3.51
D	1:2	1.6–3.3	1.0×10^{-9}	2.82

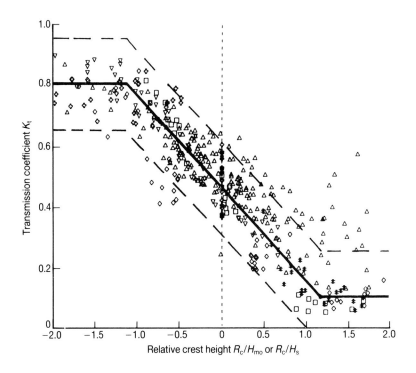

Figure 158 *Wave transmission over and through low-crested structures*

freeboard, crest width and water depth, permeability, and on the wave conditions, principally the wave period.

Hydraulic model test results measured by Seelig (1980), Allsop and Powell (1985), Daemrich and Kahle (1985), Ahrens (1987) and van der Meer (1990a) have been re-analysed for this manual by van der Meer (1990b) to give a single prediction method. This relates C_t to the relative crest freeboard, R_c/H_s. The data used are plotted in Figure 158 and the prediction equations describing the data may be summarised:

Range of validity	Equation	
$-2.00 < R_c/H_s < -1.13$	$C_t = 0.80$	(5.30)
$-1.13 < R_c/H_s < 1.2$	$C_t = 0.46 - 0.3 R_c/H_s$	(5.31)
$1.2 < R_c/H_s < 2.0$	$C_t = 0.10$	(5.32)

Figure 158 shows all results together with the equations. These equations give a very simplistic description of the data available, but will often be sufficient for a preliminary estimate of performance. The upper and lower bounds of the data considered are shown by lines 0.15 higher (or lower) than the mean lines given left. This corresponds with the 90% confidence bands (the standard deviation was 0.09).

A few comments can be made on Figure 158. The points with $R_c/H_s > 1$ and $C_t > 0.15$ are caused by a low wave height relative to the stone diameter ($H_s =$ about D_{n50}). The low wave travels simply through the crest which consists of armourstones. Transmission coefficients of 0.5 can be found in such cases. A structure under design conditions (with regard to stability) with $R_c/H_s > 1$ will always show transmission coefficients smaller than 0.1. For the range of low wave heights compared to the stone diameter and $R_c/H_s > 1$, Ahrens (1987) gave a relationship which has much less scatter than shown in Figure 158:

$$C_t = 1.0/[1.0 + X^{0.592}] \quad \text{for } R_c/H_s > 1 \tag{5.33}$$

where X is defined by:

$$X = H_s A_t / L_p D_{n50}^2 \tag{5.34}$$

with $A_t =$ cross-sectional area of structure and $L_p =$ local wave length.

Transmission through the structure may be computed by numerical models. The HADEER model described in Box 62 in Section 5.2 can give an estimate of the transmission through (not over) a permeable structure.

Another reason for the scatter in Figure 158 is the influence of the wave period. A longer wave period gives always a higher wave transmission coefficient and will be located in the upper range of the test results.

5.1.2.4 Reflections

Waves will reflect from nearly all coastal or shoreline structures. For structures with non-porous and steep faces, approximately 100% of the wave energy incident upon the structure will reflect. Rubble slopes are often used in habour and coastal engineering to absorb wave action. Such slopes will generally reflect significantly less wave energy than the equivalent non-porous or smooth slope. Although some of the flow processes are different, it has been found convenient to calculate the reflection performance given by C_r using an equation of the same form as for non-porous slopes, but with different values of the empirical coefficients to match the alternative construction. Data for random waves are available for smooth and armoured slopes at angles between 1:1.5 and 1:2.5 (smooth) and 1:1.5 and 1:6 (rock).

Data of Allsop and Chanell (1988) will be given here as well as data of van der Meer (1988a), analysed by Postma (1989). Formulae of other references will be used for comparison.

Battjes (1974) gives for smooth impermeable slopes:

$$C_r = 0.1 \xi^2 \tag{5.35}$$

Seelig and Ahrens (1981) give:

$$C_r = a\xi_p^2 / (b + \xi_p^2) \tag{5.36}$$

with $a = 1.0$, $b = 5.5$ for smooth slopes
$a = 0.6$, $b = 6.6$ for a conservative estimate of rough permeable slopes

Equations (5.35) and (5.36) are shown in Figure 159 together with the reflection data of van der Meer (1988a) for rock slopes. The two curves for smooth slopes are close. The curve of Seelig and Ahrens for permeable slopes is not a conservative estimate, but even underestimate the reflection for large ξ_p values.

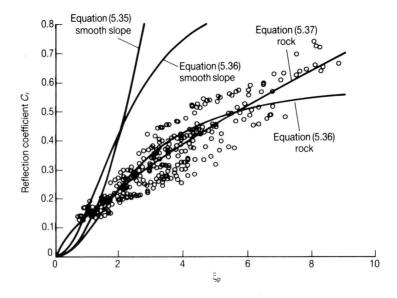

Figure 159 *Comparison of data on rock slopes*

The best-fit curve through all the data points in Figure 159 is given by Postma (1989) and is also shown in the Figure:

$$C_r = 0.14 \xi_p^{0.73} \quad \text{with} \quad \sigma(C_r) = 0.055 \tag{5.37}$$

The surf similarity did not describe sufficiently the combined slope angle–wave steepness influence. Therefore, both the slope angle and wave steepness were treated separately and Postma derived the following relationship:

$$C_r = 0.071 P^{-0.082} \cot\alpha^{-0.62} s_{op}^{-0.46} \tag{5.38}$$

where $\sigma(C_r) = 0.036$
P = notional permeability factor described in Section 5.1.1.2

The standard deviation of 0.55 in equation (5.37) reduced to 0.036 in equation (5.38), which is a considerable increase in reliability.

The results of random wave tests by Allsop and Channell (1989), analysed to give values for the coefficients a and b in equation (5.36) (but with ξ_m instead of ξ_p), are presented below. The rock-armoured slopes used rock in two or one layer, placed on an impermeable slope covered by underlayer stone, equivalent to $P = 0.1$. The range of wave conditions for which these results may be used is given by:

$$0.004 < s_m < 0.052 \quad \text{and} \quad 0.6 < H_s/\Delta D_{n50} < 1.9.$$

Slope type	*a*	*b*
Smooth	0.96	4.80
Rock, two-layer	0.64	8.85
Rock, one-layer	0.64	7.22

Postma (1989) also re-analysed the data of Allsop and Channell which were described above. Figure 160 gives the data of Allsop and Channell together with equation (5.37). The curve is a little higher than the average of the data. The best-fit curve is described by:

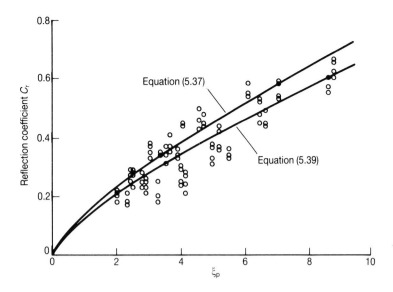

Figure 160 Wave run-up data

$$C_r = 0.125 \xi_p^{0.73} \quad \text{with} \quad \sigma(C_r) = 0.060 \tag{5.39}$$

There are no reliable general data available on the reflection performance of rough, non-porous slopes. In general, a small reduction in the level of reflections might be expected, much as for wave run-up (see Section 5.1.2.1). Reduction factors have not, however, been derived from test. It is not therefore recommended that values of C_r lower than for the equivalent smooth slope be used, unless test data are available.

Some structures may incorporate a step or berm into the armoured slope at or near the water level. This berm length, B, may lead to a further reduction in C_r. Few data are available for such configurations. Example results from Allsop and Channell (1989) are shown in Figure 161 in terms of the relative berm length, B/L_{ms}, where the wave length of the mean period is calculated for the water depth at the structure, h_s.

An alternative use of rock armouring is in a mound placed against a vertical wall. Such protection will significantly reduce reflections, protecting the wall from wave impact, and reduce any scour problems at the toe of the seawall. No general method is available to predict the reflection performance of such a structure, but see Box 47 for an example of results from a site-specific study.

5.1.3 Structural response

5.1.3.1 Introduction

The hydraulic and structural parameters have been described in Section 5.1.1 and the hydraulic responses in Section 5.1.2. Figure 139 gives an overview of the definitions of the hydraulic parameters and responses as wave run-up, run-down, overtopping, transmission and reflection and Figure 143 shows the structural parameters which are related to the cross-section. The response of the structure under hydraulic loads will be described in this section and design tools will be given.

The design tools described here will permit the design of many structure types. Nevertheless, it should be remembered that each design rule has its limitations. For each structure which is important and expensive to build, it is advised to perform physical model studies.

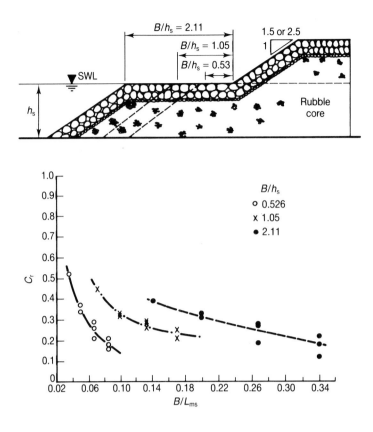

Figure 161 *Effective relative berm length to local wave length on reflection*

Figure 163 gives the same cross-section as in Figure 142, but it now shows the various parts of the structure which will be described in the following sections. Some general points and design rules for the geometrical design of the cross-section will be given here. These are:

- The minimum crest width
- The thickness of (armour layers)
- The number of units or rocks per surface area
- The bottom elevation of the armour layer.

The crest width is normally determined by construction methods used (access on the core by trucks or crane) or by functional requirements (road/crown wall on the top). Where the width of the crest can be small a minimum width, B_{min}, should be provided where (*SPM*, 1984):

$$B_{min} = (3-4)D_{n50} \qquad (5.40)$$

The thickness of layers and the numbers of units per square metre are given in Box 48. The number of units in a rock layer depends on the grading of the rock. The values of k_t which are given in the box describe a rather narrow grading (uniform stones). For rip-rap and even wider graded material the number of stones cannot easily be estimated. In that case the volume of the rock on the structure can be used.

The bottom elevation of the armour layer should be extended downslope to an elevation below minimum still-water level of at least one (significant) wave height, if the wave height is not limited by the water depth. Under depth-limited conditions the armour layer should be extended to the bottom as shown in Figure 163 and supported by a toe. Further details regarding specific structures are given in Chapter 6.

Box 47 Example of reflection from rubble placed against a vertical wall

Tests results are available to describe the reflection performance of typical examples of rubble protection to an existing seawall. Measurements of reflections for three sections are presented in Figure 162. As expected, the existing wall (section 1) has high reflections. At the higher water levels, C_r approaches 0.9, but at lower water levels, wave breaking in front of the wall has reduced the reflections to $C_r \approx 0.65$. The reflection performance of the alternative rock protection (sections 2 and 3) varies with water level, particularly with the relative position of the berm formed by the crest of the rock protection. For those water levels close to the crest level of the rock the reflection coefficient reaches minimum values with C_r around 0.20–0.30. At lower water levels the waves reflect from the armour slope, and the performance deteriorates slightly with C_r generally nearer 0.4.

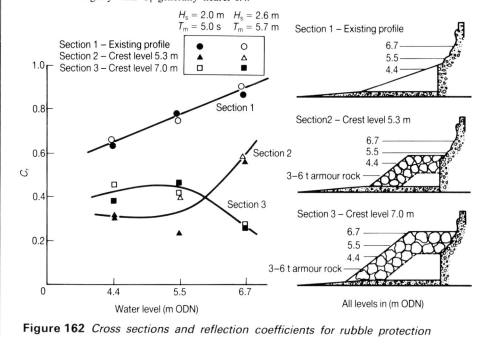

Figure 162 *Cross sections and reflection coefficients for rubble protection placed against a vertical wall*

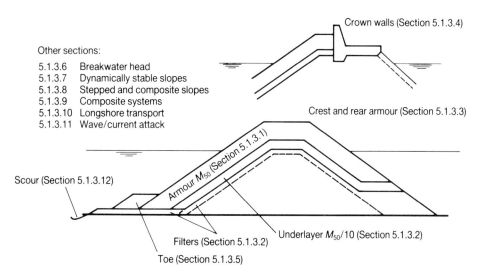

Figure 163 *Various parts of a structure whose hydraulic design is described in this chapter*

> **Box 48** Thickness of layers and number of units
>
> The thickness of layers is given by:
>
> $$t_a = t_u = t_f = n k_t D_{n50} \qquad (5.41a)$$
>
> The number of units per square metre is given by:
>
> $$N_a = n k_t (1 - n_v) D_{n50}^{-2} \qquad (5.41b)$$
>
> where t_a, t_u, t_f = thickness of armour, underlayer or filter,
> n = number of layers,
> k_t = layer thickness coefficients,
> n_v = volumetric porosity.
>
> Values of k_t and n_v for rock are taken from Section 3.2.3. In that section more data are given concerning placement.
>
> | Shape of rock | Placement | k_t | n_v |
> |---|---|---|---|
> | Irregular | Random | 0.75 | 0.40 |
> | Irregular | Special | 1.05–1.20 | 0.39 |
> | Semi-round | Random | 0.75 | 0.37 |
> | Semi-round | Special | 1.10–1.25 | 0.36 |
> | Equant | Random | 0.80 | 0.38 |
> | Equant | Special | 1.00–1.15 | 0.37 |
> | Very round | Random | 0.80 | 0.36 |
> | Very round | Special | 1.05–1.20 | 0.35 |

5.1.3.2 Armour layers

Many methods for the prediction of rock size of armour units designed for wave attack have been proposed in the last 50 years. Those treated in more detail here are the Hudson formula as used in the *SPM* (1984) and the formulae derived by van der Meer (1988a).

Hudson formula

The original Hudson formula is written as:

$$W_{50} = \frac{\rho_r g H^3}{K_D \Delta^3 \cot \alpha} \qquad (5.42)$$

K_D is a stability coefficient taking into account all other variables. K_D values suggested for design correspond to a 'no-damage' condition where up to 5% of the armour units may be displaced. In the 1973 edition of the *Shore Protection Manual* the values given for K_D for rough, angular stone in two layers on a breakwater trunk were:

$K_D = 3.5$ for breaking waves
$K_D = 4.0$ for non-breaking waves

The definition of breaking and non-breaking waves is different from plunging and surging waves, which were described in Section 5.1.1.1. A breaking wave in equation (5.42) means that the wave breaks due to the foreshore in front of the structure directly on the armour layer. It does not describe the type of breaking due to the slope of the structure itself.

No tests with random waves had been conducted, and it was suggested that H_s be used in equation (5.42). By 1984 the advice given was more cautious. The *Shore Protection Manual* now recommends $H = H_{1/10}$, being the average of the highest 10% of all waves. For the case considered above, the value of K_D for breaking waves was revised downwards from 3.5 to 2.0 (for non-breaking waves it remained 4.0). The effect of these two changes is equivalent to an increase in the unit stone weight required by a factor of about 3.5 and a very conservative change to the design approach.

The main advantage of the Hudson formula are its simplicity, and the wide range of armour units and configurations for which values of K_D have been derived. The formula also has many limitations. Briefly, these include:

- Potential scale effects due to the small scales at which most of the tests were conducted;
- The use of regular waves only;
- No account is taken in the formula of wave period or storm duration;
- No description of the damage level;
- The use of non-overtopped and permeable core structures only.

The use of $K_D \cot \alpha$ does not always best describe the effect of the slope angle. It may therefore be convenient to define a single stability number without this $K_D \cot \alpha$. Further, it may often be more helpful to work in terms of a linear armour size, such as a typical or nominal diameter. The Hudson formula can be rearranged to:

$$H_s/\Delta D_{n50} = N_s = (K_D \cot \alpha)^{1/3} \tag{5.43}$$

This equation shows that the Hudson formula can be written in terms of the structural parameter $H_s/\Delta D_{n50}$, which was discussed in Section 5.1.1.2.

Van der Meer formulae—deep-water conditions

Based on earlier work of Thompson and Shuttler (1975), an extensive series of model tests was conducted at Delft Hydraulics (van der Meer (1988a)). These include structures with a wide range of core/underlayer permeabilities and a greater number of wave conditions. Two formulae were derived for plunging and surging waves, respectively:

For plunging waves:

$$H_s/\Delta D_{n50} = 6.2 P^{0.18} (S_d/\sqrt{N})^{0.2} \xi_m^{-0.5} \tag{5.44}$$

and for surging waves:

$$H_s/\Delta D_{n50} = 1.0 P^{-0.13} (S_d/\sqrt{N})^{0.2} \sqrt{(\cot \alpha)} \xi_m^P \tag{5.45}$$

The transition from plunging to surging waves can be calculated using a critical value of ξ_m:

$$\xi_{mc} = [6.2 P^{0.31} \sqrt{(\tan \alpha)}]^{1/(P+0.5)} \tag{5.46}$$

For $\cot \alpha \geq 4.0$ the transition from plunging to surging does not exist and for these slope angles only equation (5.44) should be used. All parameters used in van der Meer's formulae (equations (5.44)–(5.46)) are described in Section 5.1.1. The notional permeability factor P is shown in Figure 143. The factor P should lie between 0.1 and 0.6.

Design values for the damage level S_d are shown in Table 31. The level 'start' of damage, $S_d = 2$–3, is equal to the definition of 'no damage' in the Hudson formula (equation (5.43)). The maximum number of waves, N, which should be used in equations (5.44) and (5.45) is 7500. After this number of waves the structure has more or less reached an equilibrium.

The wave steepness should lie between $0.005 < s_m < 0.06$ (almost the complete possible range). The mass density, ρ_{ssd}, varied in the tests between 2000 kg/m^3 and 3100 kg/m^3, which is also the possible range of application.

The reliability of the formulae depends on the differences due to random behaviour of rock slopes, accuracy of measuring damage and curve fitting of the test results. The reliability of van der Meer's formula can be expressed by giving the coefficients 6.2 and 1.0 in equations (5.44) and (5.45) a normal distribution with a certain standard deviation. The coefficient 6.2 can be described by a standard deviation of 0.8 (variation coefficient 6.5%) and the coefficient 1.0 by a standard deviation of 0.08 (8%). These values are significantly lower than that for the Hudson formula at 18% for $K_D^{1/3}$ (with mean K_D of 4.5). With these standard deviations it is simple to include 90% or other confidence bands.

Van der Meer's formulae (equations (5.44)–(5.46)) are more complex than the Hudson formula (5.43). They also include the effect of the wave period, the storm duration, the permeability of the structure and a clearly defined damage level. This may cause differences between the Hudson formula and the van der Meer formulae (equations (5.44)–(5.46)). Box 49 gives a comparison between the formulae. Nevertheless, it is more difficult to work with the van der Meer formulae since for a good design a sensitivity analysis should be performed for all parameters in the equations.

Box 49 Comparison of Hudson and van der Meer formulae

The $H_s/\Delta D_{n50}$ in the Hudson formula is only related to the slope angle $\cot\alpha$. Therefore a plot of $H_s/\Delta D_{n50}$ or N_s versus $\cot\alpha$ shows one curve for the Hudson formula. Equations (5.44)–(5.46) take into account the wave period (or steepness), the permeability of the structure and the storm duration. The effect of these parameters is shown in Figure 164.

Figure 164(a) shows the curves for a permeable structure after a storm duration of 1000 waves (a little more than the number used by Hudson). The Figure 164(b) gives the stability of an impermeable revetment after wave attack of 5000 waves (equivalent to 5–10 hours) in nature. Curves are shown for various wave steepnesses.

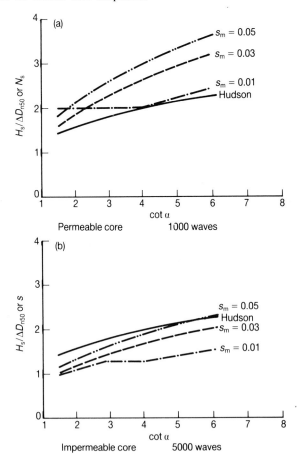

Figure 164 *Examples of stability number versus slope angle plots*

The deterministic procedure is to make design graphs where one parameter is evaluated. Three examples are shown in Boxes 50–52. To give a wave height versus surf similarity plot, which shows the influence of both wave height and wave steepness (the wave climate). The other shows a wave height versus damage plot which is comparable with the conventional way of presenting results of model tests on stability. The same kind of plots can be derived from equations (5.44)–(5.46) for other parameters (see van der Meer, 1988a).

A deterministic design procedure is followed if the stability equations are used to produce design graphs as H_s versus ξ_m and H_s versus damage (see Boxes 50–52) and if a sensitivity analysis is performed., Another design procedure is the probabilistic approach. Equations (5.44) and (5.45) can be rewritten to so-called 'reliability functions' and all the parameters can be assumed to be stochastic with an assumed distribution. The procedure of the probabilistic approach is described in more detail in Section 2.3.2. Here only one example of the approach will be given. A more detailed description can be found in van der Meer (1988b).

The structure parameters with the mean value, distribution type and standard deviation are given in Table 35. These values were used in a Level II first-order second-moment (FOSM) with the approximate full distribution approach (AFDA) method. With this method the probability that a certain damage level would be exceeded in one year was calculated. These probabilities were used to estimate the probability that a certain damage level would be exceeded in a certain lifetime of the structure.

The parameter FH_s represents the uncertainty of the wave height at a certain return period and the wave height itself is defined by a two-parameter Weibull distribution. The coefficients a and b take into account the reliability of the formula, including the random behaviour of rock slopes.

The final results are shown in Figure 165, where the damage S_d is plotted versus the probability of exceedance in the lifetime of the structure. From this figure it follows that start of damage ($S_d = 2$) will certainly occur in a lifetime of 50 years. Tolerable damage ($S_d = 5$–8) in the same lifetime will occur with a probability of 0.2–0.5. The probability that the filterlayer will become visible (failure) is less than 0.1. Probability curves as shown in Figure 165 can be used to make a cost optimization for the structure during its lifetime, including maintenance and repair at certain damage levels. Assessment of the damage S_d in prototype should be by profiles measured according to the methods described in Appendices A and 6 and compared as described in Tables 45 and 49 in Chapter 7.

Table 35 Parameters used in Level II probabilistic calculations

Parameter	Distribution	Average	Standard deviation
D_{n50}	Normal	1.0	0.03
Δ	Normal	1.6	0.05
$\cot \alpha$	Normal	3.0	0.15
P	Normal	0.5	0.05
N	Normal	3000	1500
H_s	Weibull	$B = 0.3$	$C = 2.5$
FH_s	Normal	0	0.25
s_m	Normal	0.04	0.01
C_{pl} (equation (5.44))	Normal	6.2	0.4
C_{su} (equation (5.45))	Normal	1.0	0.08

Box 50 Example of plot of H_s versus ξ_m showing influence of damage levels

The parameter whose influence is shown is the damage level S_d. Four damage levels are given: $S_d = 2$ (start of damage), $S = 5$ and 8 (intermediate damage) and $S_d = 12$ (filterlayer visible). The structure itself is described by: $D_{n50} = 1.0$ m ($W_{50} = 2.6$ tonnes), $\Delta = 1.6$, $\cot \alpha = 3.0$, $P = 0.5$ and $N = 3000$.

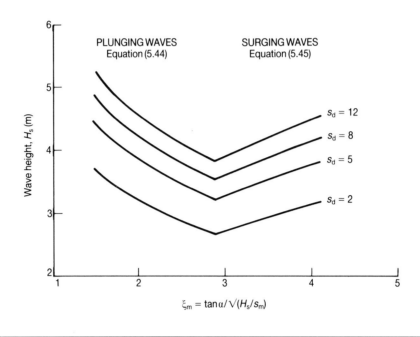

Box 51 Example of plot of H_s versus ξ_m showing influence of permeability

The parameter whose influence is shown is the notational permeability factor P. Four values are given: $P = 0.1$ (impermeable core), $P = 0.3$ (some permeable core), $P = 0.5$ (permeable core) and $P = 0.6$ (homogeneous structure). The structure itself is described by: $D_{n50} = 1.0$ m ($W_{50} = 2.6$ tonnes), $\Delta = 1.6$, $\cot \alpha = 3.0$, $P = 0.5$ and $N = 3000$.

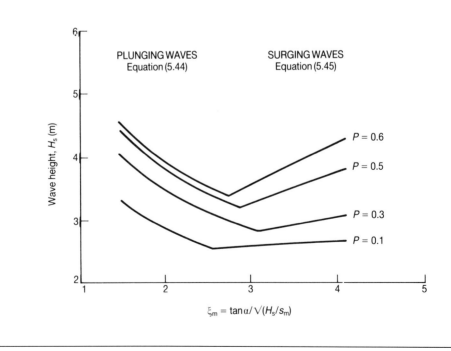

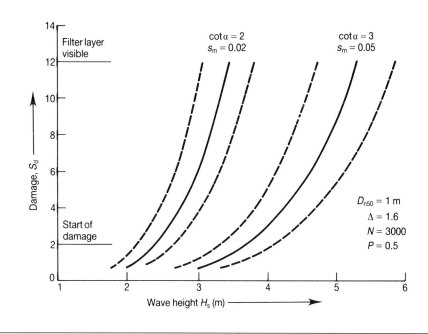

Box 52 Example of wave height—damage graph

Two curves are shown, one for a slope angle with $\cot\alpha = 2.0$ and a wave steepness of $s_m = 0.02$ and one for a slope angle with $\cot\alpha = 3.0$ and a wave steepness of 0.05. If the extreme wave climate is known, plots as shown in this box are very useful to determine the stability of the armour layer of the structure. The graph also shows the 90% confidence levels which give a good idea about the possible variation in stability. This variation should be taken into account by the designer of a structure.

Van der Meer's formulae—shallow-water conditions

Up to now the significant wave height, H_s, has been used in the stability equations. In shallow-water conditions the distribution of the wave heights deviate from the Rayleigh distribution (truncation of the curve due to wave breaking). Further tests on a 1:30 sloping and depth-limited foreshore by van der Meer (1988a) showed that $H_{2\%}$ was the best value for the design on depth-limited foreshores, i.e. that the stability of the armour layer in depth-limited situations is better described by $H_{2\%}$ than by H_s.

Equations (5.44)–(5.46) can be rearranged with the known ratio of $H_{2\%}/H_s$. The equations become:

For plunging waves:

$$H_{2\%}/\Delta D_{n50} = 8.7 P^{0.18} S_d/\sqrt{N})^{0.2} \xi_m^{-0.5} \qquad (5.47)$$

and for surging wave:

$$H_{2\%}/\Delta D_{n50} = 1.4 P^{-0.13}(S_d/\sqrt{N})^{0.2} \sqrt{(\cot\alpha)} \xi_m^P \qquad (5.48)$$

Equations (5.47) and (5.48) take into account the effect of depth-limited situations. A safe approach, however, is to use equations (5.44) and (5.45) with H_s. In that case the truncation of the wave height exceedance curve due to wave breaking is not taken into account, which can be assumed as a safe approach. If the wave heights are Rayleigh distributed, equations (5.47) and (5.48) give the same results as (5.44) and (5.45), as this is caused by the known ratio of $H_{2\%}/H_s = 1.4$. For depth-limited conditions the ratio of $H_{2\%}/H_s$ will be smaller, and one should obtain information on the actual value of this ratio. Section 4.2.3.2, Box 43, gives

Goda's method for estimating $H_{max} = H_{1/250} = H_{0.4\%}$ in depth-limited conditions, and this can be used to make an estimate of $H_{2\%}$.

Influence of grading and shape on stability

In Section 3.2.2 the block shape was discussed and the Fourier asperity roughness parameter, P_R, was given for five block shapes. Rock structures with these shapes were tested (Latham *et al.*, 1988) and the influence of reduced armour layer thickness and block shape was recently given (Bradbury *et al.*, 1990) by two modifications to equations (5.44) and (5.45) as described below:

1. The individual but random placement of stones into a two-layer armour system resulted in a reduced thickness of $t_a \approx 1.6 D_{n50}$. (See also Section 3.2.3 and Table 20 for a warning on realistic layer thicknesses.) Such armour layers gave a greater damage than for typical rip-rap layer thicknesses of $t_a \approx 2.2$ to $2.4 D_{n50}$. The damage could be predicted assuming that the power 0.2 for the $S_d/\sqrt{N}$ term in equations (5.44) and (5.45) is replaced by 0.25.
2. The coefficient 6.2 in equation (5.44), called C_{pl}, and the coefficient 1.0 in equation (5.45), called C_{su}, were used to describe the shape effects, where

$$C_{pl} = 5.4 + 70 P_R \tag{5.49}$$

$$C_{su} = 0.6 + 40 P_R \tag{5.50}$$

The values of P_R for five block shapes (see also Section 3.2.2) are:

Elongate/tabular $P_R = 0.017$
Irregular $P_R = 0.014$
Equant $P_R = 0.012$
Semi-round $P_R = 0.010$
Very round $P_R = 0.005$

With the value of $P_R = 0.010$ for semi-round rock and equant rock, the coefficients 6.2 and 1.0 can be found. Higher values of P_R indicate greater stability while very round rock gives lowest stability. Thus values of 6.2 and 1.0 will generally indicate a safe approach for prototype blocks which are usually found to have $P_R > 0.012$.

As the results are based on only a limited number of tests (1:2 slope, $t_a \approx 1.6 D_{n50}$ and impermeable core), modifications (1) and (2) above should be used only with care. However, modification (1) for reduced layer thickness will give the safer approach for structures with prototype conditions and random double-armour layer construction methods similar to those used in the model tests. In Box 53 the influence of grading on stability is discussed, where it is concluded that the van der Meer formulae should be limited in application for revetments to gradings where $D_{n50}/D_{n15} < 2.5$.

Stability against ship wave attack

Use of the van der Meer or other formulae for design against ship wave attack has not been assessed in model tests, and thus there is no definitive guidance to the designer on this subject. It is recommended that model tests be carried out for specific projects where ship wave attack is significant. For preliminary design, energy considerations indicate that ship waves of height H should be conservatively assumed to be equivalent in effect to random waves of significant height H_s at least $1.5H$.

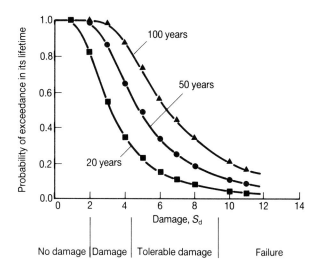

Figure 165 *Probability of exceedence of the damage level S_d in the lifetime of the structure*

Box 53 Stability of armour layers with very wide gradings

A few studies were undertaken for this manual, and one was a model study of the stability of very wide gradings with $D_{n85}/D_{n15} = 4.0$. A limited number of tests were performed on a 1:2 structure with an impermeable core (revetment). The tests showed displacement of first, small rock and then larger rock. Two tests showed large damage ($S_d = 10$–13) and a repetition of these tests with exactly the same conditions showed no damage at all ($S_d = 2$). This large scatter may be an effect of the very wide grading. Furthermore, it is difficult to obtain a good gradation over the length of the structure, and the W_{50} or D_{n50}, may change considerably along it.

Based on the model tests and the difficulties in construction of a homogeneous armour layer, it is advised *not* to use very wide gradings ($D_{n85}/D_{n15} > 2.5$) for a revetment of limited thickness (2–$3\, D_{n50}$). It may be possible to use very wide gradings for reef-type structures which consist only of a homogeneous mass of stone. Model tests are required in that case. Box 45 gave examples of narrow, wide and very wide gradings. The model tests are described by Allsop (1990).

5.1.3.3 Underlayers and filters

Structures in coastal and shoreline protection are normally constructed with an armour layer and one or more underlayers. Sometimes an underlayer is called a filter. The dimensions of the first underlayer depend on the structure type.

Revetments often have a two-diameter thick armour layer, a thin underlayer or filter and then an impermeable structure (clay or sand), with or without a geotextile. The underlayer in this case works as a filter. Smaller particles beneath the filter should not be washed through the layer and the filter stones should not be washed through the armour. In this case the geotechnical filter rules in Section 5.2.5.4 are strongly recommended. Roughly, these rules give $D_{15}(\text{armour})/D_{85}(\text{filter}) < 4$–$5$.

Structures such as breakwaters have one or two underlayers followed by a core of rather fine material (quarry run). The *SPM* (1984) recommends for the stone size of the underlayer under the armour a range of 1/10 to 1/15 of the armour mass. This criterion is more strict than the geotechnical filter rules and gives $D_{n50}(\text{armour})/D_{n50}(\text{underlayer}) = 2.2$–$2.3$. A relatively large underlayer has two advantages. First, the surface of the underlayer is less smooth with larger stones and gives more interlocking with the armour. This is specially the case if the armour layer is constructed of concrete armour units. Second, a large underlayer results in a more permeable structure and therefore has a large influence on the stability (or required mass) of the armour layer. The influence of the permeability on stability has been described in Section 5.1.3.1. Therefore it is recommended to use size of 1/10 to 1/15 M_{50} of the armour for the mass of the underlayer.

5.1.3.4 Crest and rear armour (low-crested structures)

As long as structures are high enough to prevent overtopping the armour on the crest and rear can be (much) smaller than on the front face. The dimensions of the rock in that case will be determined by practical matters as available rock, etc. If this is not the case, the preliminary approach suggested by Pilarxzyk (1990) should be used (see Box 54).

Most structures, however, are designed to have some or even severe overtopping under design conditions. Others are so low that also under daily conditions the structure is overtopped. Structures with the crest level around still-water level and sometimes far below it will always have overtopping and transmission.

It is obvious that when the crest level of a structure is low, wave energy can pass over the structure. This has two effects. First, the armour on the front side can be

---

Box 54 Protection against overtopping

No definite method for designing against overtopping is known; it is due to the lack of the proper method on estimating the hydraulic loading. Pilarczyk (1990) proposes the following indicative way of designing the splash area (see Figure 166):

a. Armour stability:

$$H_s/\Delta D_{n50} = 2.25 \; \cos\alpha_i \xi_p^{-0.5}/[1-(R_c/R_{u2\%})]$$

b. Width of protection, L_s, required in the splash area:

$$L_s = 0.2\psi T_p \sqrt{[g(R_{u2\%}-R_c)]} \geq L_{min}$$

where the parameters have their normal meanings and:

α_i = slope angle of crest or leeside slope of seawall or dike over the length L_s requiring protection,
R_c = crest height above still-water level,
$R_{u2\%}$ = 2% wave run-up level on an (imaginary) plane slope,
ψ = factor related to the importance of the structure and its function ($\psi \geq 1$).

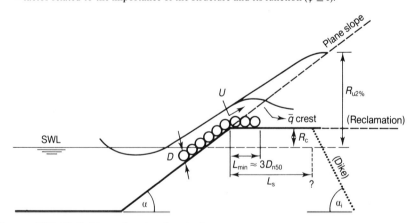

Figure 166 Definition sketch of splash area

The practical minimum width of protection (L_{min}) is equal to about $3D_{n50}$, where D_{n50} related to the armour used on the front-face slope. L_{min} can also be considered as the minimum transition length from the slope regime to the crest regime. For specific projects model studies are recommended.

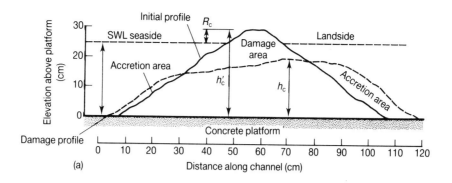

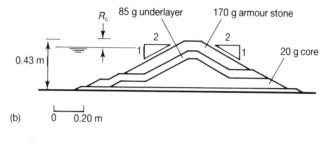

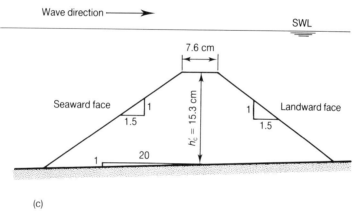

Figure 167 *Cross-sections of low-crested structures. (a) Cross-sectional view of initial and typical damaged reef profiles (SWL denotes still-water level); (b) overtopped breakwater; (c) submerged breakwater*

smaller than on a non-overtopped structure, due to the fact that energy is lost on the front side. The second effect is that the crest and rear should be armoured with rock which can withstand the attack by overtopping waves. For rock structures the same armour on front face, crest and rear is often applied. The methods to establish the armour size for these structures will be given here.

Low-crested structures can be divided into three categories, as shown in Figure 167:

1. *Dynamically stable reef breakwaters.* A reef breakwater is a low-crested homogeneous pile of stones without a filterlayer or core and is allowed to be reshaped by wave attack (Figure 167(a), Ahrens, 1987). The equilibrium crest height, with corresponding transmission, are the main design parameters. The transmission has been described in Section 5.1.2.2.
2. *Statically stable low-crested breakwaters* ($R_c > 0$). These are close to

non-overtopped structures, but are more stable due to the fact that a (large) part of the wave energy can pass over the breakwater (Figure 167(b), Powell and Allsop, 1985).

3. *Statically stable submerged breakwaters* ($R_c > 0$). All waves overtop these structures and the stability increases considerably if the crest height decreases (Figure 167(c), Givler and Sorenson, 1986).

Dynamically stable reef breakwaters (Figure 167(a))

The analyses of stability by Ahrens (1987) and van der Meer (1990(a) concentrated on the change in crest height due to wave attack (see Figure 167(a)). Ahrens defined a number of dimensionless parameters which described the behaviour of the structure. The main one is the relative crest height reduction factor h_c/h'_c. This factor is the ratio of the crest height at the completion of a test to the height at the beginning of the test. The natural limiting values of h_c/h'_c are 1.0 and 0.0, respectively. Ahrens found for the reef breakwater that a longer wave period gave more displacement of material than a shorter period. Therefore he introduced the spectral (or modified) stability number, N_s^*, defined by equation (5.9).

The relative crest height can be described by:

$$h_c/h = \sqrt{[A_t/\exp(aN_s^*)]} \tag{5.51}$$

with

$$a = -0.028 + 0.045C' + 0.034h'_c/h - 6.10^{-9}B_n^2 \tag{5.52}$$

and $h_c = h'_c$ if h_c in equation $(5.51) > h'_c$.

Equation (5.52) was derived by van der Meer (1990a), including all Ahrens' (1987) tests. The parameters are given by:

A_t = area of structure cross-section,
$C' = A_t/h'_c$ (response slope),
h = water depth at structure toe,
$B_n = A_t/D_{n50}^2$ (bulk number)

The lowering of the crest height of reef-type structures as shown in Figure 167(a) can be calculated with equations (5.51) and (5.52). It is possible to draw design curves from these equations which give the crest height as a function of N_s^*. An example is shown in Box 55.

Statically stable low-crested breakwaters (Figure 167(b))

The stability of a low-crested breakwater (overtopped, $R_c > 0$) can be related to that of a non-overtopped structure. Stability equations such as (5.44) and (5.45) can be used, for example, but in fact each stability formula can be employed. The required stone diameter for an overtopping breakwater can then be determined by a reduction factor for the mass of the armour, compared to the mass for a non-overtopped structure. The derived equations are based on specific analysis for this manual (van der Meer 1990a):

Reduction factor for

$$D_{n50} = 1/(1.25 - 4.8R_p^*) \tag{5.53}$$

for

$$0 < R_p^* < 0.052$$

where

$$R_p^* = R_c/H_s \sqrt{(s_{op}/2\pi)} \tag{5.54}$$

The R_p^* parameter is a combination of relative crest height, R_c/H_s and wave steepness s_{op}. Design curves are shown in Box 56.

Box 55 Example of stability relation for reef-type breakwater

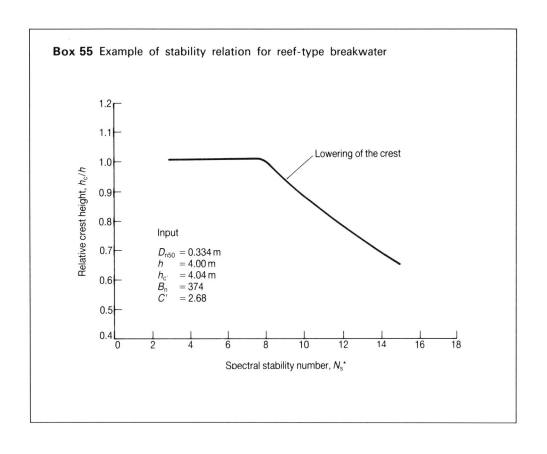

Box 56 Design curves for low-crested breakwaters ($R_c > 0$)

An average stability increase of 20% is obtained for a structure with the crest level at the water level. The required mass in that case is a factor $(1/1.25)^3 = 0.51$ of that required for a non-overtopped structure.

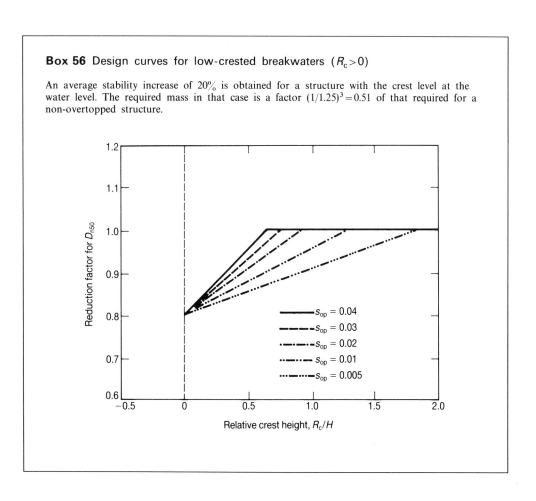

Box 57 Design curves for submerged breakwaters ($R_c < 0$)

Equation (5.55) is shown in the graph for three damage levels and can be used as design graph. Here again, $S_d = 2$ is start of damage, $S_d = 5-8$ is moderate damage and $S_d = 12$ is 'failure' (lowering of the crest by more than one D_{n50}).

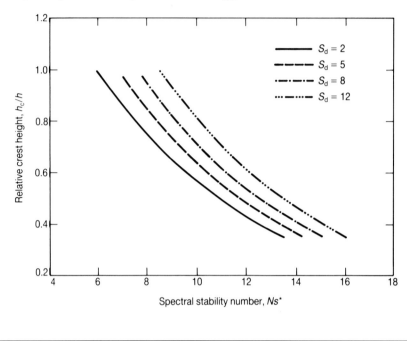

Statically stable submerged breakwaters (Figure 167(c))

The stability of submerged breakwaters depends on the relative crest height, the damage level and the spectral stability number. The given formulae are based on a re-analysis of the tests of Givler and Sørensen (1986) by van der Meer (1990a). The stability is described by:

$$h'_c/h = (2.1 + 0.1 S_d) \exp(-0.14 N_s^*) \tag{5.55}$$

For fixed crest height, water level, damage level, and wave height and period, the required ΔD_{n50} can be calculated, giving finally the required stone weight (Box 57).

5.1.3.5 Crown walls

The overtopping performance of a rubble breakwater or seawall is often significantly improved by the use of a concrete crown wall. Such elements are also used for access, to provide a working platform for maintenance and occasionally to carry pipelines or conveyors. The influence of such walls on the overtopping performance was described in Section 5.1.2.2 and the effect of overtopping on stability of the crest and rear armour was covered in Section 5.1.3.3. This section considers the loads applied to the crown wall by wave action (see Figure 168).

Wave loads on crown wall elements will depend upon the incident wave conditions as well as strongly on the detailed geometry of the armour at the crest and the crown wall itself. The principal load is applied to the front face. A second effect is the uplift force acting on the underside of the crown wall. These forces will be resisted by the weight of the crown wall and by the friction force mobilised between the crown wall and the rubble.

There is no general method of predicting the wave forces on a crown wall for all

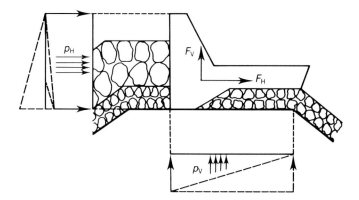

Figure 168 *Forces and pressues on a crown wall showing typical assumed pressure distributions*

configurations. There is also wide divergence between the different data sets available and the calculation methods that have been used. Model test data are available for a few examples of crown walls from studies by Jensen (1984) and Bradbury *et al.* (1968). An empirical equation has been fitted to test results for the structure configurations shown in Figure 169. This equation describes the maximum horizontal force, F_H, in terms of the wave height, H_s; peak wave length, L_p; the height of the crown wall face, h_f; the armour crest level, A_c; and empirical coefficients a and b.

$$F_H/(\rho g h_f L_p) = a H_s/A_c - b \tag{5.56}$$

Values of coefficients a and b have been derived for the sections tested by Jensen and Bradbury *et al.* and are summarised in Figure 169. Examples of the data plots are shown in Figure 170.

There are substantially less data on the uplift force, F_v, or on the forms of the pressure distribution on the front or underside of the crown wall. Examples of typical assumed pressure distributions are given in Figure 168.

A relatively safe estimate of loading may be made by assuming that the distribution of p_H is rectangular, $p_H = F_H/h_f$, and that p_v reduces from $p_v = p_H$ at the front to zero at the back. The uplift force, F_v, will then be given by:

$$F_v = (\rho g B_c L_p/S) * (a H_s/A_c - b) \tag{5.57}$$

A more conservative estimate will be given by assuming that the distribution of uplift pressures is rectangular, resulting in a total uplift force twice that calculated above. If this measure of uncertainty proves critical to the design, hydraulic model tests should be conducted to assure the stability of the crown wall.

Values of F_H and F_v may be used to calculate the stability of the crown wall against sliding. A simple equilibrium of forces may be calculated using these forces, the weight of the crown wall and a friction force. The value of the friction coefficient, μ, is generally assumed to be around 0.5. Where the crown wall incorporates a substantial key into the underlayer a value of $\mu = 0.8-1.0$ may be assumed. These values assume that the crown wall is cast in place directly onto underlayer or prepared core material. Pre-cast elements, or those cast *in-situ* onto finer material, will give lower values of μ. It is recommended that tests be conducted at large or full scale to establish more confident estimates of μ when critical to the design.

5.1.3.6 Toe protection

In most cases the front-face armour layer on the sea-side near the bottom is

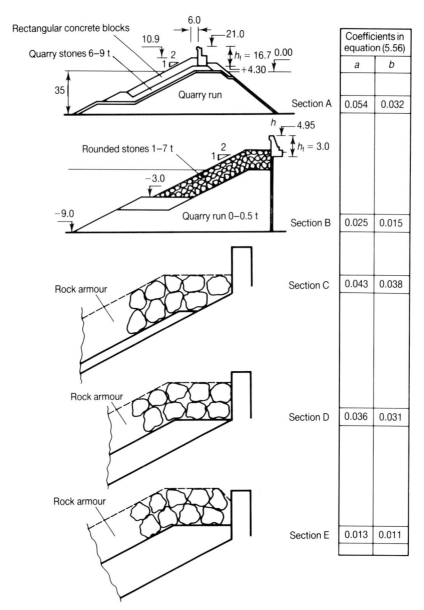

Figure 169 *Cross-section giving empirical coefficients a and b for horizontal wave forces on crown walls*

protected by a toe (see Figure 163). If the rock in the toe has the same dimensions as the armour, the toe will be stable. In most cases, however, one wants to reduce the size of the stones in the toe. Following the work of Brebner and Donnelly (1962) given in the *SPM* (1984), who tested rubble toes to vertical-faced composite structures under monochromatic waves, a relationship may be assumed between the ratio h_t/h and the stability number $H/\Delta D_{n50}$ (or N_s), where h_t is the depth of the toe below the water level and h is the water depth (see also Figure 142). A small ratio of $h_t/h = 0.3$–0.5 means that the toe is relatively high above the bottom. In that case the toe structure is more like the berm or stepped structures described in Section 5.1.3.8. A value of $h_t/h = 0.8$ means that the toe is near the bottom, and for such situations ($h_t/h > 0.5$) guidance should be sought in this section.

Toe protection to sloping front-face armour layers

Sometimes a relationship between $H_s/\Delta D_{n50}$ and h_t/H_s is assumed where a lower value of h_t/H_s should give more damage. Gravesen and Sørensen (1977) describe how a high wave steepness (short wave period) gives more damage to the toe than a low wave steepness. However, this assumption was only based on a few

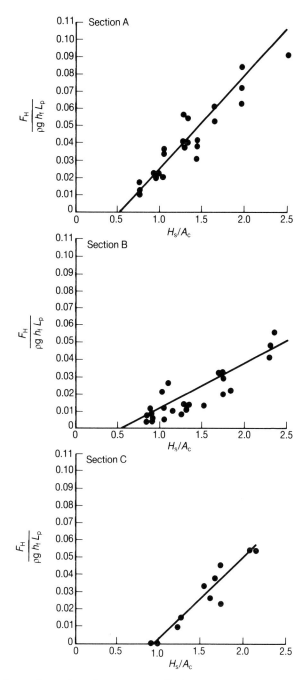

Figure 170 *Wave force data plots for crown walls in cross sections shown in Figure 169*

data points. In the CIAD (1985) report on the computer-aided evaluation this conclusion could not be verified. No relationship was found there between $H_s/\Delta D_{n50}$ and h_t/H_s, probably because H_s is present in both parameters. An average value of $H_s/\Delta D_{n50} = 4$ was given for no damage and a value of 5 for failure. The standard deviation around these values was 0.8, showing a large spreading.

A more in-depth study was carried out for this manual. The results presented in the CIAD report were re-analysed and compared with other data and Figure 171 shows the final results. Seven breakwaters (with alternatives) tested at Delft Hydraulics were taken and the behaviour of the toe was examined. The wave boundary conditions for which the criteria '0–3%', '3–10%' and 'failure, >20–30%' occurred were established. Here '0–3%' means no movement of stones (or only a

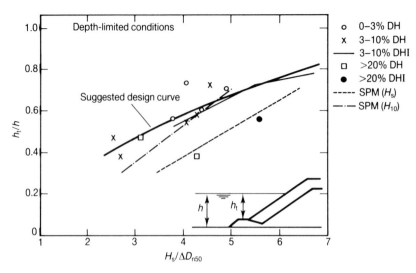

Figure 171 Toe stability as a function of h_t/h

few) in the toe and '3–10%' that the toe flattened out a little, but the function of the toe (supporting the armour layer) was intact and the damage is acceptable. 'Failure' means that the toe has lost its function and this damage level is not acceptable.

In almost all cases the structure was attacked by waves in a more or less depth-limited situation, which means that H_s/h was fairly close to 0.5. This is also the reason it is acceptable that the location of the toe, h_t, is related to the water depth, h. It would not be acceptable for breakwaters in very large water depths (more than 20–25 m). The results of the analysis are, therefore, applicable for depth-limited situations.

Figure 171 shows that if the toe is high above the bottom (small h_t/h ratio) the stability is much smaller than for the situation where the toe is close to the bottom. The results of DHI (internal paper) are also shown in Figure 171 and correspond well with the 3–10% values of Delft Hydraulics. It is interesting to note that if the higher-stability Brebner and Donnelly (1962) curve (for rubble toe protection to vertical walls) is added with $H = H_s$, the curve shows the same trend but is too low compared with the other results. If one assumes $H = H_{1/10}$ (as was done in *SPM*, 1984), the curve corresponds well with the other results and may indicate the more severe hydraulic loading when reflective vertical structures are present (see below).

A suggested line for design purposes is given in Figure 171. In general, it means that the depth of the toe below the water level is an important parameter. If the toe is close to the bottom the diameter of the stones can be more than twice as small as when the toe is half-way to the bottom and the water level. Design values for low and acceptable damage (0–10%) and for more or less depth-limited situations are:

h_t/h	$H_s/\Delta D_{n50}$
0.5	3.3
0.6	4.5
0.7	5.4
0.8	6.5

Three points are shown in Figure 171 which indicate failure of the toe. The design values given above are safe for $h_t/h > 0.5$. For lower values of h_t/h one should use the stability formulae for armour stones described in Section 5.1.3.1.

As indicated above, amplification of near-bed water particle velocities due to wave reflection in the presence of vertical structures leads to lower toe-stability numbers $H_s/\Delta D_{n50}$ (or N_s) than where a sloping rubble face is being protected. The curves of Brebner and Donnelly (1982) given in the *SPM* (1984) referred to could be used for such situations, but have the disadvantage of being derived from monochromatic rather than random wave tests, and thus lead to the problem of determining an appropriate wave height value (e.g. $H_{1/10}$) corresponding to the monochromatic H value. It is suggested instead that, for preliminary design, the results of model tests in Japan by Tanimoto *et al.* (1982) on caisson breakwaters under random wave attack should be used as described below. For rubble-mound foundations to conventional caisson breakwaters the Japanese model tests suggest that for stability, $H_s/\Delta D_{n50}$ (or N_s) should not exceed about 2. For vertically composite caisson breakwaters, the tests lead to the following values for $H_s/\Delta D_{n50}$ for the armour layer in the rubble-mound bund (Tanimoto *et al.*, 1982):

$$H_s/\Delta D_{n50} = \max\{1.8,\ 1.3\alpha + 1.8\exp[-1.5\alpha(1-k)]\} \tag{5.58}$$

where

$\alpha = \{(1-k)/k^{1/3}\} \cdot h'/H_s,$
$k = k_1 \cdot k_2,$
$k_1 = (4\pi h'/L')\sinh(4\pi h'/L'),$
$k_2 \sin^2(2\pi B/L')$

where h' denotes the depth of the berm, L' the wave length at this depth and B the berm width. In practice, values of $H_s/\Delta D_{n50}$ (or N_s) will be very close to 2, the value given for the toe stability for the foundations to conventional caisson breakwaters.

5.1.3.7 Breakwater head

Breakwater heads represent a special physical process. Jensen (1984) described it as follows:

> When a wave is forced to break over a roundhead it leads to large velocities and wave forces. For a specific wave direction only a limited area of the head is highly exposed. It is an area around the still water level where the wave orthogonal is tangent to the surface and on the lee side of this point. It is therefore general procedure in design of heads to increase the weight of the armour to obtain the same stability as for the trunk section. Alternatively, the slope of the roundhead can be made less steep, or a combination of both.

An example of the stability of a breakwater head in comparison with the trunk section and showing the location of the damage as described in the previous paragraph is shown in Figure 172, taken from Jensen (1984). The stability coefficient ($H_s/\Delta D_n$ for tetrapods) is related to the stability of the trunk section. Damage is located about 120–150° from the wave angle. No specific rules are available for the breakwater head. The required increase in weight can be a factor between 1 and 4, depending on the type of armour unit. The factor for rock is closer to 1.

Another aspect of breakwater heads was mentioned by Jensen (1984). The damage curve for a head is often steeper than for a trunk section. A breakwater head may show progressive damage. This means that if both head and trunk were designed on the same (low) damage level, an (unexpected) increase in wave height can cause failure of the head or a part of it, where the trunk still shows acceptable damage. This aspect is less pronounced for heads which are armoured by rock.

Finally, the stability of a berm breakwater head should be discussed. Burcharth and Frigaard (1987) have studied longshore transport and stability of berm breakwaters in a short basic review. The recession of a breakwater head is shown as an example in Figure 173 for fairly high wave attack ($H_s/\Delta D_{n50} = 5.4$). Burcharth and Frigaard give as a first rule of thumb for the stability of rock armour on a breakwater head that $H_s/\Delta D_{n50}$ should be smaller than 3.

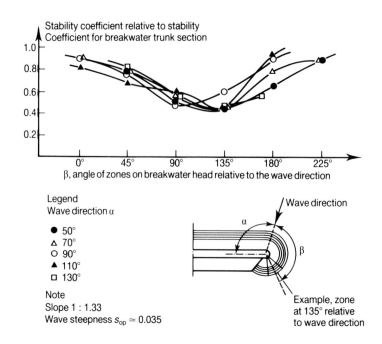

Figure 172 *Stability of a breakwater head armoured with tetrapods*

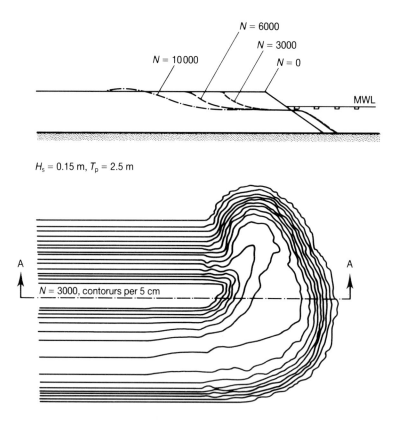

Figure 173 *Example of erosion of a berm breakwater head*

5.1.3.8 Dynamically stable slopes (berm breakwaters and shingle beaches)

Statically stable structures can be described by the damage parameter, S (see Section 5.1.1.2) and dynamically stable ones by a profile (see Figure 145). Other

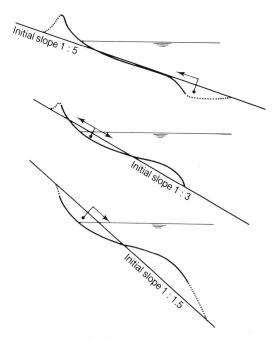

Figure 174 *Examples of dynamically stable profiles for different initial slopes*

typical profiles, but for different initial slopes to the 1:5 slope shown in Figure 145, are shown in Figure 174. The main part of the profiles is always the same. The initial slope (gentle or steep) determines whether material is transported upwards to a beach crest or downwards, creating erosion around still-water level.

Based on extensive model tests (van der Meer, 1988a), relationships were established between the characteristic profile parameters as shown in Figure 145 and the hydraulic and structural parameters. These relationships were used to make the computational model BREAKWAT on a personal computer, which simply gives the profile in a plot together with the initial profile. Boundary conditions for this model are:

- $H_s/\Delta D_{n50} = 3\text{--}500$ (berm breakwaters, rock and gravel beaches)
- Arbitrary initial slope
- Crest above still-water level
- Computation of an (established or assumed) sequence of storms (or tides) by using the previously computed profile as the initial profile.

The input parameters for the model are the nominal diameter of the stone, D_{n50}, the grading of the stone, D_{85}/D_{15}, the buoyant mass density, Δ, the significant wave height, H_s, the mean wave period, T_m, the number of waves (storm duration), N, the water depth at the toe, h, and the angle of wave incidence, β. The (first) initial profile is given by a number of (x, y) points with straight lines in between. A second computation can be made on the same initial profile or on the computed one.

The results of a computation on a berm breakwater is shown in Figure 175, together with a listing of the input parameters. The model can be applied to:

- Design of rock slopes and gravel beaches
- Design of berm breakwaters
- Behaviour of core and filter layers under construction during yearly storm conditions.

The computation model can be used in the same way as the deterministic design approach of statically stable slopes, described in Section 5.1.3.1. There the rather complicated stability equations (5.44) and (5.45) were used to make design graphs such as damage curves, and these graphs were used for a sensitivity analysis. By

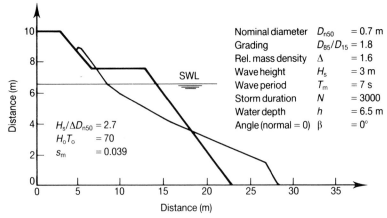

Figure 175 *Example of a computer profile for a berm breakwater*

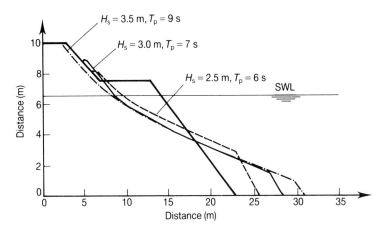

Figure 176 *Example of influence of wave climate on a berm breakwater profile*

making a large number of computations with the computational model the same kind of sensitivity analysis can be performed for dynamically stable structures. Aspects which were considered for the design of a berm breakwater (van der Meer and Koster, 1988) were, for example:

- Optimum dimensions of the structure (upper and lower slope, length of berm)
- Influence of wave climate, stone class, water depth
- Stability after first storms.

An example of the results of these computations is given in Figure 176 and shows the difference in behaviour of the structure for various wave climates.

Computations with the computational model can, of course, only be made if the model is available to the user. This is often not the case for the reader of a handbook and therefore a more simple (and less reliable) method should be given which is able to give the user a first impression (but not more than that!) of the profile that can be expected. This method is described below. First estimates of profiles of gravel beaches may also be made using the parameter profile model recently developed by Powell (1990) presented in Box 58. Berm breakwater profiles may also be estimated using the simple approach of Kao and Hall (1990) (Box 59).

Figure 177 gives the schematised profile simplified from the original profile (see Figure 145) in the BREAKWAT computational model. The connecting point is the intersection of the profile with still-water level. From this point an upper slope is drawn under 1:1.8 and a lower slope under 1:5.5. The crest of the profile is situation on the upper slope and the transition to a steep slope on the lower part.

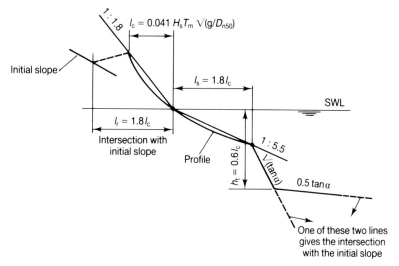

Figure 177 *Simple schematised profile of rock and gravel beaches*

These two points are given by the parameters l_c (*l*ength of *c*rest) and l_s (*l*ength of *s*tep). Of course, a curved line goes through the three points.

The connection with the upper part of the profile and the initial profile is given by l_r(*l*ength of *r*un-up). Below the gentle part under still-water level a steep slope is present, and if the initial profile is gentle (cot $\alpha > 4$) again there is a gentle slope which gives the 'step' in the profile. The transition from a steep to a gentle slope is given by h_t (*h*eight of *t*ransition). If the initial slope is not a straight line, one should draw a more or less equivalent slope, taking into account the area from $+H_s$ to $-1.5H_s$, which gives tan α. The relationships between the profile parameters and the hydraulic and structural parameters are:

$$l_c = 0.041 H_s T_m \sqrt{(g/D_{n50})} \qquad (5.59)$$

$$l_s = l_r = 1.8 l_c$$

$$h_t = 0.6 l_c$$

Steep slope below still-water level: $\sqrt{(\tan \alpha)}$

Gentle slope below still-water level: $0.5 \tan \alpha$

Finally, the profile must be shifted along still-water level until the mass balance is fulfilled. Figure 177 and equation (5.59) give a rough indication of the profile that can be expected. For $H_s/\Delta D_{n50}$ values higher than about 10–15 the prediction is quite reliable. For lower values the initial profile has a large influence on the profile and therefore the given method is less reliable. This also applies to berm breakwaters and in that case the method should really be treated as a very rough indication.

5.1.3.9 Stepped and composite slopes

The stability formulae as described in Section 5.1.3.1 are applicable to straight slopes. Sometimes structures are a combination of slopes (composite slopes) and/or have a horizontal berm below the water level (stepped slopes). Design curves will be given in this section for three types of structures. Stepped slopes were investigated by Delft Hydraulics-M 2006 (1986) and composite slopes by van der Meer (1990d).

Box 58 Powell's parametric model for shingle beach profiles

A parametric model for shingle beach profiles has recently been developed by Powell (1990) based on an extensive series of 131 model tests at HR Wallingford, UK, designed to simulate the behaviour of shingle beaches. The schematised beach profile is shown in Figure 178 and may be applied for the range $20 < H_s/\Delta D_{n50} < 250$.

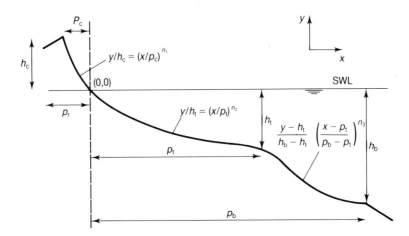

Figure 178 *Schematic shingle beach profile*

The parametric model is described by the following equations which should be read in conjunction with the definitions in Figure 178. The model is presented here to enable the designer to make first estimates. Profiles are, to some extent, duration limited and Powell (1990) suggests a method for accounting for this but also concludes that complete reprofiling will occur within only 500 waves.

Basic beach profile prediction

1. Run-up limit, p_r

$$p_r/H_s = 6.38 + 3.25 \ln(H_s/L_{om})$$

2. Crest position, p_c

$$p_c D_{50}/H_s L_{om} = -0.23(H_s T_m g^{1/2}/D_{50}^{3/2})^{-0.588}$$

3. Crest evaluation, h_c

$$h_c/H_s = 2.86 - 62.69(H_s/L_{om}) + 443.29(H_s/L_{om})^2$$

4. Transition position, p_t

For $H_s/L_{om} < 0.03$:

$$p_t \cdot D_{50}/H_s L_{om} = 1.73(H_s T_m g^{1/2}/D_{50}^{3/2})^{-0.81}$$

For $H_s/L_{om} \geq 0.03$:

$$p_t/D_{50} = 55.26 + 41.24(H_s^2/L_{om}D_{50}) + 4.90(H_s^2/L_{om}D_{50})^2$$

6. Transition elevation, h_t

For $H_s/L_{om} < 0.03$:

$$h_t/H_s = -1.12 + 0.65(H_s^2/L_{om}D_{50})^2 - 0.11(H_s^2/L_{om}D_{50})^2$$

For $H_s/L_{om} \geq 0.03$:

$$h_t/D_{50} = -10.41 - 0.025(H_s^2/D_{50}^{3/2}L_{om}^{1/2}) - 7.5 \times 10^{-5}(H_s^2/D_{50}^{3/2}L_{om}^{1/2})^2$$

6. Wave base position, p_b

$$p_b/D_{50} = 28.77(H_s/D_{50})^{0.92}$$

7. Wave base elevation, h_b

$$h_b/L_{om} = -0.87(H_s/L_{om})^{0.64}$$

8. Curve 1, crest to still-water level

$$\frac{y}{h_c} = \left(\frac{x}{p_c}\right)^{n_1}$$

where $\quad n_1 = 0.84 + 23.93 H_s/L_m \quad$ for $H_s/L_m < 0.03$

and $\quad n_1 = 1.56 \quad$ for $H_s/L_m \geq 0.03$

continued page 287

Box 58 continued

9. Curve 2, still-water level to transition

$$\frac{y}{h_t} = \left(\frac{x}{p_t}\right)^{n_2}$$

where $n_2 = 0.84 - 16.49 H_s/L_{om} + 290.16(H_s/L_{om})^2$

10. Curve 3, transition to wave base

$$\frac{y - h_t}{h_b - h_t} = \left(\frac{x - p_t}{p_b - p_t}\right)^{n_3}$$

where $\quad n_3 = 0.45 \quad$ for $H_s/L_{om} < 0.03$

and $\quad n_3 = 18.6(H_s/L_{om}) - 0.1 \quad$ for $H_s/L_{om} \geq 0.03$

Position of predicted beach profile

The cross-shore position of the predicted beach profile may be established shifting the profile until a simple balance of cross-sectional areas about the initial beach profile is established. This, of course, assumes that the net longshore transport across the section is zero.

Correction for effective beach thickness, D_B

To be applied when $30 D_{50} \leq D_B \leq 100 D_{50}$. For values of $D_B < 30 D_{50}$ the beach is destabilised. Correction $R_{c_{pc}}$ applies only to beach crest position, p_c.

$$R_{c_{pc}} = \frac{p_c(D_B/D_{50})}{p_c(D_B/D_{50} \geq 100)} = 6646 H_s/L_m (D_B/D_{50})^{-1.68} + 0.88$$

Correction for depth-limited foreshore

Correction factors necessary for positional parameters when $H_s D_w > 0.3$, and for elevation parameters when $H_s/D_w > 0.55$.

Correction factor $\quad\quad\quad R_{cd} = Par_{meas}/Par_{pred}$

where the predicted parameter value uses the wave conditions at the toe of the beach based on Goda (1985) as shown in Box 43 and D_w = depth of water at toe of beach.

Depth-limited wave height:

$$H_{s_b} = 0.12 L_{om}[1.0 - \exp(-4.712 D_w(1.0 + 15 m^{1.33})L_{om})]$$

Depth-limited wave length:

$$L_{m_s} = T_m(g D_w)^{1/2}$$

1. Upper profile limit correction

$$R_{cd} = 1.08(H_s/D_w) + 0.72 \quad \text{for } 0.3 < H_s/D_w < 2.5$$

2. Crest position

$$R_{cd} = 3.03(H_s/D_w) + 0.12 \quad \text{for } 0.3 < H_s/D_w < 2.5$$

3. Crest elevation correction

$$R_{cd} = (H_s/D_w) + 0.41 \quad \text{for } 0.55 < H_s/D_w < 2.5$$

4. Transition position correction

$$R_{cd} = 0.007(L_{om}/D_w)^{1.2} + 0.45 \quad \text{for } 40 < L_m/D_w < 130$$

5. Transition elevation correction

$$R_{cd} = 1.0 \quad \text{for } 0.55 < H_s/D_w < 2.5$$

6. Wave base position correction

$$R_{cd} = 1.08(H_s/D_w) + 1.31 \quad \text{for } 0.3 < H_s/D_w < 0.8$$
$$R_{cd} = 0.20(H_s/D_w) + 0.28 \quad \text{for } 0.8 < H_s/D_w < 2.5$$

Box 59 Berm breakwater profile model due to Kao and Hall (1990)

New guidelines for the design of berm breakwaters have recently been presented by Kao and Hall (1990) based on an extensive series of model tests at Queen's University, Canada. The guidelines are specific to a particular initial profile shown in Figure 179(a), but are useful since this profile is one that has been widely adopted, matching both typical quarry yields from dedicated quarries and natural as-constructed side slopes. The results are applicable to the range $2 < H_s/\Delta D_{50} < 5$.

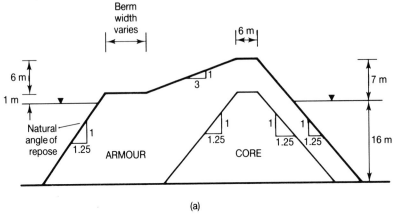

(a)

Figure 179(a) *Basic berm breakwater outline*

Kao and Hall (1990) define four basic parameters (see Figure 179(b)):

A = volume of armour stones required for stable reshaping
L = width of toe after reshaping
B = width of berm eroded
R_p = percentage of rounded stones in armour

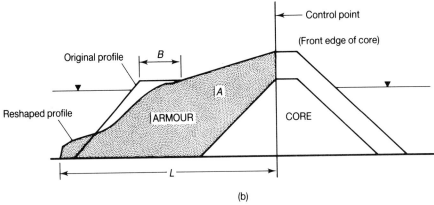

(b)

Figure 179(b) *Definition sketch for berm breakwater outer profile parameters*

The following equations relate these design parameters to wave climate, stone size, grading and shape and the values determined thereby for A and B must be considered as minima to be provided. Kao and Hall found that peak period, T_p, groupiness factor, GF, and wave steepness, s, had no significant influence on the stable profile for berm breakwaters. The equations are converted from the form originally presented by Kao and Hall in order to express them in terms of nominal diameters D_n rather than sieve sizes D. The conversion is based on the ratio $D/D_n = 0.84$ discussed in Section 3.2.2.3. For each parameter, the value based on 3000 waves is first presented followed by a correction for other numbers of waves, N.

The basic (3000) wave equations are:

$$A_{3000}/D_{n50}^2 = 104 + 29.5(H_s/\Delta D_{n50})^{1.9} + 137(D_{n85}/D_{n15}) - 18.9(D_{n85}/D_{n15})^2$$

$$L_{3000}/D_{n50} = 22.1 + 1.2(H_s/\Delta D_{n50})^{2.2} + 9.28(D_{n85}/D_{n15}) - 1.23(D_{n85}/D_{n15})^2$$

$$B_{3000}/D_{n50} = -8.7 + 0.66(H_s/\Delta D_{n50})^{2.5} + 6.32(D_{n85}/D_{n15}) - 0.90(D_{n85}/D_{n15})^2 + 5.14R_p$$

The time duration (number of wave) corrections are:

$$A_N/A_{3000} = (N/3000)^{0.043}$$

$$L_N/L_{3000} = (N/3000)^{0.042}$$

$$B_N/B_{3000} = 1 + 0.111\ln(N/3000)$$

Kao and Hall have recently found good agreement between predictions based on these equations and data obtained from prototype berm breakwaters.

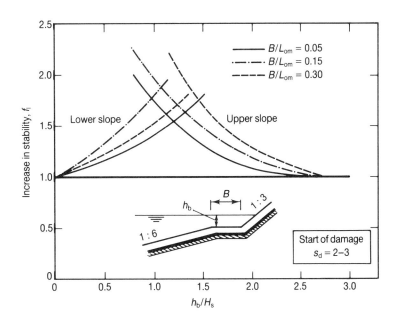

Figure 180 *Stability increase factors, f_i, for stepped or bermed slopes*

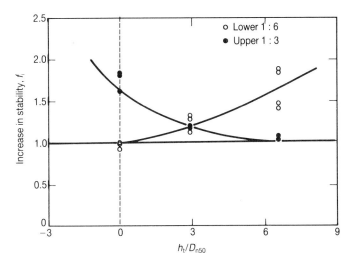

Figure 181 *Stability increase factors, f_i, for a composite slope with gradients: upper slope 1:3, lower slope 1:6*

The results are shown in Figures 180 to 182. The reference for stepped or composite slopes is always the stability of a straight slope, described in Section 5.1.3.1. The stability of the stepped or composite slope is then described by an increase in stability with regard to a similar but straight slope. This increase in stability, f_i, will be 1.0 if the stepped or composite slope has the same stability as a straight slope. It will be larger as 1.0 as soon as the step or transition of slopes has a positive effect on stability. The curves are given for start of damage, $S_d = 2-3$.

The design procedure is as follows:

- Calculate the required D_{n50} for the part of the stepped or compositive slope according to a straight slope, given in Section 5.1.3.1.
- The required D_{n50} can be calculated by dividing the D_{n50} found above by the increase in stability factor, f_i, taken from Figures 180 to 182.

Three types of structure were investigated:

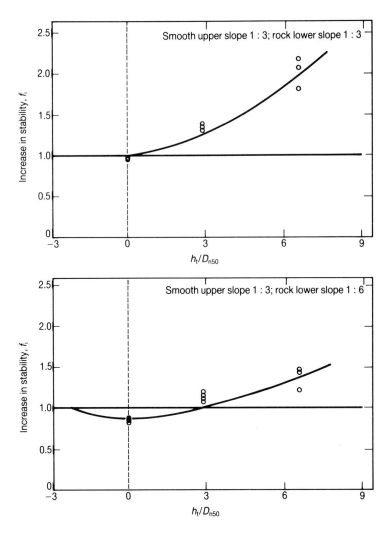

Figure 182 *Stability increase factors, f_i, for smooth upper slopes and rock lower slopes*

1. A stepped slope with a horizontal berm at or below the water level. The upper slope was 1:3 and the lower slope 1:6. The possible range of application, therefore, of the design curves given in Figure 180 may be 1:2 to 1:4 for the upper slope and 1:5 to 1:7 for the lower one.
2. A composite slope with an upper slope of 1:3, a lower slope of 1:6 and the water level at or above the transition. The possible range of application of the design curves shown in Figure 181 may be again 1:2 to 1:4 for the upper slope and 1:5 to 1:7 for the lower one.
3. A composite slope with a smooth upper slope of 1:3. This can be an asphalt slope or a placed block revetment. The lower slope was either 1:3 rock or 1:6 rock. The same possible ranges of application of the design curves shown in Figure 182 can be assumed as for the other structures.

The general trend from Figures 180 to 182 is that the lower slope increases in stability as soon as the water level is higher than the transition. The upper slope increases in stability as soon as the water level is less than $2H_s$ or $6D_{n50}$ above the transition. As soon as the transition of a stepped structure is well below still-water level, the stability of the lower slope can also be described by using the guidelines for a toe (Section 5.1.3.5).

5.1.3.10 Composite systems

The stability of randomly dumped quarried rock can often be substantially

Table 36 Indicative stability comparison for various rock-protection systems

Criterion	$\frac{H_s}{\Delta D} = F_1 \cdot F_2 \frac{\cos\alpha}{\xi_z^b} = F_1 2.25 \frac{\cos\alpha}{\xi_z^b}$			Limits	$\cot\alpha \geq 2$ $\xi_m \leq 3$	$N = 3000$ waves $P = 0.1$
System	D b	Δ	F_1	Description	Sublayer	Limits/remarks
Rock (reference)	D_n $b = 0.5$	Δ ≈ 1.65	1.0	Rip-rap (two layers)	Gr	Damage 1 to 3 stones
			1.33	Rip-rap (tolerable damage)	Gr	Damage $< D_n$ depth
Pitched stone	$D =$ average layer thickness $b \approx 2/3$	Δ stone ≈ 1.65	1.00	Poor quality (irregular-)stone	Gr	
			1.33	Good quality (regular-)stone	Gr	
			1.50	Natural basalt	Gr	
Grout	D_n $b = 0.5$ to 2/3	Δ stone ≈ 1.65	1.05	Surface grouting (30% of voids)	Gr	Avoid impermeability
			1.50	Pattern grouting (60% of voids)	Gr	$H_s < 3\text{–}4\,\text{m}$
Open-stone asphalt	$d =$ thickness top layer $b \approx 2/3$	Δ asphalt (≈ 1.15)	2.00	Open-stone asphalt	G+S	$H_s < 2\text{–}3\,\text{m}$
			2.50	Open-stone asphalt	SA	$H_s < 3\text{–}4\,\text{m}$
Gabions	d $b \approx 0.5$	Δ mattress	2	Gabion/mattress as a unit	G+S/	$H_s < 1.15\,\text{m}$ (max. 2.0 m)
	D_n	Δ stone	2	Stonefill in a basket	(G)+C	$d_{min} = 1.8 D_n$

Notes: Gr = granular, G+S = geotextile on sand, G+C = geotextile on clay.
SA = sand asphalt, S = sand, C = clay.
P_m = permeability ratio of top layer and sublayer/soil

improved by using rock as part of a system or with some additive as described in Sections 3.5.3 and 3.5.5. For some of these systems it is possible to give some rough (indicative) stability criteria, which allows the designer to make a comparison with randomly placed rock and thus a proper choice of protection. The following systems will be considered:

- Rock/rip-rap (as a reference for comparison)
- Regularly placed (pitched) stones
- Bound or grouted stone (cement grout or bitumen)
- Open-stone asphalt and other bituminous systems (asphalt concrete, sand asphalt)
- Gabion baskets and mattresses

A comparison of the hydraulic stability of these various protective systems is given in Table 36, and is based on a general empirical (approximate) formula derived by Pilarczyk (1989):

$$H_s/\Delta D \leq F_1 F_2 \cos\alpha / \xi_m^b \quad \text{(for: } \xi_m < 3, \cot\alpha \geq 2\text{)} \tag{5.59}$$

in which:

F_1 = System-determined (empirical) stability upgrading factor ($F_1 = 1.0$ for rip-rap as a reference and $F_1 \geq 1$ for other revetment systems),
F_2 = Stability factor or stability function for inception of motion of rock defined at $\xi_m = 1$,
b = Exponent related to the interaction process between waves and revetment type, $0.5 \leq b < 1$. For rough and permeable revetments such as rip-rap, $b = 0.5$. The exact values of b for various other revetments is not known, but $b = 2/3$ is a typical value,
D = Specific size or thickness of protection unit,
Δ = Relative mass density of a system-unit but allowing for any voids (as in gabions and mattresses).

For $\xi_m > 3$, the sizes calculated at $\xi_m = 3$ can still be applied. The stability factor, F_2, for rock can be more generally defined using equation (5.44). In the case of an impermeable core (i.e. sand or clay, $P \approx 0.1$) and limited number of waves ($N \approx 3000$) the following indicative F_2 values have been determined:

$F_2 = 2.25$ for inception of motion (motion 1 to 3 stones over the width of slope equal to D_n). This value is used in Table 36 as a constant reference value for the comparison of rock with other systems.

$F_2 = 3.0$ as a first approximation for maximum tolerable damage for a two-layer system on a granular filter (i.e. $S_d = 8$, filterlayer visible)

These conditions are close to the average test conditions in the past when rock and other alternative systems were examined based on Hudson's stability equation.

An important difference between unbound irregular randomly placed rock and the alternative systems concerns the behaviour of the systems after any initial movement or 'damage'. Due to the self-healing effect of rock a certain displacement of rock units can often be accepted (up to, say, $F_2 \approx 3$). In the case of alternative systems, such as pitched-stone revetments, the initial damage (i.e. removal of one block) can easily lead to a progressive damage; there is no reserve stability.

In all cases, experience and sound engineering judgement must play an important role in applying these preliminary design rules, which should be used with caution, and mathematical or physical model testing is advised to provide optimum solutions. However, the following additional advice may be useful in preparing preliminary designs of protective systems involving rock in relation to their hydraulic stability.

Grouted rock

Surface grouting is not advised where highly permeable sublayers will be present. In particular, the creation of a completely impermeable surface should be avoided because it may introduce extra lift forces (blasting effect), and pattern grouting (where only about 50–70% of the total surface is filled) is preferred. The upgrading factor achieved with grouting is very dependent on good execution, and care must be taken to ensure that the grout does not either just remain on the surface of the armourstone layer or sag completely through the layers. In areas of high wave impact, the grouted lumps themselves can be split by dynamic forces, and it is therefore recommended that this type of construction should only be applied up to $H_s = 3$ m (frequent loading) and $H_s \leq 4$ m (less frequent loading). In the latter case, for safety reasons, it is recommended to use three layers of rock in the armour layer, since if a lump of grouted stones is split and washed away, the third layer will still protect the core, being held by the overlying grouted lumps.

Bituminous systems

In the case of open-stone asphalt on a sand asphalt filter, the thickness of the system may be defined as the total thickness of both layers. For the edges of all bituminious systems, $F_1 = 2$ should be applied. Because of possible liquefaction, open-stone asphalt on a geotextile and sand combination is only recommended up to $H_s = 2$ m. For $H_s > 2$ m, a sand asphalt filter under the top layer of open-stone asphalt is recommended. The resistance of open-stone asphalt to surface erosion allows this system to be applied up to $H_s = 3$ m and, for a less frequent wave loading, up to $H_s = 4$ m.

For practical reasons, the minimum thickness of open-stone asphalt is 0.08 m if prefabricated and 0.10 m if placed *in-situ*. More common thicknesses are 0.10 and 0.15 m, respectively. Bituminous plate-systems (especially if impermeable) should also be checked for exceedance of allowable stresses and strains (bending moments) under wave pressures and uplift. Detailed calculation methods can be found in TAW (1985). Typical thicknesses of various asphalt revetments required to resist various wave loadings are given below for situations where the revetment is laid on a compacted sandbed with a slope 1 on 3:

H_s (m)	Asphalt concrete (m)	Open-stone asphalt (m)	Sand asphalt (m)
2	0.10	0.20	0.40
3	0.20	0.40	(0.80)
4	0.30	0.65	
5	0.40		

In general, the resistance of sand asphalt is limited to a wave height of 1.5 m (or $H_s \leq 2$ for less frequent loading).

Gabion baskets and mattresses

The primary requirement for a gabion or mattress of a given thickness d is that it will be stable as a unit. The thickness of the mattress can be related to the stone size D_n. In most cases it is sufficient to use two layers of stones in a mattress ($d = 1.8 D_n$) and an upgrading factor (see Table 36) in the range $\leq F_1 < 3$ can be recommended.

The second requirement is that the movement of stones in the basket should not be too high because of the possible deformation of baskets and the loading on the mesh wires. To avoid the situation that the basket of a required thickness d will be filled by too fine a material, a second criterion, related to D_n, has been formulated. The choice of, say, $F_1 = 2$–2.5 related to D_n means that the level of loading of the individual stones in the basket will be limited roughly to twice the loading at the incipient motion conditions. Thus the requirements for stability may be summarised as:

D_n (dynamic stability) (when $F_1 \leq 2.5$, and
d (static stability)   when $d \geq 1.8 D_n$

In a multi-layer gabion or mattress system (more than two layers) it is preferable to use a finer stone below the top layers (i.e. up to $1/5 D_n$) to create a better filter function and to diminish the hydraulic gradients at the surface of the underlying subsoil.

The stability formulations in Table 36 for gabions and mattresses are only valid for waves with a height up to $H_s = 1.5$ m, or for less frequent waves up to $H_s = 2.0$ m. In either case it is important that both the subsoil and the stone filling inside the gabion or mattress baskets are adequately compacted. Where the design wave height exceeds 1 m then a fine granular sublayer (about 0.2 m thick) should be provided between the mattress or gabion and the subsoil. Elsewhere it is satisfactory to place the mattress directly onto the geotextile and compacted subsoil. For practical reasons, the minimum thickness of mattresses is about 0.15 m.

5.1.3.11 Longshore transport

Statically stable structures as revetments and breakwaters are allowed to show damage only under very severe wave conditions. Even then, the damage can be described by the displacement of only a number of stones from the still-water level to (in most cases) a location downwards. Movement of stones in the direction of the longitudinal axis is not relevant for these types of structures.

The profiles of dynamically stable structures as gravel/shingle beaches, rock beaches and sand beaches change according to the wave climate. 'Dynamically stable' means that the net cross-shore transport is zero and the profile has reached an equilibrium profile for a certain wave condition. It is possible that, during each wave, material is moving up and down the slope (shingle beach). Oblique wave attack gives wave forces parallel to the alignment of the structure. These forces can cause transport of material along the structure. This phenomenon is called 'longshore transport' and is well known for sand beaches. Also, shingle beaches change due to longshore transport, although research on this aspect has always been limited.

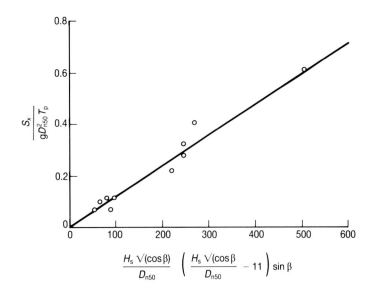

Figure 183 *Longshore transport relation for gravel beaches*

Rock beaches and berm breakwaters are or can be also dynamically stable under severe wave action. This means that oblique wave attack may induce longshore transport, which can also cause problems for these types of structures. Longshore transport does not occur for statically stable structures, but it will start for dynamically stable conditions where the diameter is small enough in comparison with the wave height. In the latter case the conditions for inception of longshore transport are important.

The *Shore Protection Manual* (1984) gives the well-known CERC formula for longshore transport of sand. This transport is related to the energy component of the wave action parallel to the coast and the approach is given by:

$$S(x) \propto H^2 c_o \sin 2\beta \qquad (5.61)$$

where $S(x)$ = material transported parallel to the coast,
H = wave height,
c_o = wave celerity = $gT/2\pi$,
β = angle of wave attack at the coast.

The longshore transport in this formulation is independent of grain size and only depends on the wave condition (wave height, period and direction).

Van Hijum and Pilarczyk (1982) have studied longshore transport on gravel or shingle beaches by random wave attack. The transport for shingle beaches is determined by bed load (rolling along the bottom) and not by a combination of bed load and suspended load, which is the case for sand beaches. Van Hijum and Pilarczyk give a formula for longshore transport of gravel beaches. This contains parameters which are a little different from those used in this chapter and a parameter which describes the refraction effects. For this manual the original data of Van Hijum and Pilarczyk (1982) were re-analysed in order to give a more simple formula with the parameters used in this chapter. Figure 183 shows the final results.

The formula for longshore transport of gravel beaches is given by:

$$\frac{S(x)}{gD_{n50}^2 T_p} = 0.0012 \frac{H_s\sqrt{(\cos\beta)}}{D_{n50}} \left\{ \frac{H_s\sqrt{(\cos\beta)}}{D_{n50}} - 11 \right\} \sin\beta \qquad (5.62)$$

The range on which equation (5.62) was established was $H_s/\Delta D_{n50} = 12-27$, i.e.

fairly large gravel in prototype. Van Hijum and Pilarczyk (1982) used the data of Komar (1969) on coarse sand to extrapolate equation (5.62) to smaller materials. They concluded that the formula could be applied up to sand beaches.

Equation (5.62) shows a dependency on the grain diameter. For small grain sizes, however, the factor 11 in the equation can be deleted and the equation can be rewritten to:

$$S(x) = 0.0012\pi H_s^2 c_{op} \sin 2\beta \qquad (5.63)$$

where c_{op} = the wave celerity = $gT_p/2\pi$. Equation (5.63) is, according to the CERC approach, given by equation (5.61), The diameter or grain size has again disappeared. The transition where the grain size no longer appears to have any influence can be given by $H_s/\Delta D_{n50} > 50$.

Equation (5.62) for longshore transport of gravel beaches indicates that incipient motion (start of transport) begins when $H_s\sqrt{(\cos\beta)} > 11 D_{n50}$. This is, however, not correct, and gives an underestimation of longshore transport for large diameters (say, $H_s/\Delta D_{n50} < 10$). It means that equation (5.62) is not valid for $H_s/\Delta D_{n50} < 10$.

Work by Kamphuis (1990) in press at the time of preparation of this manual suggests that an empirical equation of the form of equation (5.63) is possible (based on the work of a number of researchers) which covers the full range of sand and gravel transport. The equation also includes wave period, T, beach slope, m, and sediment size, D_{n50}, and indicates that transport is proportional to $D_{50}^{-0.25}$, confirming that sediment size is more important for gravel transport than for sand transport.

The start of longshore transport is the most interesting consideration for the berm breakwater where profile development under severe wave attack is allowed but longshore transport should be avoided. The berm breakwater can roughly be described by $2.5 < H_s/\Delta D_{n50} < 6$. Burcharth and Frigaard (1987) performed model tests to establish the incipient longshore motion for berm breakwaters and their range of tests corresponded to $3.5 < H_s/\Delta D_{n50} < 7.1$. Longshore transport is not allowed at berm breakwaters, and therefore Burcharth and Frigaard gave the following (somewhat premature) recommendations for the design of berm breakwaters, which are in fact the criteria for incipient longshore motion:

For trunks exposed to steep oblique waves $H_s/\Delta D_{n50} < 4.5$
For trunks exposed to long oblique waves $H_s/\Delta D_{n50} < 3.5$
For roundheads $H_s/\Delta D_{n50} < 3$ (5.64)

The various formulae and criteria presented in this section are summarised in Box 60.

5.1.3.12 Combined current and wave attack

This section deals with the stability of rip-rap under a combination of unidirectional steady flow and wave-(non-breaking) induced oscillatory flow and is directly related to the design of offshore rockfill structures, described in Section 6.4. Since the average bottom shear stress action on the bed is dominated by the wave-induced shear stress, a design method is described which is able to combine both effects. Consequently rip-rap design procedures which deal only with unidirectional flow will not be treated here.

Critical shear concept

The traditional design method for the hydraulic stability of rockfill is based on the 'incipient motion' or 'critical shear' concept. For unidirectional steady flow the initial instability of bed material particles on a plane bed is given by the Shields criterion (Shields, 1936). This essentially expresses the critical value of the ratio of

> **Box 60** Longshore transport formulae and criteria for rock and gravel
>
> Longshore transport depends on the type of structure (sand beach, shingle beach, rock beach or berm breakwater) and the wave climate, and can be described by the following ranges and formulae:
>
> *Gravel beach*
>
> $H_s/\Delta D_{n50} > 50$ up to sand beaches:
>
> $$S(x) = 0.0038 H_s^2 c_{op} \sin 2\beta \qquad (5.63)$$
>
> *Rock/gravel beach*
>
> $10 < H_s/\Delta D_{n50} < 50$:
>
> $$\frac{S(x)}{gD_{n50}^2 T_p} = 0.0012 \frac{H_s\sqrt{(\cos\beta)}}{D_{n50}} \left\{ \frac{H_s\sqrt{(\cos\beta)}}{D_{n50}} - 11 \right\} \sin\beta \qquad (5.62)$$
>
> $5 < H_s/\Delta D_{n50} < 10$: No equation
>
> *Berm breakwater*
>
> $H_s/\Delta D_{n50} < 5$:
>
> For trunks exposed to steep oblique waves $H_s/\Delta D_{n50} < 4.5$
>
> For trunks exposed to long oblique waves $H_s/\Delta D_{n50} < 3.5$ $\qquad (5.64)$
>
> For roundheads $\qquad\qquad\qquad H_s/\Delta D_{n50} < 3$

the fluid forces tending to move the particle to stabilising forces acting on a particle. The forces which tend to move the bed material particle are related to the maximum shear stress exerted on the bottom by the moving fluid and the stabilizing forces are related to the submerged weight of the particle. When the ratio of the two forces, referred to as the Shields parameter, exceeds a critical value, movement is initiated. The Shields criterion for steady flow can be expressed as:

$$\psi_{cr} = \frac{\tau_{cr}}{(\rho_r - \rho_w)gD} = \frac{\tau_{cr}}{\Delta \rho_w gD} = \text{fn}\left\{\frac{U_{*cr}D}{v}\right\} = \text{fn}(Re_*) \qquad (5.65)$$

where ψ_{cr} = dimensionless shear parameter or Shields number,
τ_{cr} = critical value of bed shear stress induced by the fluid at which the stones first begin to move,
ρ_r = mass density of the rock material,
ρ_w = mass density of sea water,
D = grain size,
U_* = shear velocity = τ_o/ρ_w,
v = kinematic fluid viscosity,
Re_* = Reynolds number based on shear velocity,
τ_o = bed shear stress.

The Shields curve is given in Figure 184. For large values of Re_* (rough bed, $Re_* > 200$), the Shields number ψ_{cr} becomes more or less constant, and from his experiments Shields concluded $\psi_{cr} \approx 0.06$. However, Shields has assumed that there is a sharp boundary between no displacement and displacement. In reality, this boundary is not very sharp due to local variations in bed shear stress, grain size and grain protrusion (see e.g. Paintal, 1971). From extensive laboratory tests Breusers and Shukking (1971) found that for high Reynolds numbers displacement of some grains begins to occur at $\psi_{cr} = 0.03$.

Because of the uncertainty about the exact value of the critical shear stress, it is also recommended that the criterion $\psi_{cr} = 0.03$ be kept for the design of rockfill to define the point at which stones first begin to move.

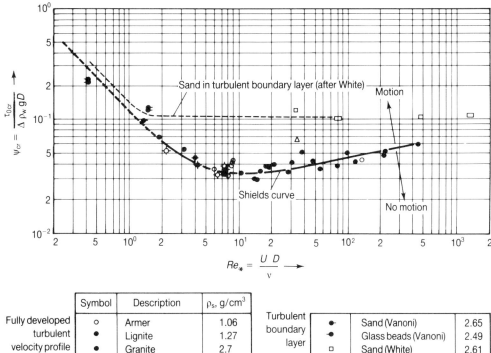

Figure 184 *Dimensionless critical bed shear stress versus shear Reynolds number*

Unidirectional flow

In steady flow the critical value of the shear stress, τ_c, acting on the bed can be computed as:

$$\tau_c = \rho_w g \frac{(\bar{U})^2}{C^2} \tag{5.66}$$

in which $\bar{U}$ is the depth-averaged current velocity and C is the Chezy coefficient. When the bed is hydraulically rough ($U_* k_s/v > 70$) the Chezy coefficient depends only on water depth, h, and bed roughness, k_s:

$$C = 18 \log(12h/k_s) \tag{5.67}$$

The Nikuradse roughness length, k_s, is an important parameter because it appears in several widely used formulas. For a flat bed the roughness length is related to the diameter of the largest grains on the surface of the bed. Van Rijn (1982) has derived from about 100 flume and field data that $k_s = 3D_{90}$, and similar values have been reported by other investigators. For engineering purposes the use of $k_s = (1-3)D_{90}$ is probably sufficiently accurate.

Oscillatory flow

The shields criterion for the initial motion has been established from experimental observations in unidirectional steady flow. For very slowly varying flows, such as tidal flows in limited water depths, the flow may, within reason, be regarded as quasi-steady. For short-period waves, such as wind waves, having a period of 5–20 s, the above quasi-steady approach is no longer justified. Various investigators have addressed the phenomenon of initial motion under wave action. Madsen and

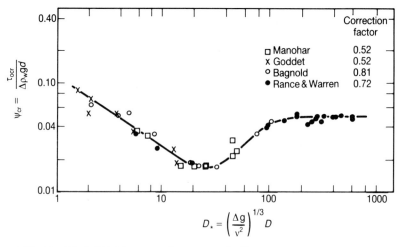

Figure 185 Modified Shields curve for unsteady flow

Grant (1975) and Komar and Miller (1975) have shown, independently, that the results obtained for the initial motion in unsteady flow are in reasonable agreement with the Shields curve for unidirectional flow if the shear stress is calculated by introducing the concept of the wave friction factor according to Jonsson (1966):

$$\hat{\tau}_w = \tfrac{1}{2} \rho_w f_w \hat{U}_o^2 \tag{5.68}$$

where $\hat{\tau}_w$ is the maximum shear stress under oscillatory flow, f_w a friction factor and $\hat{U}_o$ the peak orbital velocity near the bed, which may be determined, as a first approximation, by linear wave theory. Swart (1974) has proposed the following empirical relationship for the friction factor, f_w, which is applicable when the flow near the bed is fully developed rough turbulent:

$$\text{for } a_o/k_s > 1.57, \; f_w = \exp[-6.0 + 5.2(a_o/k_s)^{-0.19}] \tag{5.69}$$

for $a_o/k_s \leq 1.57, \; f_w = 0.3$

where a_o = amplitude of horizontal wave motion of bed = $\hat{U}_o T/2\pi$.

Because it is not certain that unsteady flow results can be adequately represented by steady flow conditions for all possible flow conditions, and because forces other than skin friction may be important under certain conditions, a *modified Shields* criterion has been established (Madsen and Grant, 1976; Sleath, 1978) using the actual results obtained for initial motion in unsteady flow by various investigators. The modified Shields function may be thought of as the ratio of the maximum fluid forces tending to move the particles to the immersed weight.

In Figure 185 the critical value of the modified Shields function is plotted together with the experimental data against the non-dimensional grain size, D_*, defined by:

$$D_* = \left\{ \frac{\rho_r - \rho_w}{\rho_w} \frac{g}{v^2} \right\}^{1/3} D = (\Delta g/v^2)^{1/3} \cdot D \tag{5.70}$$

For comparison with experimental data a correction factor has been used to take into account the fact that different investigators have used different criteria for the initial motion condition. The correction factors applied are also shown in Figure 185.

The experiments of Rance and Warren (1968), which correspond to fully rough turbulent flow conditions, refer to the initiation of motion of very coarse sediment for oscillatory flow. Rance and Warren presented their experimental data on initiation of motion in a diagram of an acceleration number:

$$A_* = \frac{a_o}{\Delta g T^2} \tag{5.71}$$

versus the relative boundary roughness, a_o/D. The acceleration number A_*, however, may alternatively be interpreted as the Shields parameter, multiplied with a factor which is a function of the relative roughness a_o/D only. In Madsen and Grant (1976) it is shown that the value of A_* can be evaluated as a function of the parameter a_o/D, if the critical value of the Shields parameter is taken as $\psi_{cr} = 0.056$ (and $k_s = D_{90}$), corresponding to the critical value of very coarse sediment. In this way, an excellent agreement with the results by Rance and Warren can be obtained.

Therefore it can be concluded that for the incipient motion of coarse material in oscillatory flow the Shields criterion for the initiation of motion can be applied when the Shields parameter is taken as 0.056 and the critical shear stress $\hat{\tau}_w$ is being evaluated according to Jonsson's wave friction concept (equation (5.68)). Where the critical shear stress is based on the *average* shear stress under oscillatory flow ($\bar{\tau}_w = \frac{1}{2}\hat{\tau}_w$) then the Shields parameter must have a value of 0.03 in order to agree with the results of Rance and Warren

Combined unidirectional and oscillatory flow

There have been very few studies on the initial motion condition in combined steady and oscillatory flow. Experimental results by Hammond and Collins (1979) suggest that if the frequency of the oscillation is very low the oscillatory flow may, at each instant, be treated as if it were steady. At high frequencies of the oscillatory flow the wave-induced boundary layer will be very thin compared with that of the steady flow, and the wave-induced and steady flow will be uncoupled. In extreme cases, with no steady current or no oscillatory flow, the condition for initial motion would be either for the wave-induced motion or for the steady component, respectively, to exceed the critical value for initial motion.

In the literature it has been suggested that for combined waves and steady current the effective velocity for initial motion should be taken as the sum of ther oscillatory and steady components of the shear stress. A formulation of the resulting bed shear stress due to combined waves and currents, which is widely applied in engineering practice, has been proposed by Bijker (1967). Further background information on this approach can be found in Sleath (1984), Herbich et al. (1984) and van der Velden (1989). According to Bijker, the maximum shear stress $\hat{\tau}_{cw}$ can be found by vectorial summation of the influences of waves and currents.

Based on the time-averaged shear stress for waves and steady current, propagating in the same direction, the following equation can be applied to evaluate the critical shear stress of the initial motion condition according to Shields:

$$\hat{\tau}_{cw} = \tau_c + \tfrac{1}{2}\hat{\tau}_w \tag{5.72}$$

where for τ_c and $\hat{\tau}_w$ equations (5.66) and (5.68), respectively, should be used. As mentioned earlier, for determining the required stable grain size D_{50} according to equation (5.65), the Shields parameter must have a value of 0.03 in order to agree with the results of Rance and Warren.

Structure slope

The above considerations were derived for a horizontal bed. Along a slope of a rockfill embankment only a part of the gravity force provides a motion-counteracting force. If the slope of the embankment is equal to the angle of repose, ϕ, of the submerged material, the stabilising force may even reduce to zero. For a side slope at right angles to the current direction, α, the reduction factor, k_d, of the critical shear stress is:

> **Box 61** Hydraulic stability formulae for seabed rockfill material
>
> *Bottom shear stress formulae*
>
> Steady flow: $\quad\tau_c = \rho_w g \dfrac{\bar{U}^2}{C^2}$
>
> Oscillatory flow: $\quad\hat{\tau}_w = \tfrac{1}{2}\rho_w f_w \hat{U}_o^2$
>
> Combined steady and oscillatory flow, propagating in the same direction:
> $$\hat{\tau}_w = \tau_c + \tfrac{1}{2}\hat{\tau}_w$$
>
> *Basic design formula—Shields criterion*
> $$D_{n50} = \frac{k_{tot}\bar{\tau}_{cr}}{\Delta\rho_w g \psi_{cr}}$$
>
> Basic input constant: $\quad\psi_{cr} = 0.03$
>
> Note: $D_{50} = 0.84 D_{n50}$ (see Section 3.2.2.3)
>
> *Other formulae required*
> $$C = 18\log(12h/k_s)$$
> $$k_s = (1-3)D_{90}$$
> f_w: for $a/k_s > 1.57$, $\;f_w = \exp[-6.0 + 5.2(a/k_s)^{-0.19}]$
> for $a/k_s \leq 1.57$, $\;f_w = 0.3$
>
> Slope effects: $\;k_d = \cos\alpha(1-(\tan\alpha/\tan\phi)^2)^{1/2}$
> $\phantom{\text{Slope effects: }}k_1 = \sin(\phi-\beta)/\sin\phi$
> $\phantom{\text{Slope effects: }}k_i = $ no general guidelines
> $\phantom{\text{Slope effects: }}k_{tot} = k_i/(k_1 k_d)$

$$k_d(\alpha) = \cos\alpha\left[1-\left(\frac{\tan\alpha}{\tan\phi}\right)^2\right]^{1/2} \tag{5.73}$$

For a slope with an angle β to the current direction, the reduction factor is:

$$k_1(\beta) = \frac{\sin(\phi-\beta)}{\sin\phi} \tag{5.74}$$

Generally, $\phi \gg \beta$ and the reduction factor of the critical shear stress for a slope in current direction can be neglected. However, in a situation with oblique current both reduction factors have to be combined:

$$k(\alpha,\beta) = k_d \cdot k_1 \tag{5.75}$$

Further considerations

Non-uniform flow conditions (e.g. local contraction effects due to the elevation of an embankment above the surrounding sea bottom) may also influence the stability of rockfill material. In such situations the actual shear stress acting on the bed may reach a much higher value as compared to the shear stress in uniform flow conditions, due to acceleration of the flow. Another phenomenon which may reduce the critical shear stress is a high level of turbulent intensities (e.g. due to flow separation, vortex shedding induced by nearby marine structures).

No general guidelines can be given as very few turbulence calculations for the type of structure under consideration have been made and turbulence influences depend strongly on specific structure dimensions. Therefore if the rockfill embankment has such domensions that the induced flow conditions can decrease rockfill (stability (large ratio structure height–water depth, steep side slopes), separate turbulence calculations should be carried out in order to quantify the applied shear stress (Box 61). These calculations require advanced numerical computer techniques.

5.1.3.13 Scour

Incidence and extent of scour

Many seawalls, breakwaters and related coastal structures are founded on sand or shingle. When the combined effects of waves and currents exceed a threshold level, bed material may be eroded from areas of high local shear stress. Close to the structure, wave and current velocities are often increased by the presence of that structure, leading to increased bed movement in this area. This commonly appears as local scour in front or alongside the structure and may in turn exacerbate any general reduction in beach levels taking place. Recent work in the UK has revealed that around 34% of seawall failures arise directly from erosion of beach or foundation material, and that scour is at least partially responsible for a further 14% of failures (CIRIA, 1986). Prevention of (or design for) local scour should therefore be a principal objective.

It should be appreciated that the main process involved in scour is always that of naturally occurring coastal sediment transport. These processes may lead to natural cycles of erosion and accretion, irrespective of the position or configuration of any structure. Such changes have, however, often be ascribed solely to the presence of the structure, and the distinction between local scour and general beach movement has often been confused. Dean (1986) has illustrated the difference between overall natural beach movement and the influence of the presence of an artificial wall in terms of onshore–offshore transport processes. Figure 186 shows normal and storm beach profiles (a) without and (b) with a vertical seawall, and Dean (1986) thereby simply explains local scour as arising from the denial to the sea by the seawall of the natural sediment sources for storm bar formation.

Accurate prediction of any beach process, including scour, requires a detailed description of the nearshore hydrodynamics and of the beach response functions. These processes lie outside the scope of this manual, and will not be treated directly here. Where local experience suggests that scour is likely, or the consequences could be particularly severe, physical and/or numerical modelling

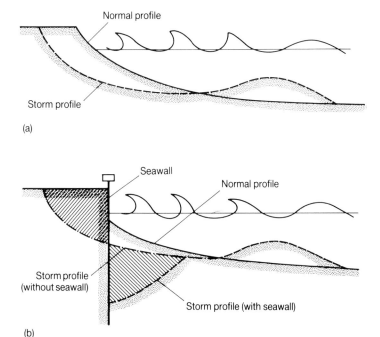

Figure 186 *Additional (local) scour immediately in front of a seawall due to storms. (a) Normal and storm profiles on a seawalled shoreline and (b) comparison with profiles on a natural shoreline*

methods should be used to quantify the effect. Again, these methods will not be described here. However, some simple estimates of the likelihood and possible extent of scour may be made from an assessment of the influence of the structure on the local hydrodynamics. The principal effects of a structure are:

1. An increase in local peak orbital velocities in front of the structure, due to the combination of incident and reflected waves;
2. Concentration of wave and todal currents along or close to the structure.

In general, the increased orbital velocities and the consequent scour can be related to the reflection coefficient, C_r, of the structure. The prediction of reflection performance has been addressed in Section 5.1.2.3. The effects of the structure on the local currents cannot be generalised in the same way, and site-specific studies may be needed.

Where scour or erosion is anticipated, particular attention should be paid to the possibility of local erosion outflanking the protection structure. On coastal revetments and seawalls, erosion effects are frequently most severe at the ends of the protection. Unless checked, such erosion may continue around the ends of the structure. This is often addressed by continuing the proposed protection well beyond the predicted erosion area, and/or tying the ends back to highe or stronger ground.

Simple prediction methods

Toe scour is the process of localised erosion occurring immediately seaward of the structure. A scour depth, d_s, may be defined as the maximum depth of scour relative to the initial bed level at the same water level/wave condition. The simple prediction methods available relate the scour depth to the incident wave conditions, the local water depth, h_s, and the structure geometry and/or reflection coefficient, C_r. These methods do not take account of the effects of angled wave attack, tidal or wave-induced currents. Although few methods include sediment side, most have been evolved for samd sizes. Prediction methods for scour on *sand beaches* may be categorised as follows:

1. Rule-of-thumb methods;
2. Semi-empirical methods based on hydraulic model tests;
3. Simple morpho-dynamic models.

The *Shore Protection Manual* (CERC, 1984) suggests that for scour under wave action alone the maximum depth of scour below the natural bed is about equal to the height of the maximum unbroken wave that can be supported by the original depth of water, h_s, at the toe of the structure:

$$d_s = H_{max} \qquad (5.76)$$

This presumably applies to vertical or steeply sloping structures. However, Powell (1987) has noted that the wave orbital velocities at the bottom of such a scour hole will still exceed those on the beach in the absence of the structure. This indicates that this simple rule may underestimate scour in some cases. Analysis of other studies suggests some other general rules:

1. For $0.02 < s_{om} < 0.04$, the scour depth is approximately equal to the incident unbroken wave height, again presumably for vertical structures;
2. Maximum scour occurs when the structure is located around the plunge point of breaking waves;
3. The depth of scour is directly proportional to the structure reflection coefficient. For structures with a smooth impermeable face, scour can be minimised by adopting a slope flatter than about 1:3. For structures faced with two or more layers of rock, steeper slopes can be adopted.

Hydraulic model tests on toe scour, from which empirical prediction methods have been derived, have been described by Herbich and Ko (1968) and Song and Schiller (1973). The studies for both these methods were restricted to normal wave attack. In both instances the derived prediction methods involve considerable

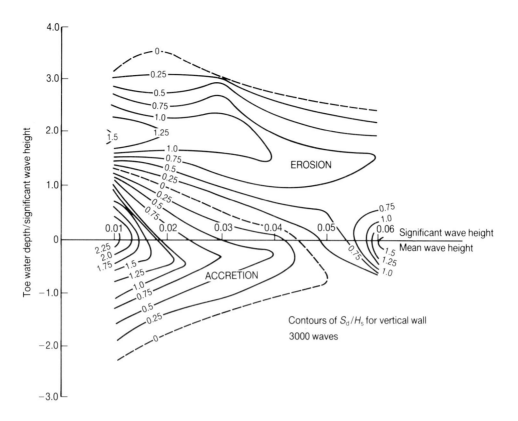

Figure 187 *Example of predicted scour depths for a vertical wall*

simplifications, and some of the trends suggested by these methods contradict the general conclusions described above.

Prediction methods for shingle beaches

Generally, scour on shingle beaches has received less attention. Powell (1989) has presented the results on wave-flume tests in the form of dimensionless design graphs. These allow the dimensionless scour depth, d_s/H_s, to be related to the mean wave steepness, s_{om}, and local water depth, h_s/H_s. An example is given in Figure 187 for vertical walls and a storm duration of 3000 waves. Correction factors are available to determine scour depths for different durations:

for
$$h_s/H_s \leq (1.0 + 25 s_{om})$$

$$d_{SN}/d_{s3000} = 0.127 \ln N - 0.03 \quad (5.77)$$

for
$$h_s/H_s > (1.0 + 25 s_{om})$$

$$d_{SN}/d_{s3000} = 0.149 \ln N - 0.21 \quad (5.78)$$

A series of correction factors are also developed by Powell (1989) to enable the scour on shingle beaches to be estimated for impermeable sloping revetments and for rock revetments. The following preliminary guidance may be of use to the designer:

1. Impermeable slopes of 1:1½ and 1:2 gave rise to no significant reduction of local scour in comparison with that given by vertical walls.
2. Impermeable slopes of 1:3 gave rise to reductions in local scour typically of the order of 25% to a maximum of 50% in comparison with that given by vertical walls.
3. Rock revetments tested showed no susceptibility to cause local scour at all and indeed tended to encourage some accretion.

Design of scour-protection measures

The principal methods of reducing or preventing the scour of bed materials can be summarised as follows:

1. Reduce forces by reducing reflections (see Section 5.1.2.4). This can be achieved by flattening the revetment slope, by adopting an energy-dissipating revetment face (e.g. rock armour) or by placing a near-horizontal berm on the revetment face at a level close to the design water level.
2. Isolate the problem area close to the structure by placing a scour-control blanket (see Section 5.1.3.6). This might consist of rockfill, preformed flexible mats or gabion mattresses.
3. Reinforce the bed foundation material by full, partial or local grouting, using cement or asphalt material (see Section 5.1.3.10).

In the design of new or rehabilitated structures the first of these options is to be preferred, since it removes the cause of the toe scour. It may also improve the performance of the structure in terms of wave run-up and overtopping. Where this is not possible, the most common method of toe protection is the provision of a rockfill blanket. The only design methods available for toe protection therefore refer to rock blankets, and they address the size of the rock needed for stability and the extent of the protection (see Section 5.1.3.6).

5.1.4 PHYSICAL AND NUMERICAL MODELLING

5.1.4.1 Physical modelling

The tools available to describe the responses of a shoreline structure, generally formulae or model test data, are limited to relatively few structural configurations, and often to a narrow range of wave conditions. In general, the structure for which data are available will often represent an idealised case, without many of the complicating features commonly found on prototype structures. The wave conditions are generally limited to normal attack in deep water and/or over a seabed of constant gentle slope.

For many practical cases the application of these to structures of complex geometry, subject to a wide range of wave conditions, will lead to unacceptably low levels of confidence in the calculated response. In such cases it is necessary to simulate the responses of the structure to the main flow processes in a physical model. Waves in nature vary in both height and period, even when the mean values remain static. As most hydraulic and structural responses are non-linear functions of wave height and period, it is important that model tests are conducted using random waves with the correct characteristics.

0Models of coastal structures are often tested to quantify armour response; wave run-up levels; and/or wave overtopping discharges. Such models will usually be constructed at an undistorted scale around 1:10–60. Where the incident wave heights are very large, it is sometimes possible to use scales up to 1:100. In each instance, the choice of scale will also be set to avoid any significant scale effects. In a few cases it has been possible to study these responses at scales between 1:1 to 1:10 in the very large wave flumes such as the Delta Flume in the Netherlands or the GWK in Germany.

The main cause of such scale effects in models of armoured structures are changes of the flow conditions within the structure, from fully turbulent conditions in the prototype to laminar or partially turbulent in the model. This will arise if the typical particle size and/or the flow velocities in the model are too small. The performance of the model will not then simulate the prototype correctly. These conditions can be avoided by ensuring that the typical Reynolds number, Re, for flow in the armour layer remains above a threshold value established from model test comparisons at various scales, including full scale. Based on comparisons between field data and model test results, Owen and Briggs (1985) suggest that values of Re as low as 3×10^3 can be allowed in the model before significant errors arise in the prediction of armour performance.

Scale effects will also affect the sizes of both air bubbles and water droplets in
the model. These will introduce some distortions in wave impact pressures
measured at small scale. These effects are not usually regarded as critical to the
model design, particularly as the model will tend to overpredict wave impact
forces.

It should be noted that physical models designed to reproduce hydraulic responses
will not reproduce the geotechnical response of the structure correctly, and cannot
therefore be used to predict geotechnical effects (see Section 5.2). This may be
overcome by using a very large model or numerical models of the geotechnical
responses with hydraulic boundary conditions measured in the physical model.

In the first stage of the design of a conventional coastal structure it may often be
most economical to model the response of the structure cross-section to normal
wave attack. The model should be built in a random wave flume, within which a
section of the approach bathymetry has been constructed to ensure that wave
conditions at the structure are reproduced correctly. Prior to construction of the
structure, the test wave conditions must be measured at the position of the
structure toe. This will ensure that the incident wave conditions are measured
without corruption from any reflections from the structure itself.

The construction of the cross-section should follow the prototype approach as
closely as possible. This is particularly important in the placement of the toe and
armour layers. The variable nature of rock armour placement will itself introduce
some variation into the test results. It is therefore essential to performance repeat
tests on the same design to quasntify possible variations in performance.

The response of rock armour (as opposed to concrete units—see Owen and
Allsop, 1983; Partenscky, 1987) to wave action is normally measured by measuring
changes to the cross-section profile and calculating the dimensionless damage area,
S_d (equation (5.10)). The model may also be used to quantify wave run-up levels
by measuring instantaneous water levels directly on the seaward face of the
structure. For those seawalls or breakwaters where some overtopping under
extreme conditions can be permitted, water overtopping the model cross-section
can be collected in a volumetric tank and measuredc directly. A wave probe
mounted on the crest of the structure can enable the number of overtopping
waves to be counted. Wave transmission over a low-crest breakwater can be
measured simply with a wave probe placed behind. The breakwater must be set
forward in the wave flume to allow space for the overtopping waves to re-
establish, be measured, and then to be absorbed on the spending beach.

The coefficient of reflection (and/or the reflection coefficient function) can be
measured during tests of overtopping and/or armour stability. Waves reflected
from the structure can be separated from the incident waves by appropriate
computer programs operating on the output of two or three wave probes placed
at least one or two wave lengths in front of the toe of the structure.

A typical model study would include a number of initial tests to allow the
performance of alternative configurations to be quantified. The final sets of tests
would then measure the performance of the optimum design. Wave conditions
used in testing would cover the range equivalent to around 50–120% of the
extreme design condition. These tests are often conducted at a single representative
water level. Where water levels vary, as under tides or surges, some tests should
be repeated at higher or lower water levels.

Where the structure is at an oblique angle to the main wave direction, or other
features cause significant three-dimensional effects, tests may be required in a wave
basin. Tests under oblique attack are particularly important in the design of
breakwater roundheads (see Section 5.1.3.7) and dynamically stable slopes (see
Section 5.1.3.8) such as for berm breakwaters (Figure 188). In the design of new
or modified harbour works, tests on the performance of new structures may
sometimes use the harbour layout model, provided that the model scale is
adequate to avoid scale effects.

Figure 188 *Model wave basin after oblique wave test on offshore breakwater (courtesy HR Wallingford)*

5.1.4.2 Numerical modelling

The basis of any numerical model is the formulation of the physical processes to be simulated. Few of the processes of wave/structure interaction of interest here can at present be described with sufficient reliability to allow the development of mathematical models. Some models have been developed to estimate wave transmission through simple permeable breakwaters (see review by Powell and Allsop, 1985). These models are restricted to very simple configurations and limited wave conditions. Similarly, a number of simple models have been developed to simulate wave run-up on impermeable slopes by Kobayashi *et al.* (1987), Allsop *et al.* (1988) and Thompson (1988). The wave equations used are only valid for non-breaking waves, and the models can generally only be run for regular waves.

An alternative form of numerical model is produced when a computer program uses empirical expressions to describe results from physical model experiments. An example of this type of model is given by van der Meer (1988a) for the estimation of dynamic slope profile changes (see Section 5.1.3.8). Such types of model will include the influence of any model scale effects although the one mentioned was verified by large scale tests.

5.2 Geotechnical interactions

5.2.1 APPROACH

Two mean components are distinguishable in the geotechnical behavior of a rock structure: the foundation and the structure, In the structure itself geomechanical processes will play a role if the structure contains rock/soils components. In the foundation, several aspects are important with regard to the functioning of the structure. The following overview shows the components considered:

ROCK STRUCTURE COMPONENTS

FOUNDATION	STRUCTURE
Seabed	Slope
Filter	Toe
Subsoil	Filter
Cavity	Core

Important information which should be collected from site includes topography and bathymetry, stratigraphy, material parameters and loading characteristics. An overview of the field and laboratory methods to obtain this information is described in Section 4.2. The following scheme explains the interrelations between the elements of geotechnical behaviour.

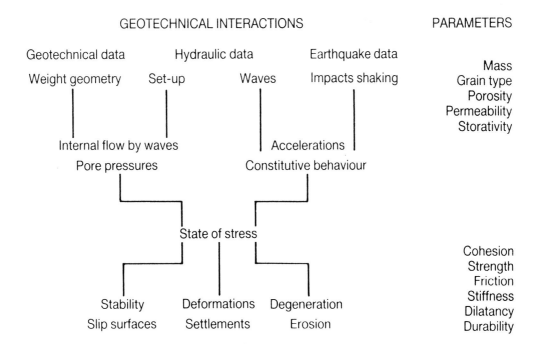

The following steps are suggested in the approach to assess the geotechnical safety of a rock structure, i.e. determination of:

- Critical cross-sections, their geometries and material description (mechanical parameter values);
- Design wave loading and in seismic areas design earthquake loading;
- Pore pressures (seepage pressures), transmission and internal set-up due to wave loading;
- Hydraulic stability of filters;
- Excess pore pressures;
- The static stability of the separate elements and the entire structure;
- The settlements;
- The dynamic stability of separate elements;
- Residual deformations;
- Long-term behaviour.

5.2.2 PARAMETERS

A provisional list of common parameters used in relevant geotechnical numerical models and formulae is presented below. The phenomena and the mechanisms to which these parameters refer are discussed in the following sections. For particular values of some of materials parameters, reference should be made to Chapter 3. For details of geotechnical site investigation and laboratory testing of foundation soils, reference should be made to Section 4.3 and Appendix 5.

Symbol	Name	Description
D	Particle size (m)	Average grain diameter ($i = 50$)
D_i	Sieve diameter (m)	Exceeding $i\%$ value of sieve curve
$\tilde{\rho}$	Specific density	
(kg/m³)	Bulk mass density	Granular skeleton
ρ_r	Specific density (kg/m³)	Mass density of rock
ρ	Specific density (kg/m³)	Mass density of water
γ	Specific weight (N/m³)	$\gamma = \rho g$ (similarly, γ' and γ)
Sh	Grain shape	Ratio maximum/minimum diameter
Rf	Surface smoothness (mm)	Irregularities in grain surface
s_u	Undrained shear strength	
R'	Equivalent roughness	Rock characterisation (Barton)
S'	Strength (N/m²)	Rock characterisation (Barton)
c'	Cohesion (N/m²)	Interlocking, cementation
n	Porosity	Volumetric density granular skeleton
K	Permeability (m/s)	Depends on D, n, Sh, Rf, gradient $\Phi_{,i}$; permeability different in laminar and turbulent conditions
ϕ	Friction angle	Depends on R', n, S', σ, D
ψ	Liquefaction	Rate of excess pore-pressure generation; measured by special test
n_c	Critical density	Porosity at zero dilatancy
G	Shear modulus (N/m²)	Shear strength skeleton
E	Bulk modulus (N/m²)	Compressive strength skeleton
σ, τ	Stresses (N/m²)	σ normal stress, τ shear stress
σ', τ'	Stresses (N/m²)	Effective stresses granular stress)
σ_c	Uniaxial compression strength (N/m²)	
p	Pore pressure (N/m²)	Pressure in the pore fluid
I	Potential gradient	$I = \nabla p/\gamma = \nabla \Phi$; Φ = potential
C	Settlement constant	
c_v	Consolidation coefficient	
C_s	Material settlement const.	
U	Consolidation ratio	Degree of consolidation (%)

Additional specific information about the above-mentioned parameters important for rock structures is discussed next.

5.2.2.1 Cohesion c'

The cohesion is related to the fact that removal of a particle at zero effective stress requires some effort. For rock and artificial armour units the cohesion is related to the interlocking. Cohesion in sand is negligible and for clay it is related to electromechanical forces at microscale. For the interlocking in rock the use of an apparent cohesion is suggested, particularly when applying a Bishop stability analysis. A value of 10–20 kN/m$_2$ for this apparent c' is suggested (Kobayashi, 1987).

5.2.2.2 Porosity n

The porosity depends on the packing of the particles (skeleton structure). An average porosity value for large, well-placed normally graded stones is about 42%. Smaller stones with a normal gradation may have a porosity between 37% and 44%. In wide-graded mixtures the porosity may be smaller (meta-structure: fines occupy the pores between the large ones). It is not practically possible to determine the porosity of existing coarse, granular rockfill in a direct manner.

5.2.2.3 Permeability K

The permeability of coarse porous media depends on the regime of flow (laminar, turbulent). A formula for laminar flow is:

$$K \approx 0.002 n^3 D^2 g/(1-n)^2 v \tag{5.79}$$

which is valid for uniform gradations with $D_{50} < 0.01$ m (ρ and v are density and kinematic viscosity of water). A practical formula for turbulent flow is:

$$K \approx 2\sqrt{(n^5 g D/|I|)} \tag{5.80}$$

and formula for the laminar/turbulent flow in rubble mounds is:

$$K \approx 1/[a/2 + \sqrt{(a^2/4 + b|I|)}] \tag{5.81}$$

with

$$a \approx c_1/(ngD_{20}) \quad \text{and} \quad b \approx c_2(1+Rf/Rf_0)/(gD_{20}\sqrt{n})$$

where c_1 and c_2 are constants depending on the Reynolds number (Hannoura and Barends, 1981). Some general values are given below:

Particle type	Range of diameter (mm)	Order of permeability (m/s)
Large stone	2500–850	1.00 (turbulent)
One-man stone	300–100	0.30 (turbulent)
Gravel	80–10	0.10 (turbulent)
Very coarse sand	3–1	0.01
Coarse sand	2–0.5	0.001
Medium sand	0.5–0.25	0.001
Sand and gravel	10–0.05	0.0001 (more than 10% sand)
Fine sand	0.25–0.05	0.00001
Silty sand	2–0.005	0.000001
Sandy clay	1–0.001	0.0000001

Some graphs concerning porous flow in coarse granular media are presented in Figure 189.

The apparent permeability is related to D^2 for laminar flow and to $\sqrt{D}$ for turbulent flow, shown in Figure 190. The flow regime also depends on the actual

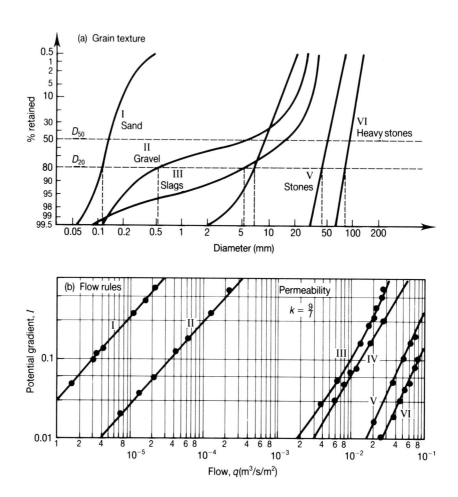

Figure 189 *Measured flow rules*

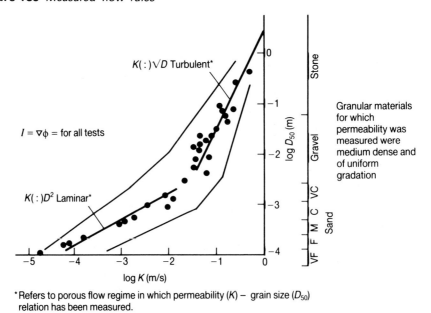

*Refers to porous flow regime in which permeability (K) – grain size (D_{50}) relation has been measured.

Figure 190 *Permeability versus grain size*

pressure gradient, which may vary in space and time (waves). Scaling problems seem unavoidable in physical modelling.

5.2.2.4 Friction angle ϕ

The granular friction angle is a material property for sands and gravels and is normally in the range of 30–45°, as given below:

	Active	Passive
Loose gravel	35°	35°
Very dense gravel	44°	41°
Loose sands	30°	30°
Very dense sands	39°	34°

Cohesionless materials dumped or discharged through water will be in a loose state so that active and passive friction will be the same. For stone/rock media the friction angle depends on various material characteristics as well as on the actual effective stress level.

In a granular skeleton of rock stones large local contact forces may occur, and at these contacts the rock may break. The internal friction depends on this process. Barton (1981) has suggested a practical empirical approach for estimating ϕ which uses the equation:

$$\phi = \phi_0 + R' \log(S'/\sigma') \qquad (5.82)$$

where ϕ_0 is the angle of repose of smooth surfaces of the intact rock, R' a roughness pareameter dependent on particle shape (see Figure 191) and packing, S'

Figure 191 *Conglomerate rock particle, polished surface (courtesy Delft Geotechnics)*

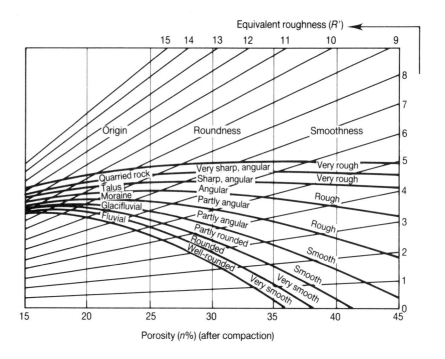

Figure 192 *Equivalent roughness R'*

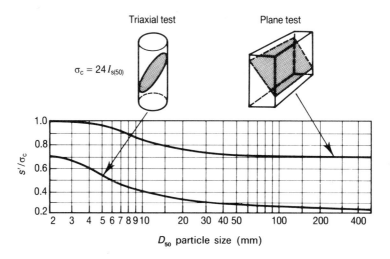

Figure 193 *Equivalent strength S'*

is a normalised equivalent strength of particulate rock and σ' the actual effective normal stress, ϕ_0 is in the range of 25–35°. The value for R' is given in Figure 192, using the porosity n and a qualitative description of particle roughness. The shape parameter P_R (Section 3.2.2.2. and Table 11) was developed (Latham and Poole, 1988) for assessment of particle asperity wear and may be used for further determination of R'. To obtain S', the typical particle size (D_{50}) and intact strength of the rock as given by the uni-axial compression strength (σ_c) are combined as shown in Figure 193. Note that applications in different strain fields (triaxial and plane strain) use different curves. The value of σ_c can be obtained from direct or index testing. The Point Load Index, $I_{s(50)}$, is recommended (see Sections 3.3.6.3 and 3.3.6.5). The effective stress, σ, can be determined by standard methods (stability or deformation models). Because the actual stress varies in a rock structure, the local friction angle will also vary. This can easily be included in a standard slope stability analysis. Changes with time occur if rock quality is poor and point contacts becomes softened (e.g. by weathering).

5.2.2.5 Dilatancy

The sensitivity of a granular material to become liquefied is expressed by the so-called liquefaction potential. This can be obtained by special laboratory tests (cyclic shear and cyclic triaxial tests) on samples from the site at various (artificial) densities. The results have to be calibrated against the *in-situ* density. A typical result is presented in Figure 194 (see also Box 67 in Section 5.2.8.10).

Excess pore pressures arise depending on the drainage capacity of the structure. Fine loose sands ($D < 300\,\mu$m) are sensitive to excess pore pressure generation. Even dense sands may show similar behaviour under cyclic loading.

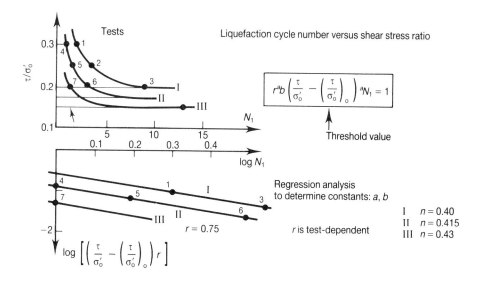

Figure 194 *Liquefaction potential (lab tests)*

5.2.3 SCOPE OF GEOTECHNICAL ASPECTS

Two facts make the geotechnical approach for rock structures different from standard geotechnics:
1. Multi-phase mechanics: air, water and rock play a role;
2. Dynamics: critical situations occur during dynamic loading.

Both aspects render the standard geotechnical methods applicable with special considerations. The behaviour of the water and the rock matrix can sometimes be separated, which simplifies the analysis pressures. The next section will explain the physics and mechanics involved.

For some phenomena a static geotechnical approach is sufficient, such as for settlements of the subsoil. For other conditions the dynamic character of the loading by waves and earthquakes is essential, which makes the analysis more complicated.

Geotechnical aspects are important in those circumstances when the construction will not function properly, referred to as failure states. The manner in which such a state can occur is called a failure mechanism. Three major geotechnical failure mechanisms of a rubble-mound structure are distinguished. Each is characterised by different aspects (submechanism):

1. *Slip failure of the slope and subsoil*
Aspects: internal shear strength

          liquefaction
          steepness of slopes
          accelerations
          excess pore pressures
          rock–(super)structure interaction

2. *Settlements of the core and subsoil*
Aspects: densification of the fill
         compression of soft subsoil layers
         squeezing of very soft subsoil layers
         collapse of underground cavities

3. *Erosion of filters and core materials*
Aspects: filter gradation
         change of material properties
         dynamic gradients
         excessive pore pressures

The state-of-the-art assessment of these geotechnical aspects mainly consists of desk calculations, varying from simple empirical formulas to sophisticated computer simulations. Physical model testing of geotechnical aspects is rarely performed because of difficulties in coping with non-linear behaviour and scale effects. A promising facility to alleviate this deficiency is geocentrifuge testing.

Standard numerical modelling, common practice in soil mechanics, will be addressed whenever applicable for rock structures. References are given for background information.

5.2.4 PHYSICAL BACKGROUND

5.2.4.1 Loading and boundary conditions

Geotechnical aspects of rock structures deals with the mechanical behaviour of rock/soil and the foundation being subjected to weight, seepage forces, pore pressures and accelerations caused by waves and earthquakes. Two types of loading are distinguished: gravity and hydraulic. Gravity loading is due to overburden by the rock mound itself and by superstructures. The loading characteristics can be easily obtained by the consideration of specific (submerged) weight and the geometry. In the case of an earthquake additional horizontal accelerations can be incorporated in a similar way as the gravity. For a realistic stress field in a ruble mound a two-dimensional stress analysis has to be carried out. Hydraulic loading by waves is obtained by measurement of the water pressures at the boundary (slope). This is achieved by special measurements in hydraulic model tests (*Breakwaters*, 1988). More information on hydraulic boundary conditions is presented in Chapter 4.

Another aspect concerns the boundary conditions with respect to the granular phase. This deals with the *in-situ* stress state, which is usually not known. The *in-situ*-stress state is important, particularly for non-linear behaviour, which is characteristic for rock/soil.

5.2.4.2 Geomechanic principles

The mechanical behaviour of a two-phase material (fluid and grains) is determined by loading (action) and by reaction, in terms of forces. In a porous medium the intergranular forces acting at the grain contacts are represented by the so-called intergranular or effective stress (Figure 195). This stress is equal to the unit-area average of all intergranular forces and has normal and tangential components. Fluid pressures in the pores represent another stress, but this has only normal components.

The reaction to a loading (applying total stresses) comprises effective stresses and

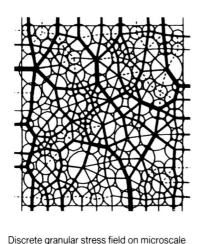
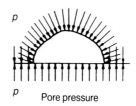
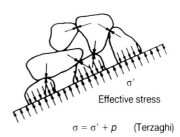

Discrete granular stress field on microscale

Pore pressure

Effective stress

$\sigma = \sigma' + p$ (Terzaghi)

Figure 195 *Effective stress principle*

pore pressures. Differences in pore pressures (pore-pressure gradients) cause flow, which in turn change the pore pressures with time. Hence, the reaction by the pore pressures to the loading changes in time. Effective stresses will also vary to compensate for the change in pore pressures. Deformations corresponding to the effective stresses are, therefore, indirectly related to the porous flow.

If the individual grain is relatively rigid compared to the rigidity of the granular skeleton and to the pore fluid, a simple concept is available to superimpose the internal stresses (Terzaghi's effective stress rule), which states for normal stress components:

$$\sigma = \sigma' + p \qquad (5.83a)$$

and for tangential stress components:

$$\tau = \tau' \qquad (5.83b)$$

Here, (σ, τ) is the total stress, (σ', τ') the effective stress, and p the pore pressure. The material response depends on effective sresses and pore pressures. The Mohr–Coulomb failure criterion, for example, states:

$$\tau = c' + \sigma' \tan \phi = c' + (\sigma - p) \tan \phi \qquad (5.84)$$

The cohesion, c', and the internal friction angle, ϕ, are material parameters which need to be measured. The pore pressure p is related to the actual flow field. The fact that p appears in equation (5.84) clarifies the importance of the flow field for the actual state of stress, particularly in rock structures.

5.2.4.3 Porous flow

Wave loading on a rock structure generates a pore-pressure field inside. For the determination of these pore pressures, and hence the effective stresses, the internal flow field must be evaluated (see above), recognising that the flow in coarse porous media is usually turbulent. In addition, variation of the water table in a porous structure causes changes in the effective stresses as a consequence of Archimedes' law. In Figure 196 an example of a model simulation of an actual porous flow field due to wave loading is presented. In this case a finite element approach (SEEP program) has been used which also includes turbulent flow (Barends, 1983).

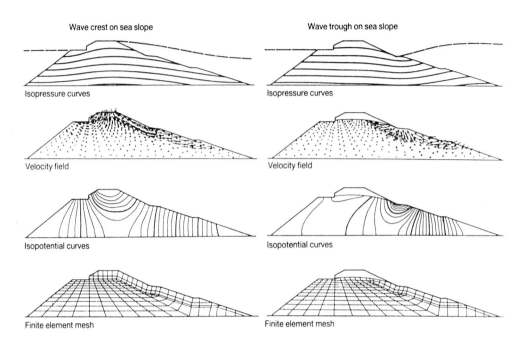

Figure 196 *Simulated transient turbulent porous flow field*

5.2.4.4 Slip failure

The effective stresses are responsible for deformations of the rock matrix. Deformations may ocur in a narrow band, visualised as a slip surface conditioned by shear strength and kinematics, or they may be found throughout the structure. The overburden of the structure may cause deformations in soft soil layers in the foundation. An example of the occurrence of slip zones and plastic zones is presented in Figure 197, showing two situations of a retaining sheetpile in a state of failure, modelled by Allersma in centrifuge tests. A different failure mode is observed dependent on the geometry (depth of the sheetpiling).

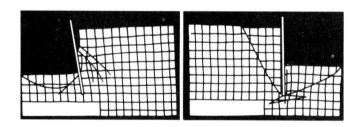

Figure 197 *Centrifuge modelling of failure*

5.2.4.5 Deformation

The deformation behaviour of rock and soil is typically non-linear and irreversible, and it is not easy to find a simple proper description of this behaviour. Field and laboratory tests are indispensable for the characterization of the soil/rock behaviour, but few complete studies about rock structure deformation in a marine environment have been published (Barends, 1981). The descriptions which follow therefore focus on simplified material behaviour.

5.2.4.6 Internal erosion/fatigue

A rock structure comprises various granular materials, fine and coarse, narrow and wide grading. Local pore fluid velocities may convey fine particles through the pores of the coarse, which may lead to internal erosion and deterioration of the structure. This process is usually described by so-called 'filter rules', which define limits of adjacent coarse- and fine-layer grain sizes and gradings (see Section 5.2.8.12).

Rockfill may degenerate with time, fall apart, or crush at the contact point. When the resulting fines are washed out, a fragile porous matrix is left behind. If the fines are not washed out, the rock structure becomes denser and behaves differently. This process can be characterised as material fatigue.

Water may also transport free sediments into the pores or voids in the structure, and when these settle out in the pores they change the hydraulic permeability significantly. This may become a serious problem under critical loading conditions (suffocation). Growth of marine organisms in pores may have a similar effect (see Section 3.2).

5.2.4.7 Liquefaction

A granular skeleton may show a volume change under shear deformation. This is called 'dilatancy' (increase) and 'contractancy' (decrease). This volume change is related to a change in the granular structure by sliding and rolling of the particles in a different composition. As a consequence, in saturated porous media excess pore pressures are generated, negative for dilatancy and positive for contractancy. At a critical density (n_c) the material is volume constrained, that is, shear deformation does not show volume changes. The critical density is a material property related to the type of particles and the grading.

The sensitivity to dilatancy/contractancy depends on the actual *in-situ* density and the critical density and on the air content of the pore water. In the ultimate state the coherence in the skeleton vanishes. Particles are free, in a state of liquefaction, as a thick fluid without significant shear resistance. The character of this phenomenon also depends on the type of loading. A sudden loading may cause a rapid decrease of pore volume and a corresponding decrease of shear strength due to excess pore pressures, in less than a second. The dissipation of the pore pressures may take some minutes or more, and in the intermediate period catastrophic consequences may have occurred. Insistent cyclic loading may cause a slow build-up of excess pore pressures with similar consequences. Both situations may occur simultaneously. This phenomenon has significant implications for the strength of rock structures where damage is sometimes initiated by wave-induced local liquefaction of the sandbed underneath.

5.2.4.8 Dynamic effects

During wave impacts and earthquakes the rapid loading induces inertial effects in the pore fluid and the porous matrix. Consequently, the related stress field and pore-pressure field vary in time and accelerations must be taken into consideration. An illustrative numerical example of the contribution of each phase (water and grains) to the energy dissipation of a wave impact is presented in Figure 198, in which the isotropic pressure histories are shown. The permeability plays a significant role, because the graphs show a completely different response when the permeability differs by a factor of 10..

An important slope-stability problem in the marine environment and for reservoir dams is that of rapid drawdown. Because it takes time for the water to seep out of

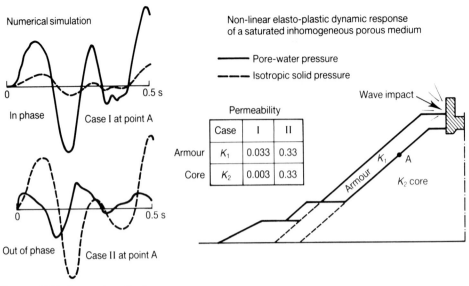

Figure 198 *Dynamic effects*

the rock structure, pore pressures will stay high for a short period and jeopardise the geotechnical stability, particularly when the slope surface is protected by an impervious cover (asphalt).

The stability of slopes subjected to earthquakes is commonly simplified by the introduction of an additional inertial force, whose magnitude is equal to the product of mass of the slice in failure and its horizontal acceleration. Usually, vertical acceleration is not considered and the horizontal acceleration is assumed constant. These inertial forces can then be taken into account in a Bishop stability analysis. A more realistic behaviour under earthquake loading can be obtained by using a sophisticated model based on a consistent stress and deformation state. Figure 199 shows the result of such an analysis (SATURN code), where the residual deformation has been calculated as the result of a complete earthquake loading.

5.2.4.9 Cavities

Large-scale subsidence can occur in sedimentary soluble rock masses, which include carbonates (linestone and dolomite) and evaporites (gypsum and salt rock). Limestone may be found with karst development, where karst is the term given to natural cave and gorge formation by carbonate solution. Areas of southern China, south-east Asia, southern Africa, north-west Europe and the south-west United States are underlain by karst. Natural solution rates are slow, around 5000 years being required for a cave of 1 m diameter to form. Vast cave systems have developed over tens of thousands of years. The largest known cavern exists in Sarawak, and is 700 m long and 400 m wide. These large cave systems can cause unexpected settlement (see Figure 200), particularly when environmental conditions are changed by man's activities.

The theory of the formation of sinkholes (sudden settlement) and dolines (slow subsidence) is well established. The initiating events may be sub-surface erosion arising from the presence of the cavities or lowering of the water table (see Figure 201). Monitoring of an inland suspect area usually consists of daily inspection on foot to look for surface cracking and cracking of structures, together with an analysis of levelling observations. In a marine environment this is not easy when often the only sub-surface tools available are bathymetric surveys and, occasionally, diving inspections. If subsidence caused by subsurface erosion is observed in time, it is sometimes possible to repair by subsurface grouting (Figure 201).

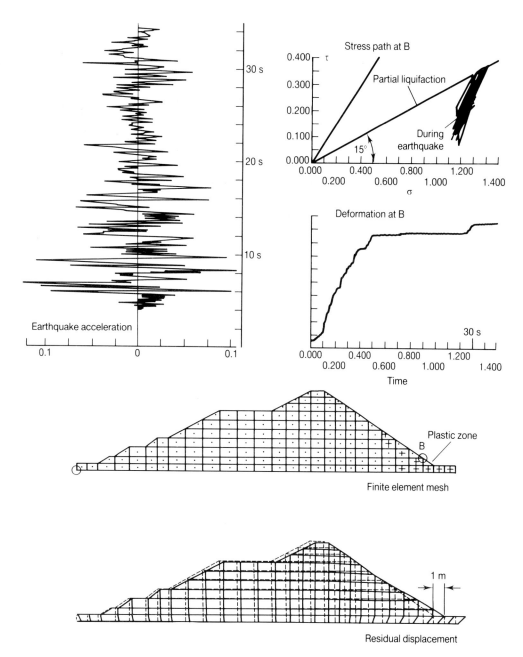

Figure 199 *Simulation of earthquake response*

5.2.5 MODELLING AND SIMULATION

The description of physical processes around, on and in rock structures with formulae, physical or numerical models, graphs or by engineering judgement is always an approximation of reality. The purpose and value of modelling is to enable the optimization of a design or a particular element of the structure by a more accurate approximation.

To assess the optimal, most reliable and economic solution one can use a probabilistic approach. In principle, the methods discussed here permit such an approach as long as the probability distribution of loading and the stochastic variation of material and geometrical parameters is known (see Section 2.4). An example of probabilistic analysis is briefly outlined in Section 5.2.6.

The simulation of fluid flow in rock structures is complicated because of non-linear

Figure 200 *A sinkhole (courtesy F. Tatsuoka, University of Tokyo, Institute of Industrial Science)*

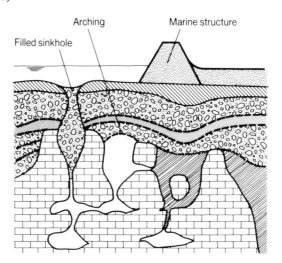

Figure 201 *Mechanisms by sub-surface cavities*

behaviour, unsteadiness, air entrainment under wave loading conditions, and interaction with the granular skeleton. It is difficult to make observations of hydrodynamic process in a porous medium because it is not transparent and, in addition to measuring the pore pressures, the response of the granular body needs to be recorded. Further problems are that the behaviour of one particular pore or grain is not necessarily characteristic of the overall behaviour, and that in model studies not all scale effects can be taken into account simultaneously.

For the simulation of aspects of geotechnical behaviour two methods are commonly used: empirical and numerical. Empirical methods are simple formulae taken from geotechnical practice and adjusted for the marine environment. Numerical methods which have become available in the last decade comprise computational models, based on schematised structures. For numerical models, scale effects do not occur, but a proper mathematical formulation of the material behaviour is a problem. Numerical models can be applied in a probabilistic way, addressing the spatial variation of parameter values together with the stochastic time-variant character of the loading.

5.2.5.1 Mathematical background of available models

The essence of numerical models is the mathematical description of a physical process. A comprehensive survey of the geotechnical behaviour of rock structures is presented here in the form of basic equations and rules. In the following section, various models based on these equations are described.

Three media are involved: air, water and rock/soil. The influence of air is important for the storativity and for dynamic impact pressures. Practical information about air effects is limited.

Pore water mechanics

For water and rock/soil, two interrelated equations of state (mechanical equilibrium) are involved. The motion of the pore fluid is controlled by an action force resulting from pore pressures, gravity, capillarity and other effects, and by a reaction force resulting from inertia and internal friction (viscosity, microturbulence, interaction with the grains). The equation of state of the pore water yields (action = reaction):

$$-np_{,i} - n\rho g z_{,i} = n\rho \dot{v}_i + n^2 \rho g(v_i - u_i)/K \qquad (5.85)$$

Pressure gradient — Gravity — Inertia — Interaction

Here, p is the pore pressure, ρ the pore fluid density, g the gravitational acceleration, n the porosity, v and u are the velocity vectors for the pore fluid and the granular matrix, respectively, and K is the hydraulic conductivity of the porous medium. The subscript i represents a spatial co-ordinate direction. If preceded by a comma it is a partial derivative along the ith direction. The dot represents a (fluid-) substantial time derivative and

$$\dot{v}_i = cv_{,t} + bv_j v_{i,j} + a\dot{u}_i \qquad (5.86)$$

Local — Convective — Interaction

in which c includes virtual mass effectys ($c > 1$) related to the unsteadiness of flow, b is the momentum distribution coefficient, and the last term represents mass effects in the flow field due to the presence of the granular phase.

The last term in equation (5.85) includes viscous effects (internal viscous boundary layers, Reynold stresses at pore size dimensions), which make the parameter K a non-linear velocity-dependent function. Many (semi-)empirical relations for this function have been suggested for coarse granular mound structures (Hannoura and Barends, 1983).

Granular matrix mechanics

The pore pressure, p, and the porous matrix velocity, u, are related to the deformation behaviour of the granular skeleton, which is described by an equation of motion for the porous matrix. This equation is described as the instantaneous equilibrium of an action force resulting from intergranular contact forces, pore-pressure gradients (drag forces) and a reaction force due to the inertia of the granular particles and indirectly the actual state of stress (non-linearity). This equilibrium is governed by the second equation of state (action equals reaction):

$$\underline{-(1-n)p_{,i}}\,\underline{-(1-n)\bar{\rho}z_{,i}}=\underline{(1-n)\bar{\rho}\dot{u}_i}\,\underline{-\sigma'_{ij,i}}\,\underline{-n^2\rho g(v_i-u_i)/K} \qquad (5.87)$$

Pressure gradient   Gravity   Inertia   Effective stress   Interaction

Here, $\bar{\rho}$ represents the density of the rock/soil, σ' is the effective stress tensor and the dot denotes a (grain) substantial time derivative. For the inertia term an expression such as equation (5.86) can be formulated.

Beside the two equations of state, the conservation of mass and constitutive relations (material behaviour) are required to complete the description. For rock granular media, the individual grains are considered incompressible. For high stresses this assumption has to be reconsidered, which may lead to other relations. The pore-water mass-conservation equation yields:

$$\underline{(pv_i)_{,i}} \;+\; \underline{(n\rho)_{,t}} = 0 \qquad (5.88)$$

In/out flow   Storage

A compressibility law for the pore fluid can be introduced accounting for the presence of entrained air. A stress–strain constitutive law describes the relation between the porous matrix stresses and the deformations (elastic, plastic, creep):

$$\sigma'_{ij}=\sigma'_{ij}(\varepsilon_{ij}) \qquad (5.89)$$

strain: $\dot{\varepsilon}_{ij}=(u_{i,j}+u_{j,i})/2$; $\varepsilon=\varepsilon_{ii}$ (volumetric strain)

These relations are complex. For practical applications, simplifications are normally made with regard to the stress–strain behaviour, i.e. linear elastic or elasto-pure-plastic.

Equations (5.85)–(5.89) represent a complete description of the mechanical process in a saturated rock/soil porous medium. Some sophisticated numerical models exist which make use of these equations (see Box 62). In most applications, simpler approaches will give satisfactory results.

Simplified approach

Simplification is possible when the dynamic soil–water–structure interaction is disregarded. Then the porous flow is independent of the porous matrix behaviour. The opposite, however, is not the case: the effective stress state still depends on the pore pressure. Introduction of the filter velocity:

$$q_i=n(v_i-u_i) \qquad (5.90)$$

and a potential function:

$$\rho g\Phi_{,i}=p_{,i}+\rho g z_{,i} \qquad (5.91)$$

transforms equation (5.85) into the well-known flow law:

$$\underline{\Phi_{,i}} \;=\; \underline{-c\dot{q}} \;-\; \underline{q_i/K(|q|)} \qquad (5.92)$$

Gradient   Unsteadiness   Forchheimer/Darcy

Box 62 Dynamic soil–water–structure interaction

The SATURN code is a computer program suited to simulate non-linear dynamic behaviour if two-phase soil structure interaction utilizing the finite element technique. A Darcy–Biot approach is used for the water–soil interaction. The soil–structure interaction includes slip and spalling and a general friction law can be applied at the interface. Equilibrium equations are solved using explicit time integration for dynamic wave propagation and implicit techniques for unsteady motion. By updating spatial positions, large strain effects are accounted for. The initial state problem for non-linear behaviour is generated implicitly to provide compatibility to the imposed non-linear dynamic problem. Available soil models are von-Mises, Drucker–Prager, Mohr–Coulomb, Critical State and Double Hardening. The pore water is compressible. The code is fully tested and applied to various complex dynamic problems (Barends, 1983).

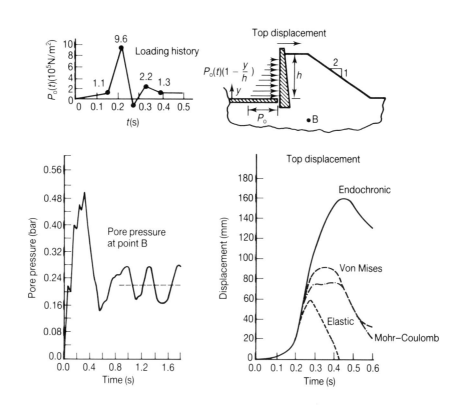

Figure 202 Pore-pressure and displacement responses to loading

Substitution into equation (5.92) and disregarding the unsteadiness yields, after some elaboration, the so-called storage equation:

$$(K/\rho g)\Phi_{,ii} = n\beta\dot{\Phi} + \dot{\varepsilon} \tag{5.93}$$

The term $\dot{\varepsilon}$ represents volumetric variation with time. It is important for soft slowly draining layers (consolidation). It is also important for loose sand deposits, which may show dilatancy. Dilatancy reflects volumetric changes under shear strain. Then, $\dot{\varepsilon} = -\gamma n\beta\psi$, where ψ is the liquefaction potential (a cyclic pore-pressure build-up rate to be determined by special laboratory tests—see Figure

194). If ε is assumed constant and the compressibility of the pore water is disregarded ($\beta=0$), a familiar equation for potential flow is found.

In the case of a phreatic surface (internal water table) the motion of this surface under time-variant loading needs attention. Local flow and storage determine the position, mathematically described by the so-called moving boundary condition. Saturation, air flow and capillarity complicate the simulation of phreatic motion. In coarse porous media the unsteadiness and the inertia effects become significant. A simple way to address this phenomenon is by averaging and integrating equations (5.85) and (5.88) over the vertical co-ordinate (like the shallow wave approach), and the disregarding the interaction with the skeleton deformation. The following equations are obtained:

$$nc_m(\bar{v})_{,t} + \bar{v}(\bar{v})_{,x} = -n^2 g(c_p h)_{,x} - n^2 gF\bar{v} \tag{5.94a}$$

$$nh_{,t} + (h\bar{v})_{,x} = 0 \tag{5.94b}$$

where h is the local water depth and $\bar{v}$ the mean velocity, n the porosity, F the internal flow resistance and c_m and c_p are correction factors for the actual mass and the vertical pressure distribution, respectively.

Simplification of the equation of state of the granular matrix is obtained by adding equations (5.85) and (5.87) while disregarding inertial terms. The result is:

$$-p_{,i} - \tilde{\rho}gz_{,i} + \sigma'_{ij,i} = 0 \quad \text{with} \quad \tilde{\rho} = n\rho + (1-n)\rho' \tag{5.95}$$

or with Terzaghi's rule: $\sigma_{ij,i} - \tilde{\rho}gz_{,i} = 0$

In a practical approach the stress field is simplified rigorously by only considering vertical stresses and these in terms of a direct relation between local stress and overburden given by

$$\sigma_{zz} = \tilde{\rho}gz \tag{5.96}$$

and by adopting, instead of equation (5.89), a simple stress–strain law, assuming negligible deformations until a failure state given by Coulomb's law:

$$t \leq c' + \sigma_{zz} \tan \phi = c' + (\sigma_{zz} - p) \tan \phi \tag{5.97}$$

where c' is the (apparent) cohesion, ϕ the internal friction angle and σ_{zz} the vertical stress.

5.2.5.2 Methods and models

A method implies a logical procedure for doing something. Here it refers to the way of applying experience gained in the design of rock structures. A method is, for example, the probabilistic approach. A model, on the other hand, could be defined as a representation made to imitate the fundamental purpose of a model being the simulation of the behaviour of a structure. A model is a basic tool in a design procedure.

The input into geotechnical models of rock structure behaviour will include loading, geometry and material properties, the output dealing with stability, rigidity and safety with regard to certain failure mechanisms. Available models can be categorised according to applicability, reliability, cost, sophistication, user-friendliness and accessibility. It is not possible to give a complete list of models, nor a detailed picture of a particular model. Some general aspects are discussed and references are mentioned.

Three types of models are distinguished:

1. Sophisticated computer models, which solve a complete problem;

2. Uncoupled computer models, which solve a porous flow problem or a deformation/stability problem;
3. Practical (hand) formulae for particular phenomena.

The two types of computer models are discussed in Section 5.2.6 and their applications in Section 5.2.7. Practical (hand) formulae and engineering experience are described and discussed in Section 5.2.8.

5.2.6 COMPUTER MODEL TYPES

5.2.6.1 Sophisticated models

Sophisticated models deal with non-linear multi-phase deformation and dynamics. Few sophisticated models are known (SATURN code). The core of the model is the solution of equations (5.85), (5.87) and formulation of non-linear behaviour and dynamic interaction, the determination of the initial state and the dynamic effect of air. Such models use a continuum approach and finite element/difference schemes.

Sophisticated models are applied in special situations: for unique and important structures, when a high degree of safety is required, and for the qualitative understanding of certain phenomena (Box 63). The dynamic energy to which a rock structure is exposed is transferred to the pore water and the granular mass. The distribution is controlled by the permeability, saturation and compressibility. The pore pressure reveals the response in the water, not in the granular part. In hydraulic engineering the granular response is often disregarded, resulting in a poor understanding of the real process and an overestimation of the functionality.

The application of dynamic models is relatively costly. It requires an extensive parameter collection (site investigation and laboratory testing). For quantitative applications the results of numerical models should be verified by physical modelling. Complete studies are rarely reported (*Breakwaters*, 1985).

5.2.6.2 Simplified (uncoupled) models

Uncoupled models simulate the porous flow without consideration of the influence of the rock/soil deformation. In a marine environment the flow generated by waves is important. The wave energy to which a rock structure is exposed generates a porous flow field with kinetic energy (fluctuations) and potential energy (internal set-up). If inertial effects in the flow in a fine granular medium is negligible, equation (5.93) applies. Non-linearity in permeability is accounted for by an iterative algorithm.

5.2.7 COMPUTER MODEL APPLICATIONS

5.2.7.1 Porous flow

In course media, a flow law according to equation (5.92) is required. A moving boundary represents the fluctuating water table inside the structure. If inertia effects are not negligible, equations (5.85) and (5.88), with a rigid granular skeleton, can be applied. For mainly horizontal flow, equations (5.94) can be used. Complicating features are the precise boundary conditions: the moving water table, overtopping, air entrainment and the formulation of the wave action on the rubble slope. At present, this boundary condition is obtained by carrying out physical hydraulic model tests with special measurements at the slope, shown in Figure 203. Few models suited for wave-generated dynamic porous flow are known, one of which is the HADEER code for open slopes (Box 63).

5.2.7.2 Deformation

The deformation of rock structures can be obtained in a semi-uncoupled fashion

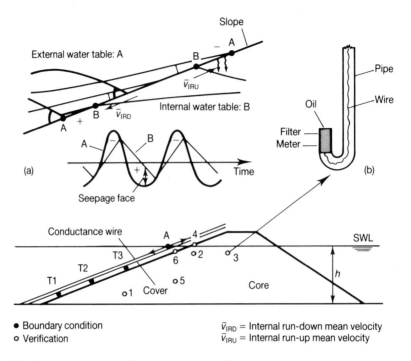

Figure 203 *Measuring hydraulic boundary conditions*

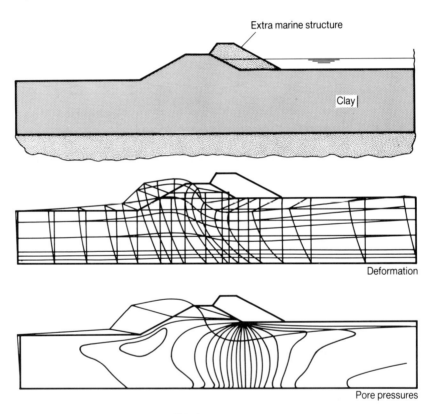

Figure 204 *Two-dimensional consolidation*

applying equations (5.89) and (5.95). The pore-pressure field must be known. Simulation of deformations of a rubble structure itself is seldom performed. Most deformation takes place during construction, and for large structures a deformation analysis is required. Settlements of compressible layers in or under the rock structure due to the overburden weight is a common phenomenon in soil

> **Box 63** Wave-induced internal water motion in a rock structure
>
> The HADEER code describes the two-dimensional water motion in a rubble structure under wave attack, including turbulence, inertia, unsteadiness and water-depth effects. The boundary condition is wave-run-up and wave pressures on the slope to be determined by experiments. The program calculates the phreatic water table by a finite-difference scheme, and then the porous flow field by a finite-element one. The result is a pressure and velocity field under a varying water table in the structure inside. Important aspects are the places of intense flow and the significant internal water-table set-up, and it can be used for wave transmission analysis (Figure 205).
>
>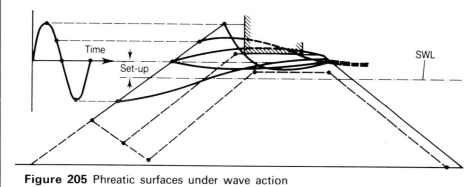
>
> **Figure 205** Phreatic surfaces under wave action

mechanics practice (consolidation). Various methods and models are available, based on equations (5.93) and (5.96). An example is presented in Figure 204, showing a river embankment founded on soft consolidating subsoil subjected to a sudden increase in river level. Retarded settlements can be expected. Non-linear behaviour must be used to model squeezing (plasticity).

5.2.7.3 Stability

The evaluation of the stability of rock slopes is a standard procedure. Usually, a failure state is stimulated by the choice of a critical kinematic system assuming continuing deformation surfaces (slip surfaces), and comparing the mobilised force with that to which the structure is exposed. Key factors are internal friction and cohesion. The stress state along the slip surfaces is determined by wave-induced pore pressures or earthquake-induced accelerations and a simplified effective stress field according to equations (5.96) and (5.97) (Bishop's method). Local excess pore pressures (liquefaction) can be accounted for by excess pore pressures or a reduction of the friction angle. Several studies have shown that this method is sufficiently accurate for rubble slopes when proper values for friction and cohesion are applied.

The models and methods oriented above can be used for a probabilistic analysis. The parameters and boundary conditions must be known in a stochastic fashion and the models must then be used in a special way. Different levels of sophistication are possible (see Sections 5.29 and Box 64).

5.2.8 PRACTICAL FORMULAE AND ENGINEERING EXPERIENCE

Practical formulae suitable for use in hand calculations are available for the analysis of the geotechnical state of a rock structure. This section indicates formulae that address the following phenomena:

- Wave penetration (structure)
- Internal phreatic set-up (structure)
- Dynamic excess pore pressures (structure/subsoil)

> **Box 64** Bishop's stability analysis
>
> Bishop's method is popular because of its convenience and its widespread simple computer codes. The most critical circle of sliding is calculated by considering the moment of weight and the shear resistance along the sliding surface. The soil mass is divided into several slices. The mechanical equilibrium of each of them is considered (see Figure 206). Horizontal forces between adjacent slices are disregarded. The error in the safety factor is 2–6%. The stability of the soil mass is expressed as a safety factor F. If F is larger than 1 the structure is safe. Usually, an extra margin is taken to account for uncertainties, and so normally $F>1.25$. The slipcircle analysis is based on effective stresses. Hence, pore pressures are important, particularly when they vary in time. The most important factors for slip stability are pore pressures and the water table, and both can be controlled by adequate drainage.
>
>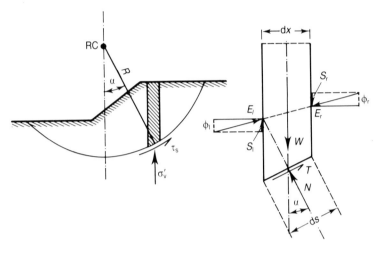
>
> **Figure 206** Principle of Bishop's method

- Slope stability
- Cover layer stability
- Earthquake effects
- Settlement of compressible soil (subsoil)
- Settlement of rockfill
- Local stability of rockfill (structure)
- Liquefaction potential of seabed sand (subsoil)
- Squeezing of very soft soil (subsoil)
- Filter rules (structure/subsoil)
- Strength of geotextile (structure)

5.2.8.1 Wave penetration

Wave loading on rock structures causes time-dependent porous flow, the time dependency being due to storage. One kind of storage occurring in coarse granular media is phreatic. This acts as a damping mechanism, attenuating the wave effect inside the structure. A formula is available which expresses the distance, λ, of penetration of wave agitation across the internal water table, assuming an unlimited structure extent (see Box 65). The significance of rapidly or slowly varying wave loading for geotechnical stability can be evaluated using this formula. For example, for a coarse sand with $n=0.20$, $K=0.0001$ m/s and $A=10$ m, the following is found:

Water waves	$t=10$ s	$\lambda=0.11$ m
Tides	$t=12$ h	$\lambda=7.35$ m
Fluvial surge (const.)	$t=3$ days	$\lambda=72.00$ m

Another useful application of the formula for penetration length of wave action is

> **Box 65** Surface wave penetration in a porous rock structure
>
> A formula is derived based on conservation of mass and on a simplified flow field. The volume of water penetrating (Figure 207):
>
> $$Q\Delta t = K \Delta t A (\Delta h/\lambda)$$
>
> equals the volume stored:
>
> $$n\Delta h [\lambda(t+\Delta t) - \lambda(t)]/2$$
>
> which gives:
>
> $$(n/2)(d\lambda)/dt) = KA/\lambda.$$
>
> Integration yields:
>
> $$\lambda = n_p \sqrt{(KAt/n)}$$
>
> where λ is the penetration length (m), n_p the parameter dependent on the loading ($n_p = 2.0$ for a sudden constant water level change, $n_p \approx 0.5$ for cyclic water level changes), K the permeability (m/s), A the drainage surface through which water seeps (m), n the effective porosity (excluding stagnant part) and t the elapsed time for a sudden constant water level change or period of cyclic loading (e.g. wave period, T). For estimates of basic parameters such as K and n see Section 5.2.2.
>
>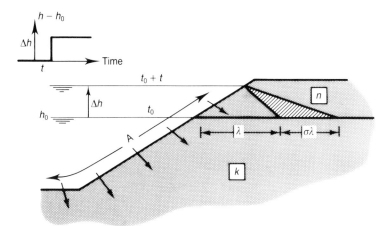
>
> **Figure 207** *Parameters used in description of surface wave penetration*

where, in coastal structure design, it is desired to absorb as much as possible of the wave water volume within the armour layer, with the objective of maximizing wave energy absorption and minimizing wave reflection. In this situation the required armour layer thickness, t_a, to achieve a substantial absorption of the wave water volume can be estimated approximately as $\lambda \sin \alpha$.

5.2.8.2 Internal phreatic set-up

The slope of a rock structure subjected to waves, swell or tides has a geometrically non-linear effect on the porous flow field generated in the structure. The non-linearities arise because the inflow surface along the slope at the moment of a high water level is larger than the outflow surface at the moment of a low water level, and because the average path for inflow is shorter than the outflow path. Hence, during cyclic water level changes, more water will enter the structure than can leave. Eventually, a compensating outflow of the surplus of water is achieved by an average internal set-up of the water level developing inside the structure.

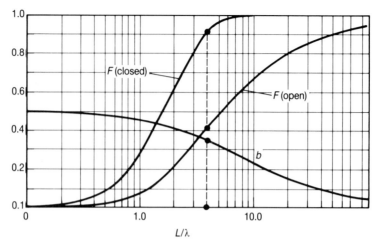

Figure 208 *Diagram for internal set-up*

The internal set-up as a consequence of cyclic water motion may be described (*Breakwater*, 1988) in terms of the following formula:

$$Z_{s\,max}/h = \sqrt{(1+\delta F)} - 1 \quad \text{with} \quad \delta = 0.1cH^2/(n\lambda h \tan \alpha) \tag{5.98}$$

where $Z_{s\,max}$ is the maximum internal set-up, h the still-water depth, c a constant depending on effects of air entrainment and rush-up/run-down ($c > 1$), H the wave height at the slope, n the effective porosity, λ the penetration length and α the slope angle. The function F is presented in Figure 208 for two cases: (1) closed lee-side and (2) open lee. In the first case the maximum set-up occurs at the lee-side. In the second the maximum set-up is found at bL, where L is the length of the inner water table (b is given in Figure 208). For the real water table inside the structure, the cyclic motion must, of course, be superimposed on the set-up. The resulting pore pressures are an essential factor in the stability analysis.

It is also worth noting (*Breakwaters*, 1988) that a breakwater with a backfill exposed to waves on a slope may develop an internal set-up which will completely saturate the core. In this situation no damping or attenuation of wave pressures in the core is possible and the backfill itself will be subjected to strong wave forces.

5.2.8.3 Dynamic excess pore pressure

A pore-pressure build-up in a rock structure is possible due to external time-dependent loading and local deformation of the granular skeleton, and dissipates by porous flow and unloading. Because excess pore pressures cause a decrease of the effective stresses and consequently a reduction in the strength of the granular skeleton, it is important in a stability analysis.

For the evaluation of a local pore-pressure build up the order of magnitude of the dissipation period has to be considered. If the dissipation period is small compared to the relevant loading period, a pore-pressure build-up will not occur. The following phenomena related to dynamic pore-pressure build-up are distinguished:

1. Phreatic storage;
2. Field storage (elastic storage);
3. Propagation of dynamic disturbance along the inner water table;
4. Dynamic pore pressures due to impact loading.

The phreatic storage effect has been discussed above.

Phenomena 2 and 3 are explained in Box 67, where it is shown how to evaluate the probability of occurrence of excess pore pressures. The evaluation of dynamic

Box 66 Internal set-up under time-variant conditions

The slope steepness (angle α) causes an extra storage of water volume every cycle in comparison to a linear situation with a vertical face (dashed area in Figure 209). This extra volume can be expressed as an average infiltration, according to: $I = I_0 \exp(-x/\lambda)$, with $I_0 = cH^2/(2\lambda T \tan \alpha)$, using the simplified formula for penetration length λ. Solving a schematic flow problem gives the expression for the average water table $(h - Z_s)$ as a function of x, where z_s is set-up of water table above still-water depth, h.

The solution is: $(h + z_s)^2 = Ax + B - \delta h^2 \exp(-x/\lambda)$, where A and B are related to boundary conditions and $\delta = 0.1\ cH^2/(n\lambda h \tan \alpha)$. Two cases are distinguished in Figure 203: closed and open lee-side. From this solution the set-up can be determined.

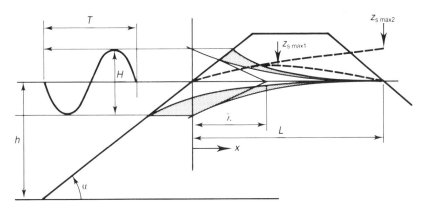

Figure 209 *Illustration of internal set-up under wave action*

Box 67 Dynamic excess pore pressures; a simple approach

Dynamic pore pressures are caused by impact loading and by the structural response. The intensity will decay due to dissipation according to: $p(t) = p_0 \exp(-Bt)$, where p_0 is sudden pressure excitation and B is a decay factor depending on the phenomenon considered. For dynamic perturbations in phreatic water $B = ng/K$ (Dracos) and for field storage $B = 3c_v/r\lambda$ (Verruijt).

Example

A rock structure is considered (see Figure 210): $n = 0.4$, $g = 10\,\text{m/s}^2$, $K = 0.05\,\text{m/s}$ (gravel), $c_v = EK/\gamma = 10\,\text{m}^2/\text{s}$ (E (Young's Modulus) $= 3$ MPa, $\gamma = 15\,\text{kN/m}^3$), average drainage path $\lambda = 4\,\text{m}$, hydraulic radius $r = V/A = 9\,\text{m}$ (volume $V = 140\,\text{m}^3$, drainage surface $A = 15.6\,\text{m}^2$). The formula yields for dynamic perturbations $B = 80$ and for field storage $B = 0.8\,\text{s}^{-1}$. A local pre-pressure build-up will not occur if the relevant loading return period exceeds $1/3B$, which yields for perturbations $0.004\,\text{s}$ and for field storage $0.4\,\text{s}$.

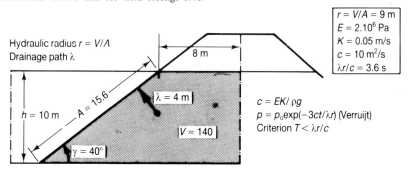

Figure 210 *Example of field storage and dissipation of excess pore pressures*

---

impact pressures (item 4) is more difficult. Some physical model results are available (Jensen, 1984), and some numerical results have been published (Barends *et al.*, 1983). In structures where it is not possible for the pore water to dissipate (venting), wave-impact loading may generate high dynamic pore pressures which can extend with high intensity throughout the entire structure and may jeopardise the stability.

5.2.8.4 Slope stability

The stability of rubble-mound slopes can be determined by a Bishop analysis (Box 64). In the marine environment the effect of water level changes is important and can cause failure (Figure 211). Three cases are considered:

1. Effects of waves;
2. Effect of rapid drawdown;
3. Effect of overtopping.

Water waves generate an unsteady flow pattern in the porous matrix (Figure 196). The major part of the flow takes place in the front slope. The effect of this porous flow on the geotechnical stability is a decrease of 20–25% with respect to the geotechnical stability factor for the corresponding still-water conditions.

Morgenstern presented stability charts to facilitate the computation of the factor of safety of earth slopes during rapid drawdown based on Bishop's stability analysis. The charts have been constructed under the assumption of a homogeneous single material with properties c' and ϕ and a slope which is constructed on a firm impervious bed. They also assume that no dissipation of pore pressures has occurred after drawdown. Some graphs are presented in Figure 212 for a slope β of 2:1 and 3:1. The heavy lines refer to $c'/\gamma H_{sl}=0.025$ and the dotted lines to $c'/\gamma H_{sl}=0.0125$, where H_{sl} is height of slope above firm impervious bed and L is the drawdown height from the top of the slope.

The stability of the rearslope is most critical during wave overtopping, when water penetrates the mound and seeps out beneath (Figure 213). This flow decreases the stability significantly, as shown in the figure. Because most rock structures have a rather steep lee slope, the effect of overtopping needs serious attention.

In the classical soil mechanics approach the conventional stability analysis assumes that peak shear strength of the soil is fully mobilised simultaneously along the total length of the potential sliding plane. This is not always the case, and instead progressive failure may occur. Particularly under certain conditions, slides in overconsolidated plastic clays and clay shales will be preceded by the development of a continuous failure surface by a mechanism of progressive failure. Such situations have to be analysed with special care.

Figure 211 *Rock slope failure (courtesy Delft Geotechnics)*

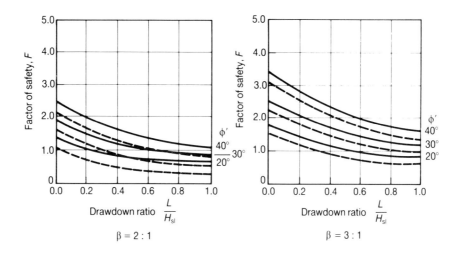

Figure 212 Stability factors during rapid drawdown

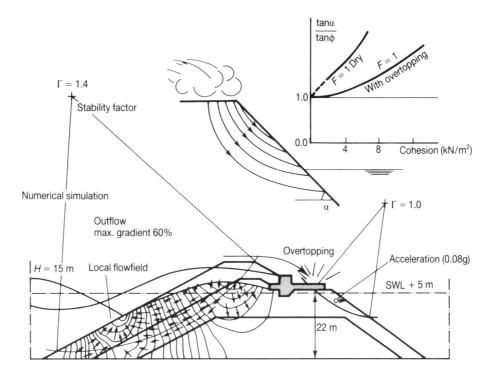

Figure 213 Porous flow during overtopping

5.2.8.5 Coverlayer stability

Coverlayer stability may be considered to have two aspects: stability against downslope sliding and against pore pressure uplift. Downslope sliding is important because the most likely shear plane is not circular, but rather along weak strata or rock-layer interfaces, where the internal friction is lower because of different stone sizes (see Figure 214). A reduction of the internal friction angle δ up to 30% with respect to the friction angle ϕ for uniform rock may be expected. Commonly, $\delta = 2\phi/3$ is applied. However, in some cases δ may be down to 0.5ϕ. For stability analysis one divides the sliding soil mass into an active wedge, a central block and a passive wedge. For thin coverlayers the wedges may be disregarded. The factor of safety is the ratio between the downslope driving forces due to weight and the active wedge and the upslope resistance forces due to friction and the passive wedge.

Figure 214 *Coverlayer failure (courtesy Delft Geotechnics)*

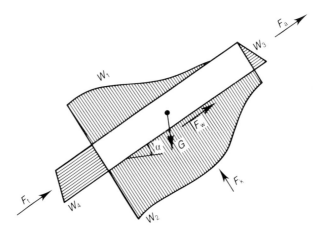

Figure 215 *Mechanics of a coverlayer*

Two types of coverlayer structures may be distinguished: with tensile strength (e.g. rock on geotextile on sand) and without tensile strength (i.e. rock on shingle). The loading on such structures may be breaking or non-breaking waves, or a steady drop in water level. The general equilibrium for the case without tensile strength (see Figure 215) is described by:

$$F_t + F_a = G \sin \alpha \left(1 - \frac{\tan \delta}{\tan \alpha}\right) + (W_e - W_4) + (W_2 - W_1) \qquad (5.99)$$

Here, δ is the friction angle in the sliding plane. The equilibrium under steady conditions is satisfied if:

$$\tan \delta > F \tan \alpha \qquad (5.100)$$

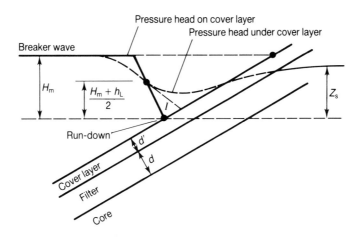

Figure 216 *Principle of coverlayer hydraulic loading*

where $F = \gamma_a/(\gamma_a - \gamma_w) \approx 2$. If there is no support from a toe structure ($F_t = 0$) then $f = 2.5$ is recommended. In general, a slope angle steeper than $\tan^{-1}(0.5)$ will not be stable, and anchoring or a special toe structure will be required. Systems with tensile strength may be useful in such situations. If the slope has an angle less than $\tan^{-1}(0.25)$ to $\tan^{-1}(0.15)$ it will be stable.

Where repeated wave loading occurs on a (relatively closed) coverlayer, significant excess pore pressures may be generated, as shown in Figure 216 (see also Box 67). The drainage capacity for such a coverlayer is expressed by the leakage height, h_L, where:

$$h_L = \sin \alpha \sqrt{(K t_f C)} \tag{5.101}$$

where C is the hydraulic resistance coverlayer ($C = t_c/K'$),
K is the permeability filterlayer underneath,
t_f is the thickness of the filterlayer,

and K' is the permeability of the coverlayer,
t'_c is the thickness of the coverlayer.

The maximum uplift pressure is approximately:

$$\Delta p \leq (H_m + h_L)\gamma_w/2 \quad \text{for} \quad h_L < H_m \tag{5.102a}$$

$$\Delta p \leq Z_s \gamma_w \quad \text{for} \quad h_L > H_m \tag{5.102b}$$

where the internal set-up, Z_s, may be calculated as described in Section 5.2.8.2.

The maximum pore pressure gradient normal to the slope is $i = \gamma_w(\Delta p)\tan\alpha/h_L$, which can be used in a stability analysis of relatively closed layers, such as pitched stone, etc., as an alternative to Pilarczyk's empirical approach in Section 5.1.3. For more details about the hydraulic behaviour under cyclic wave loading reference should be made to Section 5.1.3.

5.2.8.6 Earthquake effects

Calculation of the stability of slopes subjected to earthquakes may be attempted by considering an additional inertial force with a magnitude equal to the product of mass of the slice participating in failure and its horizontal acceleration. Usually, vertical acceleration is ignored and the horizontal acceleration is taken constant over the height. In a standard Bishop stability slope analysis the effect of the horizontal acceleration can easily be taken into account.

Horizontal accelerations are normally expressed in national codes as a seismic coefficient times the acceleration due to gravity. In earthquake areas basic regional seismic coefficients as defined by national standards will typically vary between 0 and 0.15. The appropriate national standard will also normally define other multipliers which should be applied to the basic seismic coefficient. These may include those subsoil conditions (typically, 0.8 for poor to 1.2 for hard conditions) and for importance of the structure (typically, ranging from 0.5 to 1.5, depending on the strategic importance, likely loss of life, etc.).

For coarse rubble media, the earthquake stability analysis can be carried out without taking the pore-water weight into consideration. This is permitted, because the sliding soil mass will not convey the pore water in its motion and yields an appropriate reduction of the driving momentum.

5.2.8.7 Settlement of compressible soil

When fills are applied on top of compressible soil layers the thickness of the fill to be placed must be determined so that the finally desired levels is obtained after settlement of the fill and the subsoil. Such a determination can be made using upgrading diagram such as that shown in Figure 217, which takes into account the final settlement curve of the ground level. For fills which are placed or settle partly beneath water level, a correction related to the buoyancy is needed.

Terzaghi discovered that for soil a linear relationship exists between the vertical strain ε_1 and $\ln(\sigma_1)$, with σ_1 the vertical load. As a consequence, the strain increase due to a load increase becomes:

$$\Delta\varepsilon_1 = (1/C)\ln\frac{\sigma_1 + \Delta\sigma_1}{\sigma_1} \quad (5.103)$$

The constant C is called the settlement constant, which depends on the type of soil. Typical values are as follows:

Sand $\quad C = 50$–500
Loam $\quad C = 25$–50
Clay $\quad C = 10$–25
Peat $\quad C = 2$–10

Since soil strain is almost irreversible, the soil behaviour will differ for loading less than the historic ultimate loading. The historic ultimate loading is called the transition stress. For heave and reloading under this transition stress the settlement

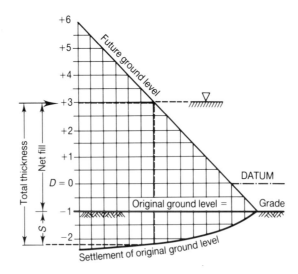

Figure 217 *Upgrading diagram*

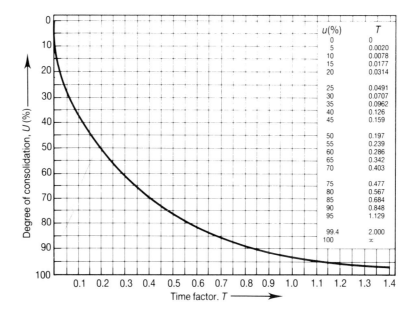

Figure 218 Consolidation ratio versus time

behaviour is characterised by a reduced valve of the settlement constant in the range $0.1C$ to $0.3C$. Loading of slowly draining cohesive soils will reveal retarded settlements due to consolidation (dissipation of excess pore pressures) and secular effects (creep).

One measure of practical importance is the average consolidation of a horizontal layer of thickness, d, drained at the top and the bottom, in time, t, which is expected in terms of the consolidation ratio, U:

$$U = \frac{\text{Compression at time } T}{\text{Final compression}} \quad \text{with} \quad T = c_v t / d^2 \text{ (dimensionless time)}$$

The general solution is shown in Figure 218. Here, c_v is the so-called consolidation coefficient, to be determined by laboratory tests. At $T=2$ the ratio is 99.4% and the process is practically complete. The corresponding time, $t = 2d^2/c_v$, is called the hydrodynamic period.

In some situations the consolidation needs to be accelerated. This can be achieved by overloading, by part removal of the soil or by vertical drains, which consist of regularly drilled vertical boreholes, generaly filled with sand that allow the water to drain horizontally as well as naturally vertically. The rate of consolidation achievable using a vertical drain system can be predetermined, the effect of the combination of horizontal and vertical dissipation being expressed by the following overall consolidation ratio:

$$U = 1 - (1 - U_h)(1 - U_v) \tag{5.104}$$

Keverling-Buisman (1941) developed a theory to divide the settlement into a primary part and a secular one. Secular effects give long-term additional settlements and the additional strains, $\Delta \varepsilon$, can be incorporated into calculations, using the relation

$$\Delta \varepsilon = (1/C_s) \log(t) \ln\left(\frac{\sigma_1 + \Delta \sigma_1}{\sigma_1}\right) \tag{5.105}$$

where C_s is a material constant.

5.2.8.8 Settlement of rockfill

The natural settlement (densification) of rockfill subjected to gravity, hydraulic and earthquake loading is sometimes important. Little is known about this effect. Earth reservoir dams are usually given an overheight allowance of 0.2–0.5% to account for it. This may seem small, but during construction, densification activities are executed and controlled carefully. In rock structures controlled densification is rarely performed.

Loose-dumped gravel and rockfill can be densified by special methods, such as the falling weight and shaking plate. The Menard method is popular and densification effects to a depth of 10 m have been reported. For gravel, a 10–15% densification can be achieved and 15% for rockfill of 10/60 kg. For large stones, however, heavy mechanical densification causes crushing.

As a first estimate, one may compare the energy of natural causes to that of machines. Densification equipment produces an acceleration of 10 g maximum. Earthquakes may produce up to 2 g in the entire structure, which will, therefore, lead to a densification of 2–4%. During earthquakes, however, internal shear failure and sliding may also occur. An example of the shakedown of a breakwater due to an earthquake by numerical simulation is shown in Figure 219.

Large wave impacts may generate in rock structures an average acceleration of the order of 1 g at most. The average densification will be of the order of 1%. Slamming waves may, however, produce locally higher accelerations and higher densifications. The resulting settlements can be observed along coastal defence structures around the storm waterline (see Figure 220).

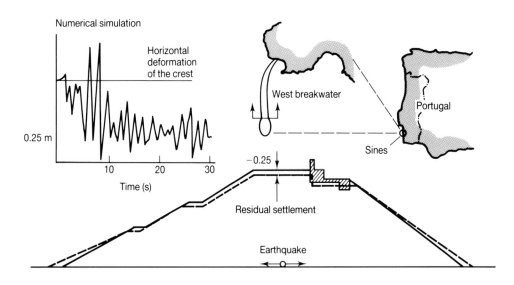

Figure 219 *Earthquake-induced residual settlements*

5.2.8.9 Local stability of rockfill

The hydraulic loading imposed on individual particles by waves, internal friction and interlocking of randomly placed stones cannot be described properly. Physical tests provide some general understanding of the stability of a stone on the surface of an armour layer subjected to hydraulic gradients. The experiments show that for a compacted granular medium, porous outflow first activates a group of coherent particles to expand, creating a loose packing, until one favourable surface particle starts moving (boiling), giving way to outflow. This surface particle seems to possess a self-healing stability potential due to the inhomogeneity which is caused by the notion itself, as the space generated by the lifting decreases the

Figure 220 *Densification by wave impacts (courtesy Delft Geotechnics)*

local pressure. This behaviour has been observed in tests on sand and gravel (turbulent porous flow). The lift can, however, be a catalyst for erosion, since a partly lifted particle can be conveyed more easily by the parallel surface flow. Further extensive information on hydraulic stability of rock is given in Section 5.1.3.

5.2.8.10 Liquefaction potential of seabed sand

Liquefaction of sands may be caused by earthquakes or waves. The earthquake-induced liquefaction is important for dams and is well described in literature on this subject.

Wave-induced liquefaction is of particular interest here. It is suggested that assessment should be according to a state-of-the-art method (Seed and Rahman, 1978), referred to as the 'uncoupled approach'. In this method the rate of pore pressure generation due to wave-induced shear stresses is assumed to be independent of the dissipation by drainage. The approach includes site

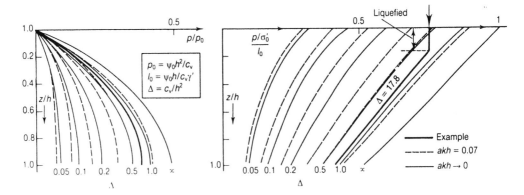

Figure 221 Excess cyclic pore pressures

investigation, laboratory testing and application of simulation models. The method proceeds by determination of:

- Seabed stratigraphy, geohydraulic properties and relevant loading characteristics;
- Static and dynamic effective stress;
- The actual liquefaction potential from laboratory tests;
- Excess pore pressures, including drainage;
- Geotechnical stability.

The liquefaction potential can be determined when the actual porosity and the critical porosity (at zero dilatancy) are known. The first is to be determined at the site and the second in the laboratory by undrained triaxial tests on samples taken from undisturbed (no flush) borings.

In the simulation the generation of excess pore pressures is compensated at the same time by drainage, which causes also a slight densification. A simple formula for the liquefaction process of a seabed sandlayer under wave loading is presented in Figure 221 (Barends and Calle, 1984; Barends, 1991) and some background information is given in Box 68.

The graphs can be used to evaluate the excess pore pressure and liquefied depth, if relevant parameters are known. The example shown in Figure 221 shows that the top 20% of the sand bed is liquefied and the remainder possesses significantly less shear strength. The application is particularly useful at the toe of a marine construction. Local failure there will subsequently jeopardise the entire structure. An example of the effect of local failure at the toe due to excess pore pressures is presented in Figure 223.

5.2.8.11 Squeezing of very soft soil

The analysis of squeezing of soft layers is rather complicated, but a simple approximate method is sufficient in many cases. This method evaluates the horizontal equilibrium of a soil column (Figure 224). The active earth pressure produces a force E_a. The resisting force is E_p and an intermediate shear force E_s. The safety against sliding is expressed by the ratio $F=(E_p+E_s)/E_a$, which can be determined by evaluation of the following formulae:

$$E_a = \int_0^h k_a(\sigma_v' + \Delta\sigma') - 2c'\sqrt{k_a}\,dz \qquad (5.106a)$$

$$E_p = \int_0^h k_p(\sigma_v') + 2c'\sqrt{k_p}\,dz \qquad (5.106b)$$

> **Box 68** Liquefaction of seabed sand under waves: a simple approach
>
> Sea waves over a seabed sand cause a stress state described by:
>
> $$\bar{\tau}/\sigma' = (m/\gamma')\exp(-kz) \quad \text{(overbar denotes amplitude)}$$
>
> where $m = \bar{p}k$ is the wave pressure steepness at the bottom, $\bar{p}$ is the pressure amplitude at the sea bottom generated by waves, k is the wave number, $\sigma'_v = \gamma'z$ is the effective vertical stress, γ' the submerged weight and τ the shear stress. The excess pore pressure is described by:
>
> $$\sigma'_v \partial^2 p/\partial z^2 = \dot{p} - \dot{p}_1, \quad \text{where } \dot{p}_1 = \psi = 0.75 f \psi' \sigma'_0 / N_1$$
>
> f is the loading frequency,
>
> where σ'_0 the static stress level and N_1 the liquefaction cycle number. The normalised rate ψ' is equal to the tangent of the relation p_1/σ'_0 and N/N_1. For $0.1 < N/N_1 < 0.9$, ψ' varies slightly with the porosity n. The value N_1 depends on the stress ratio τ/σ'_0 in the test and the threshold stress-ratio (τ/σ'_0) under which no dilatancy occurs.
>
> This approach yields $\psi = \psi_0 \exp(-akz)$, $\psi = bf\psi'\sigma'_0\{(m-m_0)/\gamma'\}^a$, where m is related to the actual wave height H and m_0 to the threshold wave height. The solution yields the excess pore pressure:
>
> $$p = \frac{2t\psi_0}{1-(ak\theta)^2}\left\{\exp(-akz) - \frac{\cosh[(h-z)/\theta] - ak\theta\exp(-akh)\sinh(z/\theta)}{\cosh(h/\theta)}\right\}$$
>
> with $\theta = \sqrt{(2c'_v T_R)}$, T_R is the duration of the storm; a and b depend on the situation (for example, $a \approx 0.3$; $b \approx 0.12$). The threshold stress-ratio may be in the range of 0.1–0.2.
>
>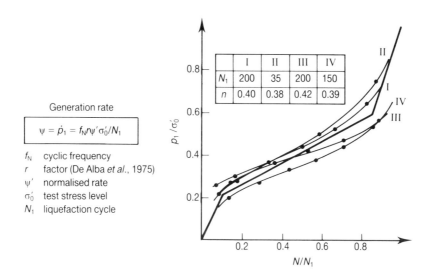
>
> **Figure 222** Dilatancy cyclic pore pressure generation in undrained cyclic triaxial tests

$$E_s = \int_0^h \tau \, dx \tag{5.106c}$$

where k_a = coefficient of active earth pressure,
 k_p = coefficient of passive earth pressure,
 σ'_v = original effective stress,
 $\Delta\sigma'$ = load increment,
 c' = cohesion,
 τ = shear stress.

The theoretical approach to the squeezing of a soft layer that has developed full plastic deformation is presented in Figure 225. The figure shows that the maximum admissible slope on top equals the apparent cohesive strength c' (Tresca model).

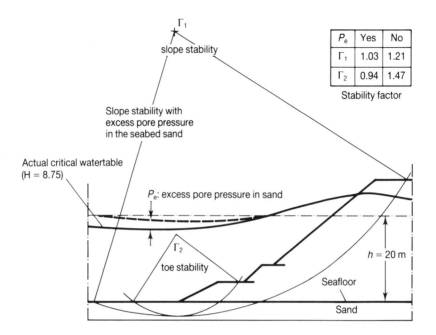

Figure 223 *Slope stability under wave action*

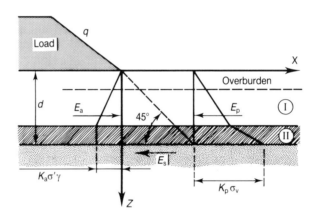

Figure 224 *Stability against squeezing*

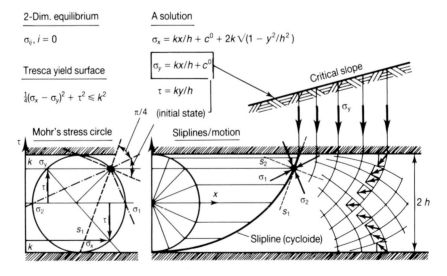

Figure 225 *Squeezing with a Tresca approach*

5.2.8.12 Filter rules

In rock structures granular filters are applied to protect against erosion by scour and migration. The design is usually based on geometric considerations: a fine grain from the bed may not pass through a pore in the coarse filter, irrespective of the hydraulic loading conditions. In practice, the hydraulic loading is relatively low, and often less strict design conditions are acceptable.

The filter stability can be undermined by mechanical and/or physical causes. The mechanical cause is due to external forces by waves, which may generate uplift pressures. The physical cause is the internal erosion (or suffocation) by migration of the finer fraction under influence of local water-pressure gradients. Such situations are characterised by filter rules, and the classic filter rules validated by physical experiments, given by Terzaghi as long ago as 1922, are related to migration and permeability, both characteristic properties of the filter.

For granular filters the rules valid for irregular-shaped particles are:

To prevent migration:
Uniform material $D_{50f}/D_{50b} < 5$
Wide graded 1. $D_{15f}/D_{85b} < 5$
2. $5 < D_{50f}/D_{50b} < 20\text{–}60$

To ensure adequate permeability:

$$D_{20f}/D_{20b} > 5$$

where the suffixes f and b relate to the filer and base (fine) materials, respectively.

Thanikachalam and Sakthivadivel (1974) have suggested filter rules which take into consideration the gradation:

1. $D_{10f}/D_{10b} < 2.50 D_{60b}/D_{10b} + 5.00$
2. $D_{60f}/D_{10f} < 0.94 D_{10f}/D_{10b} - 5.65$
3. $D_{50f}/D_{50b} < 2.41 D_{60f}/D_{10f} + 8.00$

In general, one can state that a well-densified filter suiting the above-mentioned rules functions under non-stationary loading conditions. If these rules cannot be matched due to manufacturing or non-availability of materials, a filter can be stable, but large gradients may give rise to filter degeneration. The filter stability under cyclic gradients is characterised by a critical gradient beyond which migration starts. Some typical test results are shown in Figure 226.

Additional tests or a different gradation is required if the filter is subjected to

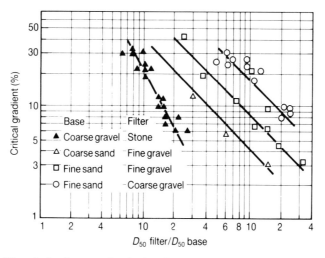

Figure 226 *Filter behaviour under hydraulic loading*

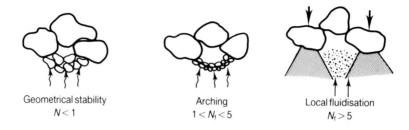

Figure 227 Mechanisms related to N_f

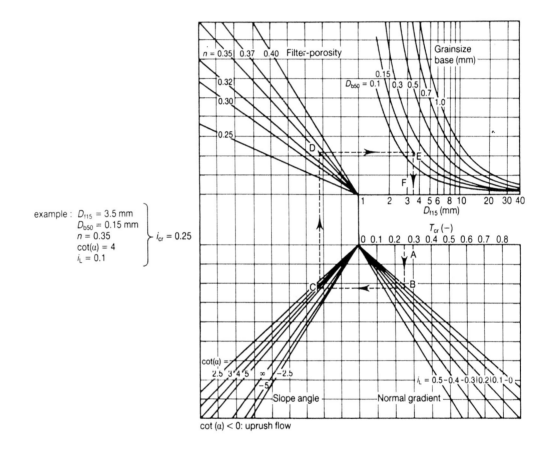

Figure 228 Nomogram for filter design

significant dynamic gradients. A gradient of at least 20% is considered significant. Extensive tests (Pilarczyk, 1984) have provided an extra norm which takes into account hydraulic loading effects: $N_f = n_f D_{15f}/D_{50b}$. This relates to various mechanisms shown in Figure 227 from which it may be concluded that under no circumstances should N_f exceed 5 and, ideally, should not be significantly more than 1.

A nomogram for the filter stability under hydraulic conditions is shown in Figure 228. Following point A (parallel gradient) through points B (normal gradient), C (slope steepness), D (porosity) and E (grainsize base material) one obtains point F (required filter material). With the incorporation of actual hydraulic loading a suitable criterion can be selected. For the application of these rules the thickness of a filter should be at least about five times D_{50f} (in order to let the gradients be a realistic average).

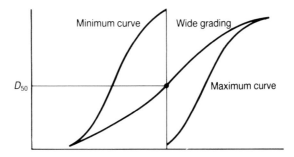

Figure 229 *A wide-graded material*

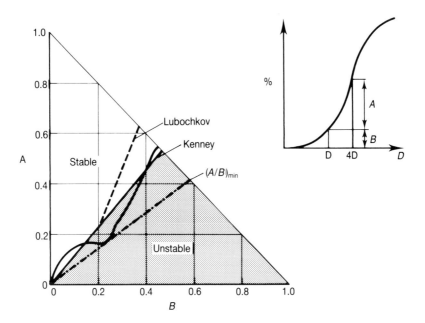

Figure 230 *Internal stability curve*

The internal stability of a widely graded filter beneath wave-stable armourstone whose grading covers several orders of grain sizes should perhaps be treated in a different way. Various tests (van der Meer, 1988a) have shown that rip-rap filter layers with a wide gradation ($D_{n85t}/D_{n15t} = 2.25$) are as stable as those with a narrow gradation as long as they are well mixed and the cover stone is of a suitable size.

One suggestion to evaluate such widely graded filters is to divide the grain size distribution curve into two curves—a minimum and a maximum curve—and to apply the rules to these artificial curves (Figure 229). As an alternative (see Figure 230), a curve may be constructed from the grainsize distribution curve by the evaluation of a factor A/B, which is related to the steepness of the gradation and the result is shown in the figure. For this curve a boundary for the internal stability has been suggested (Kenney, Lubochkov). A minimum of $1.5 A/B$ is advised. The application of this approach is validated for sand, gravel and minestone.

An example of the application of the classic filter rules is shown in the following tables, worked out for several types of granular materials, of which the grainsize distribution curves are presented in Figure 231.

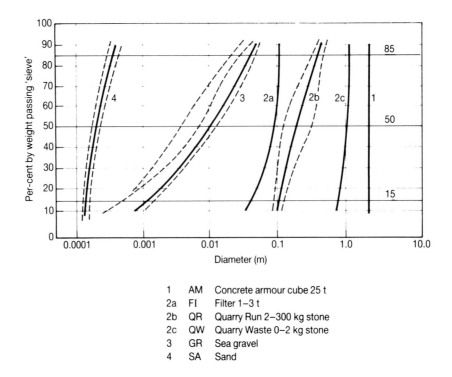

1 AM Concrete armour cube 25 t
2a FI Filter 1–3 t
2b QR Quarry Run 2–300 kg stone
2c QW Quarry Waste 0–2 kg stone
3 GR Sea gravel
4 SA Sand

Figure 231 *Grainsize distribution curves*

For the classic non-migration rules:

D_{15f}/D_{85b}	FI	QR	QW	GR	SA
FI	.	1.50	6.70	>10	>10
QR		.	0.85	2.20	>10
QW			.	0.80	80.0
GR				.	2.0

D_{50f}/D_{50b}	FI	QR	QW	GR	SA
FI	.	4.10	10.7	>10	>10
QR		.	2.60	26.0	>10
QW			.	10.0	>10
GR				.	50.0

For the adequate permeability rule and dynamic hydraulic loading norm, N_f:

D_{20f}/D_{20b}	FI	QR	QW	GR	SA
FI	.	6.92	18.0	>10	>10
QR		.	2.60	65.0	>10
QW			.	25.0	>10
GR				.	10.0

N_f	FI	QR	QW	GR	SA
FI	.	1.30	3.39	33.9	>10
QR		.	0.38	3.85	>10
QW			.	1.36	68.0
GR				.	1.25

From these tables one may compose a best-possible filter structure. The combinations QW–SA and FI–QR are not unconditionally stable. For FI–QR the condition for the permeability is critical.

5.2.8.13 Design aspects of geotextiles

A rock structure often contains layers of various materials, a filter construction, which may consist of grains (loose grains, bonded grains, packed stones) and/or fibres (synthetic, natural). Design aspects are:

Granular filters	*Fibre filters*
Advantages	
Sometimes self-healing	Small construction height
Durable elements	Tensile strength
Smaller hydraulic gradients	Relatively cheap
Good support and load spreading	Generally easily placed

Disadvantages

Larger construction height	Uncertain long-term behaviour
Preparing and mixing gradations	Gaps when base settles
Spreading composition	Easily damaged
Unclear porosity	Larger hydraulic gradients
Difficult to place accurately	

A geotextile is often employed to diminish the required filter height and easily placed. Its principal function is a separator which prevents wash-out of fines. It must prevent migration through the textile, which may be caused by 'driving gradients'. However, a geotextile will affect the pressure gradients across it. It constitutes an extra resistance to flow and decreases the permeability of the filter structure. Usually, a geotextile cannot fully replace all the required functions of the filterlayer, and a thin filterlayer may still be required on top for weight or underneath for in-place drainage.

The following criteria may be used for the evaluation of the soil tightness of geotextiles:

Non-cohesive beds

Stationary loading $O_{90}/D_{90b} < 1$ regular woven mattresses
$O_{90}/D_{90b} < 1.8$ non-regular non-woven
Cyclic loading $O_{98}/D_{15b} < 1$ no-clogging, fine graded
$O_{98}/D_{85b} < 1$ clogging, widely graded

Cohesive beds
$O_{90}/D_{50b} < 10$; $O_{90}/D_{90b} < 1$; $O_{90} < 0.1$ mm

The geotextile is a filter for the bed material and a resistance to flow. The corresponding pressure gradient can be considered as internal loading. In some situations the fines from the bed may clog in the fabric and decrease the permeability significantly, causing high-pressure gradients leading to uplift, flapping, parallel internal or 'in-plane' transport of fines and irregular settlements. An acceptable criterion is that the geotextile normal pressure gradient equals the normal gradient in the base material under design loading conditions. Since the flow through the textile can be turbulent, the precise definition of the permeability needs careful attention. Clogging may occur by silt (pollution) or biological growth (algae, bacteria). For more information, reference should be made to Geotextile (1986).

The strength of geotextiles is important when sliding of soil masses may occur through the fabric or when settlements are opposed by the geotextile. Rip-rap protection or armourstone placed directly on a geotextile can potentially damage it by sharp edges or where the fabric is anchored along the upper edge of a slope.

To determine the required strength of a fabric, a modified Bishop stability analysis developed for reinforced earth walls may be used (software packages are available) in which a horizontal force representing the geotextile strength is introduced at specified locations. The method determines (1) the required tensile force Z, (2) the width of the reinforced zone, and (3) the anchor length, L, computed by the formula:

$$L > FZ/(\gamma h \tan \phi') \qquad (5.107)$$

where F is a safety factor ($F = 1.3$), γh is the overburden weight and ϕ the soil-texture friction (30° for fine, clayey soils, 45° for granular soils, where $\phi' < \phi$).

Geotextiles can be applied to reinforce low bearing capacity soils. The textile acts as a membrane spreading loads on top so that soil stresses are in equilibrium. A Brinch–Hansen stress criterion for the subsoil together with sufficient anchoring of the textile at the sides can be applied to assess the feasibility of a particular situation.

5.2.9 APPLICATION OF PROBABILISTIC ANALYSIS

Risk analysis (see Section 2.2) improves understanding of the reliability of structures by evaluation of causes of failure and by the calculation of their probabilities. Because of the complex and iterative character of this approach, computer models are indispensable. Moreover, risk analysis requires the designer to present a logical scheme of relevant causes leading to failure of the structure. Each cause constitutes a failure mechanism for which a simulation model may exist. The description and the definition of relevant failure modes and mechanisms are important issues for further research.

Risk analysis offers a method to determine the probability of malfunctioning qualitatively and quantitatively, having in mind the minimization of the generalised cost of the structure and the maximization of its utility (Minimax) (see Section 2.3). A close inspection of the individual contributions of different failure mechanisms to the total probability of failure of a design leads to rational improvements and to research dedicated towards improvements of the weakest elements in the design.

In a complex structure such as rubble-mound rock many events (accidents) can be distinguished which may lead directly or indirectly to failure. Hydraulic, geotechnical stability and material failure by fracture or loss of position must be considered. Failure may start at the toe, near the berm, on the slope, at the crest or in the interior. It may progress into complete malfunctioning with time (accident sequence).

Failure of the structure should be carefully defined. The choice of the definition will imply the meaning of the calculated probability of failure. A clear and precise definition can be achieved in a fault tree, which states in a logical sequence the possible causes leading to malfunctioning. Evaluation of available simulation models shows that several important processes are still not covered by a model or a design formula.

An important result of a probabilistic analysis is the quantification of the contribution of the various failure mechanisms to the probability of malfunctioning. The outcome shows clearly the weak points in the considered design. With these results it is possible to evaluate which improvement is most efficient. The failure or malfunctioning of a rubble-mound rock structure due to excessive geotechnical deformations has been a subject of a published study (CIAD, 1985).

Three different mechanisms of deformation of the mound and the subsoil have been identified:

1. Slip failure of the mound and the subsoil, particularly at the toe;
2. Excessive settlements of compressible soil layers in the subsoil;
3. Collapse of cavities in the subsoil.

With regard to slip failure mechanisms the following causes of failure need to be considered:

- Limited internal shear strength
- Cyclic mobility and liquefaction of seabed sand
- Critical steepness of slopes
- Excessive wave-induced pore pressures
- Excessive earthquake-induced accelerations

To illustrate some of these slope stability aspects in relation to the probabilistic design approach the example of the breakwater considered in the CIAD (1985) study is presented in Box 69.

Box 69 Example of probabilistic analysis of a breakwater (CIAD, 1985)

In the CIAD (1985) study a particular breakwater was selected as an example for probabilistic analysis (see Figure 232). This box describes the analysis procedure and results.

Geotechnical slope ability under wave loading

Waves create an internal water table in the mound which is significantly higher than the mean water level outside. This internal set-up was determined by computer. The most critical moment was found to be when the wave trough was above the toe of the breakwater.

The waves cause cyclic pore-pressure variations which had the potential to gradually increase in the loosely packed seabed sand layer near the toe. The intensity of these excess pore pressures would ideally have been determined by adequate field measurements and sampling, special laboratory tests and evaluation of the test results with respect to the *in-situ* conditions (stress state, dissipation) determined by the computer models. These data were not available for the CIAD study, and a range of values had to be chosen on the basis of experience.

Finally, the CIAD study obtained probability of failure of the outer slope by integrating the probability of failure for a given wave class and the probability of occurrence of the wave class. This probability was determined for one year and for ten-year wave-return periods with and without excess pore pressures. The results showed that liquefaction significantly increases the probability of failure.

The local stability showed the same order of failure probability as for the total stability, and again the probability of occurrence of excess pore pressures is critical to geotechnical failure.

Geotechnical slope stability under earthquake loading

The area where the breakwater was located showed crustal tectonic activity with a medium to high seismic risk. Once a year an earthquake larger than scale III was observed in the area. Such earthquakes were able to generate horizontal ground accelerations at the site of the breakwater, the peak of which might cause significant damage, particularly if the seabed contained a sand layer that was likely to liquefy.

Historical observation showed a local peak acceleration of about 0.1 g with a return period of 80 years. Assuming a linear relation between return and earthquakes of less intensity, different classes of earthquake loading were distinguished.

The geotechnical stability of the outer slope was evaluated (and the inner slope could also have been examined). The top layer of seabed material comprised sand which was subject to earthquake-induced excess pore pressures and which might partly liquefy.

Two zones in the top layer were distinguished, one in front of the toe and one beneath the breakwater toe. The generated excess pore pressures in both these zones were clearly related to the intensity of the earthquake effect, and were estimated on the basis of experience with undrained cyclic triaxial tests in the laboratory. The estimated excess pore pressures were found to cause a considerable reduction in the shear strength of the foundation.

The stability of the outer slope was then evaluated using Bishop's slipcircle method. The results were obtained for total and local (toe) failure. The results showed that the toe collapsed easily and constituted the main part of the overall stability for the range of earthquake intensities considered. It was recognised that earthquakes of larger intensity might occur, but the breakwater would then be severely damaged in any case.

(Continued overleaf)

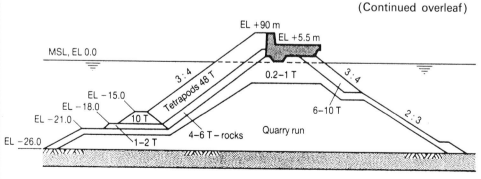

Figure 232 *Cross-section of breakwater*

Box 69 (Continued)

Probabilities of geotechnical failure

By waves in the case of a loose seabed layer (slope)	0.0072
By waves in the case of a loose seabed layer (toe)	0.0084
By waves in the case of a rigid seabed layer	0.000003
By earthquakes in the case of a loose seabed layer	0.024
By earthquakes in the case of a rigid seabed layer	0.001

The total probability of geotechnical failure could have been found by adding the (independent) partial probabilities. Clearly, the earthquake event dominated as far as geotechnical slope stability was concerned.

The analysis demonstrated that for the breakwater studied, worthwhile improvements to the foundation could be achieved by either stabilising or removing the top sand layer.

6. Structures

The integration of information on environmental boundary conditions and material availability into a functional and economic design follows the systematic procedure described in Chapter 2. Although the main steps are similar for various types of marine structures used in coastal and shoreline engineering, there will be differences in the details.

The purpose of this chapter is to describe the typical aspects related to the design and construction of five main areas of the use of rock in coastal and shoreline structures:

- Rubble-mound breakwaters
- Seawalls and shoreline protection structures, including groynes;
- Dam-face protection;
- Gravel beaches;
- Rockfill offshore structures.

For each group an overview is given of structural components, types and subtypes, relevant tools during conceptual and detailed design and typical construction and cost aspects. Practical examples of real cases of design and construction are given in order to illustrate critical points.

6.1 Rubble-mound breakwaters

This subsection describes the general structural design considerations for rubble-mound breakwaters, with considerations of the layout and the overall shape and dimensions followed by a discussion of the structural details and construction and cost aspects. Rubble-mound breakwaters are structures built of quarried rock or other stone materials. Generally, the larger rock armourstones are used for the outer layer, which must protect the structure against wave attack. Stones in this outer layer are usually placed with more care to obtain a better interlocking and consequently better stability. Although other materials (concrete, bitumen, etc.) are also used for this outer layer, this section only deals with rock armourstone.

Rubble-mound breakwaters are attractive because their outer slopes force storm waves to break and thereby dissipate their energy, causing only partial reflection. They are also extremely numerous because:

- Rock can often be supplied from local quarries;
- Even with limited equipment, resources and professional skills, structures can be built that perform successfully:
- There is only a gradual increase of damage once the design conditions are exceeded (graceful degradation). Design or construction errors can mostly be corrected before complete destruction occurs. If local wave conditions are not well known, this is particularly important;
- Repair works are relatively easy, and generally do not require mobilisation of very specialised equipment;
- The structures are not very sensitive to differential settlements, due to their flexibility. Foundation requirements are limited as a result of the sloping faces and wide base.

The concept generation, selection and detailing of a rubble-mound breakwater can be summarised by the logic diagram in Figure 233, which is an extract from part of the main logic diagram of Chapter 1. The numbers refer to the relevant parts of this section.

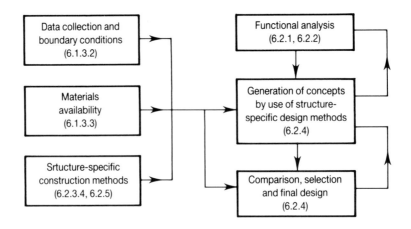

Figure 233 *Breakwater design logic diagram*

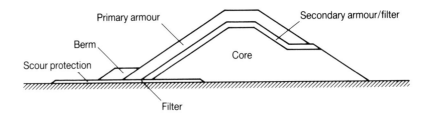

Figure 234 *Cross-section of a typical rubble-mound breakwater*

6.1.1 DEFINITIONS

As mentioned earlier, rubble-mound breakwaters are structures built of stone materials, usually protected by an outerlayer of larger armourstones or concrete blocks. Breakwaters generally serve the purpose of providing quiet water for anchorage or mooring of vessels, protected from the attack by waves and or currents. Other functions are also possible, as explained in Section 6.1.2.

A typical cross-section of a rubble-mound breakwater is shown in Figure 234 to illustrated the various components. The main body comprises the core, usually built of quarry run, one or more filter layers and the outer armour layer. The crest may be protected by the armour layer, but very often shows a concrete crown wall. The toe and scour protection at the seaward slope is needed if the breakwater is built on sandy bed material, to maintain stability of the slope in case of erosion of the seabed. Depending on the type of subsoil, the breakwater may be built directly on the seabed or on special filters, made of stone material or geotextile. In very poor foundation conditions complete soil improvement or other measures may be needed to achieve geotechnical stability. The following types of rubble-mound breakwaters will be treated in this chapter (see Figure 235).

Rubble mound with crown wall

The crown wall allows easier access and can become essential if the breakwater has functions other than simply protection against wave action.

Berm or S-slope breakwaters

An excess amount of stones is placed in a berm at the seaward slope and in an S-slope breakwater these stones are statically stable. In a berm breakwater the

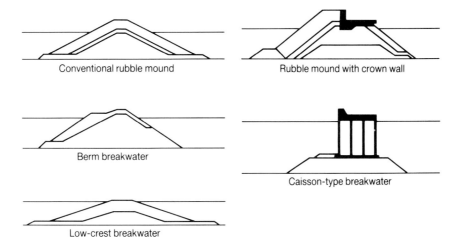

Figure 235 *Examples of various types of breakwaters*

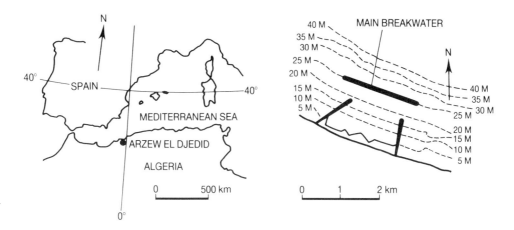

Figure 236 *Location and layout of Port d'Arzew El Djedid*

stones are dynamically stable. During more severe storms this material becomes redistributed by the waves to form a natural stable profile.

Reef-type breakwaters

These are low structures, generally detached and parallel to the shore, with much overtoping. They are composed of a mound of (graded) stones, which is allowed to develop naturally into a dynamically stable profile, as opposed to low-crest or submerged breakwaters, which are conventional statically stable rubble mounds.

Caisson-type breakwaters

These are combinations or rubble mound with caissons, where the latter may be placed on top of the mound or behind it (see Figure 235).

In most cases breakwaters are connected to the shore ('attached'). These structures comprise the root at the landward end, a trunk portion and the head at the seaward end. If composed of trunk portions with different azimuths, an elbow is found at the junctions. In some cases breakwaters are fully detached and need two heads. A well-known example of a detached main breakwater is found at Arzew-el-Djedid in Algeria (see Figure 236).

The choice between attached and detached is primarily defined by functional requirements. Where a reef-type breakwater serves to protect an eroding stretch of coastline, this structure will (often) be detached to reduce costs (unless strong longshore tidal currents are present). If protection of a harbour basin and berths is the main function, an attached breakwater is often the most logical solution. For very large ports which need two entrances at least a part of the protection is created by a detached breakwater.

6.1.2 LAYOUT

As mentioned above, the alignment of a breakwater depends on the functional requirements. The development of an optimum and cost-effective layout from the functional requirements often takes place in the early stage of a project, but is of the utmost importance for the final result. The clear differences in the functional requirements between port functions and coastal protection functions lead to different planning considerations and methods. The design of breakwater layouts for coast protection functions is discussed in Section 6.2.2. This section concentrates on discussing the development of the alignment of a breakwater as part of the overall planning process of a port.

The basic boundary conditions for the port planning process are the anticipated throughput, the type and number of vessel (for consecutive phases of development in the case of master planning) and environmental data such as wind, wave and current conditions. Once the total berth length required has been determined, the physical layout can be worked out, including the breakwater alignment.

The overall layout requires consideration of a multitude of requirements, such as port operations, inland connections, environmental impacts and flexibility for future expansion, which are not of direct influence on the breakwater alignment. Indirectly they do have influence, because they define the shape of the port, which needs to be protected. The functional requirements, which directly control the alignment, are the following:

- Adequate protection of berthed vessels against waves and sometimes wind and currents;
- Protection of access channel and turning circle for safe stopping and manoeuvring of incoming vessels and safe departure manoeuvres of outgoing ones;
- Reduction of maintenance dredging costs.

Although all three factors need to be considered simultaneously to arrive at a balanced design, different technical methods are applied to address each of the requirements. These are treated separately below.

6.1.2.1. Influence of need for berth protection

Wave penetration into a port depends on the orientation and the width of the port entrance. The actual wave disturbance inside the port at a specific berth also includes the effect of dissipation (e.g. by spending beaches or slope revetments). For ports, receiving vessels of 30 000 DWT and above, the occurrence of long waves becomes an important part of the wave disturbance, due to the sensitivity of the vessels to such waves. The basic task is to determine the expected downtime of vessels using the port, as a function of breakwater alignment. The optimum design will show a minimum for the accumulated total costs of downtime and breakwater construction.

Diffraction analysis

Methods to determine wave disturbance and vessel response vary depending on the design phase. During the initial phase of concept development and selection, wave-penetration factors are estimated by means of templates such as those given in Figure 121. The relation between the annual wave climate at the port entrance and the average annual downtime of the fleet calling on the port is shown schematically in Figure 237.

To derive the distribution of wave heights, periods and directions at Berth 1 from

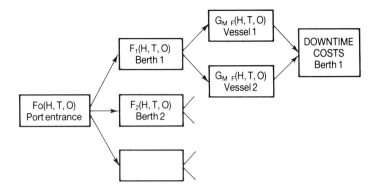

Figure 237 *Schematic relation of annual wave climate to total downtime costs*

the distribution at the port entrance is a predominantly linear problem. Hence linear diffraction models can be applied to compute wave-disturbance factors for selected values of wave period and direction and thus the resultant wave disturbance for the design directional wave spectrum being considered (see Section 4.2.2.5). Different types of models are presently available, such as source and finite element models.

Vessel response

The step from wave condition at berth to the distribution of ship motions and mooring forces (G_{MF}) is, in princple, a non-linear problem, due to the damping characteristics of the vessel in restricted water depths and the non-linear characteristics of the fenders and mooring lines. The general approach is to execute a limited number of non-linear computations to define the limits of operability for various vessel sizes at each berth location. Numerical models are operations, which compute the six motions of the moored vessel and the mooring line and fender forces in the time domain.

Hydraulic models

The application of hydraulic physical scale models for the above process is limited, although in some cases a hydrauylic model has advantages (e.g. for very complex port layouts). The main disadvantage of physical models in the large number of conditions which need to be tested in order to evaluate the full range of options and the inflexibility of these models in relation to rapid changes of geometry. Numerical models are therefore often preferred for the initial selection from the full range of options.

6.1.2.2 Influence of need to provide protection to access channel

Vessels must be able to safely enter the port (and depart from it) even under adverse weather conditions. The actual limit to these conditions depends on operational and economic considerations, and it is assumed that this limit has been defined. Subsequently, the breakwater alignment needs to be determined in compliance with the following requirements:

- Elimination of strong current gradients outside the actual entrance;
- Sufficient stopping length, so that the vessel comes at rest in the turning circle with engines at half-astern;
- Reduction of wave disturbance in the approach channel and turning circle, so that tugs can make fast to assist the vessel during stopping and can safely manoeuvre the vessel to its berth.

Again, the methods vary in complexity with the phase of the port development.

During the concept development phase empirical relations for stopping distance and wave disturbance may be used.

Fast-time ship-manoeuvring simulation

Alternatively, a computer program can be used, which simulates the manoeuvring of a vessel into a port, along a prescribed track, under influence of currents, waves and wind and with the assistance of tugs. In this so-called 'fast-time simulation' the functioning of the helmsman is schematised by a set of equations, comparable to an automatic pilot. This gives also a limitation to the method, because the output of the simulation (deviation from track, residual speed in the turning circle, etc.) is the result of a standard manoeuvre and lacks the effects of human judgement, anticipation and possible error.

Real-time ship-manoeuvring simulation

Hence in detailed design the port layout and breakwater alignment has to be checked and optimised by means of real-time simulation. In this case a human pilot and helmsman steer the vessel into or out of the port, whereby their actions are based on similar information, as is available on a real bridge; outside view, radar, speed indicator and other instruments.

6.1.2.3 Influence of need to reduce maintenance dredging costs

For ports which are designed for large, deep-draught vessels the previous two requirements will generally supersede this third functional requirement. However, for smaller ports, which are not built far out into the sea, siltation of the port entrance can become an important operational and economic factor. In these cases the cost of a longer breakwater may outweight the capitalised maintenance dredging cost. This cost optimisation leads to an alignment for which the sum of additional breakwater and maintenance dredging costs has a minimum.

The most common cause of siltation of the port entrance is littoral transport of material, which, after a period of accretion against one of the breakwaters, starts to bypass the breakwater head and deposit in the entrance. This also explains why smaller ports are more vulnerable.

When the breakwater extends into deeper water the capacity for accretion is much greater and, when an equilibrium is reached after some decades, sand will not be transported across the entrance but will deposit in deep water. In some cases, where a port is located in an estuary or inlet and an entrance channel is needed through a bar, a breakwater may be considered to reduce siltation in the channel. In such cases an additional cause of siltation is formed by the deposition of silt transported by the tidal flow, through the channel, across it or both. (It is noted that in such cases the breakwater may be low-crested, but needs to be above high water for safety reasons.)

Littoral transport and channel siltation

Methods to determine siltation volume again depend on the study phase. In the phase of concept development and selection, simple computer models are available, which provide order of magnitude figures of littoral transport. As described in Section 6.2.2.4, the effect on the shoreline (i.e. the rate of accretion at the breakwater) can be computed by a so-called one-line model, where the coast is schematised to one line (Pelnard–Considere, 1956). For a first assessment of siltation in a channel one often takes the siltation rate proportional to the depth of the dredged cut below the natural seabed. If data are available on suspended sediment in the water upstream of the channel, siltation in the channel can be assessed by calculating deposition on the basis of fall velocity of the particles.

Detailed models

During a detailed design stage further optimisation of breakwater length warrants the use of more accurate models. Within the breaker zone the coast can be schematised in more detail with so-called 'n-line models' (Perlin and Dean, 1985) while for the computation of sediment transport the distribution of velocity across the breaker zone can be estimated (Fleming, 1990a). For siltation in a channel an overall model of the port, reproducing tidal flows and sediment transport, is the most accurate tool. If density currents do not play a role, a two-dimensional horizontal model (depth-averaged flow) is adequate. If, however, density currents are important, models are needed which schematise the three-dimensional flow by defining horizontal layers or which solve the three-dimensional equations of motion (Box 70).

Box 70 Breakwater layout development

An example of a breakwater, designed for nautical reasons only, is the extension of the existing breakwater at Taichung Port in Taiwan. Due to strong northerly winds, currents and waves, the existing entrance was unsafe during some months each year, and several groundings occurred. It was decided to extend the north breakwater to improve nautical safety.

The alignment of this extension was developed in three different steps:

1. The global length required to allow arriving vessels to slow down and pass the original entrance at a reasonable speed was determined by rule-of-thumb formula for stopping distance at five times the vessel length.
2. Subsequently, five alternative alignments were compared using the fast-time simulation model SHIPMA. Standard manoeuvres for two different 'design' vessels provided an objective basis for selection. An example of the output of the SHIPMA model is shown in Figure 238.
3. The selected alignment with 850 m new breakwater was studied in detail by means of a real-time simulation. Different pilots executed repeated arrival and departure manoeuvres under a range of wind/current/wave conditions, with the aim of obtaining statistical information to determine the chance of grounding, thus establishing the limiting conditions for safe navigation.

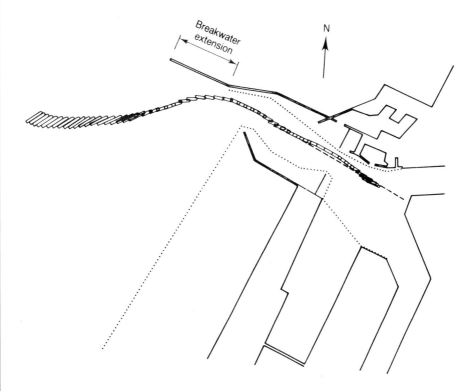

Figure 238 *Example of use of ship movement simulation model in design of breakwater extension at Taichung Port, Taiwan*

6.1.3 GENERAL DESIGN CONSIDERATIONS FOR BREAKWATER CROSS-SECTION

6.1.3.1 Cross-section concept generation, selection and detailing

An initial selection of the type or types of breakwater cross-section to be examined in more detail should be made on the basis of the functional requirements (Section 6.1.3.2), boundary conditions (Section 6.1.3.2), materials availability (Section 6.1.3.3) and construction considerations (Section 6.1.3.4). If this selection still permits alternative designs to be considered, the final choice should then be made on the basis of optimisations using cost comparison (see Section 6.1.6 and as well as Sections 2.1.5 and 2.1.7).

For the selected option or options, the required values of the main dimensions (crest height, size and thickness of primary armour and underlayers, etc.) should first be determined by using the structure-specific hydraulic and geotechnical design tools presented in Chapter 5. Actual dimensions and practical details should then be obtained from the stucture-specific considerations and rules of thumb presented in Section 6.1.4, which include constructability, availability of rock of the various sizes and gradings and the level of maintenance that is possible or preferred.

However, much of the information in both Chapter 5 and Section 6.1.4 is really only adequate for preliminary design purposes and the detailed design should still be checked in a hydraulic physical model (Section 5.1.4), making use of state-of-the-art techniques. Alternatively, uncertainties in the design formulae may be translated into (increased) safety factors, but even for small breakwaters this will quickly lead to substantial cost increases. In most cases model tests are cost effective and lead to optimisation of the preliminary design.

6.1.3.2 Data collection and boundary conditions

The main environmental conditions serving as input parameters for the design formulae and mathematical or physical models are:

- Water depth, tides and currents;
- Long-term wave statistics;
- Seabed properties (and seismic activity).

Chapter 4 provides full details of how to determine these and any conditions affecting the construction, such as:

- Short-term wave statistics and seasonal variations of importance for workability;
- Prevailing meteorological conditions (wind, temperature, visibility);

Other factors affecting the design and constructions of rubble-mound breakwaters include:

- Environmental restrictions for construction (water and air pollution, noise limitations, traffic restrictions)—Section 2.4;
- Availability of construction materials (see Section 3.1 and 3.4), equipment and labour;
- Local experience with comparable construction works;
- Infrastructional facilities (road, railways, ports);
- Facilities for future maintenance (monitoring and repair)—see Chapter 7.

The latter aspects are a factor in the assessment of the construction method and production costs, and as such determine the design concept together with stability considerations.

6.1.3.3 Materials availability

The materials for rockfill structures are supplied by quarries (Section 3.1 and 3.4), the geological characteristics of which determine the maximum size and shape of the rock blocks. Where a quarry is dedicated to a breakwater project, to obtain the required design quantities of armourstone in a conventional rubble-mound breakwater usually involves blasting away more rock than would be necessary to obtain the quantities of core materials required. This means an overproduction of certain gradations for which normally no other commercial application can be found, even when that required for concrete aggregates has been used, and this material is then effectively used as waste. The design of a rockfill structure in this situation should therefore be tailored to the expected quarry output as much as possible. Berm breakwaters (Section 6.1.4.3) offer an example of a design approach which enables this tailoring to be achieved, as described in Box 73 in Section 6.1.6. Information on assessing quarry output and on production techniques in the quarry is given in Section 3.4.

Quarry stones are often also obtained from permanent quarries (see Section 3.4), but it must be recognised that the majority of these quarries provide rock for aggregates and therefore blast for maximum fragmentation. Rock armourstone is therefore essentially a by-product in this case. Other permanent quarries may adjust their blasting patterns using fragmentation techniques described in Section 3.4.1 to maximise armourstone production.

6.1.3.4 Construction considerations

The minimum overall dimensions of the breakwater cross-section are determined by the hydraulic interactions and functional requirements. The actual dimensions follow from structure-specific construction methods. For construction with land-based equipment the crest has to to allow for the appropriate traffic. Dump trucks generally have to be able to pass the cranes, tip and turn, and subsequently pass each other.

The capacity of a crane is determined by the maximum weight of stones at the longest reach. This means that particularly the stones at the toe of the structure, and the berm, determine the type and size of crane required. Sometimes use can be made of the buoyancy of these stones, to extend the reach. Alternatively, floating equipment can be employed for these sections. If land-based equipment is selected, the type of crane to be used becomes a boundary condition for the dimensioning of the breakwater crest, as the crane must have a sufficiently wide track from which to operate. Details are given in Section 6.1.5.

The use of seaborne equipment is practical for placing at levels of 3 m below low water level and deeper. Floating cranes for higher parts of the breakwater are generally avoided because of limited workability and poor accuracy of placing. The basic decision therefore that has to be taken regarding the construction method is whether to use land-based or waterborne equipment (or a combination of the two). The main impacts on the structure are illustrated in Box 71 and can be summarised as follows:

Land based	**Waterborne**
Cross-section	
Crest elevation and crest width determined by the dimensions of cranes and trucks, and acceptable workability due to overtopping waves and spray. Flat slopes, wide berms and deep aprons present problems regarding the maximum reach of cranes.	The crest is determined by hydraulic stability and overtopping requirements. The core level is preferably 3.0 m below low water, to allow free dumping of material from barges.
Length section	
The work front is necessarily short. The various construction phases follow each other closely since the works are concentrated around the position of the crane(s).	The work front is extended over a large area to allow for sufficient manoeuvring and anchor spaces.

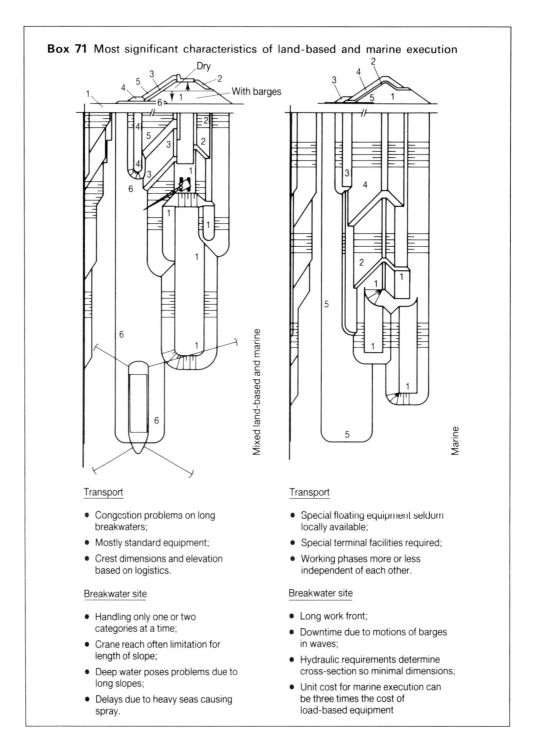

Box 71 Most significant characteristics of land-based and marine execution

Transport

- Congestion problems on long breakwaters;
- Mostly standard equipment;
- Crest dimensions and elevation based on logistics.

Breakwater site

- Handling only one or two categories at a time;
- Crane reach often limitation for length of slope;
- Deep water poses problems due to long slopes;
- Delays due to heavy seas causing spray.

Transport

- Special floating equipment seldom locally available;
- Special terminal facilities required;
- Working phases more or less independent of each other.

Breakwater site

- Long work front;
- Downtime due to motions of barges in waves;
- Hydraulic requirements determine cross-section so minimal dimensions;
- Unit cost for marine execution can be three times the cost of load-based equipment

Logistics

All supplies over the breakwater. Short breakwaters are therefore most suitable for land-based operation. Normally, the existing infrastructure can be used for transport between the quarry and the site. Work generally restricted to one front.

A loading terminal has to be provided for the barges before actual construction can start. The length of the breakwater forms no logistic limitation. Work may be initiated and proceed at different locations.

Morphology

Due to current concentrations, scouring holes at the temporary head can develop, since no extensive scour protection can be made.

A scour protection can be made well in advance to avoid the development of scouring holes.

Limiting factors

Reach and capacity of the construction cranes form a limitation to cross-section and progress of work. Breakwaters in deep water therefore become increasingly troublesome.

A water depth of at least 3–4 m is required. Sometimes work is done during part of the tidal cycle. Barges and marine equipment are generally not as widely available as land-based equipment, and their use requires specialised personnel.

Environmental constraints

Available construction time is determined by the freeboard between working platform and water level.

Available construction time is determined by the allowable motions of barges and floating equipment and downtime is generally longer than for a land-based operation. Crane pontoons are particularly sensitive in this respect. Excessive impacts between barges and pontoons, forces on the crane and risk of collision with the breakwater generally precludes crane operations in waves H_s is greater than about 0.5–0.75 m.

Damage during execution

The risk of damage is generally large, as the core and small secondary armour extends above water level. However, the length of unprotected structure is limited.

The risk of damage to core and secondary armour can be limited by keeping the top of this material at a low level. However, if damage occurs, it is generally over a large length.

Planning

The lead time for construction is determined by the quarry preparation. Mobilisation time for land-based equipment is generally short.
Construction progress depends, however, on one or two cranes, which are critical for a number of construction phases.

Lead time for construction might be long if special terminal facilities have to be constructed. The mobilisation time for marine equipment is longer, but production at critical phases can easily be increased by bringing in more equipment and working in parallel. Large stockpiles or high production rates are required for efficient operation of marine equipment.

Maintenance, repair

If the breakwater is provided with a sufficiently wide crest element, maintenance from the superstructure can be done with standard land-based equipment. Frequently this is not the case, and floating equipment is required.

Maintenance and repair can only be done with floating equipment.

6.1.4 STRUCTURE-SPECIFIC DESIGN ASPECTS

6.1.4.1 Conventional rubble mound

A definition sketch for a conventional rubble-mound breakwater is shown in Figure

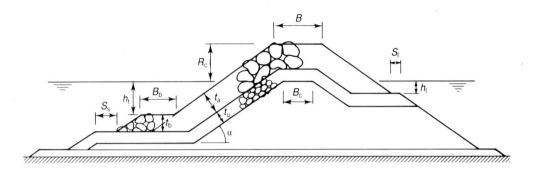

Figure 239 *Definition sketch for a rubble-mound breakwater*

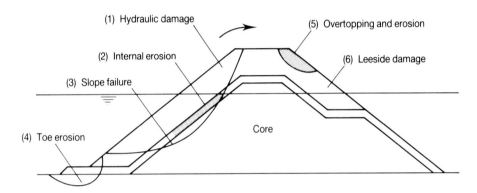

Figure 240 *Failure mechanisms of a rubble-mound breakwater*

239. The structure consists of a core of quarry run and is protected by a primary armour at the seaward slope, on the crest and on part of the leeside slope. A filterlayer or secondary armour layer may be needed between core and primary armour, depending on the filter requirements (Sections 5.1.3.2 and 5.2.8.12) and required wave protection of the core during construction. A filterlayer may also be required between the structure and the seabed. A berm is often built to support the armour layer. The typical failure mechanisms which are relevant for this type are given in Figure 240.

Having determined the main dimensions of the breakwater required to ensure an adequately low risk of failure in these modes using the design tools in Chapter 5, the following practical considerations which refer to Figure 239 should also be incorporated.

Shoulder width, S_s

The shoulder width at the sea-side, S_s, is mainly determined by placing tolerances and is generally not less than 2 m. Also S_l at the lee-side is determined by tolerances; $S_l = 0.5 * t_u$ is a practical value.

The use of seaborne equipment is practical for placing at levels of 3 m below low water level and deeper. Floating cranes for higher parts of the breakwater are generally avoided because of limited workability and poor accurac of placing. The basic decision therefore that has to be taken regarding the construction method is whether to use land-based or waterborne equipment (or a combination of the two). The main impacts on the structure are illustrated in Box 71 and can be summarised as follows:

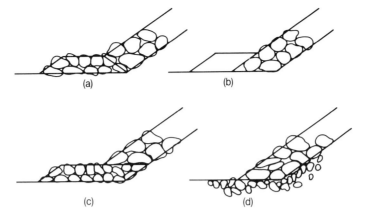

Figure 241 *Alternative arrangements of a toe berm in rubble-mound breakwater*

Berm width, B_b

The berm, width, B_b, should allow at least three stones to be placed, hence $B_b \geq 3.3 D_{n50}$. For the thickness t_b see requirements for t_a below. For the detailed shaping of the berm various alternatives are possible, as indicated in Figure 241.

Figure 241(a) poses the problem that the stones have to be placed accurately and that some erosion of the slope may fill the area between berm and slope, so that actually Figure 241(c) is obtained, which provides poor support for the lower armourstones. Figure 241(b) is a better alternative, and also permits both land-based and marine execution. Omission of a berm as in Figure 241(d) can be considered if the bedding layer contains stones that do not differ much from the size of the armourstones, so the latter can nest in the bedding layer, or is at such low level that no large forces occur ($>2*H_s$ below low water).

Crest width, B

The crest width B should be sufficient to permit at least three stones to be placed in the crest. This is a particularly important requirement if massive overtopping will take place. Hence three or four times D_{n50} is a minimum value (see Table 20 and Section 3.2.1.4 for a discussion of stone packing). The actual crest width also depends on the core crest, B_c. If the core is built out with dump trucks, B_c may be much wider than hydraulically required (see also Section 6.1.5).

Layer thicknesses

The layer thicknesses, t_a and t_u, follow from the requirement that for randomly placed stones a double layer is required to ensure that at all places the inner layers are properly protected, even after occasional washing out of individual stones. Exact layer thicknesses were discussed in Section 3.2.1.4 (see especially Table 20) and the potential need for verification test panels (Section 3.6.3.2) should be noted because of layer thickness coefficient problems in the past. For full design information on fixing the crest width and thickness of layers, see also the introduction to Section 5.1.3 and for the effect of rock characteristics, Section 3.2.3.4.

Slope angles

The slope angle, α, adopted in design for the front face depends on hydraulic and geotechnic stability considerations, but is generally not steeper than 1:1.5. This slope may be compared with the natural angle of response of material dumped under water, which can be as steep as 1:1.2. A slope adjustment to that eventually required can be made in the secondary armour layer, but it should be borne in mind that this requires a layer thickness larger than theoretically required and, as mentioned in Section 6.3.3, these stones are often the most expensive per unit volume.

The lee-side slope is generally built as steep as possible, but seldom more than

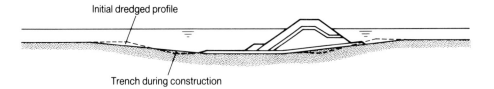

Figure 242 *Shallow-water breakwater cross-section*

1:1.5. If seismic activity is to be taken into account, the slopes are generally flattened as necessary to allow for the expected horizontal accelerations.

Crest freeboard

The minimum crest free board, R_c, follows from run-up and overtopping requirements (see Section 5.1.2), but is usually also determined by the level of the core if constructed with land-based equipment. This requires normally a level of at least 1 m above high water. With marine equipment the level of the crest can be arbitrarily chosen, recognising that all material more than 3 m below low water level cannot be simply dumped and will need to be placed by crane barges.

Berm levels

The water depth at the berm, h_t, is generally selected as 1–1.5 times the design value of H_s below low water. The exact level also depends on the weight of the berm stones. The water depth at the lee-side berm, h_l, depends on the wave attack from inside and the amount of overtopping. The effect of the latter can only be assessed in hydraulic model tests. In the determination of h_t and h_l the expected quarry output also has to be considered, so that the required volumes may be matched as closely as possible with the yield curve (Section 6.1.3.3).

Toe in shallow water

In the case of shallow water the design of the toe often presents a problem because its theoretical size brings the toe level close to still-water level. This leads to strong attack and heavier stones, which, in turn, aggravate the problem, which can be solved by constructing the toe in a dredged trench as shown in Figure 242.

Dredging of the trench in shallow water is difficult due to the transport of sediment. Side slopes in the order of 1:20 may develop, even if dredged more steeply initially. The initial trench width should allow for this.

Dredging in a marine envrironment of a sandy bottom can best be done with a trailing hopper dredger. However, such a dredger requires a depth of at least 6 m below high water. Dredging with a grab is relatively expensive due to limited production rates, and should be restricted to small volumes.

Alternatively, geotextiles can be used under the berm to limit the construction height. Intermediate layers between the berm and the seabed are replaced by the geotextile. However, the placing in waves is complicated and often requires extensive use of divers.

6.1.4.2 Rubble mound with monolithic crown wall

For a number of reasons, the introduction of a crown wall on top of a rubble-mound breakwater is a logical step:

- As mentioned above, rubble-mound breakwaters are often designed to sustain some damage, and access across the breakwater is needed for repairs;
- A crown wall with parapet may lead to a substantial reduction in the amount of stone which would otherwise be needed for a comparable conventional design (see Figure 243);

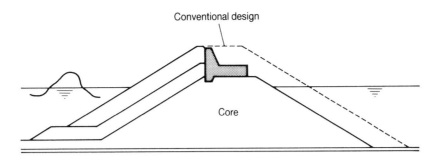

Figure 243 *Concept of a rubble mound with crown wall*

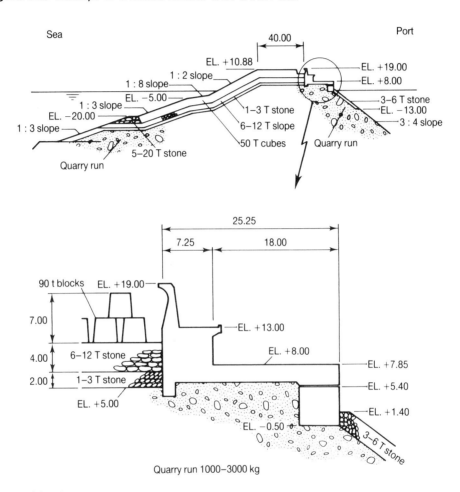

Figure 244 *Superstructure—west breakwater, Sines*

- When overtopping is allowed the crown wall may limit the width of the mound and, by its shape, protect the lee-side slope (see Figure 243).

Where berths were constructed immediately behind a breakwater it became the practice that the crown wall carried facilities for cargo loading/unloading (pipelines, belt systems) and electrical/water supply systems. As a consequence of this, the crown-wall designs became more complex and the term 'superstructure' was used to describe them (see Figure 244).

There are certain disadvantages related to a crown wall, which should be taken into account in selection and design:

- The crown wall represents a rigid element in a structure, which is flexible by nature. Uneven settlements may lead to a great problems for the elements of the crown wall and even more to transport facilities on a superstructure;

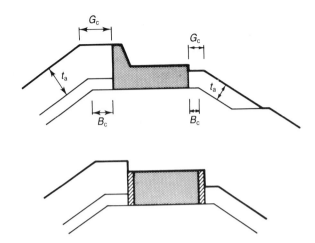

Figure 245 *Crown wall configurations*

- The tendency to increase the parapet wall in order to reduce the volume of stones leads to very large wave-impact shock forces on this wall;
- The reduction of overtopping by a crown wall gives increased attack on the armour layer;
- The crown wall increases the danger of excessive pore pressures in the mound;
- Overtopping water becomes concentrated more into a jet, and is a potential danger for the lee-side armour.

The design of crown walls should commence with an assessment of their stability, using the method and force information proviced in Section 5.1.3.4. In summary, apart from the weight of the crown wall, wave forces form the only other significant load. The horizontal force exerted on the vertical face of the crown wall is either a dynamic load (short-duration impulsive force cause by the wave front) or a quasi-static load, due to the overtopping water. Depending on the elevation of the underside of the crown wall, the mound beneath it may or may not be saturated. In the former case the wave pressures on the front will also work on the underside and may lead to a considerable uplift force. Taking into account that design forces will generally only occur at one time on a limited number of crown-wall elements, a horizontal coupling or keying is recommended.

Turning to practical details for crown walls, both L-shaped and rectangular options are available. The former is normally built in two phases: first, the horizontal slab, followed by the parapet wall. The latter can consist of precase concrete mould elements, with a fill of mass concrete (see Figure 245).

Generally, a large stationary crane is used to place any precast elements. This can be done by working back from the head or by dumping the stone of the core with the same crane. The choice of placement method has an effect on the required stability of the core during construction in view of the different exposure times. The width and elevation of the core and crown wall should, of course, be sufficient for the required heavy crane.

It is normally preferable to place the crown wall on the core and not on secondary armour, in order to avoid the higher penetration of uplift pressures under the wall, which occurs in the latter case. However, the underside of the crown wall must also be kept sufficiently above still-water level to reduce uplift forces to acceptable values for stability.

The crest element is sometimes made of a simple slab of mass concrete if the crest elevation is sufficiently high to avoid wave pressures under the slab. High parapet walls, extending above the level of the primary armour which may attract high wave-impact pressures, should normally be avoided. Where a high parapet is unavoidable, the structure must be strengthened appropriately.

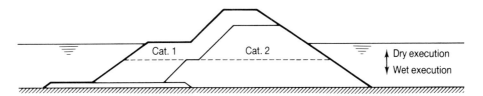

Figure 246 *Berm breakwater with two stone categories*

Adjacent to the crown wall, shoulders, B_c, have to be made (see Figure 245), the width of which follows from tolerance considerations, but should not be less than 0.5 m. To avoid the situation that some settlement may expose the crown wall to wave impacts, a horizontal shoulder of dimension G_c approximately equal to t_a must be projected.

6.1.4.3 Berm or S-slope breakwater

When rubble-mound breakwaters are subjected to extreme wave forces they develop some common characteristic properties:

- A flatter slope develops just below the still-water level (typically, about 1:5);
- Below a certain elevation no significant erosion occurs and material is deposited at a steep slope;
- In the area above still water hydrodynamic forces are reduced by the flat slope in front, leading to steeper slopes here also.

Many rubble-mound breakwaters built in the nineteenth century were designed to allow wave motion to mould the structure into its final shape. Well-known examples are the breakwaters at Plymouth (UK) and Chebourg (France), which have been damaged but never totally destroyed. Indeed, the Plymouth breakwater has remained functional over the past 125 years, which cannot be said of all the more recent structures.

Breakwaters built according to the above principle are now referred to as berm or dynamically stable breakwaters. If, however, a flatter slope is initially constructed at about still-water level with steeper slopes above and below but no deformations are allowed (static stability), one speaks of an S-shaped breakwater.

The development of the profile of berm breakwaters can be left to hydrodynamic forces, provided a sufficient volume of stone is deposited. Alternatively, one can construct as closely as possible to the appropriate dynamically stable profile. Determination of the final profile can be made using the method described in Section 5.1.3.7.

As described in the overview of structure types in Section 5.1.1.2, the ratio of the design significant wave height to the average stone size $H_s/\Delta D_{n50}$ is of the order of 3–6 as compared with conventional type breakwaters, where this ratio is about 1–4. The consequence of this difference is that the weight of the armourstones at the seawater face can be a factor 5–10 smaller than for a conventional rubble mound. Efficient use of quarry output can be made (Baird and Woodrow, 1985). With a wide gradation ($D_{n85}/D_{n15} = 2$–2.5) for the armour layer, the rest of the quarry output can be placed in the core without further separation (Figure 246).

The slope geometry and berm porosity are important parameters for the stability of these breakwaters, the mild slope section acting as a stilling basin in which the breaking wave plunges and dissipates energy. The large absorption capacity of the voluminous berm also reduces pore-water pressure gradients and uplift forces, which, when coinciding with maximum external drag forces, cause dislodging of the armourstones. In this respect also the term 'mass armoured breakwater' is used, which refers to the mass volume of armourstones at the seaward face of the breakwater.

One important point for concern for dynamically stable breakwaters are the three-dimensional effects. When oblique waves hit the breakwater, longshore transport can occur, jeopardising the stability of a breakwater, which would be stable under perpendicular wave attack. It is therefore advisable in the case of oblique waves to limit the value of the parameter $H_s/\Delta D_{n50}$ as described in Section 5.1.3.10, taking particular care with roundheads.

In summary, the most important advantages of a dynamically stable breakwater compared to a conventional one are:

- The maximum weight of individual stones is smaller. Consequently, more quarries can potentially supply the armourstone;
- The election activities in the quarry can be limited. If well designed the complete quarry output can be used;
- Wide tolerances are allowed, so most of the profile can be constructed by free dumping;
- The core can be made wide for easy traffic access.

The drawbacks are:

- Limited experience;
- Empirical data and reliable design formulae;
- No overtopping allowed, leading to a relatively high crest;
- Danger of progressive damage due to oblique wave attack, particularly at the roundhead;
- Durability of stones may be a problem due to frequent motion, but see Section 3.2.4 for a design approach to this materials problem.

6.1.4.4 Reef breakwater

Reef breakwaters are essentially composed of a large homogenous volume of stone without underlayers or core, which is allowed to reshape under wave attack (see Figure 247). Design data for the stability of reef breakwaters are given in Section 5.1.3.3. The geometric design is largely determined by the fact that marine equipment is normally required for construction. Sometimes construction is carried out with land-based equipment via a (temporary) causeway, but this approach is not favoured, as it requires substantially more materials handling.

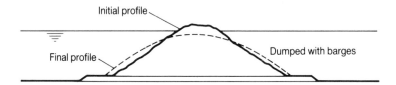

Figure 247 Reef breakwater

6.1.4.5 Low-crested and submerged breakwaters

If, alternatively, the core is covered with a statically stable armour layer, one speaks of a low-crested breakwater or submerged breakwater, depending on the water level with respect to the crest elevation. The stability enhancement due to the lower crest can be calculated using the graphs in Section 5.1.3.3. Other design considerations for these latter types do not differ in principle from those for a conventional rubble mound (Figure 248). However, in view of the low crest level, marine execution is required, mostly using crane pontoons.

> The crest height is determined by the required wave reduction. Since wave energy is transmitted over the breakwater the armour has to extend on the lee-side as well. The rate of overtopping determines the dimensions of the lee-side armour, which can only be assessed properly by model tests.

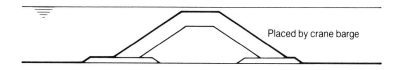

Figure 248 *Low-crested/submerged breakwater*

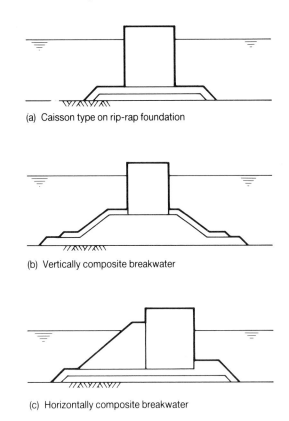

Figure 249 *Typical caisson-type breakwaters*

Reef breakwaters are generally built as part of a coastal defence scheme, and therefore are mostly constructed in shallow water, with the consequences of shallow water breakwater design and construction, as indicated earlier in this section.

6.1.4.6 Caisson-type breakwater

Caisson-type breakwaters have found wide application in some Mediterranean countries (e.g. Italy) and in Asia. Figure 249 shows the three different concepts which should be distinguished, each having its specific conditions and design criteria.

In all three sub-types rock often forms an important part of the construction, and therefore this aspect of caisson breakwaters is addressed in this manual. In sub-type (a) the use of rock is limited to a foundation layer only. Therefore this design is attractive in countries where there is insufficient good-quality rock for construction of a conventional rubble-mound breakwater. If the water depth increases, the width of the caisson also grows and the vertically composite breakwater may become an economical alternative.

In the past many failures occurred to sub-type (a) breakwaters due to very high impulsive forces caused by breaking waves. If, at a specific location, the wave conditions are such that breaking waces can occur, sub-type (c) may present a

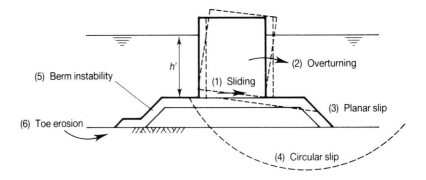

Figure 250 *Failure mechanisms of a caisson breakwater*

viable alternative. Initially, the solution was developed in Japan to protect existing caisson-type breakwaters against (further) damage. At present, it is still applied at sites where stone material is scarce and breaking waves cannot be excluded.

Caisson type on rip-rap foundation

The principal failure modes are shown in Figure 250. Horizontal sliding (1) will occur if the total horizontal wave force exceeds the friction resistance at the interface between caisson and foundation. Overturning (2) implies a rotation around point A, which effectively means such an increase of the effective stress around point A that the stones are crushed. If the horizontal slide does not occur at the underside of the caisson (for instance, by provision of keys), a failure of the upper part of the foundation may occur, called a planar slip (3). Geotechnical instability along a slip circle (4) can occur in very poor conditions of the existing subsoil. Wave impact forces on the caisson may result in liquefaction if the subsoil consists of loosely packed sand. Finally, the wave action in front of the caisson could lead to hydraulic damage of the toe (5) or erosion of the seabed outside the toe (6).

The design of the caisson stability, involving mechanisms (1)–(4), is determined by an extreme wave height–wave period combination, while the instability of the toe is linked to the occurrence of a design storm, characterised by a significant wave height. This difference is due to the fact that a single wave may cause the caisson failure, contrary to the more gradual process of hydraulic damage to a mound of stone. As a consequence, different probabilities of exceedance will be applied in defining the single wave height, H_D, for the caisson design and the significant wave height, H_s, for design of the toe.

Failure mechanisms (1) and (2) are not directly influenced by the rock foundation and are therefore not relevant within the framework of this manual. The forces on the caisson and exerted by it on the foundations are important, and therefore caisson design has been summarised below. The wave forces on the caisson can be calculated from the empirical pressure distribution developed by Goda (1974) for both breaking and non-breaking wave conditions. It is emphasised that breaking-wave conditions exclude those which can lead to high impact pressures. Based on the many failures mentioned above, it is current practice in Japan not to consider this sub-type if impacting waves may occur. A checklist has been developed which allows the designer to ascertain the dnger of so-called 'impulsive pressures'. A detailed reference for design of caissons, including this checklist, is provided by Goda (1985). A summary of the formulae for non-breaking wave pressures is given in Box 72.

The hydraulic forces determined by the formulae in Box 72 form an adequate input into the preliminary design of the caisson. This is governed by failure mechanisms (1) and (2), which lead to the following requirements:

Sliding $\qquad P < [\mu(W-U)]/F$

Overturning $\qquad M < [W.B/2 - M_u]/F$

Box 72 Non-breaking wave pressures and forces on caissons

The distribution of wave pressures according to Goda (1974) is shown in Figure 251 and the various parameters in this figure may be calculated according to the following formulae:

$$\eta^* = 1.5 H_D$$

$$p_1 = (\alpha_1 + \alpha_2)\rho_w g H_D$$

$$p_2 = \frac{p_1}{\cos_h(2\pi h/L)}$$

$$p_3 = \alpha_1 \alpha_3 \rho_w g H_D$$

where

$$\alpha_1 = 0.6 + \tfrac{1}{2}\left(\frac{4\pi h/L}{\sinh(4\pi h/L)}\right)$$

$$\alpha_2 = \min\left[\frac{h_b - d}{3 h_b}\left(\frac{H_D}{d}\right)^2, \frac{2d}{H_D}\right]$$

$$\alpha_3 = 1 - \frac{h'}{h}\left\{1 - \frac{1}{\cosh(2\pi h/L)}\right\}$$

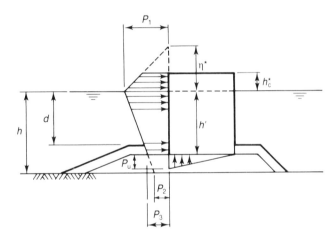

Figure 251 *Wave pressures on a caisson*

In the above formulae, H_D is taken as the maximum breaking wave height at a distance $5H_s$ seaward of the caisson (where the local depth is defined as h) and may be calculated as described in Section 4.2.3. For non-breaking waves H_D is determined for the given exceedance criterion, using the long-term distribution of significant wave heights and using the Rayleigh distribution. H_D is generally taken as $H_D = H_{max} = 1.8 H_s$.

The uplift pressure acting on the bottom of the caisson is assumed to have a triangular distribution with toe pressure, p_u, and with a heel pressure zero. Theoretically, $p_u = p_3$ holds, but the expression for p_u gives a somewhat lower value based on the experience that the actual pressures on the bottom diminish rapidly.

From these pressures, total forces and overturning moments about the base of the caisson may be derived and calculated as follows:

The total horizontal wave force, P, and moment, M_p, about rear heel of caisson:

$$P = \tfrac{1}{2}(p_1 + p_3)h' + \tfrac{1}{2}(p_1 + p_4)h_c^*$$

$$M_p = \tfrac{1}{6}(2p_1 + p_3)h'^2 + \tfrac{1}{2}(p_1 + p_4)h' h_c^* + \tfrac{1}{6}(p_1 + 2p_4)h_c^{*2}$$

in which *(Continued overleaf)*

> **Box 72** (Continued)
>
> $$p_4 = \begin{cases} p_1(1 - h_c/\eta^*) : \eta^* > h_c \\ 0 \quad\quad\quad\quad\quad : \eta^* \leq h_c \end{cases}$$
>
> $$h_c^* = \min\{\eta^*, h_c\}$$
>
> The total uplift, U, and its moment, M_u, about the heel of the caisson:
>
> $$U = \frac{1}{2} p_u B$$
>
> $$M_u = \frac{2}{3} U B$$

where P, U, M_o and M_u are as defined in Box 72 and W is the submerged weight of the caisson.

In the design of caisson breakwaters in Japan the safety factors against sliding and overturning (F) must not be less than 1.2. The coefficient of friction between concrete and foundation, μ, is usually taken as 0.6. By varying the width of the caisson, B, both requirements can be satisfied. For a given design the bearing capacity of the foundation must also be analysed, following the approach for foundations with eccentric inclined loads. The bearing pressure at the heel should not exceed values of 400–500 kPa. Subsequently, geotechnical failure mechanisms, such as planar slip (3) and circular slip (4), need to be checked using Bishop's approach, as described in Section 5.2.

For the hydraulic stability of the rubble-mound foundation, reference should be made to Section 5.1.3.5, where preliminary design formulae are given. For detailed design, model tests should normally be carried out. These allow the wave pressures on the caisson to be measured and the stability of the foundation material assessed.

Toe erosion may be caused by the high velocities which occur in the node of the standing wave pattern seaward of the breakwater. A rule-of-thumb approach suggests than an area of up to three-eighths of a wave length, measured from the vertical face, will be subject to possible erosion. The need for bottom protection depends largely on the width of the toe structure and its flexibility to follow an erosion immediately seaward of it.

Vertically composite breakwater

As indicated above, this type of breakwater becomes attractive when the water depth increases. The design has one potential danger—incoming waves are forced to break by the underwater mound and cause impulsive wave forced on the caisson. For that reason, the mound should not be too high and the width of the seaward berm should not exceed $1/20L$, where L is the length of the steepest design wave which can occur at the breakwater. This condition is included in Goda's checklist procedure for the elimination of such shock forces mentioned above. It is necessary to check the final design in model tests for a range of wave conditions to ensure that no vertical wave front hits the caisson.

For vertically composite caisson breakwaters, failute mechanisms are very similar to those for conventional caisson breakwaters, as shown in Figure 250 and the design approach for (1)–(4) and (6) is also the same. For stability of the primary armour layer protecting the berm against hydraulic damage (5), design information is provided in Section 5.1.3.5.

Horizontally composite breakwater

The mound in front of the caisson must break and absorb the wave energy effectively. In most of the examples of this type built in Japan the mound consists of one type of (concrete) armour unit, without core or filterlayers, in order to

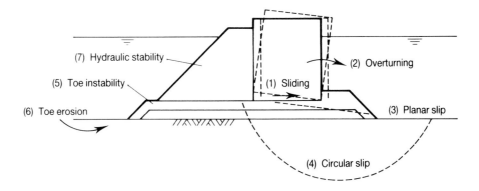

Figure 252 *Failure mechanisms of a horizontally composite breakwater*

achieve a high porosity. Because of this protection the impact wave forces on the caisson are greatly reduced. However, the same failure mechanisms apply to this type as for the previous two, since quasi-static wave pressures penetrate the mound (see Figure 252). The hydraulic stability of the mound is basically similar to that of the primary armour layer of a conventional breakwater. The coefficients will be different, however, due to the high porosity, on the one hand, and the reflecting caisson face, on the other. Japanese test results in the form of a damage coefficient K_D versus percentage damage show considerable scatter (Tanimoto *et al.*, 1982). No conclusive relation can be obtained from them and hence model tests are required to conform any particular design. In principle, this solution is feasible with rock in front of the vertical wall and thus may be very attractive from a cost point of view.

6.1.5 CONSTRUCTION ASPECTS

This section describes the construction aspects that are most relevant for rockfill structures. Typical plant capacities and execution tolerances are covered in relation to the design of the overall dimensions of the structure. For the characteristics of production and transport equipment, reference should be made to Sections 3.5 and 3.7.

6.1.5.1 Land-based operation

Placement of rockfill can be done by direct dumping by trucks or loaders, or by crane. Direct dumping of wide gradings generally involves the problem of segregation. There will be a tendency for larger stones to roll down and smaller fractions to stay on top, therefore a poor filter will be obtained at the seabed.

A land-based operation poses restrictions on elevation and width of the (temporary) crest or access. Since the access road needs to be above sea level, the structure is more vulnerable to wave erosion during execution. However, a land-based method is normally more economic than marine placing, particularly, if material is hauled directly from the quarry to the construction site.

Core construction

Core construction is done most simply by direct dumping of bulk material with dump trucks (Figure 253), generally carrying 20–50 t, often with the assistance of a bulldozer. Dump trucks require a haul track at least 4 m wide to drive, with 'passing places' of at least 7 m at intervals to allow turning and passing. The elevation of the haul track should preferably be 1.5 m above sea level, to avoid problems due to splash and spray. Only if the track area is very well protected can a smaller freeboard be applied.

Driving directly on rockfill is not possible with rubber tyres, but accessibility can be

Figure 253 *End-tipping rock to repair a breakwater breach (courtesy HR Wallingford)*

maintained by dozing fines over the surface with a bulldozer. Access on stones up to 1 t can be made in this way. If the design does not permit such an impermeable layer to remain, the fines will have to be removed afterwards by air or water jetting. This is a very costly operation, which should be avoided if not strictly necessary for stability reasons. (Clauses 12.3 and 12.7 of the model specification, Appendix 1, cover this point.)

Core material can also be placed by crane, with materials supply by dump trucks. The capacity of the crane then completely determines the progress of the work. Cranes can use a clamshell or orange-peel grab to dig into the stock of core material dumped by trucks, or work with skips or rock trays, which are filled by a loading shovel at the quarry or stockpile and carried by trucks or trays loaded at the construction site directly by dump trucks. In the former case space had to be provided for a shovel or front-end loader and a truck. In the latter, heavy cranes are required which also call for much space on the breakwater.

Of the methods discussed, the most economic is clearly direct dumping. However, this will achieve a steep side slope of approximately 1:1.25. Any core material required outside that slope line will need to be placed by crane or dump barge.

Sometimes there may be the possiblity of using sand as core material. To avoid wide cross-sectional profiles in this situation, rock bunds are placed which contain the sandfill between them. Alternatively, if economic and the prevailing wave climate permits, a wide substructure of sand (see Section 3.1.3) can be made with gentle slopes, often covered on the seaward side with a scour protection as construction progresses.

Secondary armour

Secondary armour can also be dumped by dump trucks. It needs trimming afterwards, however, to bring the material to the required profile. For smaller breakwaters such trimming can best be done by hydraulic (back-hoe) cranes, provided the slope is not too long and the stone weight not too large (less than 2 t). Figure 254 presents the characteristics of current types of equipment.

Since the secondary armour always follows the construction of the core, sufficient

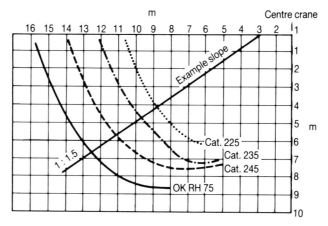

Curves show reach of actual typical current hydraulic cranes

Figure 254 *Typical equipment capacities for slope trimming*

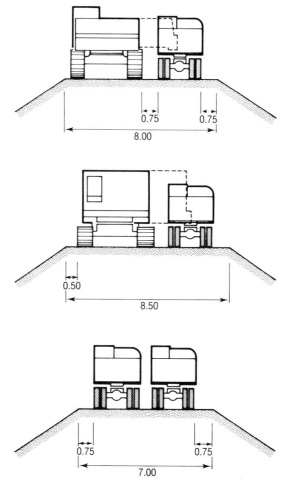

Figure 255 *Space requirements at a breakwater crest*

space has to be provided for to allow passing of dump trucks, with the (hydraulic) crane in operation (see Figure 255). Secondary armour may also be tray placed (e.g. a 2 t self-weight rock tray holding a 12 t payload).

Berm, scour protection and armour

These structural elements have in common the fact that they require cranes with

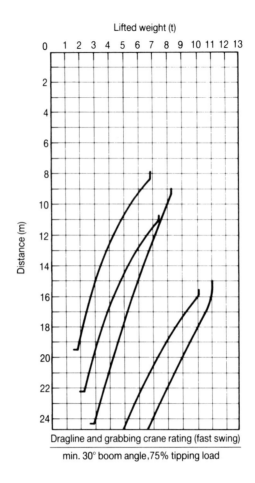

Figure 256 *Lifting capacity of common types of cranes*

sufficient reach for their construction. Generally, hydraulic cranes cannot be used, but rope-operated cranes (draglines) are required.

The lifting characteristics of some cranes are presented in Figure 256. The smaller cranes require a working platform of at least 5 m wide (no passing possible). It is important to note that the weight of a digging grab or cactus grab (for core and armour, respectively) normally used by these cranes is 50–65% of the maximum lifting capacity (Figure 257 and 258). Armourstones can be provided with eyebolts to avoid this loss of lifting capacity. If core material (for bed protection) is tray placed, the ratio of container to payload is about 1/6. Further information is given in Section 3.4.

The production capacity of cranes is determined by the rotation and lifting speed. A fast average for conventional dragline cranes is 15 cycles per hour for armour placing. For hydraulic machines this can increase to about 30 cycles per hour, but for accurate placing it may be less. This means that, in spite of higher production cost at the quarry and increased layer thicknesses, larger armourstones can still be more economic, as a result of the smaller number of units to be placed.

For larger stones, or for dumping core material with skips, trays or wire nets, much larger skip cranes are required, some typical characteristics of which are given in Figure 259. Such cranes can be used for all phases of construction, their production capacity depending on the filling capacity of skips, trays or nets by trucks or loaders.

Placing of stones is often controlled by using a grid system. With modern computer-aided design techniques the cross-section design drawing can automatically

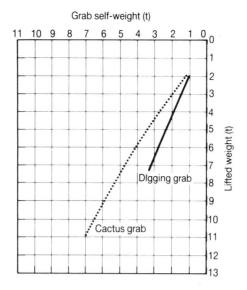

Figure 257 *Relative weight of grab versus payload*

Figure 258 *Orange-peel grab placing rock at breakwater (courtesy Aveco Infrastructure Consultants)*

be transformed into co-ordinates for the crane engineer (rotation and boom angle) at which certain volumes have to be dumped.

6.1.5.2 Seaborne transport and placing

The principal reasons for adopting seaborne (or marine) transport and placing (see also Section 6.2.3.4) include:

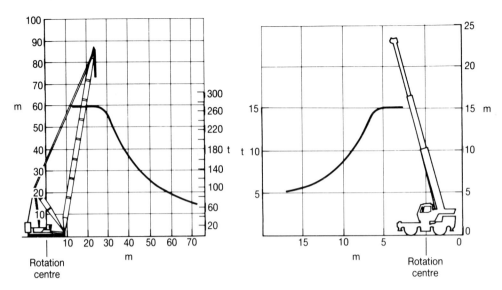

Figure 259 *Lifting capacity of two typical heavy cranes*

- Congestion problems on the breakwater with land-based plant when large volumes of stone have to be placed under water;
- Economics, depending on the quarry location (inland, ashore) and the haul distance (marine operations may also be more economic).

For breakwaters in deep water with long slopes or narrow crests, or for the placing of berms, the required reach may prevent placement using land-based plant operating from the breakwater crest. In these situations direct dumping from barges is often possible and this aspect is discussed in detail in this section. In addition, floating cranes may be used to overcome reach problems, using rock trays if appropriate, as illustrated in Figure 260.

Core construction

Several types of self-unloading barges can be used, differing only by the method of unloading (see Figure 261):

- Split barges
- Bottom-door barges
- Tilting barges
- Side-unloading barges

Commonly available self-unloading types have load capacities of the order of 500–800 t. The first three types do not allow great precision of placing, but are generally adequate for core construction.

For bedding layes scour protections and berms, flat-deck barges with a bulldozer for discharge can also be used. Capacities of such barges can be much higher, typically reaching 5000 t. For all types, strengthening of the surfaces in contact with rock is required.

The maximum construction elevation for barge-dumped core material is governed by two criteria:

- The maximum draught of the barges, plus a safety clearance for heave (vertical motion) of the barge. The highest practical level by this criterion is about 3.0 m below low water, but bottom-door barges, which require a greater clearance, will need the level to be lower;
- For 'winter construction', the need to limit the loss of material during typical storms to acceptable values, reshaping within the contours of the final core clearly being acceptable. The rate of deformation can be assessed from the hydraulic design tools on low-crested structures in Section 5.1.3.3.

Figure 260 Placing rock from floating plant using a rock tray (courtesy Aveco Infrastructure Consultants)

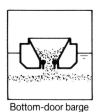 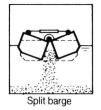  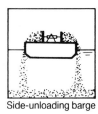

Bottom-door barge | Split barge | Tilting barge | Side-unloading barge

Figure 261 *Types of self-unloading core barges*

The barges can be loaded by conveyor systems (for 0–1 t material), or by trucks, shovel or crane. It is preferable to have a stockyard at the loading-out point in order to make the barge transport less dependent on the supply from the quarry. Loaded barges will then dump their loads at predefined locations, which form part of a grid system. For the measurement of the volume placed (see Appendix 3), there are two possibilities:

- Weight measurement, using the barge draught before and after loading multiplied by the number of barge trips;
- The *in-situ* measurement by soundings before and after construction. This sounding is required in any case to check that the core levels are according to specification.

The weight measurement is the fairest yardstick if it can be readily determined (see also Appendix 3). Soundings do not take into account bed settlements, scour or filling of scour holes at working faces. For larger stones, where there are likely to be disputed sounding results, these may be minimised by adopting the sounding survey technique given in Appendix 1. Prediction of the results of the rock-dumping process may be made using an appropriate computer simulation model (see Box 73).

Filter and toe construction

These components generally require good precision. Of the barges listed for core placement above, only the side-unloading or flat-deck ones can offer such precision. Where these vessesl often have lateral control by a Voith–Schneider propeller (or even dynamic positioning) they allow the placing of rather thin layers (0.50 m) on the seabed or on the core. Thin layers of the order of 0.5 m can be built up by multiple passages of the dumping barge to limit segregation. Alternatively, the material can be placed by a clamshell or front-end loader working from a barge. In most cases it is more economic to have this equipment on a separate barge, which remains at the placing location.

Placing of gravel-size gradings can be carried out using modern trailing suction hopper dredgers. Such hoppers are equipped with system to pump the mixture back through the suction pipe, with the draghead syspended only a few meters above the seabed. Following the dumping, the layers can also be levelled by lowering the draghead to the proper level and dozing off the high spots.

Placing of stone armour

For construction of an armour layer of relatively small rock a side-unloading barge may still be used. No defined limit can be given, since it also depends on the sea conditions in which the barge has to operate.

For armour layers of stone above 1 t or of concrete units the use of derrick barges or pontoon-mounted cranes is more common. Often specifications do not allow dumping because of the required accuracy of placing. Also, rock armour needs to be placed piece by piece in order to built it up into a proper two-layer construction.

For all seaborne construction, dependence on the weather is a common concern

Box 73 Simulation of rock dumping

The dumping of rock (e.g. for breakwater core construction) is a process characterised by a large variety of parameters influencing the final resulting profile shape and position. As a result, considerable uncertainty has always existed in regard to the quality of the final result. In order to guarantee satisfactory construction and functional performance, the historical approach has been either to dump surplus material or to employ more intensive and frequent inspection. Both options unfortunately involve extra cost.

For these reasons, a simulation model of the dumping process (STORTSIM) has been developed in the Netherlands aimed at:

1. Improving the predictability of the result of the dumping process;
2. Using, as input, practical process parameters;
3. Optimising the turning of the process control parameters;
4. Permitting comparison of the result of the dumping with the required profile shape and position to achieve the functional requirements.

The applicability of the model is dumping by any method for which the following boundary conditions are known:

- The position of the submerged dumping point(s) as a function in time;
- Bottom geometry;
- Specifications of rock material to be dumped;

Simulations are made for the discharge of a specified compartment of a barge and the corresponding layer of the structure. The final results of multi-layer dumping can thereby be simulated.

Analysis of the dumping progress

The principal question during construction of a submerged rock layer is 'what are the actual geometrical characteristics of the quantity of rock that has been dumped from the barge?' In order to answer this question the model simulates the movements of the stones released from a barge having a known position with time. The discharge characteristics of the stones are determined by the shear properties of the rock. As a result of both mutual interaction between the stones and the waves and currents, the stones will spread in directions parallel and perpendicular to the barge. Following contact of the stones with the seabed, they will spread further along it, depending on bottom slope and characteristics.

Input parameters

The model requires values to be assigned to the input parameters in the following list, which also serves as a general summary of the various dependencies in the dumping process:

1. Barge
   Distribution of rock across the barge;
   Load capacity;
   Unloading characteristics (capacity, devices, speed);
   Accuracy of measuring system on board;
   Hydraulic and dynamic characteristics;
   Positioning system.
2. Rock material
   Weight distribution (Section 3.3);
   Specific density;
   Shape;
   Porosity;
   Cohesion;
   Friction angle.
3. Dumping strategy
   Barge velocity and direction;
   Control characteristics of unloading devices;
   Number of layers.
4. Seabed
   Co-ordinate system and grid parameters for calculation;
   Bottom topography;
   Soil characteristics.
5. Environmental conditions
   Current velocity and direction;
   Wave height, frequencies and direction;
   Wind speed and direction;
   Visibility;
   Manoeuvring space and possible structures.
6. Requirements
   Layer thickness (minimum/average/maximum);
   Segregation;
   Slope angle;
   Maximum elevation;
   Maximum spreading.

(Continued overleaf)

Box 73 (Continued)

Process parameters

1. Barge position
   Positioning system;
   Vertical, horizontal, trimming and rolling motions;
   Distribution of loading;
   Way of unloading.
2. Characteristics of rock discharge
   Wet/dry material?
   Way of loading;
   Barge motions.

Results of the simulation model

Based on the specified grid, the as-laid layer characteristics can be obtained. These may be presented in tables and plots describing:

- Spatial distribution of layer thickness (cross-sections, thickness contours);
- Probability distribution of layer thickness (minimum/average and maximum values);
- Discharge from barge as a function of time.

Some examples are shown in Figure 262.

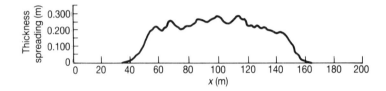
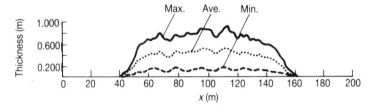

(a) Predicted cross-sectional distributions

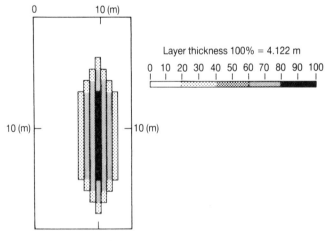

(b) Predicted contour plot

Figure 262 *Example of output from STORTSIM rock dumping simulation model*

and planning for breakwater construction should take a certain percentage of downtime into account. Often the crane, particularly if it is barge mounted, determines the maximum allowable wave condition, which is determined by the ringer mechanisms and derricks. Cranes are not normally designed to take any lateral forces such as caused by swinging loads due to barge motions, and for this reason maximum allowable tilts of 1° are not unusual.

6.1.5.3 Combination of land-based and seaborne operations

Land-based and seaborne operations may often be carried out in parallel. This may arise because of planning requirements dictating the need to operate on two or more fronts of a breakwater simultaneously. Another reason may be to reduce erosion at the land working front by placing, from barges, for bedding layers and lower sections of the breakwaters in advance of the superstructure. Island sections of breakwaters may have armour being placed by crane, standing on the island section, with supplied by barges.

6.1.5.4 Placing tolerances

Placing tolerances should be related to the size and shape of the material used, the possibilities of standard equipment, and the measuring techniques. A list of execution accuracies for profiles and layer thicknesses for core, underlayers, and armour layers are given in Appendix 1, where in some cases they are related to the size of rock being placed. Other accuracies which can reasonably be met include:

Dredging $+0\text{–}0.5\,\text{m}$
Dumping gravel $\pm 0.5\,\text{m}$ vertical $\pm 3\,\text{m}$ horizontal
Levelling with draghead $\pm 0.2\,\text{m}$

The above practical tolerances should be taken into consideration in the design, particularly at the interface between various construction activities—for instance, at transition berms (see Figure 263).

Overall tolerances for construction of rockfill layers depend on the size of the rock and the dimensions of the layers. Tolerances related to the layer thickness are important, because they eliminate the possible accumulation of opposite deviations from the design profile, which could then lead to unacceptably thin layers. Thus, in Appendix 1, Section A1.13, while recommended tolerances on profile for general usage are given in absolute terms or as a function of D_{n50}, these are combined with an overriding clause to ensure a sufficiently large thickness for each layer.

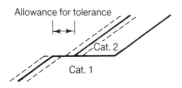

Figure 263 *Tolerances at transition berms*

6.1.6 COST ASPECTS AND PROJECT OPTIMISATION

The design of a breakwater is usually aimed at obtaining a cost minimum within the limits of the functional or environmental boundary conditions. The total capitalised costs can be separated into the following items:

1. Operational and capital costs of the total scheme directly related to the layout (extent) of the specific structure in question, but excluding the costs of the structure. Examples are downtime costs related to port accesses, operation of

berths protected by the breakwater, and the costs of protection works and wave-absorption structures in the harbour;
2. Design and construction costs (capital costs);
3. Maintenance costs;
4. Cost of repair in case of damage;
5. Extra costs imposed on the total scheme due to damage. Examples are downtime costs until completion of repair, damage to vessels and installations, etc.

It is important to appreciate that, when damage occurs, the methods, the time and the costs of repairs are dictated by the original design. For this reason, the design should also include considerations of repair of the structure (see Chapter 7). Distinction should be made between regular maintenance work and incidental major repairs. The most important elements that play a role to ease the work of maintenance and major repairs are given in Table 37.

Table 37 Relative importance of various considerations for maintenance and major repair

Consideration	Maintenance	Major repair
Maintenance prescriptions	× ×	—
Accessibility from land or waterside	× ×	×
Non-specialised equipment needed	× ×	×
Materials available	× ×	×
Funds	×	× ×

The overall approach to achieving a total capitalised cost minimum has been described in Section 2.1.7. This section therefore seeks to address those aspects specific to the on-site construction of breakwaters and to provide some examples. Production and transport cost information can be found in Section 3.4.

6.1.6.1 Production aspects of the cost of breakwaters

For a breakwater, the total production volume required from a temporary quarry opened up for the purpose of the project depends on the theoretical volume required in each size category, the losses and the fragmentation curve achieved in the quarry (see Section 3.4). Any volume requirement for a particular category in excess of the fragmentation curve will require a proportional extra production for all categories (Figure 264). The excess production for the other categories is to be considered as waste, unless other commercial use can be found, which is only occasionally so.

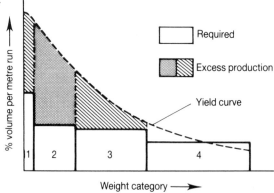

1 = filter material/aggregate for crown wall
2 = core material
3 = secondary armour
4 = primary armour

Figure 264 *Produced and required quarry output. 1 Filter material/aggregate for the crown wall; 2 core material; 3 secondary armour; 4 primary armour*

The optimum cost situation is clearly the situation where the required volumes match the fragmentation curve. Since only a small fraction of the total production provides large blocks, the required volume of these is critical and often leads to overproduction of fines. An example of this optimisation is given in Box 74 and the process is considered in more detail in Vrijling and Nooy van der Kolff (1990).

Every effort should be made to tailor the design to the anticipated fragmentation curve. It should further be borne in mind that the actual yield curve is frequently quite different from the anticipated fragmentation curves (see Section 3.6).

Rock can sometimes be obtained from permanent quarries. For aggregate quarries, which are by far the most common, the large blocks are considered as waste. Collecting these 'wastes' (preferably in standard gradings from a number of quarries) can sometimes be an economically attractive option for small projects. However, required production rates of 10 000–20 000 t per week are not unusual for a breakwater, with 50 000 t being a norm for larger breakwaters and hence often a dedicated quarry is required to meet such a demand.

6.1.6.2. Cost aspects related to activities at the site of the structure

The organisation of the works at the site of the structure is based on logistical principles. Hence the costs depend on the logistic system selected. The total volumes of material and dimensions of the structure are boundary conditions for the logistical system. For complex structures the assessment of execution method, production rates and cost optimisation can best be done on the basis of a project-simulation model, in which disturbances are introduced into the quarry and on-site construction processes in a stochastic way. (see Box 75).

An important element in the construction cost is the risk of damage to unfinished fronts during construction. This can be expressed as:

$$\text{Risk} = \text{probability of event} * \text{consequences}$$

Evaluation of 'risk' is an element in the determination of the total cost, and thus in the selection of alternatives.

The cost of handling and placing on-site with land-based equipment, working on low-water shifts with working periods of 6–8 hours are of the order of \$10/tn. The operational cost for floating equipment can be 200–300% of those of land-based equipment and productive time for floating equipment can be as low as 40–70% compared to 80% for land-based plant. Consequently, the unit rates can be at least three times more expensive when using floating equipment. However, this difference can sometimes be more than compensated for by the required smaller dimensions of the cross-section (Burcharth and Rietveld, 1987, see examples, in Section 6.1.6.3).

The size of breakwater core material can also sometimes limit the productive working time. By using heavier but more expensive grades the downtime can be reduced and savings obtained in both cost and the total construction time (see examples in Section 6.1.6.3).

Burcharth and Rietveld (1987) also carried out example cost calculations for breakwaters at Zeebrugge (Belgium) and found that the relative costs were:

Core   50%
Armour layer 15% (relative to whole structure 9%)

In this case, the armour layer formed a minor part of the total cost, but the progress of the working front of the breakwater was limited by the time needed to place the blocks, whereas ideally the progress of core, secondary and primary armour should match. Since the progress was determined by the number of blocks to be placed, the example showed a saving on overall cost when using 30 t blocks instead of 25 t due to the smaller number of blocks to be placed (see example in Section 6.1.63 and also Section 5.1.5.2).

Box 74 Matching demand for stone to quarry fragmentation curves

The actual production of quarry stone depends not only on the total theoretical volume required but also on the match of the demand curve to the yield curve of the quarry. Since average quarries produce only few heavy armourstones, this category determines to a large extent the required production. Important cost savings can be realised by designing with the best match possible.

By way of example, two breakwater sections with similar safety are compared, one a conventional type and the other a berm-type breakwater. The cross-sections and quarry yield curve are shown in Figure 265 and the stone requirements related to the quarry output in the table below.

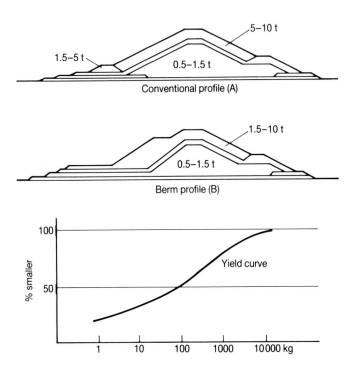

Figure 265 *Example breakwater and quarry yield curve*

		Conventional type		Berm type	
Category	Average yield (%)	Volume required (%)	Volume to be produced (%)	Volume required (%)	Volume to be produced (%)
Filter	—	11	—	11	—
Core	70	45	350	48	98
500–1500 kg	15	16	75	20	21
1.5–5 t	10	3	50	21	—
5–10 t	5	25	25	—	21
TOTAL	100	100	500	100	140

It is not unrealistic to assume that the costs for drilling, blasting and handling in the quarry are 30% of the toal costs of quarry stone in the stockpile. Since these total quarry stone costs represent 25–35% of the unit cost in the breakwater profile for both profiles, the excess production in the quarry of 400% for the conventional breakwater compared with 40% for the berm breakwater represents an extra cost of 36% in this case.

Box 75 Example of breakwater construction logistic simulation

Simplified simulation of a logistic system for the breakwater construction scheme shown in Figure 266, with three rock size categories, a dedicated quarry at 25 km and buffer stocks at both the quarry and the site.

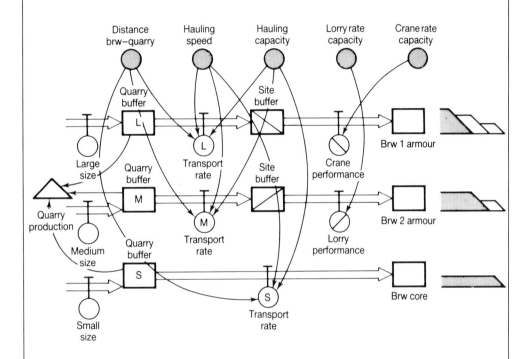

Figure 266 *Simulation diagram for breakwater and quarry operations*

Total blast volume	= 100% if one of site buffers ≤ 0, otherwise 0%
Large-size production	= 0.2 ∗ total blast volume
Medium-size production	= 0.3 ∗ total blast volume
Small-size production	= 0.5 ∗ total blast volume
Transport rates	= hauling speed/distance quarry to brw to quarry ∗ hauling capacity, or 0 if side buffer ≥ 100%
Quarry buffer	= quarry buffer + Δt ∗ (production−transport)
Lorry performance	= 100% lorry rate capacity, or 0 if site buffer medium ≤ 0 or brw 2 armour/brw core ≥ 0.9
Site buffer medium	= site buffer medium + Δt ∗ (transport−lorry performance)
Crane performance	= 100% of crane rate capacity, or 0 if site buffer large ≤ 0 or brw 1 armour/brw 2 armour ≥ 0.9
Site buffer large	= site buffer large + Δt ∗ (transport−crane performance)
Brw 1 armour	= brw 1 armour + Δt ∗ (crane performance)
Brw 2 armour	= brw 2 armour + Δt ∗ (lorry performance)
Brw core	= brw core + Δt ∗ (transport rate small)

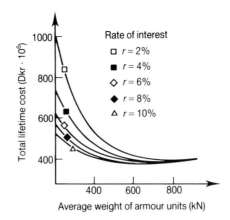

Figure 267 *The variation of expected total expenses of a breakwater armour layer during a lifetime of 100 years, depending on the armour unit weight, W, and rate of return, r*

6.1.6.3 Examples of cost optimisation in design and construction of breakwaters

As described in Section 2.1.7, optimisation takes place in three phases. Of most interest here in terms of the structural design of rock breakwaters are the second and third phases: structure optimisation and construction process optimisation, respectively. This section gives worked examples for these two phases of breakwater cost optimisation, drawing from those presented in Burcharth and Rietveld (1987).

General considerations

In the second stage of design, of the costs (2)–(5) listed at the beginning of Section 6.1.6, only design and construction costs can be calculated accurately whereas maintenance, repair, downtime costs etc. are much more difficult to estimates due to the randomness of the environmental loads and unforeseen developments within the structural lifetime, which is usually of the order of 50 years or more. Often in the design process only cost items (2)–(4) are considered, simply because the extra costs (5) are not relevant or are almost impossible to estimate. In such cases the design should represent a cost minimum as indicated in Figure 12 of Section 2.1.7.

The examples given in this section will illustrate, for instance, that overdesign of the main armour layer can in some cases give a direct reduction in the construction costs. For a quantification of the principles of cost optimisation, a stochastic model should preferably be used (Nielsen and Burcharth, 1983). An example of the use of such a model for armour layers is shown in Figure 267. Note the very flat minima which support a conservative choice of armour unit weight.

Core material

In the design much emphasis is laid on the stability of the armour layer of the breakwater. However, regarding the construction and the total costs the armour layer is often not so important. The core, toe protection and secondary armour layers of quarry rock are at least as important. This can be demonstrated with an example of the relative costs of the construction materials used in Zeebrugge, Belgium (see Table 38). Figure 268 shows the typical cross-section of the breakwater. As can be noted, the core takes almost 50% of the costs of the structure (excluding seabed preparation and protection) and the armour layer only 15%. Related to the cost of the whole structure the armour layer cost is only approximately 9%.

In Zeebrugge the core material is the cheapest grade of stone, being quarry run of 200–300 kg. This grade of stone, however, limits the workability to a significant wave height of 1.2 m, which is exceeded during approximately 20% of the time.

Table 38 Relative construction costs for Zeebrugge breakwaters

Material	Costs of materials per cubic metre of breakwater in percentage of core material
Sand	8
Gravel	67
Core 200/300 kg + 1/3 t	100
Secondary layer 1/3 t	95
Armour layer 25 t cubes	72
Berms 1/3 t and 3/6 t	120

	Costs in percentage of total/m			Structure excl. seabed preparation and protections alone (%/m)
	Materials	Execution	Total	
Soil replacement + compaction	0	92	92	
Bottom protection	139	201	340	
Toe protection (berms)	58	42	100	18
Core	204	63	267	47
Secondary layers	65	26	91	16
Armour layer	41	45	86	15
Cap construction	12	12	24	4
Totals	519	481	1000	100

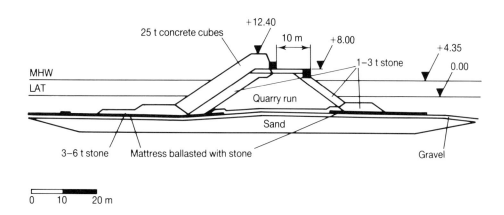

Figure 268 Typical cross-section of outer breakwaters of Zeebrugge, Belgium

Therefore it was worth trying to reduce the downtime using heavier but more costly grades of stone when the waves exceed the limit for quarry run. In this way important savings could be made, as the following example may show.

The toal costs for the construction of the core can be approximated by

$$C = Q*u + n*C_F + (n-x)*C_0$$

in which

Q = total quantity of stone (t)
u = unit rate (BEC/t)
n = total number of necessary working days
C_F = fixed costs of operations (BEC/day)
$(n-x)$ = total number of workable days, x = days of delay due to unfavourable weather; $(n-x) = Q/p$ being the daily production (t)
C_0 = operational costs (francs/day)
BEC = Belgian franc corresponding to approximately U.S.$0.026

Table 39 Example of influence of grading of core material on costs

Quarry stone grade	Limiting Wave height (m)	Exceedence%	n−x=Q/P Days	n Days	x Days	u BEC/t	ū(mean) BEC/t	C mill. BEC	C %
200/300 kg	1.2	20	1000	1250	250	600	600	15 350	100
200/300 kg + 1000/ 3000 kg	1.2 2.5	20 5	1000	1053	53	600 800	632	14 163	92
Deviations		15		−197	−197	+32		−1187	−8

With the following realistic figures:

C_F = 7 million BEC/day
C_0 = 3 million BEC/day
Q = 6.10^6 t
p = 6000 t/day

the total construction costs can be calculated as shown in Table 39. This is done for two types of stone, 200/300 kg and 200/300 kg plus 1000/3000 kg for bad weather conditions. The grade 1000/3000 kg is stable up to significant wave heights of about 2.5–3.0 m. As can be seen, a direct cost reduction of 8% (representing over a billion francs in the example) can be achieved. Moreover, 16% is saved in time, which also represents a considerable cost saving.

A breakwater is, in most cases, the first part of a harbour development and almost by definition the most critical from a construction point of view. The necessary time for the construction of breakwaters is often critical—especially on exposed locations—and may influence the total costs signficantly. The design has a great influence on this. From Table 39 it is seen that the possibility of using heavier core material during rougher sea states reduced the construction time by 16%.

Armour units

Although the armour layer forms the minor part of the total costs, the choice of its elements has a great influence because it affects not only the construction costs but also the construction period and the maintenance costs (see Figure 269). The following example illustrates the influence of armour unit weight on construction time.

In a conventional design the seaward slope of a rubble-mound breakwater is protected by rock or concrete blocks. As these blocks tend to be heavy and the placing has to be done very carefully in a predefined way, the progress of the dam front is defined by the time needed to place the blocks. Each block takes about the same time to place independently on the weight (within a fairly wide range), and thus the progress depends on the number of blocks per metre dam to be placed. A steep slope with a small number of heavy blocks permits a quicker progress than a gentler one with a great number of smaller blocks.

For instance, a slope of 1:1.5 with cubic blocks of 30 t and a slope length of 28 m has 6.1 blocks per metre breakwater. A slope of 1:2 with an equivalent block of 22.5 t needs 9.1 block per metre breakwater. With the same frequency for the placing of the armour units the second concept consumes 50% more time. This leads to a difference of 50% in construction time or would require an additional crane for placing rocks. In some cases it might even be profitable to overdesign the block weight, thus reducing the number of blocks to be placed, to save time.

These considerations tend to give preference to heavy armour units on steep slopes. Of course, heavier blocks may require a larger and more expensive crane, which puts restrictions on the argument. Moreover, a possible change in the damage sensitivity must be considered. Even with constant slope, the use of oversize armour

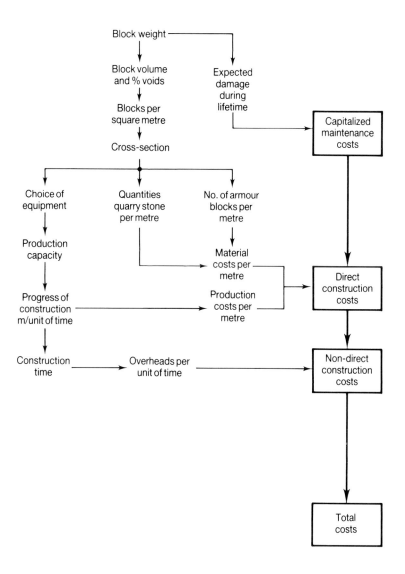

Figure 269 *Influence of choice of armour block on breakwater costs*

units can often be advantageous, as illustated in Table 40 for an example where a 25 t cube would suffice for stability reasons but overdesign to 30 t cubes gives a slightly cheaper result.

In this special case the overdesign would save 3% in money and 10% in time. On top of this might come the economic benefit from an earlier completion of the project. The construction strength/resistance, expressed in terms of significant wave height, would also be increased by a factor of $(30/25)^{1/3} = 1.063$ and the capitalised maintenance costs would be reduced to $(45/217).100\% = 21\%$.

Crest width

If land-based equipment is used for the placement of the armour the width of the breakwater should be determined with due consideration to the possibility of establishing a construction road wide enough to allow stone dumpers to pass the crane and to turn. In the case of the Zeebrugge breakwater (Figure 268) the 10 m wide crest of the completed structure is not sufficient in this respect. Therefore, either a wider structure must be designed or a construction road established at a lower level. However, a lower level means more downtime due to overtopping. A compromise was found at level +6.8 m, where the total width of the core plus the adjacent filterlayers is 13.7 m, enabling an American Hoist 11.310 crane to work and dumpers to pass (Figure 270). This lower level also reduced the reach of the crane

Table 40 Cost comparison of use of 25 t and 30 t cubes as armour units

Unit weight armour	t	25	30
Length of breakwater	m	4000	4000
Expected damage during lifetime in number of armour units		1250	221
Capitalised maintenance costs	mill. BEC	217	45
Block volume	m^3	10.4	12.5
Percentage voids in armour layer	%	45	45
Armour units per m^2 of slope		0.254	0.225
Number of armour units per m breakwater		7.1	6.3
Quantity of quarry stone per m breakwater	ton	900	900
Material costs per m breakwater	mill. BEC	0.836	0.854
Progress of construction in m per month	m/month	80	90
Production costs per m breakwater	mill. BEC	0.327	0.322
Direct construction costs per m breakwater	mill. BEC	1.163	1.176
Direct construction costs (4000 m)	mill. BEC	4652	4704
Construction time for 4000 m	months	50	45
Overheads per month	mill. BEC/month	14	14
Overheads for 50 months (4000 m)	mill. BEC	700	630
Total costs for 4000 m breakwater	mill. BEC	5569	5379
Percentage costs			
Capitalised maintenance costs		4%	1%
Direct construction costs		84%	85%
Total costs after overheads		100%	97%
Construction period		100%	90%

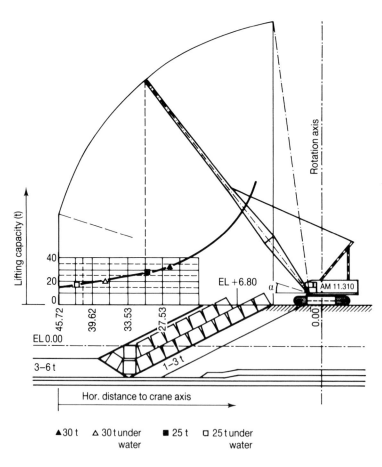

Figure 270 *Construction road, Zeebrugge breakwater*

necessary to place the concrete blocks at the toe of the slope. Note that if the fine material from the surface of a construction road fills the voids in the filterlayers it should be removed before completion of the structure.

Although there may be a specific design criterion for the crest width, it is relevant to evaluate this width as function of the accessibility for construction and future repairs. The elements to be considered are the costs of:

- Placing armour units with a land-based crane
- Placing armour units with a floating crane
- Increased or decreased crest width

Provision of good accessibility, either for land-based or floating equipment, is essential, The width of access roads and cap structures should cope with the equipment necessary and available for the placing or armour blocks. For floating equipment, the water depth and the exposure are very important factors. For each type of breakwater the sections can be evaluated by drawing the accessibility and tonne-metre graphs as shown in the examples in Figure 271.

For land-based equipment increasing tonne metres require larger cranes and wider roads. Floating equipment tends to be more expensive than land-based plant, especially on the seaward side of the breakwater, where workability is more limited. The ideal case is that local authorities or contractors have the necessary non-specialised equipment to maintain and repair the breakwater. Locally available materials should preferably be used. Rock armour or pre-cast armour blocks for maintenance can be made in the stock and the client could buy some forms for production of blocks.

The costs of placing armour units depend on the equipment, the labour costs and the production. Typical sets of equipment are listed in Table 41. As can be seen, the unit rate for placing armour from the exposed seaward side is at least three times more expensive than using a land-based crane. However, working with a land-based crane requires a crest width of at least 10–14 m for medium-size breakwaters. The answer to the question of economic width of the crest depends on each individual case but there is a strong tendency for narrow crests, as can be shown an example.

We assume a breakwater with the following properties:

- Water depth 12 m below datum
- Crest at 8 m above datum
- Armour units, modified cubes, 7 units/m breakwater
- Designer criterion 2–5% damage for design conditions

and an estimated accumulated armour layer damage over the lifetime of approximately three times the damage under design conditions, i.e.

$$3.4\% = 12\% \text{ or } 0.12 * 7 = 0.84 \text{ units/m breakwater}$$

The repair costs may be double the construction costs, but as these can be discounted over the lifetime, the unit rate to be considered for repair will not differ much from the initial construction costs. Thus for construction and repair $7+0.84$ units have to be placed. When the unit rate floating $= 4 *$ unit land rate land based, the difference equals $3 *$ unit land based placing $(3 * B)$.

This has to be compared to the costs of the width of the breakwater. When the breakwater in its centre consists vertically of 18 m quarry rock covered by 2 m concrete, 1 m of the width costs

$$18 \, m^3 \text{ rock} * R + 2 \, m^3 \text{ concrete} * C$$

where, for example, $B = \text{US\$100}$ for transport and placing of one armour unit
$R = \text{US\$35}$ for supply and placing stone per cubic metre
$C = \text{US\$100}$ for $1 \, m^3$ concrete

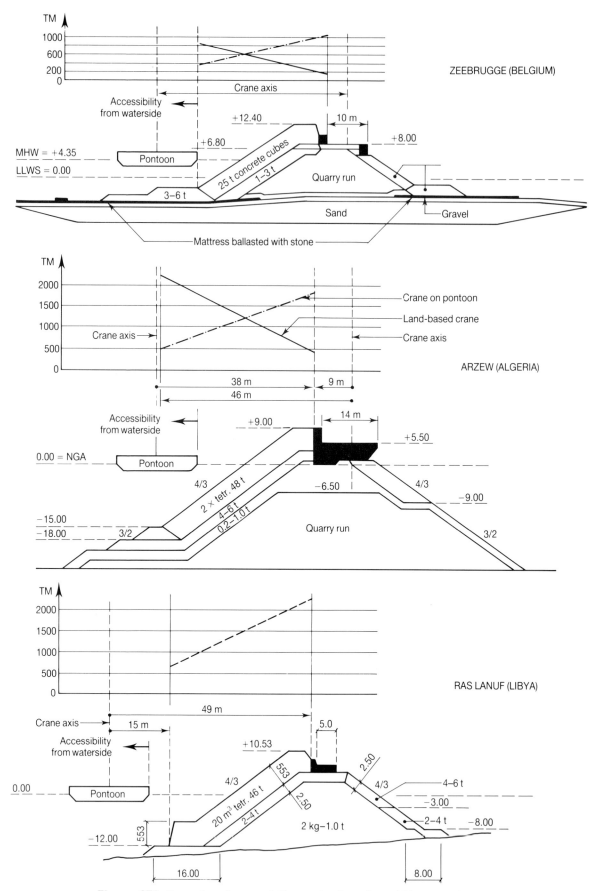

Figure 271 *Examples of accessibility evaluations for rubble-mound breakwaters at Zeebrugge, Arzew and Ras Lanuf*

Table 41 Comparison of land-based and floating cranes

Floating crane		Land-based crane	
Crane, placing armour		Crane, placing armour	
Crane barge			
Pontoon for transport (one or two units)			
Crane, loading pontoon			
Trucks, short-hauling distance		Trucks, long-hauling distance	
Crane, loading trucks		Crane, loading trucks	
Operational costs	200–300%		100%
Workability	40–70%		80%
Unit rate	300–500%		100%

B, R and C depend, of course, on local circumstances. The costs of 1 m extra width is then $18.35 + 2*100 = US\$830$. The difference in costs between floating and land-based placing is $7.84*3*100 = US\$2352$ per metre breakwater.

Thus this example shows that the difference between floating and land-based placing of armour equals roughly 3 m of width. However, for land-based placing the crest must be 10–14 m wide. The conclusion can therefore be drawn that if there is no need for a wide crest (traffic or other) the economics tend to support a narrow crest.

Referring back to the examples of Zeebrugge and Ras Lanuf shown in Figure 271, Zeebrugge has a wide crest dicatated by traffic Considerations. Ras Lanuf has a narrow one of only 5 m which, in that case, is also an economical solution. It should be noted that in the case of serious damage, generally the concrete superstructure will also be damaged (tilted seaward or pushed backwards), which sometimes makes it impossible to use it as a foundation for crane operations during rehabilitation.

6.2 Seawalls and shoreline protection structures

As with breakwaters, seawalls and shoreline protection structures built of and/or armoured with rock have a number of attractions when compared with other materials and forms of construction. These advantages include:

- Durability—most rock sources withstand wear and attrition well and are ideally suited to the coastal environment.
- As coastal rock structures are porous and generally have shallow sloping faces, they readily absorb wave energy and minimise the adverse scour consequences. In this respect, they must be seen as superior to traditional highly reflective massive vertical structures, which only tend to exacerbate beach erosion.
- The use of rock is less damaging to the limited timber resources in the earth, avoiding the use of tropical hardwoods which often involves extensive denudation of rain forests.
- Rock structures are readily modified to take account of changing environmental conditions.

In additional to the above, they also have the advantages mentioned in Section 6.1 in connection with breakwaters, particularly because:

- Even with limited equipment, resources and professional skills, structures can be built by that function successfully.
- The rock structurs are flexible, can adjust to settlements and are only damaged in a modest way if the design conditions are exceeded.
- Repair works are relatively easy and generally do not require mobilisation of very specialised equipment. If properly designed, damage may be small and repairs may only involve work to reset displaced stones.

This section concentrates on the features of seawalls and shoreline protection structures which differ from those of breakwaters, making appropriate cross-reference to the section on breakwaters where appropriate. Differences between breakwaters, seawalls and shoreline protection structures arise out of variations in function, layout, depth of water in which situated, relationship to adjacent structures and effect on coastal sediment processes.

Following a structure similar to the previous section on breakwaters, the general structural design considerations for seawalls and shoreline protection structures are given. Questions of selection of protection concept and layout are followed by discussion of armouring systems, structural details and construction cost and maintenance aspects. The section covers a range of structures, from revetments and anti-scour mats to structures designed to retain sand or shingle beaches, including conventional and fishtail groynes and offshore breakwaters and sills.

6.2.1 DEFINITIONS

Rock structures used in coastal and shoreline engineering generally have components similar to those of breakwaters described in Section 6.1.1 and illustrated in Figure 233. However, as will be seen later, they will frequently only have two gradings of stone in them because of their more modest proportions.

For this reason, unless the structure is large enough to be described as a breakwater, the outer layer is generally known as the 'armour layer' or 'coverlayer' rather than the 'primary armour layer' and the underlying rock as the 'underlayer', 'bedding layer' or core. If a third material is used in rock structures for coastal defence purposes, this will tend to be rather fine, such as sand, clay or other cliff or beach materials. The following types of trock structure will be discussed in this section.

Conventional revetment or revetted mound (dike)

A revetment has been described as a cladding of stone, concrete or other material used to protect the sloping surface of an embankment, natural coast or shoreline against erosion. In the case of interest here, where rock is the revetment material, rock may be used in its own right or with stability improvement (asphalt, pitching, gabions, mattresses, etc.). The rock may be used for protection of cliffs, sand dunes, reclamation, and existing seawalls or dikes requiring repair or renewal.

Scour protection

This comprises an almost flat layer or layers of rock placed in front of an existing seawall, cliff or sand dune to prevent further undermining of the toe of the basic coastal defence structure. It is often used in conjunction with an overlying beach- or dune-nourishment scheme, in which case the scour protection is essentially provided to guarantee the integrity of the coastal defence structure in an extreme storm situation in which the beach material may be temporarily stripped off before milder wave action allows it to recover.

Bastion groyne or artificial headland

This is a relatively short structure running seawards from the beach head, whose primary function is to interrupt the longshore transport of sediment in order to build or retain higher beach levels (and often thereby to protect an existing coast defence structure). Rock bastion groynes are typically of the order of 50 m long, but may be larger if they are terminal groynes at the end of a long ungroyned beach. A related type of groyne is the hammerhead or boot bastion which seeks to exploit a diffractive sediment retention in the lee of the groyne.

Offshore breakwater

This is breakwater, generally surface piercing (at least for most of any tidal cycle), and lying roughly parallel to the shoreline whose function is to reduce wave activity and encourage beach building at the shoreline in its lee. Sediment is transported into the lee of the breakwater by wave diffraction-induced currents. Such breakwaters generally have a length similar to their distance offshore, which is typically 200–300 m, although smaller structures are possible if required by local conditions.

Fishtailed breakwater (artificial headland)

These breakwater-like structures, which are a relatively new concept, seek to combine the attractions of offshore breakwaters with the conventional long-stop function of a groyne. Beach nourishment may be used in conjunction with such groynes to create sandy amenity beaches. Typically, these groynes may extend to 200 or 300 m offshore.

L-shaped and T-shaped breakwaters (artificial headlands)

This is a precursor of the fishtail breakwater used in situations where the tidal range is small (e.g. the Mediterranean) to create so-called 'pocket beaches' generally of sandy sediment.

Sill or submerged breakwater

Used to retain a beach of relatively mild slope on an existing, possibly steeper sloping foreshore, sills have been employed most successfully in situations where the tidal range is small. They can be used in conjunction with L- or T-shaped groynes or full-height offshore breakwaters to retain pocket beaches. They trigger the breaking of larger (most destructive) waves but have little effect on normaly day-to-day activity, so that recreational aspects of the beach are not diminished.

6.2.2 PLAN LAYOUT AND OVERALL CONCEPT SELECTION

The plan layout of a coastal or shoreline defence structure depends on its required function, planning policy decisions regarding the overall line of the coast, physical site conditions, the interrelation with adjacent shorelines, amenity/environmental requirements and benefit-cost considerations. It is also significantly influenced by the choice of material, and here it may be noted that rock structures offer significant advantages.

6.2.2.1 Functional requirements

Coastal and shoreline defences are constructed to fulfil a requirement to:

- Protect the coast against erosion; or
- Alleviate flooding by the sea.

Sometimes the defences may fulfil both functions, and can additionally be used for amenity purposes or to protect reclaimed land. In the majority of cases, protection against the effects of wave action (e.g. in terms of damage to land an property' is of prime importance. Thus the coast-protection function applies in most cases, but the flood-alleviation aspect is only of major concern when the land behind the defences is below high water.

Structures used in coastal defences can fulfil their function either directly by forming a physical barrier or line of defence or indirectly by providing protection to (or encouraging creation of) natural such as beached and sand dunes. Direct structural defences normally have to be designed to function in spite of beach behaviour. By contrast, an indirect coastal defence scheme generally functions because of the beach behaviour (as controlled by the structures provided). Of those structures defined in Section 6.2.1, conventiuonal revetments, revetted mounds and scour-protection structures are of the direct 'in-spite-of-beach' kind whereas the others (groynes offshore, breakwaters, fishtail, L- and T-shaped groynes and sills) all relate to indirect 'because-of-beach' defences. Clearly, this schematisation is simplified, and overlaps of concept and/or joint approaches occur in practical situations.

6.2.2.2 Position of shoreline

The starting point of determining the plan layout of a rock coastal defence structure (given the functional requirements for the system) must always be the existing shoreline. This is defined by the beach contours and is often taken as the high-water mark. However, the low-water mark is clearly also important as is the gradient from low- to high-water marks. Other coastal forms such as sand dunes and cliffs are also important in defining the shoreline. Frequently an existing seawall defines the present shoreline, but this shoreline may no longer be the natural position if it has been constructed on a coast which is naturally eroding or accreting.

Given the existing shoreline or line of defence, the coastal planning authority will define a policy option which will generally be one of the following:
- Withdrawal (retreat), providing new set-back sea-defence flood banks (dikes) only where necessary for safety;
- Selective erosion control, maintaining the existing defence line at key locations;
- Full erosion control, maintaining the existing line of defence;
- Seaward expansion (advance), creating more land or beach.

Selection of the appropriate policy option will be influenced by a number of factors, but the most dominant will be the benefits and costs for each option expressed in financial terms. Where residential or industrial land lies behind the shoreline it is generally possible to justify maintaining the existing line or indeed to advance it by means of reclamation. Where agricultural land is involved, the value of this is rarely high enough to justify *per se* defending an existing shoreline, unless it is low-lying and a flood-defence dike is involved. Thus the defences of a seaside town may be maintained while agricultural land between is allowed to erode (selective erosion control).

Ideally, planning policy options should be defined on a regional scale, at least for each coastal cell unit so that conflicting objectives are not sought within the same unit. A good example of regional planning is the decision of the Dutch government in June 1990 to adopt the full erosion control (*handhaven*) option for the entire 350 km of Netherlands coastline to be based on maintaining the 1990 shoreline. The main consequence of this Dutch commitment is the need for beach nourishment of some 140 km of dune coastline at a cost of 60 million guilders per year.

6.2.2.3 Layout options and design

The basic types of rock structure that can be used in a coastal or shoreline defence scheme have been defined and described in outline in Section 6.2.1 above. This section described how to determine the plan layout in each case. To illustrate some of the matters discussed, a series of figures are provided which illustrate the structures described in perspective and to a similar scale for the situation where there is a need to protect an existing concrete seawall.

Conventional revetment or revetted mound (dike) (Figure 272)

In general, the wall alignment should follow the average alignment of the beach contours or of the existing seawall which the mound is strengthening or replacing. Where the new structure is a replacement for a previously collapsed or failed structure it will be necessary to check that the wall alignment was not a constributory factor in the collapse.

Plan layout of seawalls which embody re-entrant angles or significant concave curvatures with respect to incoming wave crests can result in a focusing of reflected

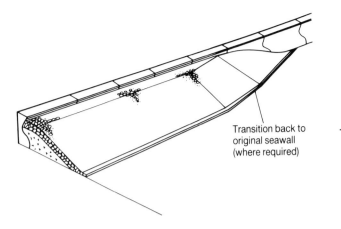

Figure 272 *Conventional rock revetments*

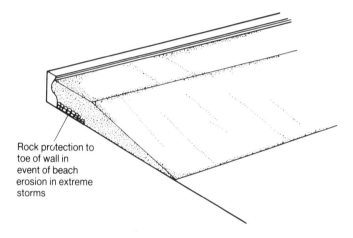

Figure 273 *Scour protection*

wave energy in a limited area offshore, with potentially deterimental effects on beach processes. Layouts of seawalls which embody significant convex curvatures can be subject to oblique wave attack on the convex curve. This can give rise to a so-called 'mach-stem' wave, which is higher than the incident waves because of the local compression of the wave front. The mach-stem wave runs along the face of the wall with potentially severe scouring effects. Convex curves can also be associated with parts of walls which protrude into marginally deeper water and can be subject to more severe wave actiuon and scour for this reason alone. Sloping rock revetments will, however, tend to ameliorate the most severe effects of concave and convex walls because of their reduced reflective and increased energy absorption capabilities.

Scour protection (Figure 273)

This will follow the alignment of the wall to be protected. It is often associated with a gravel or shingle nourishment scheme (see Section 6.4).

Bastion groyne (Figure 274)

The groyne concept is essentially based on allowing the coastline between two groynes to reorientate towards the predominant waves, thereby reducing longshore sediment transport. The exact length, orientation and spacing of a bastion groyne or groynes depends on a number of factors, including the extent of beach retention required and the size and hence slope of the beach material to be retained. For detailed advice on the principles and effectiveness of the use of groynes, the reader is referred to the *Guide on the Uses of Groynes in coastal Engineering* (Fleming,

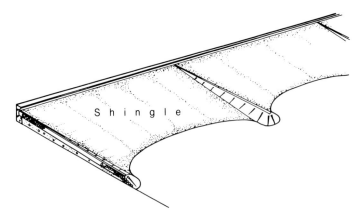

Figure 274 *Bastion rock groynes retaining shingle beach*

1990a) recently published by CIRIA. There are no simple and absolute rules for groyne length and spacing as these critically depend on local conditions (beach material, water depth, wave climate, availability of beach sediment, longshore and onshore/offshore transport regimes, etc.) It is therefore essential for experienced specialist engineers to be consulted for relevant studies, field work and execution of detailed design.

The groyne layout will relate primarily to the beach type being addressed. Four principal beach types have been distinguished for which groynes might be considered (see Figure 275):

- Shingle
- Shingle upper/sand lower
- Shingle/sand mixed beach
- Sand beach

Section 5.13 of the CIRIA report on groynes (Fleming, 1990a) also provides advice on the application of groynes to each of these beach types. Most of the advice is general but is often related to the timber groyne fields which historically were prevalent in the UK.

Important differences in performance exist between timber and rock groynes, rock groynes being more permeable, less reflective and tending to provide (unlike timber groynes) some diffractive sediment trapping in their lee. As a consequence, while groyne lengths will be similar, being largely dictated by such matters as beach slope and width, groyne spacing may be different.

For shingle beaches, groyne spacing/length ratios are determined from the range of probable nearshore wave directions which will cause a reorientation of the beach, the beach in each groyne cell tending to align itself rapidly to be parallel to the prevailing wave crests. Timber groyne spacing/length ratios of 1:1 were typically found to be most effective. With rock groynes, however, a more cuspate or crenulate beach form develops in the shingle because of the combination of groyne permeability and less diffractive capability. This allows spacing/length ratios for rock groynes to be increased to 2:1 as a typical value (Figure 276) without undue risk to the beach head. Cost comparisons between timber and rock groynes (see example in Box 79) should also take account of these spacing differences. Groyne length will be dicated by beach width and slope, the latter being typically 1:8, reducing to about 1:12 in extreme storms.

For sand beaches, when the mean grain size is of the order of 0.15 mm or less (medium-fine sands), beaches are typically extremely flat with slopes of the order of 1:100 and dominant transport mechanisms are onshore-offshore rather than longshore. Here conventional groyne fields which attempt to control the sediment movement are relatively ineffective, and concepts such as offshore breakwaters or

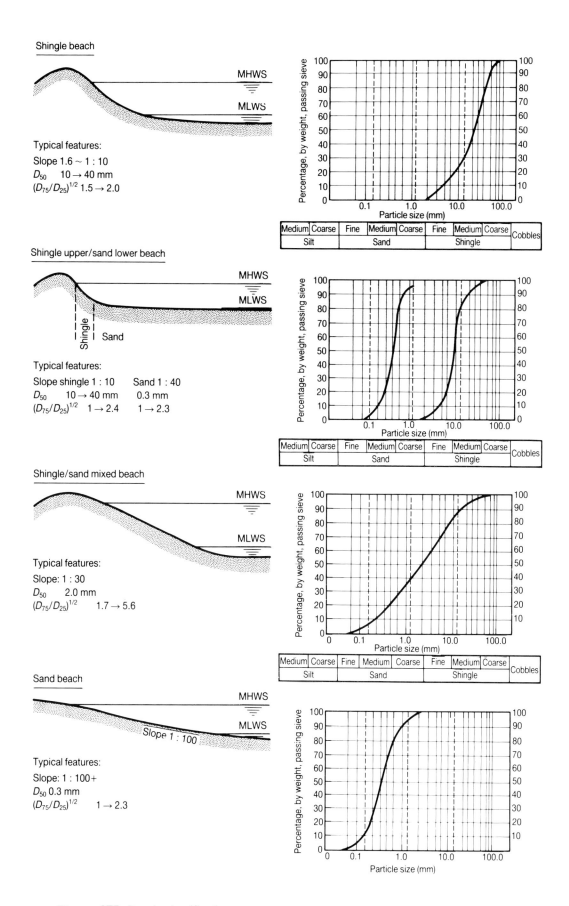

Figure 275 *Beach classification*

Figure 276 *Rock groynes on shingle beach, Barton-on-Sea, UK (courtesy J.D. Simm, Robert West & Partners)*

fishtail groynes which control the wave climate will probably be more appropriate (see below).

For medium-coarse sand beaches conventionel groyne fields are possible. Since significant current circulations will exist in each groyne bay, addition of a modest T or Y head to the groynes is desirable to reduce wave-induced rip currents against the groyne stem. Groynes also need to be long in this situation if they are to control the entire beach profile. Historically, timber groynes were rarely taken beyond mean low water of neap tides because of construction and maintenance cost constraints, and the need for sand rock groynes to have crests which follow the beach profile means that similar length constraints apply. Alternatively, groynes can simply be designed to control the upper beach profile, selecting a length of groyne which will envelop the breaker line for a moderate summer wave climate when the beachis building up. A good starting point for the spacing-to-length ratio of groynes for sand beaches is 2:1, this applying equally to timber and rock structures.

Where beaches consist of a distinct shingle bank or ridge setting on top of a sand base (shingle upper/sand lower), a groyne system should be regarded as consisting of two distinct components. For the upper beach, the groynes are designed for a shingle beach as described above. If the sand is fine to medium and the beach is flat, it is doubtful whether extension beyond the maximum drawdown point of the toe of the shingle will be beneficial. However, where the lower beach is medium to coarse sand, some groynes may be extended to act on the lower beach according to the principles described above.

Where beaches are a mixture of shingle and sand, recent investigations on the east coast of England (Fleming and Lunt, 1988) have led to the conclusion that relatively short groynes (40-50 m) are sufficient to hold the existing upper beach. The reason for this positive behaviour is thought to lie in the fact that even on a mixed material beach, the shingle elements are sorted and found more in the material on the upper sections of the beach. Retention of this coarser material appears to be sufficient to encourage the growth of a stable overall beach profile.

Special considerations are required when dealing with the last groyne in a system or with an isolated groyne on an otherwise ungroyned beach. Such terminal groynes may fulfil two functions:

1. Preserving the natural or nourished beach on the updrift side;
2. Arresting the longshore drift to prevent siltation in an inlet to a tidal estuary, creek or harbour.

The terminal groyne might deliberately be made longer and higher than other groynes in order to create a reservoir of drift material which can be mechanically transported updrift to nourish depleted beaches. In other locations it may be more important to reduce the immediate impact of downdrift erosion, resulting from retention in the groyne field of littoral material which would otherwise reach the downdrift beaches. In this situation the groynes should be made progressively shorter at the downdrift termination and boot heads pointing in the downdrift direction provided to encourage diffraction accretion in their lee. However, if the groyne bays are nourished artificially at the outset, the potential for downdrift erosion will be reduced significantly.

Offshore breakwater (Figure 277)

As mentioned above, rather than physically trapping sediment by the long-stop method, offshore or detached breakwaters operate by causing a zone of reduced wave energy behind the breakwater in which sediment will tend to deposit and crescentic beaches are thereby formed. Isolated breakwaters may be particularly useful to protect lengths of coast where erosion is occurring because net longshore transport rate is higher than elsewhere, ideally bringing the net longshore transport rate in line with adjacent coasts. Offshore breakwaters have been used with most success on coastlines where the tidal range is negligible or small. They also offer considerable attractions as opposed to groynes (see above) when applied to wide foreshores of fine sand where the dominant sediment transport mechanism is onshore–offshore.

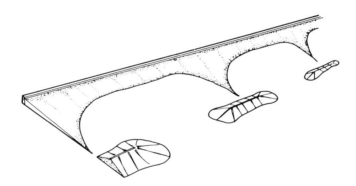

Figure 277 *Offshore breakwaters retaining sand beach*

To understand how offshore breakwaters operate, consider an expression for the volumetric rate of sediment transport in its simplest form (Fleming, 1990b):

$$Q = (H^2 C_g)_b \left(a_1 \sin 2\alpha_{bs} - a_2 \cos \alpha_{bs} \frac{\partial H}{\partial Y} \right)_b \qquad (6.1)$$

where H is the wave height, C_g the group velocity, α_{bs} the breaker angle, $\partial H/\partial Y$ is the alongshore gradient of the wave height and a_1, a_2 are empirical parameters dependent on sediment size, density, beach slope, etc. This expression recognises that there are two components driving the sediment: (1) the transport due to waves breaking obliquely to the shoreline and (2) transport by currents caused by wave height gradients. Clearly, in the case of an offshore breakwater the wave heigh graidient creates current into the lee of the structure, irrespective of the incident wave direction. These, combined with reduced wave heights, result in deposition of

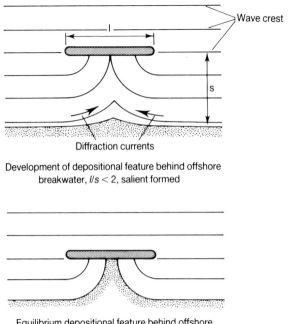

Figure 278 *Conditions for tombolo formation with decreasing relative distance from breakwater to shore*

material. There must also be a corresponding outflow which occurs immediately landward of the structure.

Thus in the absence of other influences beach material will be transported into the area to form a tombolo or salient. Depending on the dimensions of the structure and its distance offshore relative to the wavelength of the incident waves (and the gap between adjacent structures if there is more than one), a tombolo may or may not attach itself to the structure (see Figure 278, after Dean, 1988).

An offshore or detached breakwater should be located roughly at the beginning of the breaker zone, which will allow it to influence the inner half of the active littoral zone. Typically, it should be at least three inshore wavelengths from the coastline, based on a wavelength calculated at a point about one wave length from the breaker line. The length of the breakwater(s), and their distance apart if more than one is being considered, are a function of the required beach form (see Figure 278). Tombolo formation may be attractive if a 'pocket beach' structure is required (see Box 76) and in this situation offshore breakwaters can be used in conjunction with sills. In other situations less modification of the beach profile and interruption of longshore transport may be appropriate, and here typically the length may be set as being roughly equal to the distance offshore.

The spacing of the breakwaters adopted is essentially a function of the required reduction in inshore wave energy required, even though rules of thumb such as those for pocket beaches may be applied initially to assess rough proportions. This reduction in energy is affected not only by the spacing or opening size but also by the crest elevation of the breakwaters (see Section 6.2.4.4), the resulting wave energy due to wave penetration through the openings and wave transmission across the breakwaters generally being determined by calculating the two components separately and then adding their effects.

The alignment of the breakwater(s) should not necessarily be parallel to the local coastline, particularly if a single dominant wave direction, or limited spread of wave directions, exists. In the latter case it may sometimes be appropriate to set the line

Box 76 Design of pocket beaches

Berenguer and Enriquez (1988) published a useful set of design equations for pocket beaches where the tidal range is less than 1 m based on an analysis of 24 Spanish beaches such as Rihuete Beach, Murcia, shown in Figure 279. With reference to the definition sketch (a), the following relations were obtained:

$$A_0 = 2A_1$$
$$A_1 = 25 + 0.85S$$
$$X \cdot B_0 = 2.5 A_1^2$$
$$S_p = X \cdot B_0 - (\pi A_1^2 / 2)$$
or
$$S_p = 0.37 X \cdot B_0$$

where S_p is the maximum stable surface area of the beach

In addition, a graphical relationship was developed between the cross-sectional surface area of the gap $(S \cdot d_g)$ and the dimensions of the resulting beach $(A_1^2 \cdot D_m^{1/2})$, as shown in (b).

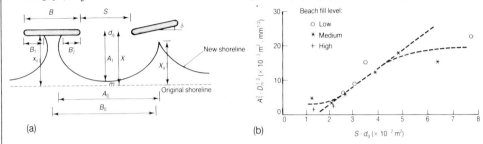

Figure 279 *Rihuete Beach, Murcia, Spain, after creation of pocket beaches (courtesy, Director Gen. Ports & Coasts, Spain)*

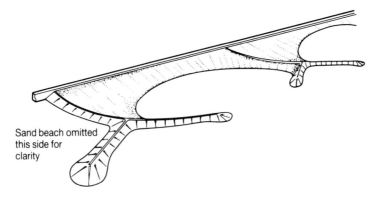

Figure 280 *Fishtailed breakwaters retaining sand beach*

of the breakwater(s) parallel to the wave crests, so long as practical (construction and cost) considerations make this possible.

None of the rules of thumb that have been developed for determining the relative dimensions and positioning of offshore breakwaters can be applied to coastlines where there are high tidal ranges and/or high tidal currents. Clearly, any shore parallel current that can pass between the breakwater and the beach can negate the wave-induced current effect and flush the material from behind the structure. This can be reduced or eliminated by making a connection between the offshore breakwater and the beach either by a causeway or a submerged reef-type structure. In many cases the former may be built as part of the temporary works to facilitate construction so that the additional costs are relatively small. This type of structure development leads into the consideration of the alternative of artificial headlands.

Fishtailed breakwater (artificial headland) (Figure 280)

The fundamental difference between a groyne and an artificial headland is that the latter is a more massive structure designed to eliminate problems of downdrift erosion and promote the formation of beaches. While these structures may take a number of different forms, their geometry is such that as with the offshore breakwater, wave diffraction is used to assist in holding the beach in the less of the structure. The fishtailed breakwater is a particular development of this concept, largely due to Dr P. C. Barber in the UK (Fleming, 1990b).

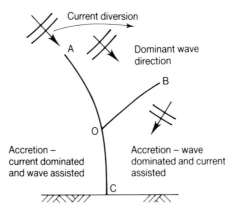

Figure 281 *Basic geometry of the fishtailed breakwater*

The concept of the fishtailed breakwater is to combine the beneficial effects of the groyne, offshore breakwater and tombolo and reduce the undesirable influences of the separate structures. The basic plan shape of a fishtail breakwater is shown in Figure 281. The breakwater arms OA and OB act as wave energy dissipators while the arm AOC provides the interceptor to alongshore drift. Thus the updrift beach is formed by normal accretion processes associated with a groyne while the downdrift beach is formed by those associated with an offshore breakwater.

The arm AC, which acts to intercept and divert offshore, alongshore and tidal currents to minimise beach erosion, is curved in plan so that the axial alignment at A is normal to the streamline of the diverted alongshore and tidal currents. The axial alignment of C is generally normal to the shoreline. The curvature of COA is designed to minimise wave reflection effects on the concave side of the breakwater and an consequent scouring. With its shallow sloping sides and porous structure (see Section 6.2.4.5), the arm OA also encourages storm waves to diffract onto the structure and to pump sand into the angle between COA and the shoreline. The arm OB is orientated approximately parallel to the most severe storm wave crests. It is located in plan sufficiently inshore from A to allow wave distance to transform out of the current field. The length OB is partly dependent on the length of OC, but mainly dependent upon achiving the desired wave diffraction effects.

The overall breakwater dimensions are thus interdependent and depend on wave height, direction and period, tidal range, beach morphology and the extent of required influence. The distance of A offshore depends on the length of coast the breakwater is intended to influence, but should be greater than three inshore wave lengths as well as less than half of the width of the active littoral zone.

Wide, shallow-sloping roundheads (see Sections 6.2.4.4 and 6.2.4.5) are provided at A and B and have two functions:

1. To improve the efficiency of the structure in diffracting waves, thus reducing their energy and assisting in natural beach accretion;
2. To provide a transition between seabed and breakwater arm, reducing the tendency for wave reflection and helping to prevent scouring of the seabed by tidal currents.

The fishtail breakwater can be expected to influence the beach in a number of ways. There is usually a small steepening of the beach gradient due to current reductions caused by the breakwater. The beach may form a crenulate bay form if the wave conditions are dominantly from an oblique wave direction. However, more often multi-directional conditions exist and more complex geometries will evolve. Where there is a high tidal range and a varied wave climate the beach will be constantly changing in plan level and gradient. An example of the application of fishtail breakwaters is shown in Box 77.

L- and T-shaped breakwaters (artificial headlands) (Figure 283)

L- and T-shaped breakwaters have been widely adopted in areas of small tidal range such as the Mediterranean to enclose sandy 'pocket beaches', the layout design of which has been described in Box 76. The only difference from a pocket beach created by offshore breakwaters is that the land link is provided in advance, thereby reducing the immediate quantity of sand nourishment required to create the beach.

Sill of submerged breakwater (Figure 283)

Sometimes used in conjunction with L- and T-shaped breakwaters, sills can be adopted in areas of small tidal range to retain a beach of relatively mild slope known as a perched beach. Box 78 provides a beach design approach and should be combined as appropriate with the guidance provided in Box 76.

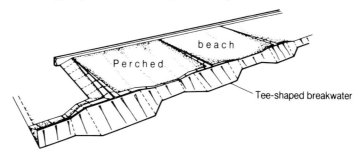

Note: This concept applicable where tidal range does not exceed about 1.0 metres.

Figure 283 *Sand beach retained by T- or L-shaped breakwaters and offshore sill*

Box 77 Clacton Sea Defence Scheme, UK

Problem

The seawall protecting a 4 km length of seafront along a low-lying area of the Essex coast in the UK was in urgent need of repair. Wave action had also eroded the protective beach in front of the wall, exposing the sheet piling which was being worn away by shingle abrasion. Reduced beach levels increased water depths and hence wave heights at the wall. The total effect was to undermine the stability of the wall, to increase wall damage due to wave overtopping and to produce tidal currents close inshore which were increasing transport of beach material.

Solution

The chosen solution was to reconstruct the seawall at Jaywick, carry out renourishment of the beach and build rock-armoured breakwaters (or fishtail groynes) to protect the new beach from future erosion. The function of the fishtail groynes were:

1. To protect the most vulnerable points of the new beach against direct wave action, as seen in Figure 282;
2. To modify the oncoming waves into a less oblique direction; i.e. to reduce the longshore drift;
3. To divert the tidal currents a sufficient distance offshore to prevent them from moving material drawn down from the recharged beach.

Benefit and costs

The total cost was £11 million and construction was phased over the period 1986–1988. The completed scheme protects an area of 200 ha from flooding to a 1 in 1000-year standard. The calculated benefit-cost ratio was 1.4:1.

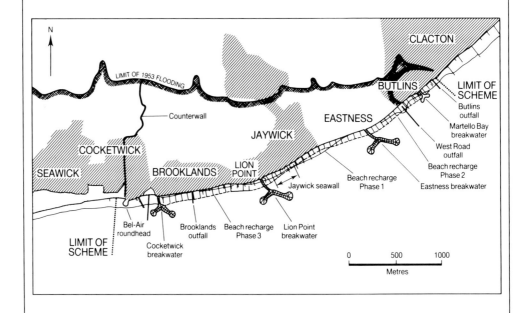

Figure 282 *Clacton sea defence scheme (courtesy Anglian Water, UK)*

6.2.2.4 Effect of selected concept/layout on beaches/coastlines

Clearly, those 'because-of-beach' concepts discussed above will influence the beach in the area being protected more or less in the desired way. However, there will be influences updrift and downdrift of the area for which the structures are being designed, unless this area is a complete and closed coastal process unit. Such influences will also exist in the case of 'in-spite-of-beach' concepts (seawalls and

> **Box 78** Design of beach using rock sill concept
>
>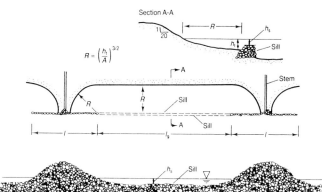
>
> (a) Definition sketch for rock sill and associated sand beach of relatively mild slope (Dean, 1988)
>
>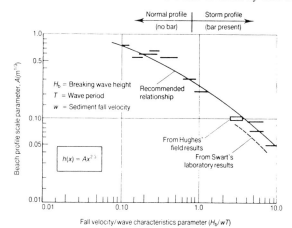
>
> (b) Correlation of equilibrium beach profile scale parameter, A, with combined sediment/wave parameter, H_b/wT (Dean, 1986)

scour mats), but in these cases there will also be a more or less unplanned influence on the beach in front of the structure.

Influences of seawalls and scour mats

Many concerns have been expressed relating to the influence of seawalls, and those have been well summarised by Dean (1986) in a useful table reproduced here (Table 42). Dean (1986) also gives methods of quantitatively assessing erosional impact and required mitigative beach nourishment.

A particular concern in the case of seawalls is always the influence that reflections from such a wall may cause in terms of local scour in front of the wall. However, to quote Kraus and Pikley (1988), an excellent summary of the state of knowledge on the subject:

> Laboratory studies, theoretical developments, and qualitative field observations all point to the conclusion that slanting seawalls and permeable, rubble mound seawalls, which have smaller reflection coefficients than vertical walls, suffer less local scour than vertical or near-vertical walls. In some cases, 'softer' structures

Table 42 Assessment of some commonly expressed concerns relating to coastal armouring (Dean, 1986)

Common		Assessment
Coastal armouring placed in an area of existing erosional stress causes *increased* erosional stress on the beaches adjacent to the armouring	TRUE	By preventing the upland from eroding, the beaches adjacent to the armouring share a greater portion of the same total erosional stress
Coastal armouring placed in an area of existing erosional stress will cause the beaches fronting the armouring to diminish	TRUE	Coastal armouring is designed to protect the upland, but does not prevent erosion of the beach profile waterward of the armouring. Thus an eroding beach will continue to erode. If the armouring had not been placed, the width of the beach would have remained approximately the same, but, with increasing time, would have been located progressively landward
Coastal armouring causes an acceleration of beach erosion seaward of the armouring	PROBABLY FALSE	No known data or physical arguments support this
An isolated coastal armouring can accelerate downdrift erosion	TRUE	If an isolated structure is armoured on an eroding beach, the structure will eventually protrude into the active beach zone and will act to some degree as a groin, interrupting longshore sediment transport and thereby causing downdrift erosion
Coastal armouring results in a greatly delayed post-storm recovery	PROBABLY FALSE	No known data or physical arguments support this
Coastal armouring causes the beach profile to steepen considerably	PROBABLY FALSE	No known data or physical arguments support this
Coastal armouring placed well back from a stable beach is detrimental to the beach and serves no useful purpose	FALSE	In order to have any substanial effects to the beaches, the armouring must be acted upon the waves and beaches. Moreover, armouring set well back from the normally active shore zone can provide 'insurance' for upland structures against severe storms

appear to have mitigated local scour and allowed the beach to respond in a manner similar to that of a natural sand beach over the erosion and recovery cycle associated with impacts of storms.

The influence of rock scour mats is restricted to times when such mats are exposed above the beach profile. Their influence in such situations is not explicitly known, but their energy-dissipating capability must reduce some of the erosive power of the incident waves and perhaps encourage sediment to be retained as wave climate reduces after storms.

Influence of groynes and breakwaters on updrift/downdrift beaches

The extent of influence of such structures depends on a number of factors, including the erosional stress in the area, the longshore transport rate and the extent of beach nourishment associated with the scheme. However, there will normally be an interruption of greater or lesser extent to the longshore transport proccess, leading to updrift accretion and downdrift erosion, and the associated beach plan form can be estimated for example using one-line numerical models developed according to the principles set out by Pelnard-Considiere (1956) and discussed in Fleming (1990a), or the *N*-line models recently developed by Perlin and Dean (1985).

Beach nourishment provided in association with a groyne field, artificial headland or offshore breakwater scheme can have several advantages:

1. The effects of downdrift erosion can be mitigated;
2. Longshore transport rates can be maintained at the appropriate levels right across the frontage; and
3. The influence on adjacent beaches can be minimised, while still retaining the desired new beach profiles on the frontage.

6.2.2.5 Environmental/amenity considerations

These considerations will have a significant influence on the concept and layout selection for a coastal defence scheme and should always be considered as an integral part of the selection process. The following factors might require consideration:

Amenity value of beach

At the simplest level this factor will influence the degree of access provided to the beach, which may involve construction of purpose-built paths or steps, replacing or stabilising the rock structures as appropriate with concrete/asphalt. At a more fundamental level, amenity value may ditate. the whole coast defence concept if swimming, sunbathing and other beach activities are the prime concern. Indeed, as with the Mediterranean pocket beaches, amenity use may be the main reason for constructing the breakwaters/groynes/sills in creating such beaches with rock structures, consideration must be given to the possibility of seaweed in the lee of such structures which may give, at least in the short term, undesirable odours as it decays.

Amenity value of rock structures

Potential amenity use of coastal rock structures includes walking, sunbathing and fishing (see Figure 284). Groynes are particularly popular for walking (because of the 'walk to the end of the pier mentality') and for fishing, as they provide immediate access to deep water. However, because of their popularity, careful consideration must be given to safety matters. Rigid adherence to hydraulic requirements may in fact lead to a structure which is unsafe for people to walk on and if this is essential, access to the structure should be forbidden. However, the adventurous spirit of the beachgoer should be acknowledged and it may therefore be better to avoid over-smooth stone and too open a stone packing arrangement, so that walking and clambering over the stones can be relatively safe. The main problem encountered with rock structures is leg injuries caused while scrambling

Figure 284 *Use of rock groyne for fishing, Melford-on-Sea, UK*

Figure 285 *Offshore breakwater, Rhos-on-Sea, UK (courtesy C. Durrant)*

around the rocks, but this has to be compared with the risk of being injured or drowned falling from the top of a vertical wall or barrier groyne. Offshore breakwaters and large groynes are also popular in terms of providing shelter for small pleasure and fishing craft, even though they may not have been deliberately designed for this purpose (see Figure 285).

Visual impact

The visual impact of coastal rock structures is inevitably considerable but may have attractions in providing a 'landmark' for the area. Nonetheless, however constructed, on close inspection they seldom look elegant, but, once completed and their surface has weathered, they eventually blend in well with the marine environment. This blending in is often assisted by their capacity to encourage beach material to build by absorbing wave energy. This resulting accretion may often substantially cover the rock structures and, when combined with a growth of vegetation on the upper slopes, can render such structures much more acceptable than they may have been immediately after completion of construction.

Ecological value

Generally, the direct ecological consequences of rock coastal defence structures are positive, and include:

- Growth of vegetation;
- Colonisation by crabs, fish and other marine life which is attractive to both sports and commercial fishing interest;
- Provision of additional roosting areas for wading birds, thereby encouraging bird life.

The primary negative consequence is normally covering up of some life form or source of food, but this is rarely so widespread as to cause irreparable harm..

A further indirect value of rock structures such as groynes and breakwaters is that they avoid the use of the main alternative material, timber, which for durability reasons normally must be a troprical hardwood. Avoidance of denudation of the world's resources of rain forest is now understood to be a prime environmental objective and use of rock rather than timber can assist here.

Archaeological/geological interest

Rock structures may be valuble if they preserve items of archaeological interest. However, conflict may arise if archaeological/geological/palaeological features are covered up or erosion which exposes fresh geological features is reduced. Such conflicts will apply, however, whatever material is used for the coastal defence scheme.

Construction aspects

Environmental restrictions on construction operation may influence the concept selection. These may include requirements to restrict water and air pollution, noise and traffic. Traffic restrictions may dictate supply of rock to the site by sea (see Section 6.2.3.4).

6.2.3 GENERAL CONSIDERATIONS FOR CROSS-SECTION DESIGN OF ROCK COASTAL STRUCTURES

Having selected a particular coastal defence concept and layout for evaluation/ detailed design, the next step is to determine the cross-section of the structure. This may be more or less constant along its length or may vary in dimensions and even in materials.

Like the concept and layout, the cross-section will be determined by various functional requirements (Sections 6.2.2.1 and 6.2.3.1); physical and planning policy boundary conditions (Sections 6.3.2.2 and 6.2.3.2); amenity and environmental considerations (Section 6.2.2.5), including local experience of comparable construction projects, materials availability and supply (Sections 6.2.3.3 and 6.2.3.4); constructions considerations and local experience (Section 6.2.3.5 and 6.2.5); and maintenance considerations (Sections 6.2.3.6). As with breakwaters, if these considerations still permit alternative designs to be considered, then a final selection can then be made on the basis of cost, taking into account the resultant benefits (see Sections 6.1.6, 2.1.5 and 2.1.7).

This section discusses some of the factors mentioned above which constrain selection. It also described the design and selection of the armour layer which is of general application to all coastal structures. Detailed dimensioning and practical details for each of the coastal defences concepts discussed in this chapter is then presented in Section 6.2.4, which provides further design guidelines in addition to those presented for breakwaters in Section 6.1.4.

Overall hydraulic and geotechnical design of the structure should be carried out using the design tools presented in Chapter 5, taking account of the potential failure modes (see particular Figure 20 in Chapter 2 and Figure 240 in Section 6.1). Detailed designs should wherever possible, be checked in a hydraulic physical model (Section 5.1.4.1), and in the case of dynamically stable structures this is essential. Alternatively, uncertainties in the boundary conditions/design formulae may be translated into increased safety factors (for example, in the case of small bastion groynes). For structures of any significant size, model tests will be cost effective and lead to optimisation of the preliminary design.

6.2.3.1 Functional requirements of cross-section

In addition to the primary functional requirements (Section 6.2.2.1), secondary functional requirements for each of the elements of the cross-section of the structure

Figure 286 *Seawall rehabilitation in progress, Morecambe, UK*

should be determined along the lines of the elements presented in Table 3. Quantification of these requirements should be based on the guidance in Chapter 5 and carried out in consultation with the appropriate planning authority. For example, in the case of a seawall a permitted maximum safe overtopping discharge can be determined by relating the information in Box 46 to the appropriate planning policy requirements.

6.2.3.2 Physical boundary conditions

The definition of the hydraulic physical boundary in terms of winds and waves, water depths, tides and currents, coastal sediment processes and seismic activity was described in Chapter 4, together with geotechnical boundary conditions. In addition, the definition and importance of exposure zones given in Section 2.1.1 should be carefully noted in cross-section design. It is sufficient at this point to remind the designer that coastal and shoreline protection structures are generally constructed in shallow water, and at some time during the tidal cycle will be located in the surf zone and subject to breaking waves. For this reason, they are also generally located in the area of maximum sediment activity. Tidal range and timing of lowest tides may also be a crucial factor in construction (see Sections 6.2.3.4 and 6.2.3.5).

The form of the cross-section may also be influenced by any existing structures present along the shoreline. This is particularly true when carrying out rehabilitation of existing seawalls using rock. If it often cost-effective not only to protect the old structure but also to incorporate it into the overall concept. The rehabilitation of the Morecambe Seawall, UK (shown in construction in Figure 286), where the wave wall at the rear of the promenade has been retained to reduce overtopping rates and the promenade now acts as a beam at the top of a sloping rock revetment, is a good example of this approach.

> **Box 79** Cost comparison of timber versus rock for bastion groynes
>
> The cost of rock-bastion groynes employed in a conventional groyne field can be compared directly with that of comparable timber, vertical-barrier groynes. On a groyne-for-groyne basis, the cost of rock-rubble groynes is roughly a function of their volume while that of vertical barrier groynes is related to their elevational area. Small rubble groynes are almost always cheaper than their timber or steel counterparts. Figure 287 (after Tyhurst, 1986) shows a simplistic exercise applied at Christchurch, UK, based on 1983 tender prices, which showed that rubble groynes were economic on a groyne-for-groyne basis up to a height, H, of 5 m.
>
>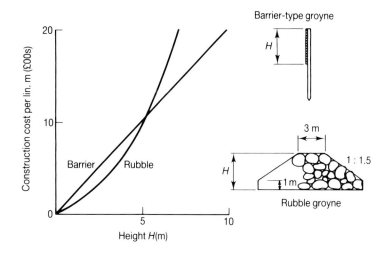
>
> **Figure 287** Rock and timber groyne cost comparison by cross-section, Christchurch, UK
>
> This cost comparison, however, neglects other factors which tend to make rock structures even more economically competitive when examined on a whole-scheme, whole-life cost basis:
>
> 1. The use of groynes in a groyne field may permit a wider groyne spacing (see Section 6.2.2.3) and thus fewer groynes and less cost per unit length of frontage;
> 2. Rock groynes are more durable than timber groynes and replacement will be required much less frequently (say, at least at double the time interval for comparable timber groynes);
> 3. Rock groynes will generally require less maintenance than timber ones.

6.2.3.3 Materials availability

Rock is a widely available material for applications in seawalls and shoreline protection structures. It will normally be cost competitive in comparison with other materials (timber, concrete) in addition to its technical advantages. This is seen, for example, in the case of applications to groynes in the UK, where it has a significant cost advantage over timber (see Box 79) as well as its technical advantages (see Section 6.2.2.3).

Occasionally, seawalls and shoreline protection structures constructed in rock justify the opening up of a temporary, dedicated quarry, either where the project is very large or where it is located in a remote area (see Sections 3.1 and 3.4). However, in most cases in Europe, the quantities of rock required are such that supply from existing local commercial quarries is the only cost-effective option.

If supply is obtained from a commercial quarry, the designer must consider which quarries could be selected by the contractor for economic supply. The range of options available may vary considerably, depending on the exact location of the site. For example, on the east coast of England there are a limited number of local, inland quarries, but seaborne supply across the North Sea from Scandinavian and other commercial quarries is very competitive. On the west coast, by contrast, seaborne supply is not attractive due to the long sailing route from European commercial quarries, and local supply from small, carboniferous limestone and igneous quarries is much more attractive. Selection of appropriate quarries for supply of rock to the Morecambe Bay Coast Protection Scheme, described in Box 80, is an interesting example of the latter.

Having determined which quarries are economically viable for supply, the designer can then decide what durability constraints in terms of rock quality (see Sections 3.3 and 3.4) he wishes to impose. These durability criteria must obviously be realistic, but may be set in such a way as to exclude certain low-durability rock-producing quarries or some faces within an individual quarry.

In a resource evaluation, the quantity of rock required for a project is also a significant factor and if it is large and the locally available quarries are small, it may lead to a requirement for rock to be drawn from more than one quarry. In this situation, marginal-durability rock may be acceptable for use in the core and underlayers of the structure.

Design of the armour layer needs particular evaluation of materials aspects if the prevailing environmental conditions, materials availability and economic considerations dicate the adoption of dynamically stable armour layers using small rock. Alternatively, use of rock material systems may be appropriate. Techniques such as asphalt grouting, gabion mattresses or stone pitching may be asopted. Armour layer design using these various alternatives is discussed further in Sections 6.2.3.7 and 6.2.4 below.

6.2.3.4 Material supply

Material supply, discussed in general in Section 3.4.4, can be by land-based or waterborne plant, irrespective of whether the *construction* equipment is land-based or waterborne. Considerations which will affect the decision as to which of the options to use will largely be economic but may also be environmental.

An important, practical constraint on seaborne supply is water depth and tidal range. Clearly, the water depth must be adequate for a barge containing rock to be floated into position and currents sufficiently low that it can deposit its load there. If the tidal range is sufficient, a technique that is frequently adopted is to dump the rock as close to the shoreline as possible at high tide and then recover this rock with land-based plant at low tide and move it to its final position. Use of seaborne plant in this way does, of course, presuppose that a cheap, local land-based source of rock is not being used.

Where import of rock to the construction site by lorries or dumper trucks is being considered, then the environmental impact of the noise and disturbance of the lorry traffic on the local residents must be taken into account. It is recognised that it is often the last 100 m or so of transport by road which poses the greatest problems to a contractor in this respect, particularly when the coast-defence works being considered are at a seaside town.

6.2.3.5 Construction considerations

As for breakwaters, the fundamental decision in construction of seawalls and shoreline protection structures is whether to use land-based or seaborne plant and the comparison of advantages and disadvantages given in Section 6.1.3 also applies here. However, there is clearly a tendency for many coastal works to use land-based plant, particularly since access to any part of the shoreline above low-water mark is normally possible using land-based plants and becomes progressively more difficult with seaborne plant, especially where the tidal range is small and draught restrictions rapidly come into effect. However, where the tidal range is large, it may be appropriate to commence the lower levels of construction using seaborne plant before transferring to land-based plant for the higher levels of construction.

Box 80 Quarry section, Morecambe Bay Protection Scheme, UK

The Coast Protection Scheme for Morecambe requires the construction of wave-reflection walls, rock armour protection walls and fishtail groynes and breakwaters over a period of 10 years. Between 300 000 and 500 000 t of rock were estimated to be required, of which about 60% would be armourstone.

Lancaster City Council, the body responsible, decided to investigate possible quarry sources in advance of inviting tenders for construction of the various phases of the project so that supply could be restricted to approved nominated quarries supplying stone of adequate quality.

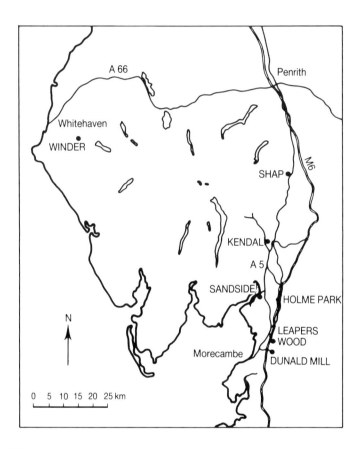

Figure 288 *Quarry locations evaluated for Morecambe Bay Coast Protection Scheme, UK*

Initially, four quarries were investigated (see Figure 288): three limestone quarries at Leapers Wood, Holme Part and Winder and a pinkish granite quarry at Shap. Rock sources were generally found to be of adequate quality, except for the base level face of the Leapers Wood quarry, were the small block size and the presence of solution features which might have a deleterious effect on integrity led to this face being excluded. The only other rock to be excluded, in spite of its excellent quality, was the Shap granite, primarily because its pink colour was not consistent with local rock formations. The Winder quarry was retained despite its long road-haulage distance from Morecambe in case a contractor might wish to supply the rock to Morecambe by sea.

For the second phase of the project a further three limestone quarries at Dunald Mill, Sandside and Kendal were investigated for addition to the nominated list of quarries. While all quarries were of adequate quality since they were all owned by the same producer, only the most local, Dunald Mill, was added. However, for the phase three works the other two were added, primarily because storms in February 1990 led to damage to parts of the existing coast-protection frontage not scheduled for rehabilitation until later phases of the project and to a consequent need to accelerate the construction programme.

If excavation is required for the construction, water depth and tidal range will also affect which excavation can be carried out in the dry and which will always be under water and for which, consequently, shallower side slopes will be necessary. The designer should also take account of underwater placing difficulties with geotextiles when specifying these as linings to underwater excavations. If geotextiles are specified, the geometrical arrangements should make their placing as simple as possible (e.g. by permitting unrolling of a weighted roll of geotextile down a uniform slope).

Another non-insignificant factor for work which needs to be carried out in the dry is the time of the day that mean low-water spring tides occur. For any given location on a coastline, this will always be at approximately the same time of the day. Hence locations where mean low-water springs take place in the middle of the day will generally offer wider scope for construction in the dry than at those where mean low-water springs are at 06.00 or 18.00 hours.

6.2.3.6 Maintenance considerations

These aspects are dealt with in some detail in Chapter 7. However, in terms of selection of the appropriate type of cross-section for the seawall or shoreline protection structure in the conceptual design process, it is important to recognise that maintenance will be a small-scale operation and therefore will be carried out most economically by land-based plant. Thus the access provided from the shore by the structure layout will have a crucial influence on the design approach to the armour layer. If an offshore breakwater has been selected or post-construction access to the structure will be impractical, the designer will need to provide an appropriate factor of safety on the design of the armour layer so that maintenance is never really likely to be required. If the structure shore access is possible with appropriate land links (even if only low-tide access for maintenance plant across the land links is possible), then a less conservative design of armour layer may be adopted.

6.2.3.7 Selection and design of armour layer (revetment)

The selection and design of the armour layer for a coastal defence structure will depend on a number of factors/choices:

1. Whether the armour is to be statically or dynamically stable;
2. If the armour is to be statically stable, the factor of safety required (related to uncertainty in design—see introduction to Section 6.3.3).
3. The required durability/lifetime of the armour;
4. The availability of different materials and materials systems;
5. The potential failure modes, given site-specific conditions.

Dynamically stable armour layers may be adopted if rock of adequate durability is available and nourishment, if necessary, is feasible. The most likely situation for adoption of a dynamically stable structure for coastal defence works is in offshore breakwaters, where a berm or reef breakwater concept may be adopted (see Section 5.1.3.3 and 5.1.3.7 for details and design tools for these concepts).

If statically stable armour layers are adopted, then a choice betweem simple rip-rap discussed in Section 5.1.3.1 and one of the many stabiity improvement systems described in Sections 3.5 (materials aspects) and 5.1.3.9 (stability aspects) must be made. The potential failure modes of these alternative rock-based revetment systems are summarised in Table 43 by Pilarczyk (1990), and the designer must take account of these in design as well as environmental availability, cost, construction and maintenance aspects in making the selection. Understanding of failure modes is, of course, normally only an aid to design out failure. It should be appreciated, however, that where failure does occur it is almost always due to one of three reasons:

1. Undersizing of armour elements due to underestimation of the wave climate;

Table 43 Critical modes of failure of rock-based revetment systems (after Pilarczyk, 1990)

Type of coverlayer	Critical failure mode	Determinant waveloading	Strength
Rip-rap	Initiation of motion Deformation	Max. velocity Seepage	Weight, friction Permeability of sublayer/core
Gabions/(sand, stone, cement) mattresses, including geotextiles	Initiation of Motion Deformation Rocking Abrasion/corrosion of wires UV	Max. velocity Wave impact Climate Vandalism	Weight Blocking Wires Large unit Permeability, including sub-layer
Placed blocks, including block mats	Lifting Bending Deformation Sliding	Overpressure Impact	Thickness, friction, interlocking Permeability, including sub-layer/geotextile Cabling/pins
Asphalt	Erosion Deformation Lifting	Max. velocity Impact Overpressure	Mechanical strength Weight

2. Improper dealing of the toe, transition elements and crest of rock structures (see Section 6.1.4)
3. Improper construction technique and placing (see Section 6.2.5)

An example of revetment damage can be seen in the failure of standard Australian Gold Coast revetments due to subsidence and overtopping described by Smith and Chapman (1982). Figure 289 shows a typical damage sequence for these revetments. The improvements recommended by Smith and Chapman (1982) to diminish the risk of such failures are shown in Figure 290.

In general, underestimation of wave climate should be avoided if the guidance provided in Section 4.1 is followed and advice on correct detailing of the key elements of toes, transitions and crests is given in Section 6.2.4 below (see also Section 6.1.4). In the case of seawalls or dikes more involved failure criteria must be considered to ensure that functional failure does not occur (see Figure 20 in Chapter 2 and the more detailed discussion and fault trees given in Section 6.2.4.1).

In making a selection between the alternative rock revetment systems it should be appreciated that the more open systems such as simple rip-rap are more likely to have at least a few individual blocks moving under storm wave attack, but, unlike relatively closed or grouted systems, are not subject to bursting pressures and, because of porosity, absorb energy. Typical maximum wave heights and life expectancies for the different systems may be summarised as follows:

	Maximum significant wave height, H_s(m)	Typical life expectancy (years)
Conventional rip-rap	Limited only by stone weight available	30–100
Pitched stone (including basalt columns)	4–5	30–100
Grouted stone	3–4	20–50(?)
Open-stone asphalt	2–4	Up to 30 proven
Gabions (see Box 81)	1.5–2	5–20

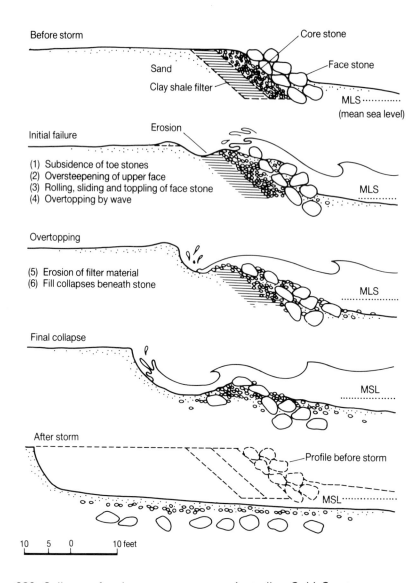

Figure 289 *Collapse of a rip-rap revetment on Australian Gold Coast*

Further information (other than that in Section 3.5) on the use of gabions and mattresses is given in Box 81 on grouted stone and asphaltic revetments in van Herpen (1990) and TAW (1985).

Revetment armour thicknesses must follow the guidance on minimum layer thicknesses given in Section 5.1. The revetment system must also appropriately match the underlying permeability/pore size of the material being protected, with appropriate permeability/pore size transitions using filter layers and geotextiles. Less permeable system may thus appear attractive in situations where the underlying material is of low permeability. However, wave reflections from (and run-up on) such permeable systems are likely to be higher and avoidance of increasing such factors will often be more important (e.g. to reduce beach scour and flooding due to overtopping).

6.2.4 STRUCTURE-SPECIFIC DESIGN ASPECTS

This section provides detailed guidance and rules of thumb for designing all the various structureal elements of the structure types discussed in this section particularly:

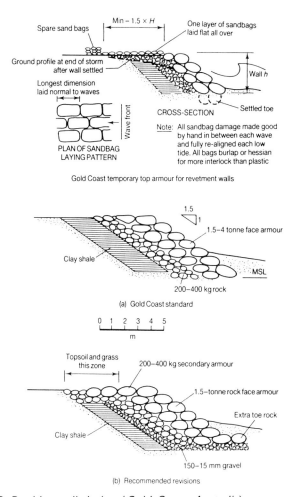

Figure 290 *Boulder wall design (Gold Coast, Australia)*
$H_s \approx 3.5\,m$, $T = 8\text{--}18\,s$, slope 1 on 1.5

- Side slopes and front face profile
- Toe stability and scour prevention
- Crest height
- Transitions

6.2.4.1 Conventional revetment or revetted mound (dike)

The cross-section of a seawall or dike protected with rock armour or a rock-based system as described in Section 6.2.3 above will depend on the actual situation and the functions required of the revetment. Given this dependency, it is nevertheless possible to identify a number of basic concepts on which site-specific solutions can be built. These basic concepts are illustrated in Figures 292 to 295, which show typical solutions for coast protection, sea defence, reclamation protection and seawall rehabilitation.

Sometimes the existing arrangement at a particular site may impose severe geometrical constraints on a solution involving a revetment or revetted mound. This is particularly true where an existing seawall and/or roadway/promenade is to be rehabilitated (Figure 293) and incorporated into the final solution. Fortunately, rock offers flexibility in this situation because of the wide range of sizes and densities available, a factor which proved invaluable in the sizing of armourstone in the case study described in Box 82.

Box 81 Use of gabions and mattresses

Wire gabions and mattresses filled with stone may be used in a conventional sloping revetment. Guidance on stability in this situation is given in Section 5.1.3.9 and information on materials and construction of gabions and mattresses in Section 3.5.

Gabion boxes are also commonly used to form gravity-type walls, which may be tied back into the soil mass as a composite structure. This form of structure, like other stone revetment systems, can accommodate settlement or consolidation of the foundation after construction. Gabion walls should be built with the front face at a slight angle (10:1) to the vertical, to allow for future forward horizontal tilting as settlement takes place. This battering of the wall may be achieved either by sloping the entire foundation or by stepping back individual units. The gabion wall concept is illustrated in Figure 291.

Careful attention should be placed on provision of an effective filter behind the gabion wall to prevent leaching out of the fine material it retains. Detailed design advice and experience of similar applications should be sought from a gabion manufacturer.

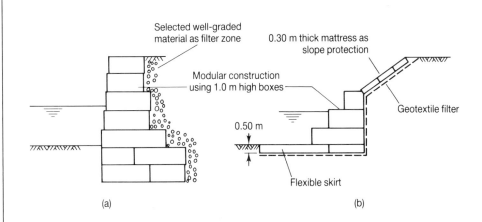

Figure 291 (a) Gabion box wall structure; (b) combined box wall/mattress structure

In approaching the cross-section design of a particular rock seawall the principal functional failure criteria can be summarised as:

- Flow under or through wall
- Flow over wall
- Damage to revetment (facing)
- Geotechnical instability

any of which could potentially lead to a breach in the wall and consequent erosion/flooding. These have been well summarised in fault trees prepared for *Seawall Design* (Thomas and Hall, 1990) and represented here as Figures 296 to 299 with some small modifications.

Geotechnical stability and flow under or through a mound comprising (or faced with) rock can be assessed using the information supplied in Section 5.2. Particular care in the case of reclamation schemes (Figure 292) should be paid to the question of flow of air and water through the revetment/revetted mound, as there have been many examples of failures involving leaching out of reclamation material and consequent sinkhole formation arising from poor filter design and failure to

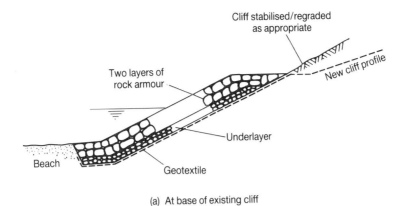

(a) At base of existing cliff

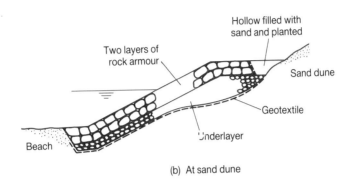

(b) At sand dune

Figure 292 *Coast-protection revetments*

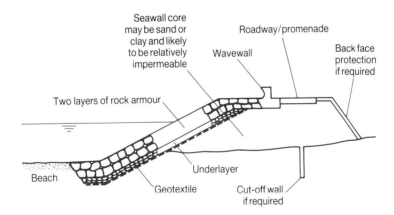

Figure 293 *Sea-defence revetment*

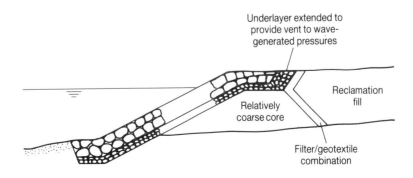

Figure 294 *Land-reclamation revetment*

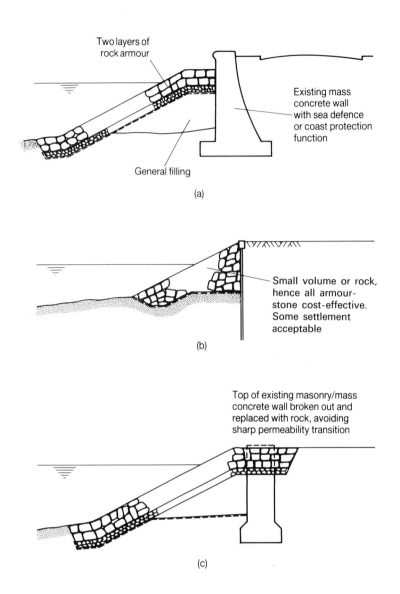

Figure 295 *Typical revetments for rehabilitation of existing vertical seawalls*

properly vent fluctuating wave pressures. Flow over the crest and armour stability can be assessed using informatiuon in Section 5.1.

Side slopes, crest level and front-face profile of revetment

To design side slopes, crest level and front-face profile for a seawall (Figure 300) a number of factors must be taken into account, given the physical boundary conditions evaluated in accordance with Chapter 4:

1. Required slope for armour layer hydraulic stability (see Section 5.1.3);
2. Required slope, crest level and width of berm(s) for limiting run-up/overtopping to acceptable values (see Section 5.1.2 with acceptable overtopping rates given in Box 46);
3. Required slope, etc. to ensure that reflected wave heights (see Section 5.1.2) are kept to acceptable values;
4. Required slope, crest level and width of berm(s) to ensure adequate stability against geotechnical slip failure (see Section 5.2);
5. Cost considerations for overall volumes of material (which increase with shallower side slopes and crest levels) and volumes of armouring (which reduce per unit area as side slopes becomes shallower).

> **Box 82** Case study in sizing armourstone for a seawall rehabilitation
>
> *The brief*
>
> Severe beach erosion had badly exposed the cut-off piles at the toe of a concrete stepped apron over a considerable length, threatening the integrity of the structure, and strengthening works were urgently required. Site investigation showed that the whole structure was founded on very weak soil and that there was a risk of failure through rotation if excavation was not severely limited.
>
> *Design step 1*
>
> A rock armour wedge along the toe was proposed to prevent further erosion and encourage beach build-up. The armourstone weight was initially calculated on the basis of a slope angle of 1 in 3 and a rock density 2.6 t/m³. The resultant D_{n50} was too great to allow two-layer armour construction without either
>
> 1. Raising the top of the wedge so high that dislodged rock could fall on and damage the existing concrete; or
> 2. Excessive excavation.
>
> Inspection suggested that a reduction of about 25% in D_{n50} would allow two-layer armour to be used.
>
> *Design step 2*
>
> There were two possible ways of reducing the D_{n50}:
>
> 1. By lowering the slope angle. Consideration of van der Meer's equation for plunging waves (equation (5.44)) indicates that D_{n50} is proportional to $(\tan \alpha)^{1/2}$. Hence to reduce D_{n50} to 0.75 times its original value, $\tan \alpha$ would have to be reduced to $(0.75)^2$ times its original value. Thus the new $\tan \alpha$ required is $(0.75)^2 \times 0.33 = 0.187$, equivalent to a new slope angle of 1 in 5.3. This had two disadvantages:
>   (a) It would extend excavation into the most probable rotational toe area; and
>   (b) It would require approximately 25% more material, with associated costs.
> 2. By increasing the specified density. Consideration of van der Meer's equation indicates that D_{n50} is proportional to $(1/\Delta)$. Hence to reduce D_{n50} to 0.75 times its original value a new Δ was required given by $\Delta_{old} * (1/0.75)$. By definition of the rock mass density ρ, this gave a revised required rock density, ρ_r, given by
>
> $$\rho_w = (\rho_{r\,old} - \rho_w) * (1/0.75) + \rho_w$$
> $$= (2.6 - 1.03)/0.75 + 1.03$$
> $$= 3.12 \, t/m^3$$
>
> This would require approximately 20% more tonnage, but carried no additional risk, while handling weight was reduced by about 50%.
>
> *Final design solution*
>
> In fact, a source of supply of rock with 3.1 density was available at a similar price per tonne, so solution (2) was adopted on both engineering and economic grounds.

Of the above factors the setting of side slopes and crest level to limit overtopping given the hydraulic boundary conditions is often one of the most dominant. To illustrate this point, Box 83 summarises Dutch practice for determination of the crest height of a dike. The flexibility of rock construction will permit relatively inexpensive raising of crests subsequently if required by possible future higher predictions for rises in mean sea level.

Revetment crest detailing

Having selected an appropriate crest level and thereby an implicitly acceptable rate of overtopping, it is important to correctly detail the crest. Correct crest detailing will include the following:

1. Adequate protection against erosion of crest, or an area such as a cliff face, above or behind the revetment, including avoidance of loss of soil support to the revetment. If rates of overtopping are high, use of rip-rap may be appropriate and the design approach suggested in Section 5.1.3, Box 54 may be adopted. If rates of overtopping are small, much lighter forms of protection such as a grass mat or sward may be appropriate for the crest. In this case, an appropriate transition in surface roughness and porosity is essential between the two systems, such as (in the case of a rip-rap to grass sward transition) the use of cell blocks,

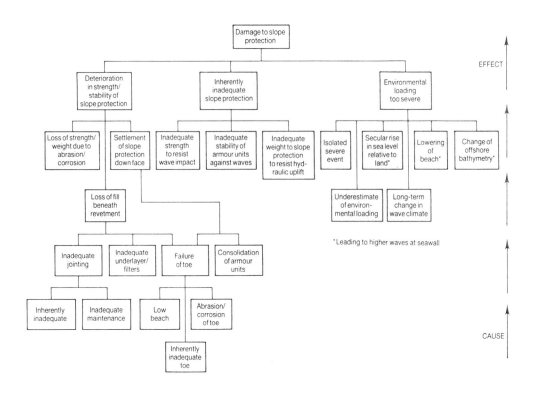

Figure 296 *Fault tree: events leading to flow under seawall*

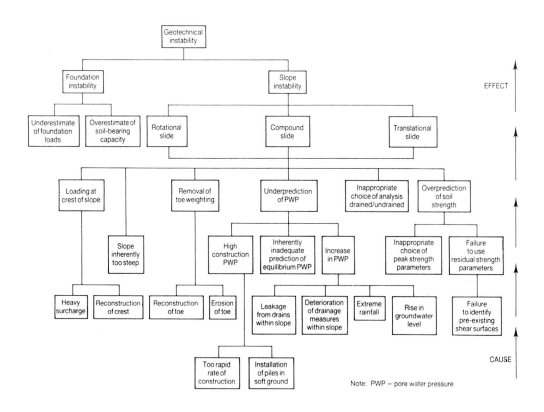

Figure 297 *Fault tree: events leading to flow over seawall*

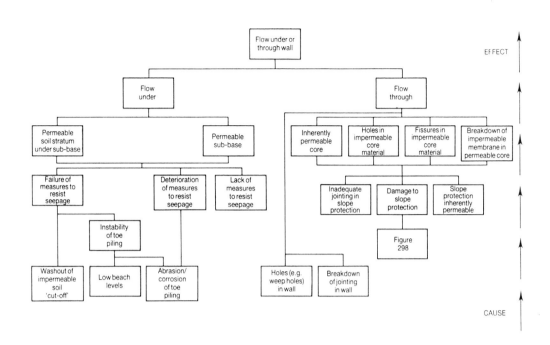

Figure 298 *Fault tree: events leading to damage to slope protection*

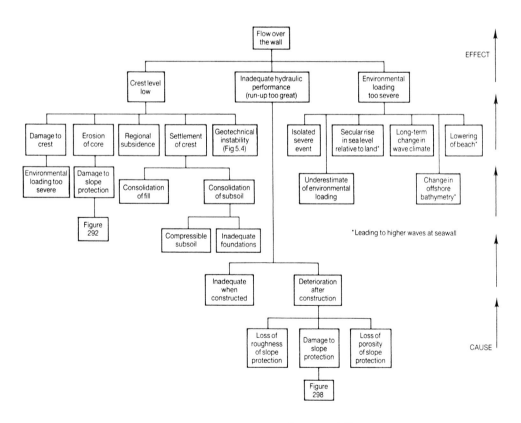

Figure 299 *Fault tree: events leading to geotechnical instability*

geogrids or other systems which permit associated use of vegetation. For more details on the use of vegetation in such engineering applications reference should be made to Hewlett *et al.* (1987) and Coppin and Richards (1990). Allowance should also be made in the detailing of protection for post-construction differential and total settlement;
2. Adequate drainage of overtopping water using the appropriate hydraulic design methods and paying careful attention to levels and falls;
3. Adequate protection against the effects of traffic (vehicle, human, livestock) and vandalism.

Revetment toe detailing

In order for a revetment toe to function successfully it must extend down to a level such that either it cannot be undermined or it is sufficiently flexible to drop down to a new lower level without damage. Revetment toe design must therefore be based on estimates of the lowest anticipated beach level, the anticipated local scour depth, S_d, calculated as described in Section 5.1.4, and the required stable armour weight.

Stable armour weights under wave action may be assessed as described in Section 5.1.3.6. Where currents are involved, stable weights may be estimated using the theory presented in Section 5.1.3.11 and scour depths using an appropriate scour theory for steady flow or from model testing. Where currents in excess of 1 m/s occur at the same time as waves, it is recommended that the armour layer thickness be increased by a factor of at least 1.3.

Estimation of lowest beach level and local scour depth, S_d, will depend on an understanding of coastal sediment processes, combined with the beach geology at founding level. Three main types of beach may be identified for toe detailing purposes:

1. *Resistant strata at foundation level*: where relatively erosion-resistant strata (including rock) are exposed or occur at a reasonable depth below beach level. Here the toe may be founded on the stratum with little risk of undermining. The armour layer must be keyed into the stratum at a minimum depth of $0.5D_{n50}$ to ensure that sliding of the layer will not occur. This will normally involve dredging, ripping or blasting an appropriate trench into the stratum (see Figure 302). Sometimes, in the case of a very hard rocky foreshore where creation of a trench may be too expensive, the option of casting a concrete nib dowelled into the rock against which the toe of the armour layer may sit may be considered.
2. *Limited-resistance strata at foundation level*: In many instances, clay or other strata with limited resistance to erosion occur at a shallow depth below the beach and erode only slowly when uncovered. Founding in such strata may therefore be considered as offering low undermining scour potential and a detail such as that in situations (a) or (b) in Figure 303 may be appropriate with the armour layer keyed into the resistant stratum at a minimum depth D_{n50}.
3. *Beach or other erodible material at foundation level*: Here the founding level must be based on th predictions of beach-level variation and local scour depth alluded to above, and the risk of undermining, depending on the reliabiity of prediction and the degree of conservatism adopted, tends to be higher. In such situations it is worth assessing whether the hydrodynamic conditions prevailing indicate a situation of potentially low, moderate or severe scour. To assist in this judgement it is worth noting the following:

(a) Situations of severe scour potential occur where wave downrush on the structure face extends to the toe and/or the wave is breaking near the toe due to shallow water conditions. Such situations may arise where the water depth at the structure toe is less than twice the height of the maximum expected unbroken wave that can exist in that water depth, and where the structure has a high reflection coefficient, which is generally true for slopes steeper than about 1 in 3;

(b) Situations of lower scour potential occur with slopes milder than 1 in 3 and where the water depth is greater than twice the maximum wave height. Here most of the wave force will be dissipated on the structure face rather than the scour apron.

Given the assessment of scour potential, an appropriate toe design may be selected using the guidance in Figure 303. The minimum thickness of armour layer in the toe should be $2D_{n50}$, but the requirements in Figure 303 often dictate a thicker layer to provide some factor of safety to the design. Particularly with underwater construction (case (d)), adequate material must be present to provide the necessary volume of rock for a 'launched' apron. In all situations of potentially severe scour, if a geotextile is used as a secondary layer it should either not be extended over the whole width of the apron (and thus permit a flexible edge in the armour of at least 1 m in case of undermining) or it should be folded back, and then buried in cover stone and sand to form a Dutch toe.

Note that the conservatism of the apron design (width and size of armour rock) depends on the accuracy of the methods used to predict the wave and current actions and to predict the maximum depth of scour. For specific projects where the revetment is to be founded in erodible material a detailed study of scour in the natural bed and near similar existing structures should be constructed at the planned site, and model studies should be considered before determining a final design. In all cases, experience and sound engineering judgement play an important role in applying the above design rules, which are also largely based on experience rather than systematic prototype or model testing.

Joints and transitions in revetments

However well designed a revetment cross-section may be, the whole revetment is only as strong as its weakest section, and thus particular care is required when designing transitions either along the length of the revetment or with existing or different structures or revetment types. Experience has shown that erosion or damage often starts at such joints and transitions.

Different treatments may be required for protection of different parts of the cross-section and may include the following: toe protection, lower slope protection in the area of heavy wave and current attack, upper slope protection (for example, a grass mat), and protection of any berm provided to reduce run-up or as a maintenance road. Different materials and construction methods may be used for these parts and hence careful attention must be paid to the joints between them.

Similarly, a new slope protection may need to be connected at one end to an

Figure 300 Rock revetment at Buckhaven, Fife, UK (courtesy HR Wallingford)

already existing protective construction which involved another protective system, as illustrated in Figure 272. Here again careful attention is needed, including avoiding sharp angles and curves (see Section 6.2.2.3).

In general, joints and transitions should be avoided as much as possible by treating cross-sections and entire coastal cell units in a unified manner, but if they are inevitable the discontinuities in behaviour introduced (e.g. in load-deformation characteristics, permeability, etc.) should be minimised and high-quality construction employed.

It is difficult to formulate more detailed principles and/or solutions for joint and transitions. The best way is to combine the lesseson from practice with some physical understanding of the systems involved. As a general principle, the transition should be of a strength equal to or greater than the adjoining systems. Very often it needs a reinforcement in one of the following ways:

- Increase the thickness of the coverlayer at the transition;
- Use concrete edge-strips or boards to prevent damage progressing along the structure.

One specific example of a transition arises in the case of a pipeline normally used for land drainage which must pass through the revetment. Here it may be appropriate to use concrete or asphalt grout to locally increase the stability of the rock armour and underlayer around the pipe and use the stabilised area to provide an appropriate haunching to the pipe.

Top edge and flank protection are needed to limit vulnerability of the revetment to erosion continuing around its ends. Care should be taken to ensure that the discontinuity between the protected and unprotected areas is as small as possible (use a roughness transition) so as to prevent undermining. For example, open cell-blocks or open blockmats (eventually vegetated) can be used as the transition from a hard protection to a grass mat.

With flank protection, extension of the revetment beyond the point of active erosion should be considered but is often not feasible. In such situations terminal groynes and protective flanking walls cut into existing land perpendicular to the line of

Box 83 Dutch practice for determination of dike height (Pilarczyk, 1990)

The height of a dike was, for many centuries, based on the highest known flood level that could be remembered, but the real risk of damage or the probability of flooding were unknown. Little was known about the relation between the cost of preventing flooding and that of the damage that might result from flooding.

In this century it was found that the frequency of occurrence of extremely high water levels and wave heights could be described adequately using probability theory, despite the risks involved in extrapolating beyond observed events. Thus after the 1953 flooding disaster in the Netherlands, having studied flooding risk frequency in relation to economic aspects, it was decided to base the design of all sea dikes on a storm-urge water level (including wind set-up) with a return period of 1 in 10 000 years.

In addition to the design flood level, several other elements play a role in determining the design crest level of a dike see Figure 301 (a)):

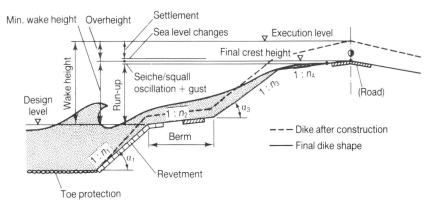

Figure 301(a) *Determination of dike height*

- Wave run-up (the 2% exceedence value is applied in the Netherlands) which depends on wave height and period, angle of wave approach, roughness and permeability of the slope, and profile shape (see Section 5.1.2);
- An extra margin to the dike height to take into account seiches (oscillations) and gust bumps (single waves resulting from a sudden violent rush of wind); this margin in the Netherlands varies from 0 to 3 m for the seiches and 0 to 0.5 m for the gust bumps depending on location;
- A change in chart datum (NAP) or a rise in the mean sea level in the Netherlands (until now assumed to be roughly 0.25 m per 100 years but now increased to an average value of 0.6 m per 100 years);
- Settlement of the subsoil and the dike-body during its lifetime (Figure 301(b)).

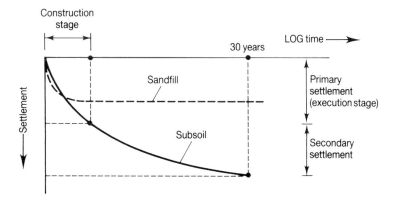

Figure 301(b) *Dike settlement as a function of time*

The combination of all the above factors defines the freeboard of the dike (called the wake height in Dutch). The recommended minimum freeboard is 0.5 m.

The above dike height is commonly combined with a minimum crest width of 2 m and slopes of 1:3 on the landward face and between 1:3 and 1:5 on the seaward one.

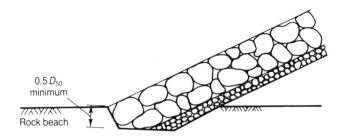

Figure 302 *Toe detail: revetment on rock beach*

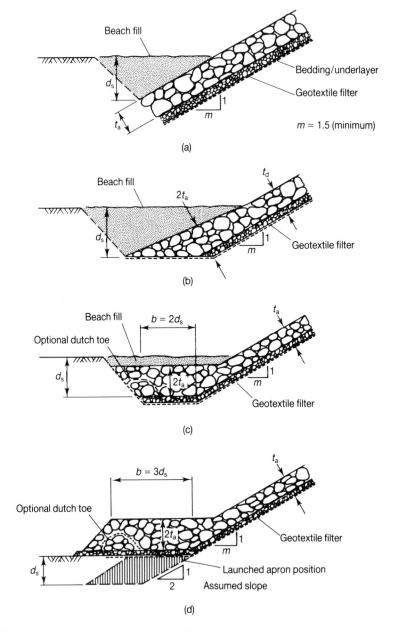

Figure 303 *Revetment toe protection—erodible beach (after US Army Corp of Engineers, 1985; SPM, 1984). (a) Low scour potential sites—construction in the dry; (b) low to moderate scour potential sites—construction in the dry; (c) moderate to severe scour potential sites—wet construction possible; (d) moderate to severe scour potential sites—construction underwater (launching apron concept). Notes: (1) for calculation of t_a see Section 3.2, Table 20; (2) for calculation of d_s Section 5.1.3.12; (3) for moderate to severe scour situations, replacement of stone after extreme events may be necessary.*

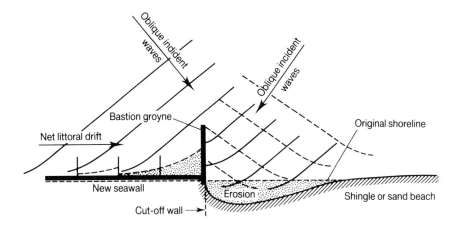

Figure 304 *Erosion at seawall termination*

defence may be required to protect against erosion (see Figure 304, Thomas and Hall, 1990), but these often only provide a temporary solution and require extension from time to time to match the rate of erosion/accretion.

6.2.4.2 Scour protection of vertical wall

Scour protection of a vertical wall by constructing a rock-armoured apron in front of it involves many considerations similar to those in the protection of caisson breakwaters discussed in Section 6.2.4, and the stability of the anti-scour armour may be assessed in a similar way using the information in Section 5.1.3.5. Design principles are otherwise generally similar to those for revetments and revetted mounds.

6.2.4.3 Bastion groyne

The cross-section of a simple, small bastion groyne may be made up of a single grading of armourstone which may be wide ($D_{n85}/D_{n15} < 2.5$—see Section 3.3 and Box 45). A typical grading might be, say, 1–9 tonne with the larger rocks set to one side during construction for placing on the outer part of the groyne, thus giving additional protection. For somewhat larger structures, modest bedding stone layers may be introduced (see Figure 305, from Barber and Davies, 1985).

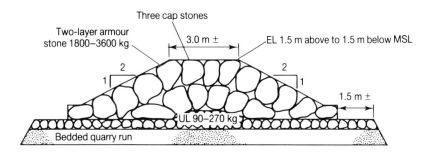

Figure 305 *Example of groyne cross-section from Atlantic Coast, North Carolina, USA (this groyne is exposed to hurricanes)*

The level of complexibility of the cross-section will be a function of site accessibility and maintenance resources available (see Chapter 7). A single, armourstone sized, grading placed straight on the beach may incur some settlement and a consequent need to add further rock in the future. However, the capital cost savings involved may be cost effective if replacement rock can be readily placed.

Figure 306 *Rock groyne on shingle beach, Barton-on-Sea, UK (courtesy J.D. Simm, Robert West & Partners)*

The crest level should generally follow the existing or proposed nourished or trapped beach profile. This will, of course, vary with a summer/winter, storm/calm beach profile resulting from onshore/offshore sediment movement. However, the crest should not normally exceed the maximum beach level expected at any position. Sometimes it may be appropriate to keep the crest level constant, particularly for short groynes, and here it should not normally exceed the height at which a storm ridge would exist at the site (see Figure 306).

The suggested crust levels and profiles should ensure that beach material is not unnecessarily retained on one side of the groyne, thereby starving the downdrift beach. The selected profile will influence the zones of worst wave attack (see Figure 307) and particular care should be taken when addressing the hyraulic stability of these areas.

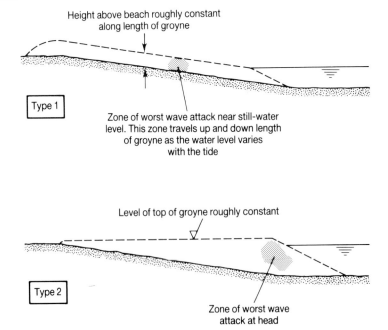

Figure 307 *Alternative longitudinal profiles of rock groynes*

Side slopes of simple groynes may be largely dictated by economy, and slopes as steep as 1 in 2 or even 1 in 1.5 are used. The primary advantage of flatter slopes (say, 1 in 3 to 1 in 4) is the reduced wave reflection that arises and the increased diffractive capability to encourage sediment to build in the lee of groynes. Particular attention should be paid to the transition between rock groynes and existing impermeable hard defences. It is advisable to ensure a proper transition in permeability/porosity. A relatively economic way of achieving this transition is asphalt grouting of a small area of the groyne armourstone immediately adjacent to the hard defence.

6.2.4.4 Offshore breakwater

Breakwater design and construction has been discussed in Section 6.1. When applied for a coast protection function as described in Section 6.2.2.3, the failure mode evaluation and cross-sectional design described in Section 6.1 may be followed. However, their application for coast protection, where they are required to encourage beach build-up, may have more influence over cross-sectional shape than

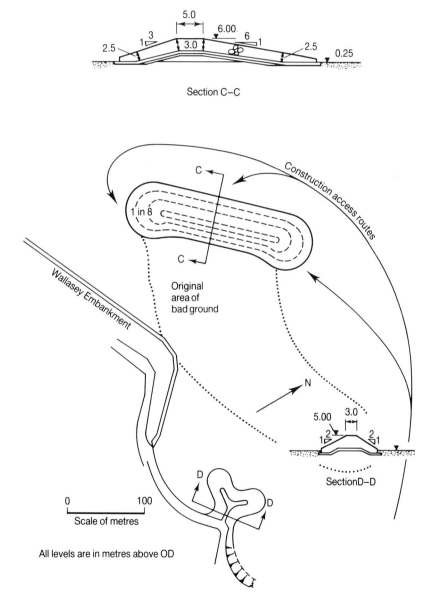

Figure 308 *Offshore breakwater, Leasowe Bay, UK*
$H_s \approx 3.0$ m, $T_2 = 5$–7 s, armour $D_{n50} = 1.0$ m

pure stability considerations. A typical example from Leasowe Bay, UK, is shown in Fig. 308. Typically, the outer face of such breakwaters should have slopes greater than about 1 in 3 or 1 in 4 if it is desired to reduce reflective scour and increase energy dissipation. The rear face can be steeper as indeed can the front face so long as any reflections and scour can be accommodated. Crest levels will be set by overtopping/transmission criteria or, in the case of dynamically stable reef breakwaters, by wave structure interaction.

Roundhead design

A critical element in offshore breakwater design is the provision of wide, shallow-sloping roundheads. The shape of this may be designed using the following rule-of-thumb approach due to Dr P. C. Barber.

1. Determine the range of wave length field about a wave length offshore from the roundhead.
2. Set the overall roundhead diameter at 1 to $1\frac{1}{2}$ times the wave lengths in the wave length field determined, above.
3. Set the slope of the roundhead to ensure a proper transition between the main breakwater and the beach and to avoid inducing horizontal circulations. This may be achieved by setting the slope at between 1 in 6 and 1 in 12; slopes steeper than 1 in 6 run the risk of inducing horizontal circulations although in practice this has been adopted by other designers to reduce the cost of such structures.
4. The transition between breakwater and beach can be smoothed still further by the introduction of a spending apron of bedding/underlayer stone, as in the case of the breakwater constructed at Leasowe Bay, UK (see Figure 309, from Barber and Davies, 1985).

6.2.4.5 Fishtailed breakwater

The design of the cross-section of the various parts of a fishtailed breakwater involves a combination of the concepts discussed above for groynes and offshore breakwaters. For a description of the various parts of the breakwater, reference should be made to the basic geometry diagram (Figure 281). The principles of cross-section design are illustrated in Figure 310 by a fishtailed breakwater for the Clacton-on-Sea sea defence project, UK, described earlier in Box 77.

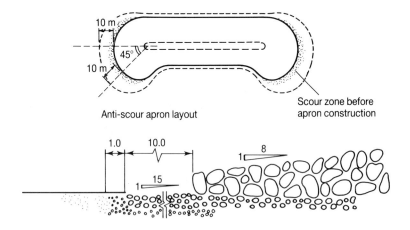

Figure 309 *Transition details, Leasowe Bay offshore breakwater, UK*

Roundheads

The design principles for roundheads are as described for offshore breakwaters in Section 6.2.4.4 above.

Land link (OC—Figure 281)

The crest of the land link, which prevents tidal currents from flowing behind the main arms of the structure and eroding the beach, is, as with groynes, normally set to follow the profile of the beach. Side slopes are typically set at about 1 in 2, again as for conventional groynes.

Downdrift outer arm (OB—Figure 281)

As the main function of the downdrift outer arm is to intercept storm waves and protect the downdrift beach from direct wave action, its crest is set generally above high-tide level but with a fall from O to B to assist in a smooth transition back to beach level. This point is well illustrated in Figure 310 (from Biss and Craig, 1987) for the Clacton breakwater.

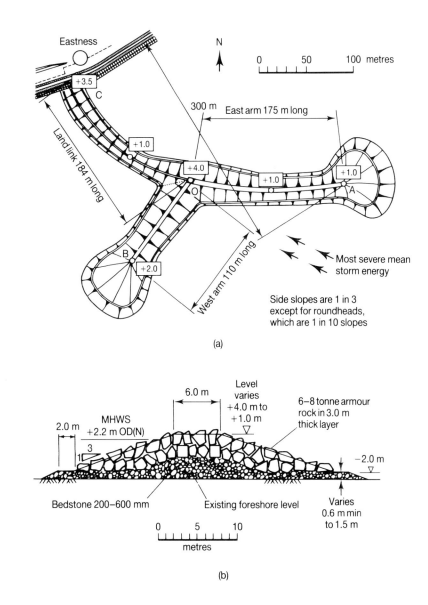

Figure 310 *Eastness Breakwater, Clacton-on-Sea, UK. (a) Plan; (b) typical cross-section. $H_{max} \approx 2.5m$, $T_2 = 6–8s$*

Side slopes for the downdrift outer arm should normally be set at about 1 in 4 on the outer face exposed to the prevailing storm waves, but can be reduced to 1 in 3 on the more sheltered inner face. Slopes as steep as 1 in 2 should be avoided, as these will cause undesirable reflections and will not have the required energy-dissipating properties to assist in sand accretion.

Updrift outer arm (OA—Figure 281)

As the primary function of the updrift outer arm is to intercept alongshore and tidal currents and divert them sufficienly far offshore to minimise beach erosion within the protected cell, the crest level of this arm can be lower, tidal currents being most severe in the mid-tide range. Indeed, if crest levels are set too high, undesirable silt patches between AOC and the shoreline may form.

The approach formulated by Barber to set the crest level of this arm, which will generally be between MHWN and MHWS, is as follows:

1. Obtain or estimate a joint probability distribution of waves and water levels (see Section 4.1), together with the associated currents for each of the principle directions of wave attack (using refraction analyses as appropriate).
2. Determine for each water level/wave direction combination a sediment number which describes the total sediment transport energy in each situation. Barber's sediment number is calculated as: Bijker (1967) bed shear stress (τ_b) × depth mean velocity ($\bar{U}$) × water depth (h). This has been written in a dimensionless form by Simm (1989) as:

$$N_{BS} = \frac{U_*^2}{\Delta . g . D_n} \left[1 + \tfrac{1}{2} \frac{(\xi U_0)^2}{U} \right] * \frac{\bar{U}.h}{v}$$

or, since $U_* = \bar{U} g^{1/2}/C$, where C = Chezy coefficient (see Section 5.1.3.12):

$$N_{BS} = \frac{\bar{U}^3 h}{\Delta . D_n . v . C^2} \left[1 + \tfrac{1}{2} \frac{(\xi U_0)^2}{\bar{U}} \right]$$

3. Assess a water level below which the majority of the total sediment transport energy falls and set the crest at this level over at least the outer half of OA, allowing the crest level to rise from the midpoint of OA towards the higher level of O already determined by the crest level calculation for the downdrift outer arm OB. The side slopes of the updrift arm should be set at between 1 in 4 and 1 in 3, depending on the degree of shelter. As with the downdrift arm, slopes as steep as 1 in 2 should be avoided.

6.2.4.6 L- and T-shaped breakwaters

These may be designed as conventional breakwaters. Their application to areas of limited range such as that shown in Figure 311 at Malaga, Spain, in the Mediterranean, means that crest level definition is relatively straightforward. Side slopes for the outer L or T will be between 1 in 3 and 1 in 6 for the reasons discussed for offshore and fishtail breakwaters above. The land link arm can have side slopes as steep as 1 in 2 except for the outer face of arms not having a pocket or protected beach either side of them.

6.2.4.7 Sill or submerged breakwater

Armour stability, side slopes and cross-sectional details of sills or submerged breakwaters can be assessed using the information in Section 5.1.3.3. Armour

Figure 311 *Pedregalejo Beach, Malaga, Spain (courtesy Director-General Ports and Coasts, Spain)*

stability will be strongly influenced by the selected crest level, and this in turn will be determined by the required profile of the beach to be retained. Insufficient information is available to give clear guidance on crest level in relation to beach profile, and model tests are always advised. However, based on an assessment of

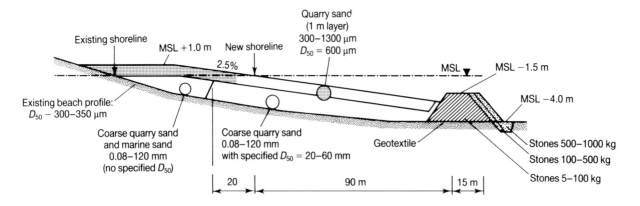

Figure 312 *Sill and perched beach design, Lido di Ostia, Rome, Italy*

work by Beil and Sorenson (1988), a starting point for design may be to set the crest level, such that the height, $h - h_s$, of the sill crest above the original beach level is about twice the height, $h_t - h_s$, of the sill above the final beach level (see Box 78 for definition of terms).

Documented applications of sills are limited, the concept being relatively new, but a model-tested design for a sill and perched beach at Lido di Ostia, Rome, is shown in Figure 312 (Toti et al., 1990).

6.2.5 CONSTRUCTION ASPECTS

Construction considerations in the design of rock structures have been addressed above (Section 6.2.3.5) in terms of land-based/seaborne construction, and the influence of tidal range and tidal timing on the need for wet or dry working. A comprehensive discussion of construction techniques in relation to breakwaters will be found in Sections 6.1.3 and 6.1.5. Much of this is also directly relevant to coastal defence structures, but it should be appreciated that coastal structures differ from breakwaters in a number of ways:

1. They are often smaller in scale, certainly in terms of their cross-section; and thus
2. They are generally constructed using smaller-scale equipment, typically track-mounted hydraulic-arm vehicles with a suitable handling attachment (see below).

Because of the above, the large crane and dragline equipment discussed in connection with breakwaters is less frequently used for coastal works. However, rock-transport vehicles will be similar to those for breakwaters, and thus similar crest-width constraints to those illustrated in Figure 259 will apply.

Unlike the situation with deep-water breakwaters, because access in the dry is possible in most coastal situations, work can proceed on more than one front without resorting to seaborne plant. However, caution is necessary, as with breakwaters, to ensure that excessively large areas of core or underlayer/bedding layer are not exposed at any one time, thereby reducing the likely extent of damage in the event of a storm. This constraint may only need to be rigorously applied in stormy seasons of the year, noting that in the waters around north-west Europe, storms may arise at any time. If the stormy season is particularly severe, work may have to be stopped during this period and full protection applied in all areas potentially subject to damage.

6.2.5.1 Handling and placing of armourstone

Armourstone, which in the context of handling and placing can be taken to cover any individually handled piece of rock, may be lifted directly from delivery trucks or taken from stockpiles of tipped stone placed adjacent to the placing equipment.

Figure 313 *Orange-peel grab placing rock (courtesy J.D. Simm, Robert West & Partners)*

Considerable care should be taken in selecting an appropriate handling attachment to the crane or hydraulic equipment being employed. If a crane or dragline is being used, then some form of cactus or orange-peel grab is probably the best available. When hydraulic-arm placing equipment is being used, as is most common, three main options are available.

Orange peel or cactus grab (Figure 313)

If fitted with a powered turntable, this provides the operator with most flexibility in placing armourstone to a desired position and orientation. Thus a powered-turntable orange-peel grab will generally achieve the tightest packing density and should perhaps therefore be used where amenity and safety considerations (Section 6.2.2.5) are important. Non-powered orange-peel grabs allow less control and the grapple is probably to be preferred.

Grapple or 'Bofors' grab (Figure 314)

A fixed-arm grapple or 'Bofors' grab provides the operator with less control over the orientation of armourstones than the powered orange-peel grab. Its attraction lies instead in its ability to permit positive placing, including pushing and easy pick-up from stockpile of armour. For this reason (see Chapter 7. Table 50) it is often also preferred for maintenance and repair work. The grapple will tend to lead to a relatively open texture to the armour layer with a higher porosity. If energy dissipation is more critical than amenity/safety, this increased porosity is to be preferred and thus the grapple should be employed in such situations.

Hydraulic bucket (Figure 315)

For small projects the readily available bucket mounted on a hydraulic vehicle is sometimes used, although is not recommended. The disadvantage of the hydraulic bucket is that, once stones have been placed, they are very difficult to remove again, and hence smooth profiling of the final surface and close packing is normally

Figure 314 *Grapple placing rock (courtesy J.D. Simm, Robert West & Partners)*

Figure 315 *Hydraulic bucket placing rock (courtesy J.D. Simm, Robert West & Partners)*

difficult. However, in all cases the quality of the resulting armour layer is very much dependent on the skill of the individual machine operator.

If it is necessary for quality control or payment purposes to weigh individual rocks, then each of the types of handling equipment mentioned above can be fitted with a load cell for this purpose. To date, load cells have been most accurate when used with orange-peel grabs but have also been employed with a useful degree of success with grapples and buckets.

Control of surface profile during construction to the tolerances in Appendix 1 can be carried out using:

1. Conventional boning rods;
2. Plastic tape sight lines set a fixed distance above the desired final profiles combined with a suitable traveller with a D_{m50} hemispherical foot as described in Appendix 1 or measuring rod; or
3. Level and staff surveys, using the $0.5D_{n50}$ diameter hemispherical foot described in Appendix 1, combined with paintain coloured marks on stones to guide the operator (one colour if the stone is above the desired profile and another if it is below).

When placing armour layers which may have a wide grading, larger stones may be selected and used as 'key stones'. These will be placed roughly flat to avoid unnecessary protuberance out of the armour layer and then the smaller stones wedged between them.

6.3 Dam-face protection

This separate section on protection to the upsteam faces of dams is provided because of special environment geometric and structural constraints which the protection of earthfill water-retaining structures imposes. The section does not attempt to give any guidance on the other aspects of the design of earthfill dams, such as the consideration of fundamental geotechnical stability, as this is outside the scope of this manual. It is assumed that dam engineers reading this book will be competent in such fields, and will only need guidance on the face-protection aspects. However some of the geotechnical considerations discussed in Section 5.2 may be of interest to the dam designer.

The purpose of the section is therefore to provide guidance to the engineer concerned with the consideration of a quarried rock protection of the upsteam face in three major ways:

1. To identify the sections of this manual that are of reference or help;
2. To identify areas of difference between dams and inland lakes, and seashore and coastal conditions;
3. To offer guidance or comment in the areas of difference.

The section will take the reader through those relevant elements of the overall logic diagram (Figure 2) but the dam engineer should refer to Chapter 2 for an overview of the design process.

6.3.1 PROTECTION CONCEPT SELECTION

6.3.1.1 Planning-policy type boundary conditions

Functional requirement

Normally, this is primarily to protect the dam face against erosion by wave action. A secondary benefit is protection against erosion from rain.

Acceptable risk (see also Section 2.1.2.2)

A dam is unlikely to collapse as the result of a single storm destroying the protection. However, maintenance, both its practicality and its continuity over the life of the structure, needs careful assessment. A firm policy on maintenance needs to be fully documented and incorporated into the dam's operation manuals.

Although risk of ultimate collapse of the dam can be low, the hazard posed is often largel It is important therefore that the face protection be fully considered in the hazard and risk assessments carried out for the whole structure.

Environmental (see also Section 2.4)

The provision of a dam and its lake are themselves a major environmental impact. A full Environmental Assessment is likely to be called for to examine these impacts. The use of armour rock facing is likely to form a relatively small element of the project but would be included in any study. The need for early discussion and consultation cannot be stressed too highly.

The following aspects of the face protection may need special attention:

1. For Europe generally, the site is likely to be part of or close to a Site of Special Scientific Interest (SSSI) or Area of Outstanding Natural Beauty (AONB) and possibly affected by the Ramsar Convention. This will have a number of constraints on the design, including the appearance of the facing.
2. The creation of a reservoir frequently has a major impact on bird and fish communities (often this can be beneficial) and on archaeological sites. It can also cause significant human disruption. The facing can form an important access to the reservoir water for amenity activities such as fishing.
3. The source of rip-rap and associated sand and crushed stones are likely to be major environmental issues during construction, both ecologically and with regard to human impact. Special, dedicated quarries are likely to be required, unlike many seashore defences, and these may be some distance from the dam.
4. Because of (3) above, material transport, often by road, is likely to have a large community impact.

Economic

The availability of suitable rock, its transport and distance of transport, and the method of placing are all significantly different in character to most sea-defence systems. The costs both in capital and in maintenance are therefore also likely to be different.

6.3.1.2 Material sources

There is normally no possibility of transporting rock by water for a dam. Road transport becomes the normal method and rail is only rarely appropriate. Consequently, the rock sources available may be constrained by economic and environmental reasons to be relatively close to the structure. In addition, the rock may be obtained from quarries opened up for the project, usually primarily to supply material for aggregates, filters or fill, and not dedicated to the supply of large stone only. In such cases, a thorough geotechnical appraisal of the potential sources is required, with trial excavation whenever possible. Careful assessment of achievable rock sizes, grading and quantities is necessary. Chapter 3 gives extensive information on such matters and advice on selection of appropriate material sources.

Although this manual primarily considered natural rock, Sections 3.5 and 5.1.3.9 give general advice on the use of concrete as artificial rock, industrial by-products and stability-improvement systems.

Stability-improvement systems

This manual primarily addresses randomly placed rock armour. Dams are, however, often faced with shaped blocks of natural rock (pitched stone), concrete setts or large concrete slabs. Some indications of the stability improvement that may be achieved with pitched stone and other stability-improvement systems for rock are discussed in Sections 3.5, 5.1.3.9 and 6.2.3.5.

Use of industrial waste

The use of industrial waste (see Section 3.5.2) needs critical examination if considered for any dam. The presence of soluble (even at low level) poisonous substances must be carefully and conservatively examined. Unlike a river or sea protection, the water in contact with the facing is changed very slowly. There are no tides or currents to carry away even low concentrations of potentially harmful chemicals. The effects of the lake water with respect to acidity or to low salt content on such materials also needs consideration to assess critically the risk of disintegration of the material with time.

6.3.2 GENERAL DESIGN CONSIDERATIONS

6.3.2.1 Materials—specification and testing

The specification of rock for face protection is covered in Appendix 1, Sections A3.3 and A3.5. Parameters for rock shape, grading and size/weight relationships are given together with guidance on blasting to produce the required rock sizes and grading (Section A3.5.2).

With respect to dams, the following points may be relevant. Others may arise that are area- or site-specific (this feature of dams frequently arises):

1. The specification should take account of the likely material sources. This can lead to:
   (a) A reduced number of different tests required—the tests are material-specific;
   (b) A better matching of desirable properties and sizes to those physically available. Use of stability-improvement system (see Sections 3.6 and 6.2.3.5) may be appropriate.
2. Maintenance of a dam face is usually difficult. Despite 1(b) above, a higher specification standard may be necessary to limit maintenance.
3. Use of weaker rocks may not be advisable.
4. Specification and testing must allow for affects such as dissolution of the preferred materials by reservoir water. For example, limestone may be unsuitable for upland lakes due to attacks from acidic run-off from peats or from conifer forests. Measurement of natural examples of limestone dissolution are discussed in Jennings (1985), where rates to 1–10 mm per year for peaty waters have been recorded in conditions less turbulent and therefore milder than at a reservoir margin subject to waves. Similarly, but no less significant, waters of natural pH and low salts content may leach sandstones.
5. Temperature ranges inland are greater than on coasts. Freeze–thaw stability may therefore become a particularly important parameters in some fissured and porous rock types especially if used at high altitudes or in cold climates. The magnesium sulphate test may prove more valuable for assessing resistance to weathering of rock used on reservoirs in a dry climate (see Section 3.4.7).

6.3.2.2 Physical site conditions and data collection

A discussion of physical site conditions and data collection relevant to dam-face protection design is given in Chapter 4, particularly Sections 4.2.2.4 and 4.2.2.7. The major common differences betweeb dam and sea defences with regard to data collection are:

Figure 316 *Wave action at Carron reservoir and dam. (a) Wave attack on dam (Fetch 3.2 km, U=18.5 m/s, H_s=1.2 m, return period=2 years); (b) downstream spray during wave attack (courtesy Central Regional Council, Scotland)*

1. The reservoir does not exist (or not in its final form) at the time of data collection and hence predictions of waves (Figure 316) must be made from wind data.

2. The presence of the dam and reservoir can modify the local wind behaviour.
3. Tidal cycles do not normally exist and lake level changes usually take place comparatively slowly, except under storm conditions, when they can rise rapidly. Exceptions do occur. For example, tidal barrages experience tidal cycles and pumped storage reservoirs can fill and empty at similar rates to tides.

Wind climate for wave climate prediction

The subject of assessing wind climate for wave climate prediction (Section 4.2.2.7) is described in general terms in Section 4.2.2.4. However, for most dams, it is necessary to predict the likely wind strengths from limited data, despite the fact that wind speeds are the only tool for assessing the future wave climate. The following steps are put forward as a basis for obtaining wind data and the design wind conditions. Some engineering judgement will be involved in the latter.

Good data of the general area can usually be obtained from:

1. National meteorological offices. These sources are available for many countries worldwide but are inevitably of varying accuracy and are dependent on local records. Data usually include both actual records (including extreme event estimates or measurements) and statistics;
2. Records of events, rarely collated statistically, which are often obtainable from airfields or airports local to the area of interest;
3. Occasional inferential data on extreme events can be obtained from historical records, although the reliability of such data is generally not good for the site of the structure. Nonetheless, weather records and local weather phenomena should be included in any collection of data. In many places, local winds of great strength can blow for periods of a few minutes to days, examples including the plateau winds of South America and the siroccos of the Mediterranean. Inferential data can often be obtained from publications (e.g. Institution of Civil Engineers, 1978) or organisations within a country that are concerned with hydraulic or meteorological events. This may include government meteorological organisations, universities, and quasi-government and private research or engineering organisations.

Having obtained suitable data, the design wind conditions should be established as follows:

1. Determine the design return period of wind and wave storms that the dam face will be required to resist. In view of the difficulty of maintenance of most dam faces and the long life of such structures, it is usually advisable to design for long return-period storms and it may be advisable to consider several return periods to determine the sensitivity of the wave size to such variations.
2. Determine the range of wind directions likely to produce severe wave action at the dam.
3. Determine mean hourly wind speeds for the chosen return periods and directions from available data and modify these for duration and topographic effects. Allowance for duration can be made using the correction factors in Table 27. It should be appreciated that the storm duration required to produce the maximum wave height is usually shorter than an hour, fetches often being relatively small even in large reservoirs.

Allowance should also be made for topographic effects at the proposed site. It is recognised that surrounding hills and valleys can have a marked influence on the local wind in both strength and direction. Local knowledge and comparison with similar projects may provide guidance. In addition, the influence i some areas of the topography on such winds as the sirocco will need consideration. Attention to their effect both on records and on the proposed structure require attention. Examples are katabatic winds flowing down valleys close to the sea and also across wide, major plateaus.

Hydrological data

The study of floods, their size and their routing through a reservoir are outside the scope of this manual. However, account should be taken of the result of such a study with regard to:

1. Rate of reservoir rise—especially first filling, when rip-rap facing can settle and more significantly;
2. Maximum flood levels—the highest water level at which wind waves may be generated.

Other data

It is frequently observed that topography influences local winds and directions. The topography of the reservoir below water level is usually well documented, but not above this level. It is advisable to obtain topographic data of the dam and reservoir. This may include the whole catchment on upland water supply reservoirs. In the UK, Ordnance Survey maps at 1/25 000 scales may be suitable. Information on earthqukes will also be required for consideration of the whole dam and would be obtained for separate studies.

Hydraulic stability of rock dam-face protection (rip-rap)

A typical design for rock dam-face protection is illustrated in Figure 317. In recent years in the UK the work by Thomson and Shuttler (1976) has been extensively used for the hydraulic design of such protection. This has led generally to much larger rock sizes being specified than would have been the case using the older Hudson approach, which was derived for highly porous coastal mounds and breakwaters. The work of Thomson and Shuttler has now been confirmed in relation to the influence of porosity by van der Meer (1987, 1988a), who has expanded it by a further extensive series of tests. The resulting findings and methods for sizing armour are given in Section 5.1.3.1. When using van der Meer's formulae (equations (5.44) and (5.45)), dam engineers are recommended to use a permeability factor P of 0.1 for face protection, unless they can justify a higher value. Section 5.1.3.2 also discusses the requirements for underlayers and filters and other sections of Chapter 5 deal with features that also may be of interest to dam engineers, such as run-up and overtopping.

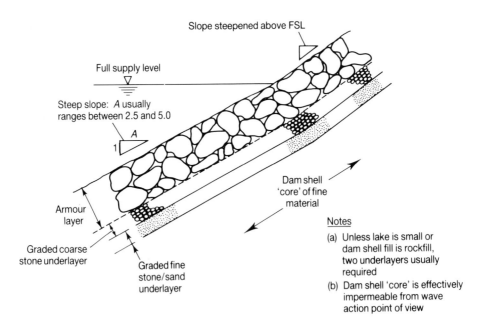

Figure 317 *Typical protection to dam face*

Geotechnical design

For dams, the geotechnical considerations of the protecting face only form a small element in the overall geotechnical appraisal and design of the structure. It should be appreciated that geotechnical dam design is outside the scope of this manual and that the sections on Geotechnical Data Collection (Section 4.3) and Geotechnical Interactions (Section 5.2) are written primarily for coastal sea-defence systems. However, some aspects of the design guidance given in Section 5.2 will be of interest and use to the dam designer, particularly those relating to wave penetration, settlement of rockfill, filter rules and designing with geotextiles.

6.3.3 SPECIAL DESIGN CONSIDERATIONS

In addition to the general hydraulic and geotechnical aspects of rock face protection there are some other special design considerations which may need to be addressed. Some of these are very site-specific, but those most commonly arising include the following.

Surface set-up

Except for shallow lakes, this is unlikely to be of any significance on most impounding reservoirs and can be ignored.

Hydrological variations in water level

The likely variation in water level due to reservoir operation and flood inflows can have a significant effect on the protection provided. It can:

1. Limit or extend the elevation range over which protection is provided;
2. Allow variation in stone size with elevation (due to reduced fetch at lower reservoir levels, for example);
3. Lead to reduced risk of consecutive wave attack at one level—if reservoir levels fluctuate regularly;
4. Result in increased risk of consecutive wave attack at one level—if reservoir levels are held sensibly constant (e.g. gated spillway reservoirs).

Seiche

Seiche effects will need special consideration along with other earthquake influences. These are comparatively long-period events, and, as waves, will have a much less severe effect than storm waves of comparable height. However, it may be necessary to take into account their influences on water level when combined with wind waves.

Overtopping

Dams are normally designed to not be overtopped even under extreme flood conditions. However, the joint probability (Section 4.2.8) of maximum wind storm with maximum flood levels needs to be considered in certain cases. The Institute of Civil Engineers (1978) suggested for the UK that 1 in 10-year return wind speeds be used for wave height computations associated with maximum flood levels, indicating that the risk of maximum wave and maximum flood coinciding is considered low in the UK. However, with proper data, a full joint probability analysis of wind and flood could be carried out to determine the actual risk more objectively, and may be advisable in countries with less extensive historical records.

Wave refraction

There are cases on record where waves approaching a dam at oblique angles have been refracted and caused damage that otherwise was not expected. In unusual-shaped reservoirs the effects of refraction, even in relatively deep water, should be investigated.

Wave reflection

It is possible that wave reflection from nearby land forms or structures such as spillway weirs may increase wave heights adjacent to the dam.

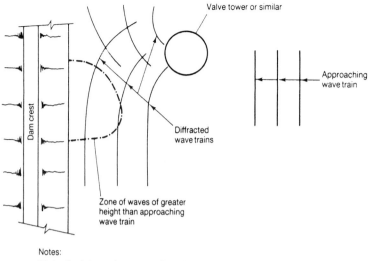

Figure 318 *Diffraction patterns. Notes: (1) Any fixed obstruction upwave of dome face will cause diffraction. (2) Zone of increased wave height and interaction with dam face depends on wave length and location of structure*

Wave diffraction

There is growing evidence that the presence of offshore structures such as valve towers and bellmouth spillways causes wave diffraction that may lead to local increase of wave height and consequent damage. Such effects should be taken into account, as the increase in wave height can be of the order of 10–20%. Figure 318 illustrates the general pattern of diffraction around a tower close to a dam.

At present there are no very simple measures that can be used to predict diffraction patterns from an isolated structure. The diffracted wave activity will be dependent on the shape of the structure, its size relative to the incident wave length and its orientation.

The *SPM* (1984) gives graphical methods of calculating the diffraction effects from structures such as harbour head entrances. These methods are appropriate for structures which are rectangular or near-rectangular in shape, and whose width (in the direction normal to the wave direction) exceeds about five wave lengths. It will be necessary to use the principle of superposition to establish the effect behind such an obstacle, as diffraction will occur round both sides of the structure. In situations where the rectangular obstacle is less than about five wave lengths there will be significant interaction between the diffracted wave fields emanating from the ends of the structure, and linear superposition should not be used. The methods used to predict diffraction in this case require the evaluation of special functions. For some configurations, diffraction diagrams, similar to those in the *SPM* (1984), are given in Smallman *et al.* (1989).

For a circular obstacle the diffracted wave field can be calculated using the method given in Mei (1983), which is based on early work by McCamy and Fuchs (1954). Calculations employing the methods described above will give the user an estimate of the effects of diffraction and enable engineering judgement to be applied to estimate:
1. The local increase (if any) of rip-rap protection required;
2. The distance from the dam, and size in relation to the wave length, of a tower that minimises or eliminates the chances of increased wave height.

It must be emphasised that the above approach must be used with caution and as a tool for engineering judgement, not precise design. Hydraulic model tests of individual designs may be warranted.

Landslides

There have been many instances of landslides into reservoirs causing large waves that overtop the dam, and separate geotechnical analysis and wave analysis will be required for such events. Although outside the scope of this manual, the designer and operator must be alert to the possibility.

Earthquakes

Earthquake effects, particularly movement of the dam against and away from the reservoir water, would form part of the studies of the stability of the whole structure. They are not therefore covered directly by this manual. (See also 'Seiche' above.)

Upstream slope angle

It must be kept in mind that on a dam the slope angle of the upstream face is controlled to a large degree by the overall design, particularly stability. There is little opportunity to use slopes that, from a protection aspect, are optimal. There may, however, be a possibility of introducing berms (as beaches) or of locally steepening or flattening slopes at some levels.

6.3.4 SPECIAL CONSTRUCTION ASPECTS

Other parts of this manual which discuss construction should also be used for general guidance with respect to dams. These include Sections 6.1.3.4, 6.1.5, 6.2.3.5 and 6.2.5 and Appendix 1, Section A1: Part C.

However, a dam, with its inland location and usually with no lake against it during construction again raises special features, many site-specific. Broadly, the differences between dam-face protection and other applications of rock discussed in this manual are that a dam is comparatively high with a dry upstream face during construction, and is of intermediate slope between conventional coastal revetments and coastal anti-scour blankets. The differences mean that gravity provides little help in placing rock, while working on the face is usually impractical and unsafe. These considerations lead to the following non-site-specific factors to be taken into account in placement of rock, which are illustated in Figure 319:

1. The method of placing must be clearly specified. As it is difficult to ensure a performance-based system, method specification is often used. Ideally, this should be based on fields trials at the site using the proposed materials, to ensure compatibility between plant used and conditions. It may be possible to include these trials in a preliminary construction contract and incorporate into the main works a specification based on them.
2. The armour and its underlayers will almost universally be placed from the surface of the dam as the dam is constructed. Access back to the face after construction of the main dam body is normally not feasible.
3. Because of placing difficulties on a dam and the need to avoid, as far as possible, the requirement for any maintenance (see Section 6.3.6), it will be necessary to take extra care that the armour layer is placed to achieve good interlock between pieces to ensure integrity of the facing. Similarly, underlayers and filters will require careful placing and compaction, and care to avoid disturbance prior and during armour layer placing (see Figure 317 and Appendix 1, Section A1: Part C).
4. Placing of rock and its underlayers should follow closely the level of the general fill. Both slight lead and slight lag have proved successful.
5. Rock should not be pushed uphill into place, nor any distance, when being placed by tracked or wheeled vehicles.
6. Pushing downhill a short distance, forcing the stone into the previously placed

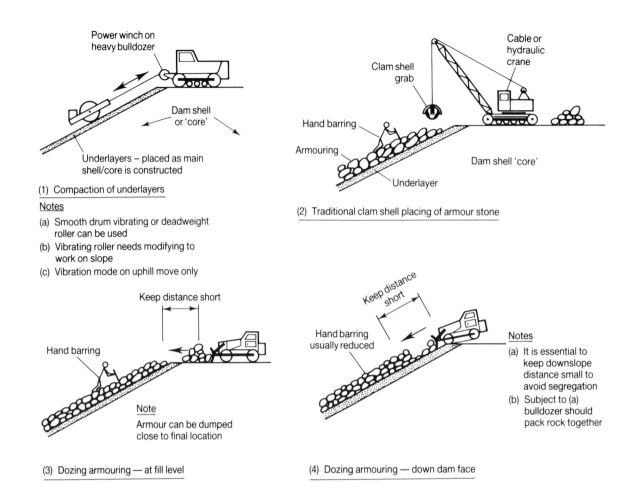

Figure 319 *Placing armourstone on dam faces*

protection, has been used successfully. Moving long distances must be avoided to prevent segregation.

7. Placing by grab and barrier by hand can be successful.
8. Compaction of underlayers can be achieved by heavy vibrating rollers operating on the slope of the dam face. Both vibrating and deadweight modes have been used successfully. However, conventional compaction in horizontal layers may be preferred, depending on the method of placing stone protection.
9. Secondary placing by crane and hand barring should be considered for achieving a uniform face free of open or loose areas.
10. End tipping from lorries or dumpers will lead to a poorly placed uneven and loose facing.

6.3.5. MEASUREMENT AND COST ASPECTS

As with sea- and coast-defence systems, measurement for payment coupled with assurance of compliance with a specified thickness has always been difficult. Appendix 3 provides guidance on this which, apart from placement through water, will generally be applicable to dams. Section 3.6 covers quality control. This again, apart from rock placed in water and as core material, is generally applicable to dams. The verification test panel approach to the layer thickness problem (Section 3.6.3.2) is commended as a way of avoiding measurement payment disputes.

6.3.6 MAINTENANCE ASPECTS

The main chapter in this manual on maintenance (Chapter 7) gives a background

> **Box 84** Maintenance of UK water-retaining dams
>
> The Reservoir Act 1975 applies to all water storages in the UK. All procedures must comply with the requirements of the Act and instructions given under it. All water storages above 25 000 m³ capacity are covered. Regular inspection and supervision are both required to be carried out in prescribed ways at stipulated intervals by engineers appointed under the Act. Also prescribed are various records, certificates and reports that have to be maintained or undertaken. All these are aimed at preventing incidents that may affect the safety of the public.
>
> It is normal practice for the construction engineer, the inspecting engineer and the supervising engineer under the Act to cover in their reports and documentation matters of general maintenance as well as safety. This usually forms part of their general brief.
>
> As a consequence of the Act, standards required in operation and maintenance are, of necessity, of a high level. Also, under the Act, the engineers will, with respect to their various roles, have instituted maintenance strategies. Application of the strategies given in Chapter 7 will have therefore to be framed within the needs of the Act and from practice arising from both this Act and its predecessor.

and philosophy for all shoreline and coastal engineering, which is also applicable to dams. The selection of the correct maintenance strategy at the outset is perhaps the most important aspect in relation to face protection for water-retaining dams. This is because access to the face of a dam is normally difficult and repair works therefore pose special problems. These difculties at the earliest stage possible in the design process should be considered and agreed policies on their solution adopted and written into the operation and maintenance manuals. This may involve approaching the design on a no-failure or low-statistical-risk-of-failure basis. Where a 'no-maintenance' requirement conflicts with practical constraints on the size and availability of the required armourstone alternative methods of face protections may need to be adjusted. Stability-improvement systems for rock (see Sections 3.5 and 5.1.3.9) may be worth considering in this context. The additional care in placing discussed in Section 6.3.4 may pay off if it avoids loose areas which may otherwise be subsequently damaged.

6.4 Gravel beaches

A gravel beach can be defined as a layer of gravel at the bottom surface both below and above the water level, which is attacked by waves. In many places such beaches exist in nature, but in others they are constructed artificially. The word 'gravel' is understood to include both rounded and angular small stones. 'Shingle' is more often used for marine rounded stones. Because gravel has a broader definition, this word will be used here.

The main difference from the design of the previous structures described in this book is that gravel is allowed to move. The elements of the other structures are supposed to stay in their place (statically stable) but gravel may have a large displacement both parallel and perpendicular to the coast. The movement perpendicular to the coast should be such that the layer as a whole stays more or less in its place (dynamically stable). The movement parallel to the coast may require renourishment.

There are two main differences with sand beaches. Firstly, the slopes are steeper, which makes it possible to have a greater height differences over a smaller distance. Second, gravel moves less than sand for a given wave climate, which makes it more stable under wave and current attack. In Section 5.1.1.2 the difference with sand beaches is expressed in the $H_s/\Delta D$ value. For sand this value is generally larger than 500 and for gravel generally smaller than 500.

The use of gravel beaches can be divided into the following four main purposes:

Recreational beaches

Existing gravel beaches are frequently used for recreation, and erosion may reduce these. Structures in the vicinity of the beach, such as harbour moles, may affect the beach stability. Renourishment with gravel can restore the beach to its original form. New recreational beaches can be constructed at locations where sand would be unstable in view of the steepness of the foreshore, or when there is no sand available. Important parameters in the design of a beach are the volume of gravel required and that available.

Protection of sandy slopes

The main difference between sand and gravel beaches is the equilibrium slope: gravel allows steeper slopes, both under water and on the dry beach. At some locations there is no room for gentle slopes, at others, less material and costs are involved when steeper slopes are made. In such cases a gravel layer gives a cheaper solution, if gravel is available near to the site. Examples are a dike of sand, an artificial area of sand bordering on the water or an artificial island of sand. A layer of gravel from some metres below the lowest water level to some above the highest water level makes a steep slope possible. An important parameter in this case is the layer thickness.

A different use of gravel is to protect a waterfront consisting of fine materials where tidal estuaries have been recently closed off from the sea. After the closure, waves may attack sandy coastlines at certain water levels. The application of gravel at the water level can reduce or halt the erosion in this situation. It can be combined with a detached breakwater (see below).

Protection of structures or beaches

A third use of gravel beaches is to reduce erosion and/or wave attack at structures, dunes or beaches. First, a gravel beach can be built against a structure (e.g. a seawall or a dune). The waves will first run up the beach and be reduced when they reach the seawall or dune. In addition, the waves cannot attack the toe of the seawall or dune because of the gravel beach.
Second, a gravel structure in the form of a detached breakwater can be constructed in the water in front of a beach to reduce the wave attack.

Energy absorption

In lakes or harbour areas vertical walls at the bank can give wave reflection. This leads to increasing ship motions and hampers port operations, especially when this reflection causes resonance in harbour basins. A rather steep gravel beach, absorbing almost all wave energy, may significantly reduce this phenomenon and consequently the operational downtime of the harbour.

The type of gravel to be used for beachfill depends on the hydralic loads and on the use of the beach. Where the beach will also be used for recreation, normally only rounded gravel (shingle) will be acceptable. In this situation the hydraulic requirements and the recreational requirements may sometimes conflict. For the mining aspects of gravel, such as borrow methods and equipment, reference is made to Sections 3.1 and 3.4.

6.4.1 DESIGN PROCESS

In the design the same main approach is valid as given in Section 1.2 (Figure 1 and 2). Often, the construction of a gravel beach will be only one solution within a wide range of alternatives that may vary from an unprotected sandy beach to a concrete seawall.

The design process for a gravel beach is very similar to that to be applied in the design of a sandy beach, a special manual on which has been published by CUR (1987). Planning and design of beach-nourishment schemes, computation methods, design parameters, execution methods, quality assurance and environmental aspects are dicusssed thoroughly in this manual, which may also be used in the design of a gravel beach. Reference should also be made to the discussion on beach design and control in Section 6.2.2.

An important factor in the design of gravel beaches is the reliability of the prediction of the beach development after completion. Such hydraulic and morphological aspects are more important in the case of gravel beaches than for rock structures with smaller values of $H_s/\Delta D_{n50}$, while the structural aspects are more straightforward.

The following steps can be distinguished in the design process of a gravel beach:

1. Definition of the characteristics (grain sizes, etc.) and the behaviour of the beach area in the past;
2. Definition of the ideal material for the new beach, taking into account the rate of loss of material in the short and long terms;
3. Definition of the availability of gravel, including the production costs;
4. Definition of the new beach profile, taking into account hydraulic aspects (overtopping, toe erosion, etc.), execution methods and costs;
5. Definition of any additional measures that may improve the stability of the gravel beach, such as groynes (see Section 6.2.2).
6. Optimisation of the beach design with respect to type of gravel to be used, life expectancy, application of additional protection of the beach, risks of damage to the beach and consequent damage and costs.

6.4.2 COASTAL PROCESSES

Both for the definition of the behaviour of the area before the construction of the gravel beach and for the prediction of the developments of the beach after completion, a thorough investigation should be made of the hydraulic and morphological processes that play a role in the coastal zone. Coastal processes are discussed further in Section 6.2. Relevant hydraulic parameters are described in Chapter 4 and Section 5.1.1. For design of gravel beaches the daily distribution of these hydraulic parameters is as important as the extreme values. In fact, the combined statistical distribution of water levels and wave parameters (height, period, direction) is required for a proper evaluation of the coastal processes.

The wave and water level statistics, together with the bathymetry and grain characteristics of the beach material, are the most important parameters for the determination of longshore transport rates and hence of the erosion rate of the beach (net transport out of the area in question). In Section 5.1.3.10 design formulae are presented to determine the longshore transport of gravel as a function of the wave characteristics. Further information may be found in van Hijum and Pilarczyk (1982), Morfett (1988), Chadwick (1989) and Kamphuis (1990). As knowledge of the coastal processes is limited, as if often also the knowledge of the hydraulic parameters (wave statistics), care and engineering judgement is needed in the application of any transport formula.

Because of these uncertainties in sediment transport calculations, the actual development of the coast in the past should also be considered in detail. This historical development can be used to determine the reliability of the calculation method and the input parameters by means of hindcast calculations of coastline development. In all cases, not only a mean value, μ, of the predicted coastal transport has to be determined but also the range that may be present. From this range an estimate can be made of the standard deviation, σ, of the longshore transport.

The loss of gravel from a breach section is determined by the differences in longshore transport between the upper boundary and the lower boundary of that section (S_1 and S_2). During the year the wave direction and, as a consequence, the transport direction of the gravel may change. Therefore S_1 and S_2 will also vary.

Integrated over the year the following erosion, E, is found:

$$E = \int S_2\, dt - \int S_4\, dt = S'_2 - S'_1$$

Both integrated transports S'_1 and S'_2 have their own mean value μ and standard deviation σ. The mean value of the erosion, E, can be found by simply subtracting S'_1 from S'_2.

Assessment of the standard deviation of the erosion is more involved since the standard deviation of S_1 and S_2 consists of correlated elements (e.g. the accuracy of the deep-water wave climate) and of statistically independent elements (e.g. the effects of the local bathymetry). In the calculation of standard deviation of erosion the independent elements in the standard deviations of S_1 and S_2 will tend to lead to a wide range around the mean value of the erosion rate, such standard deviations being added using $\sigma_{Ei}^2 = \sigma_{1i}^2 + \sigma_{2i}^2$. In this way the standard deviation of the erosion rate can become as high as the mean value or even higher.

The longshore transport on a beach may be influenced by coastal structures such as harbour moles. An updrift structure may cause erosion downdrift, when no supply occurs. A downdrift structure may lead to an accumulation of gravel, which can easily be recirculated on the beach. Where there is no structure downdrift, the longshore transport may move outside the area of interest normally moving the finer material first. The environmental impact of such movement should be assessed, for instance, on navigation channels or on the ecology (see Section 2.4).

6.4.3 GRAVEL BEACH PROFILE

In the short term, immediately after construction, a major reshaping of the profile will take place. The extent of the initial reshaping depends on the differences between the profile 'as placed' and the equilibrium profile of the gravel beach. When the gravel is mainly placed above the waterline this short-term adaptation may appear to be a very serious erosion, particularly since it will occur immediately after construction. In reality, there is no loss of material, gravel simply being transported downwards along the slope for form a stable profile from seabed to crest. In the long term, fluctuations of the beach profile will also occur which will depend on the prevailing wave and water level conditions. The work of Powell (1990) suggests that such changes occur rapidly (within 500 waves).

Calculations of gravel beach profiles may be made using the profile models of van der Meer (1987, 1988a, b) or of Powell (1990) decribed in Section 5.1.3.7. Apart from the grain size of the gravel (D_{n50}), the profile is determined by the wave conditions and the still-water level. The range of fluctuations in the dynamically stable profile can be determined by calculating the beach profile for several typical wave and water level conditions.

The thickness of the gravel layer on the seabed subsoil should be such that the whole range of calculated profile fluctuations can be met in the gravel layer itself without exposure of (parts of) the subsoil to the wave action. Where the gravel beach is also subject to a long-term erosion because of gradients in the longshore transport, this condition should not only be met immediately after construction but also after a number of years at the end of the design life of the beach (see Figure 320).

Special attention should be paid to the crest and toe levels of the gravel fill. Where sufficient gravel is provided, the crest level will adjust itself naturally to coincide with a run-up level, $R_{u2\%}$, that will minimise overtopping (see Section 5.2.1.1 for details). It is therefore not always necessary to raise the top level of the gravel fill

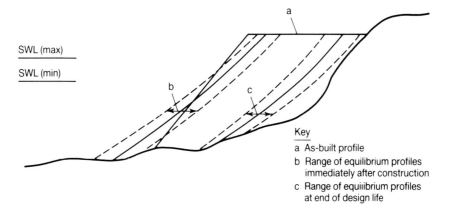

Figure 320 *Ranges of simplified gravel beach profiles (vertically exaggerated scale)*

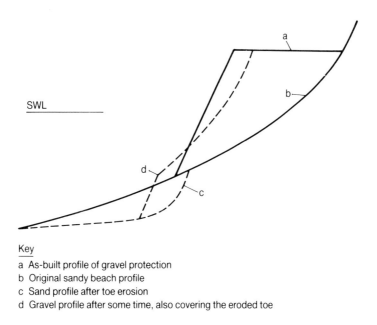

Figure 321 *Self-healing process of toe gravel beach (vertically exaggerated scale)*

as high as the highest run-up that can be expected during the lifetime of the gravel beach, so long as an adequate cross-sectional volume of gravel is provided.

Sometimes it will not be economic to provide a volume suitable for the most extreme conditions and, depending on he infrastructure in the zone immediately behind the gravel beach, overtopping and even erosion of the backfill may be acceptable during such extreme conditions. In such a situation the only criterion is that the washing out of backfill material should not damage the overall stability of gravel layer. If, however, the stability of the backfill itself is one of the main reasons for providing the gravel protection, the full volume of gravel should be provided or other protective measures (e.g. underlying rock scour blanket) provided.

For the toe level of the gravel similar considerations are valid. When the gravel is placed on a sandy bottom, erosion of the toe can be accepted, provided that the volume of gravel is sufficient to cover the eroded toe (see Figure 321).

6.4.4 CONSTRUCTION AND COST ASPECTS

The main features and costs of constructing a gravel beach are the transport stages

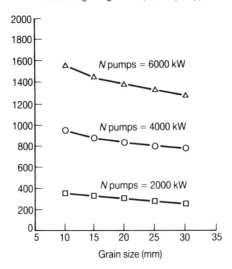

Figure 322 *Maximum discharge length for gravel pumping*

from borrow area to beach and their costs. Therefore the method of transport (discussed briefly in Section 3.4) will be discussed here in more detail.

The gravel is generally first taken by a dredger from the seabottom. Geneally, this will be a trailing suction hopper dredger (TSHD). The TSHD dredges the gravel from the bottom into its hold. For large projects TSHDs of 5000–10 000 m^3 capacity are the most economical. Dredges which pump the gravel into another ship are also available.

Next, the gravel has to be transported close to the required site. This travel distance makes a great contribution to the costs of a project.

Once at the site, small amounts can be brought ashore in a harbour and carried by road to the site. For large contracts, the gravel is dealt with in two principal ways:

1. The gravel is dumped and later pumped ashore by another dredger, typically a cutter section dredger (CSD); or
2. It can be pumped ashore from its hold by the TSHD itself by a pipeline. Occasionally, the lowest part of the underwater beach can also be directly dumped.

Because of the energy needed to pump the gravel ashore, the pumping distance is generally restricted. Figure 322 gives a preliminary indication of the pumping distance as a function of pumping capacity and the mean grain size of the gravel.

One difference between pumping gravel as compared with sand is that the maintenance costs of both pumps and pipelines are much higher. The pumps wear quickly and have often to be replaced frequently. Pipes also wear more quickly, especially rubber floating pipelines. In general, the costs of pumping gravel per cubic metre are several times higher than those of pumping sand per cubit metre.

In the intertidal area the pipeline has to lay on the bottom, as a so-called 'sincker' pipeline. Because the pipeline leaves the dredge at the water surface with a floating pipeline, a special method is necessary to bring the pipeline gradually to the bottom. Generally, this method is related to the way the ship is kept in its correct position. Several combinations of ship-positioning methods and pipeline arrangements may be used:

1. The ship is positioned on anchors. The pipeline is on the surface kept in place by the ship. Over some distance the buoyancy material around the pipeline is

Figure 323 *Artist's impression of gravel beach construction at Seaford, UK in 1987 (courtesy Zaner Dredging)*

reduced, thereby reducing the buoyancy and allowing the pipeline to gradually slope down to the seabed.
2. The ship is moored to a pile construction or an anchored pontoon. The pipeline goes from the pile or the pontoon to the sincker pipeline at the bottom.
3. The ship is kept in position by dynamic positioning, but in this case the energy required for such positioning cannot be used for pumping. In the same way as a case (1) above, the pipeline slopes to the bottom, where an underwater pontoon or anothe structure fixes the connection between the floating and the sincker pipeline. An example of this working method is given in Figure 323.

An important cost factor is the maximum wave height and the maximum wind speed, at which the connection between ship and pipeline can be maintained. It is important to choose a time of the year for gravel beach construction when waves and wind are favourable.

On the beach the gravel comes out of the pipe and forms a gravel hill. This hill is

> **Box 85** Construction details for Seaford gravel beach-nourishment project, 1987
>
> An impression of the operation of a beach-renourishment project at the English south coast (Seaford) is given below:
>
> | | |
> |---|---|
> | Hopper capacity | 8000 m^3 |
> | Total power installed | 11 474 kW |
> | Hopper pump ashore power | 2 × 2270 kW |
> | Pipeline diameter 0.8 m | |
> | D_{50} | 15 mm |
> | Amount of gravel brought on beach | 1 500 000 m^3 |
> | Production rate average | 75 000 m^3/wk |
> | minimum | 70 000 m^3/wk |
> | maximum | 90 000 m^3/wk |
> | Pumping distance | 1250 m |
> | Water depth at begin pipeline | 11.5 m |
> | Tidal range | 3 m |
> | Maximum waves | 4 m |
> | Waves, limiting operations | 2 m |
> | Maximum wind | Force 8 (Beaufort scale) |
> | Wind, limiting operations | Force (Beaufort scale) |
> | Maximum storm-surge water level | CD + 6 m |
> | Time to connect ship to pipeline | 30 min |
> | Project time | 8 months |

then rapidly spread by bulldozers over the area around the end of the pipe. If the spreading distance becomes too large, pipes are connected to a new location, where the procedure is repeated.

If levelling is used, experience indicates that the beach profile can be made with an accuracy of ±0.1 m. Unlevelled or underwater beaches will have less accuracy. Details of construction of a typical gravel beach are given in Box 85.

6.4.5 MONITORING AND RENOURISHMENT

Large amounts of gravel or considerable layer thicknesses will give protection for a long time, but the costs are high. More frequent renourishment with less material will often be a more economic option. The timing of any subsequent renourishments can be determined on the basis of monitoring data.

For a proper analysis of the behaviour of the nourished gravel it is important to know properly the initial condition before nourishment. Just before the gravel is nourished, a detailed survey (in-survey) of the whole area should be made in an intensive network, obtaining levels at points on a square grid 10 m × 10 m. The cost of this detailed survey will be small compared to the project cost and will make it possible later to know the layer thickness in every location.

After placement of the gravel a new survey (out-survey) should be made in order to determine the layer thickness at every point so that subsequent erosion can be measured. The in- and out-surveys can also be used to determine the total fill volume and the payment to the contractor. After completion, further surveys should be made on a regular basis to determine any regression of the coastline or any weak thin spot in the gravel layer. These weak spots can be surveyed more frequently. Surveys are also recommended after severe storms.

Once the layer thickness at any point reduces to zero, any subsequent erosion process depends on the characteristics of the original seabed. Where this seabed is sandy, further erosion will take place and a scour hole will develop. Cases where underlying chalk has eroded have also been recorded. However, such holes will partially be filled in by gravel from the surroundings. In this way, an ongoing erosion of the subsoil is normally prevented: a gravel beach is normally self-healing. Local erosion of the subsoil during the large cross-shore gravel movement which takes place during severe storms may not necessarily therefore be critical, as the original shape of the profile will be re-established during moderate daily conditions after the storm.

The main situation where self-healing will not occur is when the erosion of the subsoil is caused by a gradient in the longshore gravel transport. Occurrence of holes in the subsoil in such a situation will normally indicate that the gravel layer has reduced to a critical thickness and renourishment must then be considered.

The working method for beach renourishment is similar to that explained in Section 6.4.4 for initial beach construction: small volumes can be transported by road, but larger amounts over a wide beach section normally have to be renourished by hydraulic fill (dredging, transport to the shore and pumping through a pipeline to the beach). The costs of renourishment per cubic metre are comparable with the original costs of the beach nourishment.

6.4.6 COST OPTIMISATION

Three main elements that determine the costs of the nourishment operation must be considered:
1. The location of the borrow area and the grading of the material in it;
2. The weather conditions during construction;
3. The volume of gravel required per metre beach length.

6.4.6.1 Influence of location of borrow area and grading of borrow material

To evaluate this factor, the grain size and the sand content of the gravel must be compared with the cost of the transport distance and the cost of pumping (which depends on the depth of water at the borrow area). If there is a high sand content in the gravel, this sand (apart from that lost in dredging) will also have to be transported to site. Although this sand content may reduce pipeline wear costs at site, the sand will only have a limited resistance to erosion. Thus the total nourishment volume to be provided will need to be larger. This extra volume can be estimated using (James, 1975; Fleming, 1990a) an appropriate overfill ratio as described in Box 86. Overfill ratios will also need to be calculated when comparing borrow areas with fine and coarse gravel as well as the relative pipeline wear that the gravel will cause.

From the above the main borrow area and borrow material grading cost considerations can be summarised as:

- Dredging depth;
- Transport distance;
- Gravel grading and sand content which in turn influence pipeline wear and the required overfill ratio.

Sometimes in the optimisation the royalties sought by the owner of the gravel also have to be taken into account as well as environmental aspects and the time needed to obtain the required permits and dredging licences.

6.4.6.2 Site conditions during construction

If fair weather conditions are only available in a short period an optimum has to be found between the cost of working in unfavourable conditions and that of planning in that short period. Recreational use of the beach, that may affect the production rates in the summer season, should also be considered.

Box 86 Estimation of beach-nourishment losses

The losses that can be expected to occur from an area that has been renourished can be estimated roughly using a method based on the composite grainsize distribution of both the imported material and the indigenous beach material. The method is based on comparisons between the respective grainsize distributions defined on the phi scale:

$$\phi = -\log_2(D) = -3.322 \log_{10}(D)$$

where D is the grain diameter (in millimetres). Grainsize distributions on natural beaches often exhibit a log-normal form and the imported form and the imported material is assumed to be similar. These distributions are defined by two principal parameters:

1. The phi mean (μ) is a measure of the location of the centroid of the grainsize distribution. This can be defined in a number of ways, for nearly log-normal distributions

$$\mu = (\phi_{84} + \phi_{16})/2$$

where ϕ_{84} and ϕ_{16} are the eighty-fourth and sixteenth percentiles;

2. The phi sorting or phi standard deviation (σ), a measure of the spread of grain sizes about the phi mean and for log-normal distributions, can be approximated by:

$$\sigma = (\phi_{84} - \phi_{16})/2$$

Comparison between indigenous material, subscripted n, and the imported material, subscripted b, are made by considering phi mean difference $= (\mu_b - \mu_n)/\sigma_n$, and the phi sorting ratio $\sigma_r = \sigma_b/\sigma_n$

Using the above parameters and defining an overfill ratio as the ratio of the volume of the material that must be placed and the required design volume, the predicted required volume of fill can be determined with the aid of Figure 324 (James, 1975).

Figure 324 is split into four quadrants. Quadrants 1 and 2 represent regions where the imported material is more poorly sorted than the indigenous material. Quadrants 1 and 4 represent regions where the imported material has a finer phi mean than the indigenous material. Points which lie in quadrants 2 or 3 will generally result in a stable fill. Those lying in quadrant 1 will result in stable fill for some combinations, but losses could be large. Points lying in quadrant 4 generally indicate an unstable fill. While this figure serves as an approximate guide, values of the overfill ratios should only be used for comparative purposes. The technique makes no allowance for the effects of groynes and should not be used to obtain absolute values.

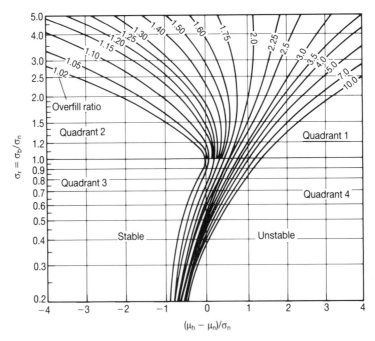

Figure 324 *Isolines of overfill ratio, R, versus phi mean difference and phi sorting ratio*

Note that the calculated overfill ratio should not necessarily be applied to the entire fill volume when reclamation as well as beach building is involved. The relative proportion will depend on the rate of filling and the extent of the active beach zone.

6.4.6.3. The volume of gravel required per metres beach length

The volume of gravel to be dumped depends on three items (see Figure 320 and Box 86):

1. The range of beach profiles to be expected as a function of the hydraulic conditions;
2. The required overfill ratio (depends on grading of gravel);
3. The reserve for long-term erosion by longshore transport during the lifetime of the structure.

None of these items are known exactly beforehand. For this reason, not only should the most likely required volume of gravel (the mean or μ-value' be determined but also the standard deviation, σ.

As stated earlier, the standard deviation can be as high as the μ-value. Consequently, the $(\mu+\sigma)$-value of the volume of gravel may be much higher than the μ-value.

Selection of the μ-value for renourishment volume will, of course, impose a significant risk of failure and, in some situations where only a very limited risk is acceptable, selection of the $\mu+\sigma$ or $\mu+2\sigma$ will be necessary. In many cases, however, a high erosion rate of the beach will not immediately lead to high consequential damages to waterfront structures, and here the design can be based on a gravel volume close to the μ-value.

Regular monitoring of the beach profile after the first nourishment may provide reliable information on actual erosion rates. This information must, of course, be evaluated before planning future renourishment activities. Other aspects that determine the optimum volume of gravel are:

- The method to be adopted in spreading and levelling the gravel;
- The conditions of the foreshore: steep slopes may lead to gravel losses to deep water when a large volume of gravel is dumped at the shoreline;
- The availability of borrow areas at the time of nourishment and thereafter;
- The availability of dredging equipment: high mobilisation costs will seriously influence the risks that can be accepted and/or the nourishment interval to be chosen;
- The available of funds.

6.5 Rockfill in offshore engineering

This section describes the design and use of rockfill in offshore engineering for bottom protection, particularly to:

- Protect and stabilise pipelines and transmission lines; and
- Provide bottom and scour protection for offshore structures in general.

Protection of pipelines and transmission lines is often necessary because possible incidents may result in:

- The release of the contents of the pipeline, cause serious environmental impact;
- High repair costs;
- A loss of income in the period between the accident and the final repair;
- A shorter lifetime of the pipeline or cable.

During operation, pipelines can encounter the following hazards:

- Hydrodynamical forces due to the action of waves and currents;
- Overstressing and vibration caused by freespans. These can be the result of scour of the seabed or fast morphological changes of the seabed (sandwaves);
- Dropping or hooking of ship anchors;

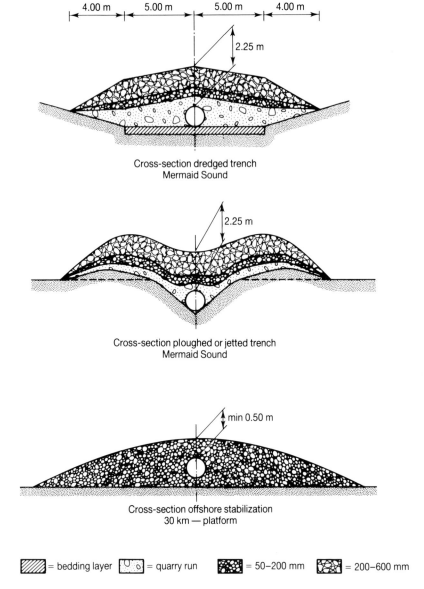

Figure 325 *Examples for a pipeline rock cover protection/stabilisation*

- Hitting or hooking of fishing gear;
- Dropped objects (containers, tools), especially in the vicinity of platforms;
- Buckling, caused by thermal expansion of the pipeline;
- Waxing within the pipeline due to a temperature drop along the pipeline;
- Decreasing viscosity of the transported crude oil, also caused by a temperature drop along the pipeline.

The pipeline can be protected in several ways:

- Placing of a rockfill cover (Figure 325);
- Lowering below the seabed by pre- or post-trenching or fluidisation;
- Self-burying of the pipeline, sometimes fitted with vortex spoiler strips, in a soft or loose-packed subsoil;
- Placing of a wide variety of mattresses;
- Placing of artificial seaweed;
- A combination of above-mentioned techniques.

Besides the protection of pipelines, scour protection is often necessary around the legs or concrete walls of platforms, due to the risk of erosion and consequently instability cause by wave and current action.

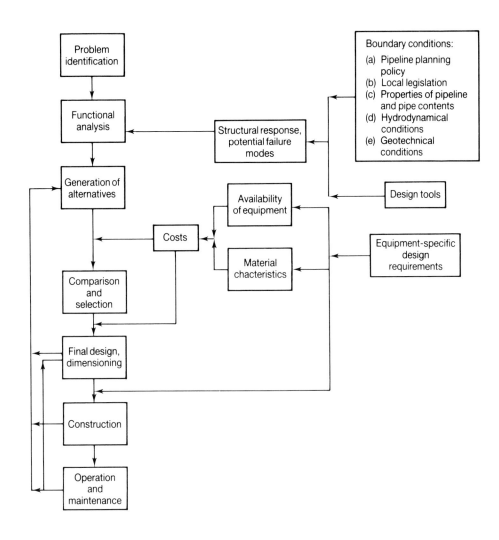

Figure 326 *Simplified logic diagram of the design process*

A simplified logic diagram of the design process for bottom protection is given in Figure 326 and the principal design considerations are discussed in Section 6.5.1 together with the main physical processes involved. These processes will be described by simple formulas or models which enable the optimisation of a design of the structure. Some models are too complicated to describe in this relatively short section and in such cases references are given to publications where a more complete description of these models can be found. Section 6.5.2 then discusses construction aspects together with their relation to the detailing of the structure.

6.5.1 PRINCIPAL DESIGN CONSIDERATIONS

6.5.1.1 Design approach

The traditional design of an offshore rockfill structure is based upon a deterministic approach. The structural dimensions are calculated by choosing a limit state condition with respect to the loading of the structure which correponds to a certain characteristic strength of the structural elements. Where possible, the selected design loading is usually extracted from a statistical data analysis.

As mentioned in Section 2.4.1, a limitation of this approach is that loadings below the design value do not contribute to the development of damage. One can see this as a limitation of this approach, but because of the small cross-sectional area of an offshore rock structure (approximate range $10–20\,m^3/m^1$) and the high costs of repair, almost no development of damage caused by loadings below the design value can be allowed. The required structural dimensions are therefore calculated to ensure that the structure is just stable.

6.5.1.2 Hydraulic stability of rockfill

Offshore-placed gravel and rockfill structures must be stable under the action of steady state (tidal and wind driven) and wave-induced currents. In Section 5.1.3.11 the traditional design method based on the 'critical shear' or 'incipient motion' concept has been described.

With the use of these equations, if applicable together with a slope angle-induced critical shear stress reduction factor, a minimal required grain size, described by the median diameter, D_{50}, can be found. If the rockfill embankment has such dimensions that the induced flow conditions can decrease rockfill stability, separate turbulence calculations should be carried out in order to quantify the shear stress exerted.

6.5.1.3 Geotechnical stability

A second step in designing an offshore rock structure is the analysis of the possible geotechnical mechanism of failure. The major mechanisms are (Section 5.2):

1. Slip failure of slope and subsoil;
2. Settlement of core and subsoil;
3. Erosion of soil and rock material.

Because these mechanisms and related aspects have been treated extensively in Section 5.2, only some aspects in relation to offshore structure will be mentioned here.

The safety factor against the occurrence of slip surfaces can be calculated using Bishop's method. The relatively small construction height and practical maximum slope angle of 1 in 2.5 are the main reasons that, in most cases, overall stability is assured. Internal settlement of the gravel embankment may occur under the densifying influence of wave-induced orbital motions. However, a decrease in porosity is not likely to exceed 4%, a typical average decrease being 1–2%.

Erosion of rock material has been treated in Section 6.5.2.2 by giving a method of determining the required grain size that is stable under the hydraulic design conditions. Erosion of seabed material may occur if a critical current velocity, induced by a local water pressure gradient, is exceeded at the boundary between filter and subsoil material. This criterion is expressed in the classic filter rules. A construction designed according to these filter rules is also stable under non-stationary (cyclic) flow. Scour of the original seabed next to the gravel structure will be induced by increased turbulence. The most important governing parameters are the slope steepness of the structure, the ratio between the amplitude of the water displacement and the construction width, and the amount of sediment in suspension. To minimise the amount of erosion to take place, it is general practice to reduce the slope of the gravel or rock structure to a maximum of 1 in 2.5.

6.5.1.4 Impact of falling objects

In the near-vicinity of offshore platforms, semi-submersibles or other places where loads are handled above the water surface, there is always a chance of an object being dropped by accident. Dropped objects such as drilling equipment, containers and anchors may cause serious damage to pipelines and electrical and optical cables laid on the seabottom.

Protection of these lines is therefore required. Because trenching in the near-vicinity

Table 44 Frequency of dropped objects with regard to weight (accidents per 100 crane-years

Dropped objects	Mass m (t)	On deck (freq.)	Overboard (freq.)	Impact energy (kNm)
Cargo up to 1 t	1	1.15	0.43	12–260
Cargo container up to 5 t	5	2.03	0.76	62–1320
Crane-block	0.5	0.26	0.10	130
Crane test weight	55	0.26	0.10	2750–15 000
Crane boom fall	10	0.53	0.20	20
Drill collar/casing	3	0.97	0.36	170–740
Bundle of pipes/casings	7.5	0.88	0.33	180
Hydrill/preventer	5	0.26	0.10	640
Riser/conductor section	10	0.18	0.07	2640
Mud pump	32	0.09	0.03	800
Well slot cover plate	4	0.18	0.07	50
Winch	25	0.09	0.03	450–5000
Cable drum	16	0.09	0.03	2800
Diving bell	3	0.09	0.03	30
Life raft	1	0.09	0.03	0
All		7.15	2.67	
Percentage of all		73	27	

of a platform is mostly not possible or allowed (or too expensive), protection is usually provided by means of placing rock protection. The necessary protection (i.e. the cover height of the rock layer) can be calculated by determining the generated impact energy of the falling object and the energy-absorbance capacity of the protection. The kinetic (impact) energy of a falling object with mass, M, and velocity, V, is equal to:

$$E_k = \tfrac{1}{2} M V^2 \qquad (6.2)$$

In most cases the falling objects will reach a constant velocity, called the equilibrium velocity, V_e, which is determined by:

$$V_e = \left(\frac{2 g \Delta I}{A_s C_d} \right)^{1/2} \qquad (6.3)$$

where Δ is the relative buoyant of the object, I the volume of the object, A_s the cross-section area and C_d the drag coefficient, which is dependent on the Reynolds number and the shape of the falling object.

Data from the Veritec Worldwide Offshore Accident Databank showed 81 crane accidents with dropped loads (of which 22 loads dropped overboard) on fixed UK platforms during the six-year period 1980–86. In Table 44 the generated impact energy has been given for the objects falling overboard and by presuming that the equilibrium fall velocity has been reached.

Because the total number of crane-years behind these accidents is known (825) it is possible to calculate the accident frequency per 100 crane-years. Based on the data used for Table 44, a distribution of the registered impact energies for the objects falling overboard can be derived (in total, 2.67 crane-incidents per 100 crane-years), as shown in Figure 327.

The frequency of incidents (number of occurrences per time) with which a pipeline or cable will be hit is then derived from the product of the probability that a dropped object will hit the pipeline or cable and the frequency of incidents with dropped loads. Based upon a permissible probability of failure, the required exceedance probability of the design impact energy can be determined and hence the required energy absorbance capacity of the pipeline protection.

Little is known about results of research into the mechanics of impact on loose-

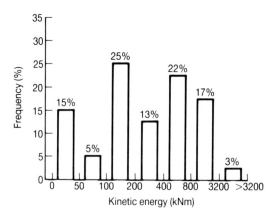

Figure 327 *Distribution of kinetic energy of falling objects just before impact (kNm)*

packed material. As a first method of approach, the system of rock cover and falling object is modelled as an ideal spring-impact model, thereby neglecting viscosity and damping effects. The impacts energy absorbance capacity, E_c, defined as the energy that is absorbed by the rock cover before the dropped object is physically touching the pipeline or cable, can be written as:

$$E_c = \int_0^p R\,dz \qquad (6.4)$$

where p is the total penetration depth and R the resistance by the rock material as a function of the depth z and the shape of the falling object. If the velocity vector of the object during impact is orientated vertically, the resistance can, for example, be described by the Terzaghi equation for the maximum bearing capacity (Lambe and Whitman, 1969). Calculations based upon the above theory show that a 1 m gravel layer offers protection against spherical falling objects with an impact energy up to 300 kNm.

In Heuzé (1990) an overview is given of experimental and analytical results of projectile penetration into geological materials, with the emphasis on rock targets. Comparison of several calculation methods and test data showed that the prediction for the rate of penetration vary considerably from one method to another. It is also showed that the applied frictional force is very uncertain but important aspect of the penetration process. Further, at velocities of up to a few hundred metres per second(!) penetration is most dependent on shear strength, and penetration depth for rock appears to scale linearly with the ratio of the penetrator's weight to its cross-sectional area. For a better understanding of the behaviour of impact into loose-packed layers under water, further testing and research is necessary.

6.5.1.5 Dragging anchors and fishing-gear

Dragging anchors

A traditional anchor is constructed to dig itself into the seabed by its flukes when the anchor chain is pulled. A wide variety of anchors is now available and a distinction should be made between:

- Standard ship anchors; and
- Work anchors with high holding power (HHP anchor).

The main differences between these two types are the holding power and the burial depth. The HHP anchor is defined as being able to have three times the maximum holding-power capacity, with the same anchor weight, of a standard anchor. The holding power is greatly dependent on the soil characteristics, the fluke area and the burial depth, which depends strongly upon the fluke-shank angle and the soil tyupe, and may be up to 10 m in soft soils. Literature studies on the behaviour of

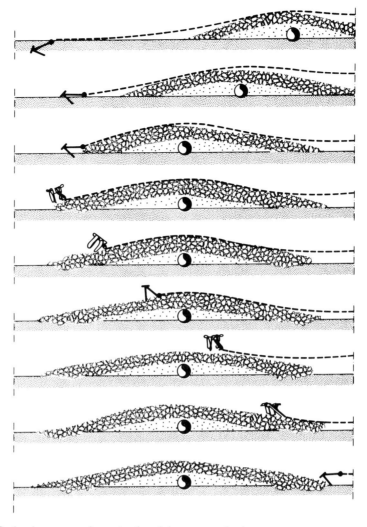

Figure 328 *Anchor on rock protection lying on seabed*

anchors have been performed by Koster (1974) and Visscher (1980). For anchors used on merchant vessels the required holding power is lower, and the burial depth is therefore smaller. About 98% of the world's cargo fleet is equipped with anchors which do not penetrate into the seabed more than 2–2.5 m.

From these data it seems obvious that pipelines and cables cannot be protected against dragging anchors by trenching only. However, wherever the probability of an anchor being dropped near a pipeline or cable is acceptably low, any protective measures against anchor damage can be omitted. At some places this probability can be much higher. In areas of frequent actvity by construction barges, supply vessels, etc. (for example, around exploration/production platforms or areas with heavy shipping) it might be advisable to protect cables and pipelines by a cover of suitable selected rock.

Depending on the layout of the rock structure, two mechanisms can lead to breaking out of an anchor:

1. A rock protection lying on the original seabed causes a change of the angle of the anchor chain, resulting in a vertical uplifting force (Figure 328);
2. A rock protection lying in a trench causes instability of the anchor due to uneven loads on the anchor flukes.

Normally, a combination of these mechanisms will determine the behaviour of the anchor when approaching and/or penetrating into the rock protection.

In the past some model and prototype tests have been performed (Schäle, 1962;

Boodt, 1981; Seymour *et al.*, 1984) to produce minimum requiremens with respect to rockfill protection structures. These tests confirm the theory that pipelines can be protected against dragging anchors by using a rock protection. However, the test series cited were limited in number and were set up to investigate a special condition. Therefore more extensive investigations are needed for deriving general rules for designing the required rock protection.

Fishing gear

Fishing gear from otter trawls (trawl doors) and beam trawlers (beam and trawl shoes) can cause serious damage to pipelines and cables on the seabottom (ICES, 1980). Freespan lengths of pipelines are potentially particularly dangerous, as the lines are likely to be hooked by fishing gear. In extreme cases the fishing vessel may even be pulled down.

A sound solution to protect cables or pipelines against fishing gear is a rockfill cover. This coverlayer should be able to withstand the horizontal impact loads, which depend mainly on:

- Shape and mass of trawl board
- Trawling speed
- Direction of pull
- Seabed conditions
- Protection of cable or pipeline

The average total weight of a trawl board is about 500–2000 kg and the trawl speed is usually 3 and 5 knots. This corresponds with an impact energy 0.5 and 6 kNm.

A gently sloped gravel structure will deflect the trawl board so that only part of this energy has to be absorbed by the rock profile. The penetration into the rock profile will be negligible with these relatively small impact energies. A rock cover of 0.5 m will be sufficient in all cases.

6.5.1.6 Freespans

Rapid morphological changes of the seabed (for example, large sand and mud waves) can result in partial exposure of an originally buried pipeline or in large freespans.

Spanning of a pipe can cause the following problems:

- Overstressing of the pipe due to its unsupported weight over the length of the freespan and, more seriously;
- Vibration of the pipe due to the oscillating wave velocities, introducing fatigue problems;
- The line becoming unprotected against dragging anchors and fishing gear.

Among other technical solutions (e.g. (re)trenching of the pipeline or placing of (block)mattresses over the pipeline) a well-designed rock protection placed over the pipeline can prevent the forming of freespans. The dimensions of the rock structure should be designed in such a way that the structure is large enough to follow the changing adjacent seabed without disintegrating and are therefore dependent on local conditions.

6.5.1.7 Upheaval buckling

Hydrocarbons produced from marginal offshore fields are usually transported at high pressure and high temperatures. The compressive stresses induced in the pipeline due to thermal expansion and internal pressure can lead to upheaval buckling. Resistance to upheaval buckling is normally provided by soil, gravel or rock cover offering enough vertical and horizontal support. A number of observed buckling cases in recent years, which were caused by inadequate backfill cover, have forced oil companies to reconsider the problem of upheaval buckling more thoroughly.

In this section an introduction to the buckling problem is given. Because of the complexity of the problem and the relatively new research developments, the section is written more as a literature review than as a technical manual on how to deal with the problem.

Theoretical modelling

In the literature the pipeline problem has been addressed by various authors. Historically, the upheaval phenomenon had been considered to be analogous to the vertically stability of railroad tracks under solar heating. In Hobbs (1984), Boer *et al.* (1986) and Richards *et al.* (1986) this analysis procedure for track buckling is used, assuming that the uplift resistance, which is composed of the weight of the pipe and that of the cover, is constant and that the foundation of the pipeline is rigid.

Pedersen and Michelsen (1988) described a mathematical model which includes the non-linear behaviour of the pipe material, the non-linear pipe–soil interaction and the geometric non-linearities caused by large deflections. Consistent with this model, a simplified approach applicable for a pipe in the pre-buckling stage was presented by Pedersen and Jebsens (1988), who studied the effects of time-varying temperature loadings and non-linear pipe–soil interactions in more detail. They concluded that the classical upheaval buckling analysis as described above is not conservative for imperfect pipelines. As a result, a new design procedure was proposed by Nielsen *et al.* (1988), based on limiting the uplift movement of the imperfect pipe to the elastic deformation of the cover. This design procedure, in combination with the mathematical model as presented in Pedersen and Michelsen (1988) and Pedersen and Jebsens (1988), can be used to determine the required uplift resistance.

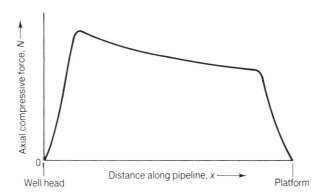

Figure 329 *Axial compressive force along a typical pipeline*

An important parameter in the analysis of upheaval buckling is the axial compressive force. At the subsea well head and production platform the pipeline is usually provided with expansion loops which result in zero axial load. Along the pipeline, surface friction forces between pipeline and subsoil and cover are mobilized until axial load reaches a level at which the pipeline is completely restrained (Figure 329). The completely restrained axial compressive force, N_0, at a distance, x, is:

$$N_0(x) = \bar{\alpha} E A_s \delta T(x) - v D^2 \delta p \pi / 2 \tag{6.5}$$

where $\bar{\alpha}$ is the coefficient of thermal expansion, E is Young's modulus, A_s is the pipeline cross-sectional area, δT is the temperature change per unit length, v is Poisson's ratio, D is the pipe diameter and δp is the internal pressure difference per unit length.

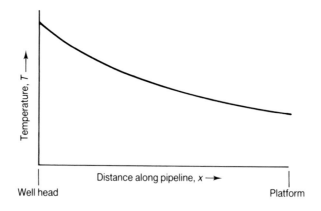

Figure 330 *Temperature profile along a typical pipeline*

Due to heat loss, the temperature and consequently the axial compressive force will normally vary along the pipeline (Figure 330). The heat loss of buried or covered pipelines is greatly influenced by the thermal properties of the cover material and surrounding soil.

Where the protection consists of fine to medium sand or gravel a fair impression of the heat loss and the resulting temperature drop of submarine pipelines can be obtained by the determination of the conductive movement of heat through the granular material. However, in situations where the cover consists of high-porosity media such as loosely pack coarse gravel or rock, the convective movement of heat is also important. Conventional heat-loss models, based on conduction only, generally underestimate the heat loss for such situations. Boer and Hulsbergen (1989) described a numerical model which can be used to compute the heat loss and resulting temperature drop of buried and covered pipelines.

Depending on the local axial load, the required resistance against upheaval buckling can be determined. Only the vertical break-out force has to be considered to determine the locally required cover height, as the friction force between pipeline and the soil/cover hardly influence the reponse in the pre-upheaval buckling stage. To give a detailed descriptions of the buckling model is beyond the scope of this manual. However, full details of the model can be found in Pedersen and Michelsen (1988).

Empirical input

It will be clear that quantitative information about the axial friction and uplift resistance of the cover is essential for practical analysis of submarine pipelines under substantial temperature changes. The pull-out mechanism of an infinitely long shallow horizontal pipeline with diameter D and a cover with a submerged bulk density γ_r and height H is illustrated in Figure 331. The maximum pull-out force P is usually written as:

$$P = \gamma_r DH(1 + fH/D) \qquad (6.6)$$

In this simple empirical formula, f represents a factor for specific geometrical and geotechnical characteristics. Geotechical literature on pull-out forces mainly refrs to horizontal anchor plates in fine granular soils with horizontal upper boundary, therefore it cannot be applied directly to a pipeline covered with rock or gravel. This limitation and the lack of verified methods to calculate pipeline pull-out force were important reasons for carrying out full-scale pipeline pull-out tests (Boer et al., 1986; Schuurmans et al., 1989). The test results indicate that the friction factor, f, varies between 0.6 and 1.0. For identical cover properties and embankment geometry, only a small tendency for decrease of f with increasing H/D values has been found.

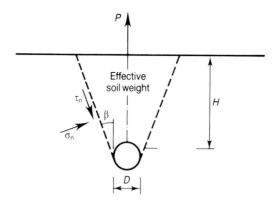

Figure 331 *Pullout mechanism for a shallow pipe*

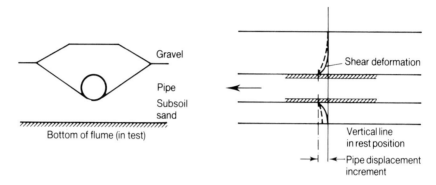

Figure 332 *Shear displacement pattern for a horizontal pipe pulled longitudinally*

If a horizontal pipe, buried in a gravel embankment with constant cross-section, is pulled in a longitudinal direction, the horizontal force will be transmitted to the surrounding soil by means of shear stresses. The transfer of shear stress along the external pipe surface must be accompanied by shear deformation in the grain skeleton (differential displacement), as shown in Figure 332. Ultimately, the shear stresses are transferred to the subsoil. The pulling force at which maximum shear stress is reached strongly depends on the friction behaviour between the soild and the pipe and thus the roughness characteristics of the pipe.

To determine the maximum friction force, full-scale pipe friction tests have been carried out. In its most simple form, the maximum horizontal force per unit length can be expressed as follows:

$$\hat{P}_H = \pi D \mu_H \bar{\sigma}_n \qquad (6.7)$$

in which μ_H is the coefficient for horizontal friction and $\bar{\sigma}_n$ is the average effective stress normal to the pipe surface, which can be expressed in terms of the cover height, H, and the submerged bulk density, γ_r, of the cover material.

Apart from the effect of the height of the cover above the pipe, the test results showed that the type of coating has a remarkable influence on the friction coefficient, μ_H. Furthermore, μ_H tended to decrease with increasing cover height, H, for pipes in a trench. This decrease may be introduced by an arching effect against the trench slopes. For a pipe with an external polyurethane coating, the μ_H value for the average pull-out force was of the order of 1.04. However, because of the considerable influence of the type of coating material this figure must be considered to be only indicative.

In view of the buckling problem, the temperature in the pipeline should be as low

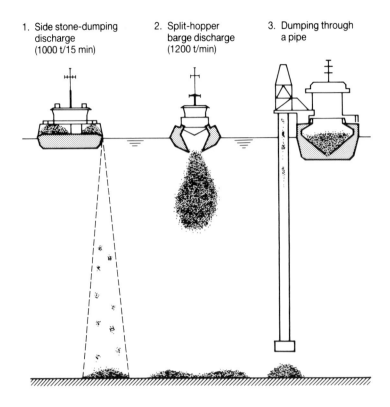

Figure 333 *Common methods of dumping stone offshore*

as possible. On the other hand, it should not drop below a certain minimum level if oil-handling problems (such as waxing and decrease of viscosity) are encountered. In an integrated approach the pipeline cover and coating can be utilised to optimise the temperature profile along the pipeline.

6.5.2 CONSTRUCTION ASPECTS

6.5.2.1 Construction methods

There are at least three methods of dumping rock or stones offshore (Figure 333):

1. From a side stone-dumping vessel or barge. The load is dumped slowy and each stone may be considered to fall individually for the purpose of evaluating the fall velocity.
2. From a split-hopper barge. After the bottom gap of the barge exceeds a certain limit, the load is dumped in a short time as one large mass. The mass of stones stays together in a 'cloud', resulting in a fall velocity exceeding the equilibrium fall velocity of each individual stone.
3. From a vessel through a (flexible) pipe in order to achieve greater accuracy in deeper water.

Dumping from a side stone-dumping vessel

An individual stone will pass through the following states:

1. Through air until it hits the water surface. The stone will accelerate from zero velocity to fall velocity:

$$V_0 = (2gh)^{1/2} \qquad (6.8)$$

where

g = gravitational acceleration (m/s^2),
h = height from release point to water surface (m).

2. After hitting the water surface, the stone will either increase or decrease its initial velocity, V_0, to the equilibrium fall velocity, V_e.
3. The rock or stone will hit the bottom with velocity V_e. The following equation may be used to obtain the equilibrium fall velocity, V_e, in water:

$$V_e = (4/3 \, g \Delta D)^{1/2} \tag{6.9}$$

in which

V_e = equilibrium velocity (m/s),
Δ = relatively density of rock (−),
D = grain diameter (m)

To obtain an impression of the impact of a stone with a certain diameter it may be helpful to know that the impact in water is comparable with that of the same rock falling in air from a height of approximately its own diameter.

The loading capacity of side stone-dumping vessels varies between 350 and 3500 t. They discharge their loads by means of sliding shovels placed on the loading deck and the shovels push the rock gradually over the side of the vessel. For a side stone-dumping vessel of 1000 t the actual dumping time is approximately 15 min. Depending on the local water depth, the dimensions of the dumping profile and the dimensions of the rock, the vessel can either keep station or track along or over the pipeline.

Dumping accuracy (height, width and location) is dependent upon the water depth, the wave response of the vessel, the vertical current velocity profile and the size and specific density of the rock material. Therefore dumping is usually carried out around slack tide. Accuracy of the dumping height can be improved by dumping several layers of rock, depending on the total required structure height and the rock size. Accuracy can also be improved by use of a computer simulation model (see Box 73).

All dumping activities are supported by sophisticated positioning and survey equipment, as positioning inaccuracies may result in a significant 'loss' of material. Positioning is achieved by a roundabout anchoring system (usually six anchors), a combination of two anchors and two sideways-oriented thruster propulsion units or a dynamic positioning system using computerised thruster propulsion.

Dumping from a split-hopper barge

A split-hopper type barge dumps its load in less than a minute! As a result, a cloud of stones and water will reach the bottom with a velocity of two to three times the equilibrium velocity of an individual falling stone. The impact of the split-hopper barge dump is very heavy and may result in damage to a pipeline or cable, particularly in freespan sections.

Moreover, the dumped material usually shifts sideways after hitting the bottom, leaving only a small quantity at the desired location. Therefore split-hopper barges are usually not utilised for work requiring accurate placing of stones, such as the protection or stabilisation of pipelines or cables in deeper water. The sphere of work of a split-hopper barge is pre-eminently the dumping of large quantities of gravel or rock in shallow water (approximately twice the maximum draught of the barge) where accuracy has no high priority.

Dumping by means of a (flexible) fall-pipe vessel

This system guides the rock to a level several metres above the seabed and is therefore especially suitable for accurate dumping in deeper water (over 50 m). A further advantage is that the dumping operation is not restricted to the period around slack tide.

The system consists of a vessel from which a (flexible) pipe can be lowered, ending several metres above the seabottom. The end of the pipe can be positioned using either an independent working propulsion unit or a free-moving Remote Operating Vehicle (ROV), both fitted with equipment capable of making pre- and post-dump surveys. The dumping material is transported by means of a system of hoppers and conveyor belts into the fall pipe. While tracking along the pipeline at a constant speed, the rockfill is placed over the pipe.

Dumping accuracy is dependent only upon the positioning of the lower end of the fall pipe relative to the pipeline. The vertical movement is controlled and restricted by a heave-compensating system. A thruster unit enables the operator to control and correct the horizontal displacement.

Essentially, there are two different systems in operation using a fall pipe:

1. *Dumping through a flexible fall pipe* (*Figure 334*). The flexible fall pipe consists of chains, which implies that the chute is permeable. The velocity of the falling rock is limited by the equilibrium fall velocity, which is 1–2 m/s for 2–6 inch rock. During the passage from seasurface to seabottom dust will be washed out, making a clear view possible and an undisturbed performance of survey equipment near the bottom. This enables video recording during the dumping operation. A ROV mounted on the lower end of the fall pipe can be equipped with the necessary monitoring and survey equipment. Because of the limited mass of the flexible chute, a limited propulsion power at the ROV is also required for the positioning of the lower end of the fall pipe.
2. *Dumping through a rigid pipe*. The fall pipe system comprises a string of polyethylene pipe joints, bolted together and suspended from a large pipe tower situated over a moonpool. The string is lowered just above the seabottom and is guided by a monitoring and thruster unit situated at the end of the fall pipe. A separately working ROV is required to monitor and control the work.

6.5.2.2 Impact of dumped rock

From the evaluation of the different construction methods it can be concluded that if rock is dumped by means of a side stone-dumping vessel or a flexible fall-pipe vessel the fall velocity of the stones will be limited by the equilibrium velocity. It is evident that the resistance of the pipelines and cables against the impact of the falling stones must be ascertained.

Full-scale as well as laboratory tests have been performed in the past with rock of 50–150 mm in order to determine possible damage to coatings of steel pipes, flexible flowlines and cables. It was concluded that rock dumping on pipelines, flowlines or cables will not lead to damage for normally coated pipelines, which have a coating of more than 1 mm.

6.5.2.3 Survey

In order to control and document the results of the dumping operations during the various stages of the works, surveys must be carried out. These can be made either from the dumping vessel itself or from a separate one, possibly equipped with a ROV fitted with multiple sensors.

A wide range of survey equipment is available, and usually the following (sub)systems form part of the survey system (see also Appendix 6):

- Surface-positioning system
- Sub-surface-positioning system
- Gyro compasses
- Scanning profilers
- Depths sensors
- Video systems
- Computer systems

Figure 334 *Dynamically positioned, flexible fall-pipe vessel Trollnes (courtesy ACZ)*

Usually, the navigation computer on board is the heart of the survey system to which can be interfaced, among others, the surface-positioning system(s), the ROV ship-positioning system, the vessel gyro compass and the ROV gyro compass and scanning profiler system. The ROV may be fitted with underwater cameras and a scanning profiler system. The ROV ship-positioning system provides the ROV position relative to the vessel.

Prior to the commencement of the survey operations it is usually necessary to perform checks in order to ensure that the overall system provides data to the required standards. The results of those checks or calibrations must be recorded in order to monitor significant changes in value over time. The following three types of survey can be distinguished.

Pre-survey

The pre-survey has, in general, two purposes:

- To establish the exact 'as-found' co-ordinates of the dump area;
- To establish the pre-dump seabed profile for later assessment of the dumped quantities, dump height and dump dimensions.

A bathymetric survey grid with pre-established intervals will be convered for this pre-survey. The dump area will be contained within this grid and cross-sectional/longitudinal profiles produced from the logged echosounder data for later comparison with final survey profiles.

Intermediate survey

This survey is carried out for appraisal of the dump dimensions.

Post-survey

After completion of the dumping a post-survey will be made over the same bathymetric grid area as the pre-survey. The results will be compared with the pre-survey data in order to confirm that the dumped profile is in accordance with the client's specifications.

The post-survey techniques are basically the same as those of the pre-survey. Preferably, cross-sectional and longitudinal profiles will be produced at approximately the same locations as the pre-survey.

6.5.3 COST ASPECTS

The construction cost of an offshore rockfill structure can be divided into purchase, transportation and placing of the rockfill material and the required surveying activities of the offshore structure. The amount of rock to be purchased is, next to the minimal required structure dimensions (i.e. minimum cover height on top of the pipe, minimum structure width and length of the pipeline), dependent on the accuracy of the dumping method. Dumping accuracy is smallest (i.e. losses are greatest) when using a split-hopper barge and greatest with a fall-pipe vessel. Rock losses will grow with increasing water depth when using the side stone-dumping vessel and split-hopper barge. Therefore for every type of rock dumping the minimum required amount of rock for a specific job can be estimated.

Based on the required rock quantity, for each rock-dumping vessel (with a certain loading capacity, sailing speed, survey facilities, etc.) the duration and expected costs of the rock-dumping works can be calculated. This is done by assessing the expected number and duration of a cycle time, which can be separated into loading of the rock, sailing to site, system set-up, rock dumping, pre-, intermediate and post-survey, system recovering and sailing back to the quarry. Workability restrictions caused by wind and wave conditions differ for each vessel and type of rock dumping.

7. Maintenance

Maintenance of rock structures in coastal and shoreline engineering should be viewed as comprising all those activities that are required to be carried out on a periodic basis after construction to ensure that a structure performs to an acceptable standard during its lifetime. A maintenance programme will, therefore, include the following essential elements discussed in this chapter:

1. Inspection and monitoring of environmental conditions and structural response;
2. Appraisal of monitoring data to assess compliance of performance with predetermined standards, which may themselves vary through the lifetime of the structure (e.g. due to secular trends in water level or wave climate, or developments in understanding of hydraulic or geotechnical processes and interactions);
3. The repair or replacement of components of a structure whose life is estimated to be less than the overall structure or of a localised area which is assessed to have failed. The latter is sometimes viewed as 'maintenance' in a narrower sense and therefore will be termed 'repair' or 'replacement' for the purposes of this chapter.

The activities to be carried out will be based on a maintenace policy developed at the design stage in a management view of the whole-life performance and costs of the rock structure in question. This design overview of maintenance, discussed in Chapter 2, will have taken account of likely maintenance requirements and costs for plant, labour and materials. The policy developed will also have considered access for monitoring and maintenance and likely owner engineering and financial resources for carrying out both practical monitoring and repair activities as well as interpretation of structure performance.

Development of a maintenance policy should be followed at the design/construction stage by production of a maintenance manual. This is always merited for rock structures in view of their flexible (and sometimes even dynamic) nature. In many situations, particularly where the structure has been designed to minimise maintenance, the manual may be very simple. However, it must contain the minimum necessary guidance on techniques and criteria for the three basic elements of the maintenance programme listed above. It will also set out the interrelations of the various activities involved, a guide to which is given in the flow chart in Figure 335. This figure also shows where information on the three principal types of activity in a maintenance programme can be found in this chapter.

7.1 Monitoring

Rock structures respond to the destructive mechanisms of wave and tidal action by changes in the profile of the structure and in the size and shape of its component parts. The alterations in shape can also be due to loss or changes to foundation of core material. Typically, rock may be displaced, abraded, fractured or even dissolved. Any quantiatative description of the state of the structure must be able to identify these different responses, which may take place either gradually or catastrophically after a major storm. It must also be able to identify the environmental forces driving the responses.

A regular monitoring programme for both structure and environment allows the designer or owner to plan repair and replacement activities with a good understanding of failure mechanisms and damage trends. Gradual deterioration of the armour layers or foundations may be unnotices without the aid of monitoring programme, and may ultimately result in the failure of the armour layers or in

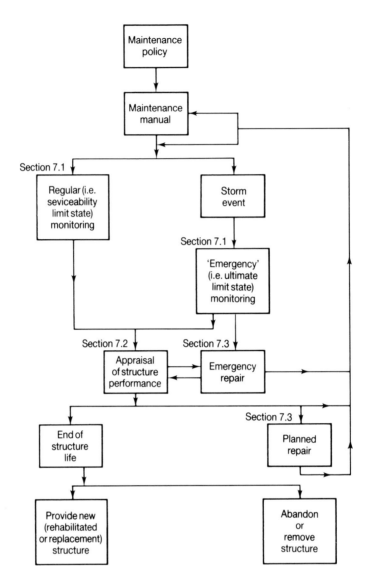

Figure 335 *Maintenance programme flow chart*

unacceptably large settlements. By permitting comparison of measures of the state of a rock structure at a number of points in time, a structural monitoring programme allows these changes to be identified at an early stage, thus enabling the appropriate maintenance action to be carried out.

7.1.1 TYPES OF MONITORING

There are two principal areas of monitoring: measurement of structural state and measurement of environmental loading conditions. Structural state monitoring concerns the performance of the fabric of the structure and its foundations. Details of the measures of structural state which may be monitored, together with appropriate survey techniques, are given in Table 45. Fuller details of the survey techniques mentioned in the table are given in Appendix 6.

Formal structural state monitoring is worth supplementing with walkover surveys carried out to record, with the aid of photographs, the overall condition of the structure, including any obvious rock movements, changes in profile, etc. This type of survey must remain very subjective and had only limited use in a detailed quantitative assessment of a structure. However, if fixed reference points can be

Table 45 Measures of the state of a rock structure

Aspect of structure state	Survey technique (full details in Appendix 6)
Level I: Location Two to 10 points on the structure measured in relation to a well-established grid and datum levels. Redundant survey points need if major movement possible.	Conventional survey techniques (any settlement markers should be installed during construction)
Level II: Geometry Outer surface description, related to level I survey points.	Conventional survey, using profiling techniques recommended in the model specification, (Appendix 1) but with profiles at wider spacings (say, 20–30 m). For underwater surveying, bathymetric techniques may give useful information
Level III: Composition Position and attitude of each piece of rock armour, including unstable pieces. Position and size of major voids and exposures of core or underlayers.	Armour-degradation inspection techniques Comparative photography Photogrammetry For underwater surveying, side-scan sonar techniques may be used in co-ordination with diver surveys to identify features on sonar traces
Level IV: Element composition Shape and size of armour rock, including any fractures.	Armour-degradation inspection techniques

Note: Levels III and IV armour degradation inspection techniques are difficult to implement on wide gradings and/or gradings with D_{15} less than about 0.3 m. The techniques, however, work extremely well on large 'single-sized' rock armour ($D_{15} > 0.3$ m; D_{85}/D_{15} in the range 1.2 to 1.4), including the 'heavy' standard gradings in Appendix 1.

established on the structure and checked at the time of the survey, walkover surveys carried out by experienced personnel can form the basis of a suitable monitoring programme.

Environmental monitoring concerns the external loading on the structure and the effect the structure has on the total environment. Details of environmental conditions or loadings which may be monitored, together with appropriate monitoring techniques, are given in Table 46. For a more detailed discussion of environmental data collection, however, the reader is referred to Chapter 4 and for information on instrumentation, to Appendices 4 and 5. In addition, for critical structures, the hydraulic/geotechnic response to wave loading may be worth monitoring and some brief suggestions on measurements and techniques that might be adopted are given in Table 47. The monitoring methods selected should related to the potential failure modes for the structure in question and, in particular, to those which have been identified as the most likely (see Section 2.2 for a discussion on failure modes and fault trees).

7.1.2 FREQUENCY OF MONITORING

As with the monitoring of methods, the frequency of monitoring should be predetermined in relation to the risk associated with particular failure mechanisms, structural elements, foundation conditions, exposure conditions and design criteria. This frequency may be different for different types of monitoring. It could also increase during the life of the structure, reflecting the structure's reducing resistance to failure as it degrades with time and the approaching need to carry out repair (Figure 336, de Quelerij and van Hijum, 1990).

The typical behaviour of rock structures commences with the individual pieces of rock within a rock structure tending to settle and pack during their early life and become more stable. Changes are then often minor for some years. Later, the degradation of individual pieces of rock may become significant and/or

Table 46 Measures of environmental conditions or loadings

Environmental condition or loading	Monitoring survey technique (full details in Chapter 4 and Appendices 4 and 5)
Water level	Tide board, visually inspected Data from nearest local tide recording stations Use of surface elevation monitor (step gauge or resistivity gauge) recordings, if available
Wave climate	Seabed pressure meter (robust and cheap). Surface elevation monitor mounted on robust support (e.g. pile, or triangulated scaffold tube arrangement) Wave rider buoy or similar (will be expensive to maintain for long periods) Hindcasting analysis for storm events using wind records
Wind climate	Standard anemograph device (depending on correlation between wind and wave direction may be a useful way of assessing directionality of wave climate)
Bathymetry and beach topography	Below high water, standard bathymetric techniques posible. Above low water, conventional land-survey techniques may be used or photogrammetry from aerial photography Land-based photography of waterline from fixed positions gives useful assessment of low to high water beach form
Stress in foundation	Pressure pads
Pore pressure in foundation	Piezometers (simple standpie or, for continuous measurements, vibrating wire electronic recording devices may be used)

Table 47 Measures of hydraulic/geotechnic response of rock structure to wave loading

Aspect of hydraulic/geotechnic response	Monitoring survey technique
Wave run-up	Parallel steel wire resistivity guage (survival a problem)
Wave transmission (for breakwaters)	Appropriate wave gauge at reat of breakwater
Mound pore pressures	Piezometers installed within mound with automatic recording facility

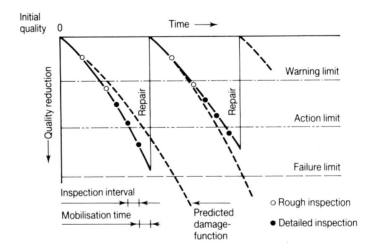

Figure 336 *Preventative condition-based maintenance*

Table 48 Frequency of planned monitoring above low-tide level

Location of whole structure	6 months visual—12 months survey
Geometry of the structure	12 months
Position of individual rocks	12 months
Shape, size, condition of individual rocks	12 months
Foundations, scour, etc.	6 months visual—12 months survey

environmental conditions become more severe, and then changes in the rock structure may be more rapid. This overall behaviour under what could be described in limit state terms are serviceability conditions may be overriden by a very severe storm event and the behaviour of the structure in the response to such an ultimate limit state loading.

It is important to commence the monitoring of new structures by establishing a 'base-line' set of information regarding the structure and its environment at the time of construction and during the construction guarantee period. This information should be stored in a form which allows for its retrieval in future years for comparison purposes. The recording of details about the structure during the construction stage is also required as a check against the assumptions and details established at the design phase. Information recorded should include not only basic geometry survey data of profiles, etc. but also records of any failures of the rock elements during the construction period. The basic monitoring information should be documented by the contractor and handed over to the owner or designer by the end of the guarantee period.

The frequency of structure state monitoring adopted thereafter during the life of the structure depends on a number of factors, principally the following:

- Location on structure
- Type of construction
- Design risk levels
- Exposure conditions
- Foundation conditions

Two basic frequencies of monitoring will be involved, one related to normal conditions or the serviceability limit state and the other to extreme conditions or the ultimate state. Monitoring related to the serviceability state should take place on a planned basis at frequencies identified at the design stage. Ultimate state monitoring will only occur during the following severe storm events, the minimum wave conditions for which should be stated at the design stage.

Monitoring frequency above low-tide level

For the frequency of planned monitoring in relation to normal conditions Table 48 indicates the recommended maximum period between planned monitoring inspections for a rock structure above low-tide level. After the first few years of life of a structure, satisfactory performance may indicate that adequate monitoring will still be achieved if the detailed surveys are reduced in frequency from 12 to 24 months or even longer. However, the experience of authorities responsible for monitoring is that there is a possibility of surveys being forgotten if they are not carried out every year. There is also the danger with such infrequent surveys that rapid deterioration near the end of the life of a structure will not be detected before a failure occurs.

For structures which have elements below the lowest tide level it should be recognised that the effects of wave exposure will not be as great. However, the influence of local currents will have to be considered when developing a monitoring frequency for elements of a rock structure which are permanently submerged.

Monitoring frequency below low-tide level

Unless special circumstances exist, it is recommended that the submerged elements of a rock structure are fully inspected at least every five years and after extreme storm events. In addition, the regular annual monitoring of the upper sections may indicate possible problems on the submerged section of a structure worth further investigation. Instrumentation can also be introduced at the construction stage into structures which are partially permanently submerged, to allow certain performance aspects (e.g. foundation settlement) to be monitored during the more frequent inspections of the upper structure.

7.2 Appraisal of structure performance

Having instigated a comprehensive monitoring programme, a designer or owner needs to document all the monitoring information and carry out a review and/or analysis of the data which will form the basis for decisions regarding the need for and extent of maintenance works. Each structure state monitoring report should be assessed in relation to the environmental conditions recorded during the inspection and compared with previous reports. Typical assessments which may be made are listed in Table 49, and enable a continual evaluation of design and performance criteria.

Table 49 Outputs from comparison of measures of the state of rock structures over a period of time

Aspect of structure state measured (see Table 45 for details	Output from comparison of structure state at a number of points in time
Level I: Location	Settlement of foundation
	Change in alignment
Level II: Geometry	Consolidation of structure
	Comparison of slope profiles enables overall armour damage parameter, S (see Chapter 5), to be determined
	Scour damage
Level III: Composition	Loss or movement of armour rocks
	Overall sliding of armour layers if this has occurred
	Voids requiring emergency/planned repair
Level IV: Element composition	Rounding of rocks and loss of material, enabling revised assessment of D_{n50} with the design wave climate, or measured wave climate, or revised design wave climate from wave measurements, allows re-assessment of armour stability parameter $H_s/\Delta D_{n50}$ using equations in Chapter 5. Comparison with design and measured damage parameter, S, is also possible

It is also useful after each monitoring survey to make an assessment of the remaining life of the structure in relation to each potential damage pattern so that the overall safety may be monitored. The time scale of the deterioration effect can usefully be presented by constructing a kind of 'measuring watch'. On this watch (see Figure 337) both the actual damage level and the damage limits (warning limit, action limit and failure limit) can be presented on a time scale (de Quelerij and van Hijum, 1990). The watch provides the managing organisation with a simple overview of the relative safety condition of the rock structure in relation to the relevant damage pattern, together with an idea of the remaining safe lifetime.

To provide a good presentation, the 'measuring watches' of all relevant damage patterns for the rock structure can be collected together into one control panel, as illustrated for a seawall in Figure 338, thereby providing a complete overview of its overall actual and near-future safety condition.

Having completed the assessment, owners or designers may decide to opt for one or more of the maintenance actions indicated in Figure 335. Their options may be briefly listed as:

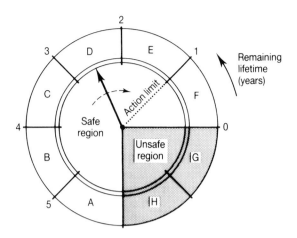

Figure 337 *Measuring watch for a selected damage pattern*

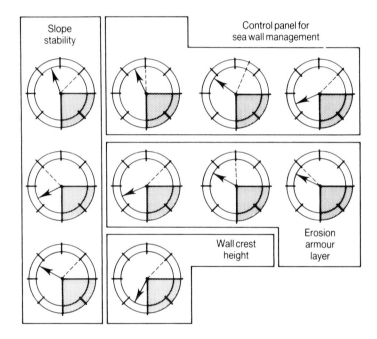

Figure 338 *Safety control panel for seawall management*

1. Do no repair/replacement work and await next planned monitoring report.
2. Do no repair/replacement work but instigate additional future monitoring of structure state and/or environmental conditions.
3. Carry out further detailed inspection before making decision.
4. Undertake temporary or 'emergency' repair/replacement works.
5. Undertake permanent repair/replacement works.
6. Instigate development of a new (rehabilitated or replacement) structure.
7. Instigate abandonment/removal of the structure.

The decisions made in response to a monitoring report should be set against performance and failure criteria which have been established at the design stage. It must be recognised, however, that performance and failure criteria can change as technical understanding develops and as the requirements for the function of a structure alter due to changes in use, safety standards or Environmental Assessment standards (see Section 2.6).

7.3 Repair/replacement construction methods

The repair construction methods adopted for any particular rock structure will be strongly influenced by decisions taken during the design and project appraisal phase prior to initial creation of the structure. Financial considerations will be one aspect, the maintenance budget set obviously having a large influence. Generally, it will be uneconomic to so underdesign the structure that when damage occurs it passes the armour layer and reaches the underlayers or core. Repairs in this situation are more in the nature of a major rehabilitation and consequently costly. Economic maintenance procedures will therefore gradually be confined to repairs to the armnour layer.

Having set the financial boundary conditions, the main considerations then become availability of (and access for) materials and plant. Since armourstone is a re-usable commodity, in many cases of shoreline protection structures where suitable access is possible, repair works will only require dislodged stone to be retrieved and placed back into the face of the structure. In other cases fresh armourstone will be required for repairs. Here, if provision for suitable access has been made at the design stage and there are no financial constraints, the stone can be imported as required. More usually, however, to import small quantities of additional armourstone is very expensive, particularly if the source is remote. In addition, access for haulage trucks after construction is complete may be difficult or impossible. In this case, stockpiling of spare material at the site as part of the main initial construction operation should be considered. This rock can be placed in a mound or, if it is necessary to soften the impact of a stockpile of large pieces of rock, buried beneath fine material or used to mark access roads. Occasionally, none of these options is practially or financially feasible, and it may be appropriate to consider asphaltic grouting or one of the other stability-improvement systems discussed in Chapters 3 and 6.

Table 50 Construction equipment for repair of rock armour layers

Method	Handling attachment	Comment	Access
Tracked hydraulic excavator	Bucket	Positive pick-up and placing. Limitations on placing and moving, tendency to drop stones	Suitable on soft beaches also used to track over large stone on crest of breakwater with aid of experienced banksman
	Fixed-arm	Positive pick-up and placing	As above
	Orange peel grab	Non-positive pick-up and placing, difficult to pick up individual stones from face	As above
Wheeled hydraulic excavator	Bucket grapple grab	As above	Only suitable where hard access available, stone relative small and short reach
Crawler crane	Orange peel grab	As above, for handling attachment, slower than excavator	Suitable for remote areas of a structure where hard access available above or below
Jack-up on pontoon with crane or excavator	Bucket grapple grab	As above	Suitable for sites which do not dry and access is not available along the breakwater
Reinforcing failed sections—cranes, excavators, boilers	Buckets and skips	Applicable to areas where the importing of large rock is difficult	Concrete or asphalt grout used to reinforce sections need good access close to structure
Use of block and tackle	Chains and lifting eyes	Suitable where rock does not have to be moved far across the structure. Attractive as military training exercise	Access for large plant materials difficult

Access for plant to, along and around the structure for handling and placing armourstone for repairs is of prime importance, and should be considered carefully when detailing or dimensioning new rock structures. Access for initial construction is very often gained along the underlayers of a partly completed structure or by purpose-built and expensive temporary works, but these forms of access are no longer available for repair works. Even concrete crown walls, sometimes included partly to make provision for future maintenance, are only of value in this regard if it can be guaranteed that they will not be damaged by storms.

Equipment commonly used for repair of armour layers is listed in Table 50 together with comments on its suitability and potential access constraints. Handling attachments are also discussed and it is worth highlighting that, for repair work, the positive pick-up and placing capability of a fixed-arm grapple often makes this the favoured attachment. This is because the grapple tends to allow more rapid working by enabling easy picking up of rock from a stockpile or from a position already in the structure. It can also be used to push rock into position, if necessary.

Appendix 1: Model specification for quarried rock applications in coastal and shoreline engineering

PART A—GENERAL

A1.1 Subject and Field of Application

This model specification provides definitions and requirements for quarried stone materials and construction, together with rules for inspection. The specification applies to quarried stone for applications in coastal and shoreline engineering. Italicised text does not form part of the specification but is only included for guidance to the specifier.

A1.2 Definitions

A1.2.1   Quarried stone: broken natural stone which is coarser than road stone.

A1.2.2   Graded quarry stone: quarried stone which is graded by sieve sizes or by weight of the stone.

A1.2.2.1   Fine-graded quarry stone: a grading which is determined with the aid of sieve sizes.

A1.2.2.2   Light-graded quarried stone: a quarried stone grading which is determined by weight or size of stone for weights less than 300 kg.

A1.2.2.3   Heavy-graded quarried stone: a quarried stone grading which is determined by weight for stones of at least 300 kg.

A1.2.3   Stone fragment: a piece of stone in a grading with a lesser weight or size than the extreme lower class limit (see Figure A1) for that particular grading class.

A1.2.4   Effective mean weight, W_{em}: the arithmetic average weight of all blocks in a sample excluding any stone fragments.

A1.2.5   Load of quarried stone: the quantity of quarried stone per unit of transport.

A1.2.6   Client: the person or organisation who commissions the construction project.

A1.2.7   Designer: the engineer responsible for the design of the works and (frequently) for supervision of its construction. The engineer is employed by the client and may be given powers to specify and accept or reject materials supplied or works constructed.

A1.2.8   Producer: the organisation which operates the production of rock materials from the quarry.

A1.2.9   Contractor: the organisation which undertakes construction of the structure, including quarried rock materials.

A1.2.10   Supplier: the organisation which supplies, including transporting, the rock from the quarry to the construction site. This may be independent or a subsidiary activity/organisation of the producer or contractor.

A1.3 Data to be Supplied

A1.3.1 DATA TO BE SUPPLIED BY THE CLIENT WITH THE ORDER

(a) The designation of the grading;
(b) Additions or variations to requirements agreed in advance of construction;
(c) In case the quarried stone has no grading in accordance with these standards, the requirements for the grading thereof;
(d) Whether the quarried stone is intended for a layer to be grouted (see Section A1.7);
(e) If relevant, the name of the producer and the name and location of the quarry, quarries or other sources of rock which the contractor may use in the works. (Note: The Client should nominate a quarry or quarries if this is the most convenient way of ensuring the rock quality required but should make certain that no economic alternative sources are thereby excluded.)
(f) A reference to this standard by specifying 'in accordance with the Model Specification for Rock Armour Materials and Construction as published by CIRIA/CUR'.

A1.3.2 DATA TO BE SUBMITTED BY THE SUPPLIER AT DELIVERY

A consignment of quarried stone must have, for each load, a certificate of origin submitted by the producer. The following data must be included on the certificate:

(a) The delivery date;
(b) The name of the ship or the designated number of the road or rail transport unit;
(c) The name of the producer;
(d) The designation of the grading;
(e) The name and location of the quarry or other source where the grading has been produced;
(f) The weight of the load;
(g) A reference to this standard by mentioning 'in accordance with the Model Specification for Rock Armour Materials and Construction as published by CIRIA/CUR'.

PART B—GEOMETRIC AND MATERIALS REQUIREMENTS

A1.4 Gradings

A1.4.1 GENERAL

Quarried stone gradings can be defined on the basis of functional demands and producibility. A number of distinctive standard gradings exist. For standard fine gradings, the four classes are designated by sieve sizes as follows: 30–60 mm, 40–100 mm, 50–150 mm and 80–200 mm. For standard light gradings, the six classes are designated either by weight or by size (and average weight) as follows: 10–60 kg, 60–300 kg and 10–200 kg, or 200–350 mm, 350–550 mm and 200–500 mm. For standard heavy gradings, the four classes are designated by weight as follows: 0.3–1 tonne, 1–3 tonne, 3–6 tonne and 6–10 tonne. The gradings to be used in the works are selected from these fourteen standard grading classes unless circumstances dictate that alternate gradings must be specified.

Alternative gradings based on weights may be specified in the following cases:

1. For extremely heavy rock armour greater than 10 tonnes;
2. For heavy rock armour gradings in the range 0.3–10 tonnes where the calculated

design W_{50} falls inconveniently in relation to the standard gradings and where the need to use heavier rocks than those theoretically required has a definite cost penalty, even when the reduced number of rock pieces required is taken into account;

3. *For wide gradings that can be produced at considerable economic saving compared with standard gradings and which are acceptable to the designer in terms of their function;*
4. *Where a quarry is being developed for the specific breakwater, dam, or coastal protection project (temporary dedicated quarry) and economics dictate minimum wastage of quarried materials.*

A1.4.2 DETERMINATION OF GRADINGS

For fine gradings, the sample is selected according to the sampling procedure in Sections A1.8.2.3.1 and A1.8.2.4. A size distribution analysis is performed using sieves and is reported as described in Appendix 2, Section A2.1.

For all heavy gradings and for light gradings designated by weight, the sample is selected for analysis according to whichever sampling procedure in Sections A1.8.2.3.2 and A1.8.2.4 has been designated as most appropriate. A weight-distribution analysis is performed and reported together with the effective mean weight using those weight class intervals described in Table A5 of Appendix 2, Section A2.2:Part 1 for the cumulative plot. For light gradings designated by size, the sample is selected according to procedure in Sections A1.8.2.3.4 and A1.8.2.4 and the size distribution and average weight are determined according to Appendix 2, Section A2.2:Part 2.

A1.4.3 FINE GRADINGS

The requirements for each of the four standard fine grading classes is defined by the minimum and maximum acceptable proportions of the total weight of the sample which will pass through the standard sieve sizes used in the particle size analysis as indicated in Table A1 and described in Appendix 2, Section A2.1.

The particle size distribution of the sub-sample referred to in Section A1.8.2.3.1 must be determined in accordance with Appendix 2, Section A2.1. The arithmetic mean of the particle size distributions from each subsample must satisfy the requirements in accordance with Table A1 for the grading class designation..../....mm.

Table A1 Fine-grading classes

Designation of grading class	30/60 mm		40/100 mm		50/150 mm		80/200 mm	
Sieve (mm)	Cumulative weight passing sieve as percentage of total weight							
	min.	max.	min.	max.	min.	max.	min.	max.
250							95	100
180					90	100	50	100
125			90	100			10	50
90					10	50		
63	90	100	10	50			0	10
45	30	70			0	10		
31.5	0	20	0	10				
22.4	0	10						

A1.4.4 LIGHT GRADINGS

Each of the three standard light grading classes designated by weight must fulfil:

Table A2 Light-grading class requirements

Standard light gradings designated by weight

Grading class designation (kg)	Class limit definition, W_y (kg)				Effective mean weight (kg), W_{em}	
	Extreme lower (ELCL) $y<2\%$	Lower (LCL) $0\%<y<10\%$	Upper (UCL) $70\%<y<100\%$	Extreme upper (EUCL) $97\%<y$	min.	max.
10–60	2	10	60	120	20	35
60–300	30	60	300	450	130	190
10–200	2	10	200	300	30	90

Standard light gradings designated by size and average weight

	Class limit definition by square hole (mm)				Average weight retained on L hole, $\bar{W}_L$ (kg)					
	% y by weight passing				Rock density, ρ_r (t/m³)					
	EL hole $y<2\%$	L hole $0\%<y<10\%$	U hole	EUCL hole $97\%<y$	<2.5		2.5–2.9		>2.9	
					min.	max.	min.	max.	min.	max.
200/350	100	200	(350)	400	20	40	25	45	25	50
350/550	250	350	(550)	650	115	180	130	200	145	240
200/500	100	200	(500)	550	45	80	50	90	55	100

The 10–60, 60–300 and 10–200 kg classes are approximately equivalent to the 200/350, 350/550 and 200/550 mm classes, respectively. The sizes given in brackets are those of the equivalent UCL or U hole and are not requirements but for consistency are used for designating the class of grading.

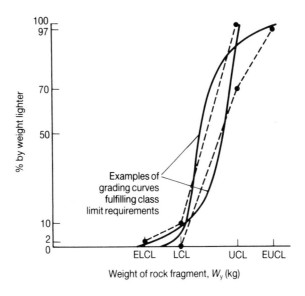

Figure A1 *Explanation of the grading class limits for a standard grading*

1. The class limit definitions; and
2. The effective mean weight limits given in Table A2 and determined in accordance with Appendix 2, Section A2.2:Part 1, where the convention for defining the classs limits and cumulative weight curves is illustrated in Figure A1.

The weight distribution and the effective mean weight of the stone pieces in the sample referred to in Section A1.8.2.3.2 must be determined in accordance with Appendix 2, Section A2.2:Part 1. The results must satisfy the requirements in Table A2 for light gradings designated by the grading class....kg.

Each of the three standard light grading classes designated by size and average weight must fulfil the requirements for:

1. The class limit definitions; and:

2. The range of average weight of stones given in Table A2 using the test method described in Appendix 2, Section A2.2:Part 2.

The results must satisfy the requirements in Table A2 for light gradings designated by the grading class.../...mm.

The size-limit specification of light gradings is only appropriate for underlayer applications, whereas the weight-limit specification is always necessary for cover layers. Any of the three standard light gradings specified by size limits but tested and certified by their equivalent weight designated grading class would be deemed acceptable for all uses as the weight definition is taken as primary.

A1.4.5 HEAVY GRADINGS

Each of the four standard heavy grading classes must fulfil both the class limit definitions and the effective mean weight limits given in Table A3 (see Figure A1). The weight distribution and the effective mean weight of the stone pieces in the samples referred to in Section A1.8.2.3.2 must be determined in accordance with Appendix 2, Section A2.2:Part 1. The results must satisfy the requirements in Table A3 for heavy gradings designated by the grading class.../...tonnes.

Table A3 Heavy-grading class requirements

Grading class designation	Class limit definition, W_y (tonnes)				Effective mean weight, W_{em} (tonnes)	
	Extreme lower (ELCL)	Lower (LCL)	Upper (UCL)	Extreme upper (EUCL)		
(tonnes)	$y<2\%$	$0\%<y<10\%$	$70\%<y<100\%$	$97\%<y$	min.	max.
0.3–1	0.2	0.3	1	1.5	0.54	0.69
1–3	0.65	1	3	4.5	1.7	2.1
3–6	2	3	6	9	4.2	4.8
6–10	4.0	6	10	15	7.5	8.5

A1.5 Shape

All results for shape determinations must refer to samples taken in accordance with Section A1.8.2.3.3. The shape specification refers to percentage by weight for light and fine gradings and percentage by number of stones for heavy gradings.

The shape specification may be chosen from either of the following (but option 1 will be appropriate in most circumstances):

1. The quarry stone sample shall not contain more than 5% of stones with a length to thickness (l/d) ratio greater than 3;
2. The quarry stone sample shall not contain more than 50% of stone with a length to thickness (l/d) ratio greater than 2 and no stone with l/d greater than 3;

where the length, l, is defined as the greatest distance between two points on the stone (e.g. diametrically opposite corners of a cuboidal block) and the thickness, d, as the minimum distance between two parallel straight lines through which the stone can just pass. The sampling is according to Appendix 2, Section A2.3 using at least 50 pieces taken at random from stones greater in weight than the Extreme Lower Class Limit (see Figure A1).

Any of the above specification clauses may be supplemented with the following sentence.

Blocks of quarry stone in heavy gradings showing clear signs of significant edge or corner wear or of severe rounding shall not be accepted.

A1.6 Rock Quality Requirements

A1.6.1 GENERAL

All results for rock quality tests reported in Section A1.6.1 to and including Section A1.6.5 shall refer to samples taken in accordance with Section A1.8.2.3.3.

A1.6.2 DENSITY

The density specification and requirements may be chosen from the following:

The average density of quarry stone used for armour/core/underlayers must be:

1. At least $2600 \, kg/m^3$ with 90% of the stones having a density of at least $2500 \, kg/m^3$;
2. At least $(x) \, kg/m^3$ with 90% of the stones having a density of at least $(x - 100) \, kg/m^3$;
3. Less than ...... kg/m^3

for sampling, testing and reporting in accordance with Appendix 2, Section A2.6 using ten density determinations, each determination being carried out on a different randomly selected stone where such stone shall have a volume of at least 50 ml and if any stone is too large, a representative part of at least 50 ml shall be taken.

The designer may wish to select different density requirements for different applications. For armourstone, it will always be from (1) or (2), with the actual density selected to suit hydraulic conditions and known local availability, but there is no need to set a density requirement higher than that in (1) from durability considerations. A requirement (3) may be appropriate in order to ensure the cost advantage of readily available low-density rocks, but this should only be used for core materials.

A1.6.3 WATER ABSORPTION

The water absorption specification and requirements may be chosen from the following:

The average water absorption of quarry stone must be:
1. Less than 2% and the water absorption of nine of the individual stones less than 2.5% *for rocks required to be of good quality*;
2. Less than 6% and the water absorption of nine of the individual stones less than 8% *for rocks where marginal quality is acceptable for durability purposes (see Section 3.2.4 of the main text for further details)*;

for sampling, testing and reporting in accordance with Appendix 2, Section A2.7 using ten water-absorption determinations, each determination being carried out on a different randomly selected stone where such stone shall have a volume of between 50 and 150 ml and if any stone is larger than 150 ml, a representative part of between 50 and 150 ml shall be taken.

A1.6.4 RESISTANCE TO WEATHERING

In addition to the water-absorption requirement in Section A1.6.3, resistance to weathering may be specified using one or more of the following:

1. Magnesium sulphate soundness for sampling, testing and reporting in accordance with BS 4639, Part 1, Appendix B:1984 ysing either the 63–125 mm sizes of crushed rock and a 50 mm sieve to determine losses or the 20–10 mm sizes of crushed rock and an 8 mm sieve to determine losses, *whichever (according to Section 3.3.7 of the main text) is deemed appropriate*, must be:

(a) Less than 2% *for rocks required to be of excellent quality such as in severe conditions in hot, dry climates where porous sedimentary rocks are used* (*see Section 3.2.4 of the main text*);

(b) Less than 12% *for rocks required to be of good quality*;

(c) Less than 30% *for rocks where marginal quality is acceptable for durability purposes* (*see Section 3.2.4 of main text for further details*).

2. Average freeze–thaw weight loss for sampling, testing and reporting in accordance with Appendix 2, Section A2.4 must be:

    (a) Less than 0.5% *for rocks required to be of good quality*;

    (b) Less than 2% *for rocks where marginal quality is acceptable for durability purposes* (*see Section 3.2.4 of the main text*)

    and there must be no crack or cracks developed during testing for at least 19 out of 20 randomly selected stones containing no visible cracks and with test sample preparation in accordance with Appendix 2, Section A2.4. If the water absorption according to Appendix 2, Section A2.7 does not exceed 0.5% for each of the ten pieces, this is taken to mean that requirement 2(a) has been met.

3. Deleterious secondary minerals shall not be present. For all rock types, this is taken to be indicated by Methylene Blue absorption values of less than

    (a) 0.7 (g/100 g) *for rocks required to be of good quality*;

    (b) 1.0 (g/100 g) *for rocks where marginal quality is acceptable for durability purposes* (*see Section 3.2.4 of the main text*)

    for sampling, testing and reporting in accordance with Appendix 2, Section A2.10. If the rock is of basaltic type, a further requirement is that there shall be no occurrences of Sonnenbrand effect in the first 20 stones tested or no more than one occurrence in the first 40 stones tested (for sampling, testing and reporting in accordance with Appendix 2, Section A2.8). In the case of a rock categorised by an experienced geologist as a fresh granite or rhyolite, or a pure quartzite, limestone or gneiss, this condition may be taken to have been satisfied.

Of the above three test requirements, checking for item (3) must always be carried out. Of the remaining two tests, checking for item (1) is more important in hot, wet and, especially, hot dry climates, whereas checking for item (2) is more important in cooler climates even if only occasionally subject to freezing.

A1.6.5 RESISTANCE TO IMPACT AND MINERAL FABRIC BREAKAGE

The resistance to impact and mineral fabric breakage may be specified using one or a combination of the following:

1. Average fracture toughness in the general planar direction of the most pronounced layer should any visible anisotropy exist and for sampling, testing and reporting in accordance with ISRM 1988 recommended method at Level I must be

    (a) At least $0.8\,\mathrm{MPa\,m^{1/2}}$ with the average minus the standard deviation of the fracture toughness at least $0.6\,\mathrm{MPa\,m^{1/2}}$ *for rocks where marginal quality is acceptable for durability purposes* (*see Section 3.2.4 of the main text*);

    (b) At least $1.4\,\mathrm{MPa\,m^{1/2}}$ with the average minus the standard deviation of the fracture toughness at least $1.2\,\mathrm{MPa\,m^{1/2}}$ *for rocks required to be of good quality*

    where at least six valid test results obtained from cores originating from at least four randomly selected stones have been used to calculate the final result and all valid results have been used to calculate the average and standard deviation.

2. Average point load index in the planar direction of the most pronounced layering should any visible anisotropy exist and for sampling, testing and reporting in accordance with the ISRM 1986 recommended method must be

    (a) At least 1.5 MPa with the average minus the standard deviation of the point load index of at least 1.0 MPa *for rocks where marginal quality is acceptable for durability purposes* (*see Section 3.2.4 of the main text*);

    (b) At least 4.0 MPa with the average minus the standard deviation of the point load index of at least 3.0 MPa *for rocks required to be of good quality*;

where at least ten valid test results obtained from pieces originating from at least ten randomly selected stones have been used to calculate the average and standard deviation, the largest and smallest valid test results obtained being excluded from this calculation. *In practice this will therefore mean that at least twelve valid test results will be required.*

3. Wet dynamic crushing value for sampling and testing in accordance with Appendix 2, Section A2.5 must be
    (a) Less than 30% *for rocks where marginal quality is acceptable for durability purposes (see Section 3.2.4 of the main text)*;
    (b) Less than 20% *for rocks required to be of good quality*;
   where the sample for analysis consists of six randomly taken stones or representative parts thereof for which the weights may not differ more than 25% and each stone weighs at least 0.5 kg.

A1.6.6 RESISTANCE TO ABRASION

The resistance to abrasion may be specified using one of the following:

The mill abrasion resistance index for sampling and testing in accordance with Appendix 2, Section A2.9 must be:
1. Less than 0.015 *for rocks where marginal quality is acceptable for durability purposes (see Section 3.2.4 of the main text)*;
2. Less than 0.004 *for rocks required to be of good quality*;
3. Less than 0.002 *for rocks required to be of excellent quality*,

where the test sample is prepared according to options 1 or 2 as given in Appendix 2, Section A2.9.2.

A1.6.7 BLOCK INTEGRITY

Blocks from heavy gradings must be free from visually observable cracks, veins, fissures, shale layers, styolite seams, laminations, foliation planes, cleavage planes, unit contacts or other such flaws which could lead to breakage during loading, unloading or placing.

The designer should also specify a requirement, as suggested in Table 21, for the drop test described in Appendix 2, Section A2.11, or select whatever non-standard test is considered appropriate to assess block integrity more objectively (see Section 3.4.8 of the main text). If the drop test is selected the following wording may be adopted:

In addition, the drop test breakage index calculated based on appropriate sampling and testing as described in Appendix 2, Section A2.11 shall be less than ...%.

The percentage to be inserted is still a matter of research but a provisional figure of 5% is suggested.

A1.6.8 MODIFICATION TO REQUIREMENTS FOR ROCK ARMOUR PROPOSED BY CONTRACTOR

The contractor may propose modifications to the rock quality requirements specified in this clause provided that the performance of the rock armour in terms of armourstone weight remaining after ... years in service, as assessed using the Rock Degradation Model (*main text, Section 3.2.4.2*), is similar to that indicated by the quality and size parameters specified. Modifications in the size of armourstone may be necessary and any modified gradings should be selected where possible from the standard gradings given in Section A1.4 *and designed following guidance in the main text.*

A1.7 Impurities

Quarried rock shall not contain visually observable or chemically detectable impurities or foreign matters in such quantities that these are damaging for the constructive application of the quarried stone or for the environment in which the quarried stone is applied.

Quarried rock intended for use in a layer to be grouted with bitumen or cement (see Section A1.3.1(d)) shall not be covered with visually observable clay or other adhesive soil.

Other than problems of visual cleanliness which can satisfactorily be assessed by the visual inspection incorporated into this clause, particular problems in bonding of bitumen to certain rock types arise. This is particularly so, for example, in the case of flint and bauxite. If the designer is uncertain about the bond characteristics of a material offered, he could consider use of a stripping test such as those discussed in Collis and Fox (1985). However, these tests are designed to assess the resistance of bitumen aggregate mixtures to the removal of aggregate pieces in road pavements and it is, therefore, difficult to correlate the results with the environment of a grouted rock revetment.

A1.8 Inspection

A1.8.1 LOCATION

A1.8.1.1. At or near site

The requirements in Sections A1.4–A1.7 inclusive apply to inspection at or near the site.

A1.8.1.2 At quarry

The interpretation of inspection results shall take into account the possible influence of storage, loading, transporting and unloading on the quality requirements in Sections A1.4–A1.7 inclusive.

It is preferable that quality control procedure be concentrated at the quarry to prevent the unnecessary transportation of sub-specification materials. Deterioration in quality and block size due to handling can often be assessed by sampling before dispatch from the quarry and after delivery to site. Suitable guidance criteria at the quarry can then be derived which when applied will normally ensure the specified requirements in Sections A1.4–A1.7 inclusive at the site.

A1.8.2 SAMPLING

A1.8.2.1 General

The samples of the grading of quarried stone to be inspected shall be taken at random and must be representative. The sampling inclusive of transport and transfer of the samples shall be carried out in a careful manner so that breakage is limited to a minimum.

The pieces of one stone which, according to observation, were broken during sampling, will be considered to comprise one stone at the inspection.

Depending on the particular quality to be investigated, it can sometimes be difficult to carry out a satisfactory representative sampling of the quarried rock. Particle size distribution and contamination can, for example, vary appreciably between locations for each batch as a result of loading, unloading and storage.

Sampling should be by one of the methods described in this standard (see Section A1.8.2.4). The choice of the most appropriate method is at the specifier's discretion while the execution of the sampling by experienced personnel will be of great benefit.

A1.8.2.2 Homogeneity of the batch

When, on the basis of visual judgement of the quarried rock batch to be inspected, non-homogeneity or possible non-homogeneity of the batch is considered to exist with regard to one or more of the relevant qualities, that batch has to be divided into parts supposed to be homogeneous. Sampling must for those qualities be carried out on the supposedly homogeneous divided parts.

When one of the divided parts does not satisfy the requirements, according to the results of the inspection, then the whole batch of quarried stone is taken to be unsatisfactory.

If separation of the divided part(s), which does (do) not satisfy the requirements, is possible without difficulty, it can be agreed to regard the remaining part of the batch as a separate batch.

A1.8.2.3 Size and composition of samples

A1.8.2.3.1 Samples for determining particle distribution

For the determination of the particle distribution of a fine-graded quarry stone, at least six sub-samples shall be taken if the sampling takes place from a stockpile or a ships's load. In all other cases the number of sub-samples has to be at least three.

The numerical value of the weight in kilograms of each sub-sample must be at least equal to the numerical value of the upper limit in millimetres of the designation of the grading concerned if that upper limit is less than or equal to 100 mm. The numerical value of the weight of each sub-sample in kilograms must be at least twice the numerical value of the upper limit of the designation if the upper limit is greater than 100 mm.

A1.8.2.3.2 Samples for determining weight distribution

For the determination of the weight distribution of the light- or heavy-graded quarry stone at least six sub-samples shall be taken if the sampling takes place from a stockpile or a ship's load. In all other cases the number shall be at least three.

The sub-samples including all the rock fragments together constitute one sample. This sample must contain at least 200 pieces of stone heavier than the extreme lower class limit of the designated grading class (see Sections A1.4.4 and A1.4.5).

When the determination of the weight distribution concerns a ship's load containing less than 200 pieces of stone, the whole load is taken to be one sample.

A1.8.2.3.3 Samples for determining shape and rock quality

The numbers of stones of specified sizes constituting samples for rock shape and quality tests are enumerated in Sections A1.5 and A1.6.

The stones should normally be chosen at random from the sub-samples which have been taken for the particle and weight distributions. Where such samples are not available, the stones shall be taken at random from the batch to be inspected. If the chosen pieces of stone are too large for the test descriptions in force, it will be necessary to break from each stone a representative piece of the required dimensions.

A1.8.2.3.4 Samples for determining grading designated by size and average weight

At least four sub-samples shall be taken if sampling is from a ship's load or from a stockpile. In all other cases, the number shall be at least two.

The sub-samples, including all rock fragments, together constitute one sample. This sample must contain at least 100 pieces of stone retained on the L square hole of size given in Table A2 for the designated grading class.

A1.8.2.4 Method of operation

A1.8.2.4.1 General

Ensure that during sampling the grab or other extraction equipment is filled to a minimum such that the degree of filling does not adversely affect the representativity of the sample taken or part thereof.

A1.8.2.4.2 Sampling from a belt conveyor

Prior to sampling material on the belt conveyor, let the belt transport for a period sufficient to ensure that deviations from the composition of the material possibly present due to the starting up of the installation will not be shown in the sample. For sampling from a belt conveyor a sample of a sufficient quantity of material should be taken by catching it from the end of the belt or by stopping the belt and then taking material from the belt. Catch the material from the end of the belt in a manner to ensure that, from the cross-section of the material flow, material is taken from each point for equal periods of time. When sampling from a stationary belt, take the sample from a point marked on the installation.

Take the required number of sub-samples at approximately equal intervals along the whole batch.

A1.8.2.4.3 Sampling from a silo

When sampling from a silo, take a sample by catching a sufficient quantity of material from the silo. For sampling from a silo account must be taken, in connection with particle size reduction and segregation, of the degree of filling of the silo and the speed of loading into and extraction from the silo. Take the required number of sub-samples at approximately equal intervals from the whole batch to be sampled.

If during the sampling, segregation is observed, the number of samples should be adjusted accordingly.

A1.8.2.4.4 Sampling from a stockpile

When sampling from a segregated stockpile, take a sample of sufficient quantity from the material which is being taken from the stockpile. Take, for this purpose, the contents of one or more loads of a wheel loader, lorry or any other transport or transfer method employed.

Simulate the removal of material from the segregated stockpile if, at the instance of sampling, no material is undergoing routine removal. Before taking the sample, make several extractions of material from the stockpile so as not to distort the sample contents with segregation effects associated with initiation of stockpile extraction.

When sampling from a non-segregated stockpile, take a sample as indicated for a segregated stockpile or by taking a sufficient quantity of material from a random location which is easily reached with the equipment available.

A1.8.2.4.5 Sampling from floating equipment

For sampling prior to the unloading of the segregated load take adequate quantities of material from the locations shown in Figure A2 at the surface of the load, with the aid of the unloading equipment. For the sampling of a non-segregated load the samples must be taken as is indicated for a segregated load or by taking an adequate quantity of material at random or evenly (Figure A3) distributed locations at the surface of the load, with the aid of the unloading equipment.

When sampling during unloading, take for each sample an adequate quantity of material with the aid of the unloading equipment. Take the required number of sub-samples at approximately equal intervals from the whole of the load to be sampled.

During unloading of a ship load a fragment rich quantity of material on the ship bottom may be generated by segregation due to unloading. This material can be sampled and inspected separately in order to implicate this material accurately in the determination of the particle or weight distribution.

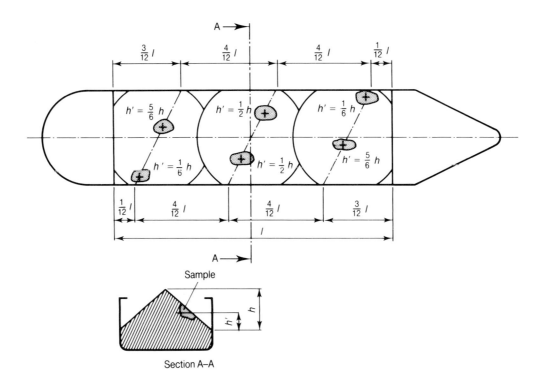

Figure A2 *Sampling locations in the load on floating equipment*

A1.8.2.4.6 Sampling from wheeled transport

For the sampling of a load of quarried stone let the load be tipped out partially or completely in a manner which produces an evenly distributed long pile. Take the required number of sub-samples from across that pile by removing at random or at equally distributed locations an adequate quantity of material, while avoiding the possible segregated material at the start and finish of the pile (Figure A4). Take the material in long strips over the full width of the pile or in equal numbers of half strips from the left- and right-hand side of the centre line of the pile.

A1.8.2.4.7 Splitting of samples of light and fine gradings

If the sample collected for the inspection of compliance with the requirements in Sections A1.4–A1.6 is too large, take a part of the sample according to one of the methods described below.

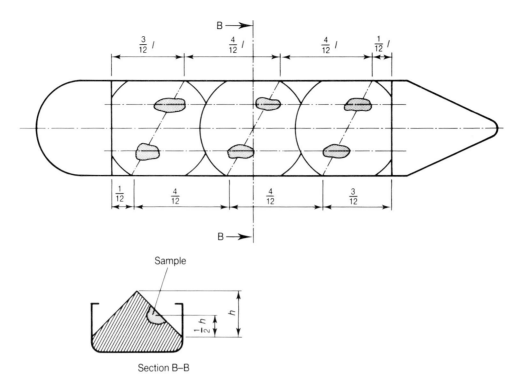

Figure A3 *Sampling locations in a non-segregated load on floating equipment*

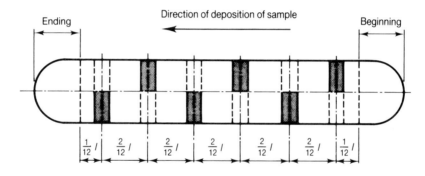

Figure A4 *Sampling locations in a spread-dumped load*

When depositing a sample, take into account the splitting to be carried out by spreading the sample appropriately. Dump the sample to be deposited and to be split over one or more buckets in a manner which allows for possible segregation. When dumping material from a wheel loader bucket, catch all the material from an imagined cross-sectional width of the bucket content in the sample bucket(s). The width of diameter of the (sample) buckets must be at least twice the sieve dimensions of the largest piece of stone.

If so desired dump the sample, which is to be deposited and to be split, over one or two vertically set plates, which will create separation planes. Proceed further in accordance with the work methods presented in the following description, utilising wires representing the imaginary vertical separation surfaces. Stretch a wire as a separation line over the sample already deposited to indicate the desired demarcation into two approximately equal parts. Where segregation has taken place in one direction of the deposited sample, place the wire in the same direction (Figure A5). Remove all material where all pieces of stone or the majority are placed to one side of the imagined vertical plane projected by the wire.

When, for the division of the deposited sample, less than half of the total sample is

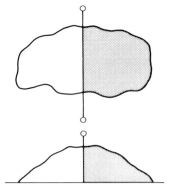

Figure A5 *Halving a sample by means of a separation plane*

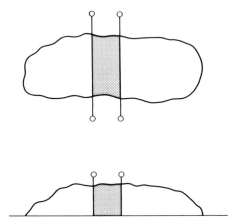

Figure A6 *Dividing a sample with two separation planes*

required, stretch two parallel wires as dividing lines over the sample, so that the desired part of the sample lies between the two separation lines. If the complete sample has been segregated in one direction, stretch the wires in the same direction (Figure A6). Take all the material from the strip between the imaginary two vertical planes between the wires, with all pieces of stone which are completely or for the largest part between the two planes. If so desired, where no segregation of material has taken place, material to be taken can be limited to half the separated strip.

Take a sample, which consists of a not too large number of stones, by a random collection of the necessary number of stones. Take the stone pieces at random by choosing them blindfolded by lottery numbers or by selecting stones at predetermined but irregular intervals.

A1.8.3 TRANSPORT AND IDENTIFICATION OF THE SAMPLES

For the transport of a sample, precautions shall be taken so that no material is broken or lost and that the sample is not contaminated. A sample shall be accompanied by a certificate drawn up by the person responsible for taking the sample. The certificate shall include the following information:

- A reference to this standard;
- The name of the producer and location of the quarry or other source where the broken stone is produced;
- The description and class designation of the grading;
- The number of stone pieces in the sample;
- Details about location and method of sampling, including the date when the sampling took place;
  The name of the sample taker.

PART C—PLACEMENT/CONSTRUCTION REQUIREMENTS

A1.9 Access to Works

If the client or designer has special or necessary requirements for, or restraints on, access routes and/or modes or on times and/or rates of delivery of rock materials, these should be stated here.

A1.10 Transport and Handling

Quarried rock which will be placed in the works in bulk shall be transported and handled in such a manner as to minimise segregation of the rock.

A1.11 On-Site Inspections

The contractor shall provide all facilities for any on-site inspection, categorisation, approval/rejection activities on materials indicated in Part B of this specification. (*The designer should list here in detail the facilities he will require.*)

A1.12 Placement

A1.12.1 CORE MATERIAL

Core material shall be placed to the positions and slopes indicated on the drawings and in accordance with the method and sequence of construction approved by the client/designer. Quarried rock core material shall be placed to ensure that the larger rock fragments are evenly distributed and the smaller rock fragments serve to fill the spaces between the larger rock fragments.

Subject to the written approval of the designer, core material may be dumped and tipped to the natural slope of the material and left untrimmed provided that the core is built up to the dimensions shown on the drawings with the material specified for the layer next overlying the core and placed in accordance with the method for the overlying layer.

In considering whether to give approval to this course of action, the designer must consider whether replacement of core material by material appropriate to an armour layer or underlayer will adversely affect the porosity or shear strength of the structure, taking account of all the functional requirements, including energy dissipation, shear strength and structure-related water motions.

Alternatively, the dumped material shall then be dressed out to the required slope and position by placing.

A1.12.2 PLACING OF MATERIAL IN ARMOUR LAYERS AND UNDERLAYERS

Quarried stone in underlayers and armour layers shall generally be placed individually in accordance with Section A1.12.3. However, placement in bulk in accordance with Section A1.12.4 is permitted in the following cases:

1. For material in the fine and light gradings as defined in Sections A1.4.2 and A1.4.3;

2. Widely graded rip-rap materials as described in the note to Section A1.4.1;
3. Heavy gradings used in underlayers.

A1.12.3 INDIVIDUALLY PLACED QUARRIED STONE FOR ARMOUR LAYERS AND UNDERLAYERS

Individually placed quarried stone shall not be dropped or tipped into position, but shall be placed piece by piece into the structure to achieve a minimum 'three-point support' and be stable to the lines and levels shown on the drawings. In exceptional circumstances, and only if approved by the client, rocking stones may be stabilised by packing sandbags of dry concrete mix under them. Stones shall be tightly packed together so as to achieve as near as possible a target weight of stone placed of....t/m^3 with a tolerance of ± 0.1 t/m^3 and shall not be placed so that they can rock or obtain their stability on a plane by frictional resistance alone prior to placing further stones.

Target tonnage concept.

The designer must select an appropriate target tonnage. For example, for carboniferous limestone of specific gravity 2.68, experience has indicated that an appropriate figure is 1.6 t/m^3. Reference may also be made to Section 3.2.3.4 in the main text for further guidance. Where no satisfactory information is available, a trial section of the works may be used to define an appropriate target tonnage (see Box 39 in the text).

Where the contractor proposes a modification to the specified requirements in accordance with Section A1.6.8 he should not only specify any revised rock density he wishes to use but also advise the revised target tonnage that results using the adjustment factor approach given in Appendix 3, Section A3.4.3.

For work above low-tide level, sufficient fine material on the surface of already-placed stones (including rocks within the layer being placed) shall be removed from those areas where surface contact will arise between the stone being placed and those already placed to ensure sound bearing and interlock between stones. The contractor shall make due allowance for the removal of such fine material which, in the case of coastal structures, includes beach sediment.

Any void below the finished profile level (as defined in Section A1.13) in excess of the armourstone D_{n35} size shall be filled with an appropriate stone or stones. Determination of the acceptability of any void shall be by means of a test sphere or cage of diameter D_{n35}.

A1.12.4 BULK-PLACED QUARRIED STONE FOR ARMOUR LAYERS AND UNDERLAYERS

The stone shall be deposited carefully to minimise disturbance to any already-placed rock and to avoid damage to any existing structures such as pipelines. Use of rock trays is permitted.

For placing above mean low water neap (MLWN) tide level, the stone need not be compacted but shall be placed to grade to ensure that the larger rock fragments are uniformly distributed and the smaller rock fragments serve to fill the spaces between the larger rock fragments in such a manner as will result in the resulting structure being well keyed, densely packed and of the specified dimensions. Hand placing or barring will be required only to the extent necessary to secure the results specified above.

For placing below MLWN tide level, techniques such as end tipping and dumping from barges may be used, providing the work is organised in such a way as to minimise segregation of the stone grading and to ensure the specified dimensions or weight per unit area.

Any void below the finished profile level (as defined in Section A1.13) in excess of the armourstone D_{n35} size shall be filled with an appropriate stone or stones. Determination of the acceptability of any void shall be by means of a test sphere or cage of diameter D_{n35}.

A1.12.5 PROTECTION OF PLACED MATERIALS

Each placed layer shall be protected by the subsequent layer (as indicated on the drawings) as soon as possible after placement, leaving a maximum length of each material of ... metres and a maximum height of ... metres unprotected, in order to minimise wave damage in the event of storms during the construction period.

The designer should select appropriate values for these dimensions, taking account of the risk of damage and the cost of repair. The values should not be unnecessarily restrictive on the contractor's method of working, nor should they allow unnecessary risk of costs in potential damage during storms and the consequent delay to the construction programme.

A1.12.6 DISTURBANCE TO PREVIOUSLY PLACED MATERIALS

Material eroded by wave action or other cause shall be made good before placing the appropriate protective layer. However, in respect of core material, if authorised in writing by the designer, the core may be built up to the dimensions shown on the drawings with the material specified for the layer next overlying the core and in accordance with the method for this overlying layer.

Notwithstanding the above, the contractor shall take all reasonable care to avoid disturbing a previously placed layer by avoiding dropping or other potentially disturbing placing methods.

A1.12.7 TEMPORARY HAUL ROADS AND TRACKS

Where the contractor wishes, or the designer or client requires, that a temporary haul road or track be created within or on the structure, it shall be constructed of free-draining local beach material if available and suitable for this purpose, or of other free-draining material approved by the designer. Such material shall be removed in accordance with Section A1.12.3 before placing subsequent layers. Where placed along the crest of a completed structure, the haul road material shall be sufficiently removed in order to expose the upper layer of stones to between one third and one half of their depth measured from their upper surfaces. Any stones laid flat to facilitate haul road construction shall be reset to the requirements of Section A1.12.3.

The designer or client should only instruct haul roads to be built where the nature of the rock and plant is such that there is a risk of extensive crushing or splitting of the construction materials. Minor chipping, even if extensive, is unlikely to significantly affect the rock grading.

A1.13 Paylines and Surface Profiles

A1.13.1 TOLERANCES

Rock materials shall be placed to the levels, dimensions and slopes (paylines) shown

Table A4 Vertical placing tolerances for rock materials

Depth of placing below low water	Bulk placed rock of grading where		All armour layers and individually placed rock with gradings where W_{em} greater than 300 kg	
	Effective mean weight (W_{em}) less than 300 kg	W_{em} greater than 300 kg (not armour layers)	On individual measurements (m)	Design profile to actual mean profile (m)
Dry, i.e. above low water	±0.2 m	+0.4 m −0.2 m	±0.3D_{n50}	+0.35D_{n50} −0.25D_{n50}
Less than 5 m	+0.5 m −0.3 m	+0.8 m −0.3 m	±0.5D_{n50}	+0.6D_{n50} −0.4D_{n50}
5–15 m		+1.2 m −0.4 m		
Greater than 15 m		+1.5 m −0.5 m		

Notes
1. Tolerances apply even if gradings being placed have not been selected from the standard gradings in Clause 4.
2. All tolerances refer to the design profile to actual mean profile unless stated otherwise.

on the drawings and, when the surface profile is measured using the techniques specified in Section A1.13.2, shall comply with the vertical tolerances in Table A4:

Notwithstanding the tolerances in Table A4, the following shall apply to armour layers:

1. The tolerances on two consecutive mean actual profiles shall not be negative.
2. Notwithstanding any accumulation of positive tolerances on underlying layers, the thickness of the layer shall not be less than 80% of the nominal thickness when calculated using mean actual profiles. Where an accumulation of positive tolerances arises and is acceptable to the designer, the position of the design profiles will need to be adjusted to suit.

A1.13.2 SURVEY TECHNIQUE

Measurements shall be carried out using a probe with a spherical end of diameter $0.5D_{n50}$. For a land-based survey this will generally be connected to a staff or EDM target; for an underwater survey it will generally be a weighted ball on the end of a sounding chain.

Measurements shall be carried out at the following intervals across the measurement profile:

1. Fine and light gradings 1 m
2. Heavy gradings $0.75 \times D_{n50}$

Measurement profiles shall be at intervals along the length of the structure (breakwater, seawall, etc.) approved by the designer. These will generally be every 10 m, but may need to be more frequent where the profile is changing rapidly or on tight-radius curves. No layer shall be covered by a subsequent layer until the profile of the former layer has been approved by the client/designer.

Further details of survey techniques are given in Appendix 6.
For underwater surveying of large breakwaters where very thick armour layers are used as part of optimal use of quarry output, survey control using narrow-beam echo-sounders may provide adequate control, even given that the survey error is likely to be

of the order of ± 0.6 m. The tolerances specified in Section A1.13.1 can clearly be relaxed in this situation.

A1.13.3 NOMINAL STONE DIAMETER, D_n

The nominal stone diameter, D_n, in the above sections shall be calculated as the cube root of the volume of the stone. The volume shall be calculated by dividing the mass of the stone by the saturated surface dry density. Where a numbered subscript is given to D_n, this refers to the percentage by weight of stones in the grading having a smaller nominal stone diameter.

A1.14 Disposal of Surplus, Unsuitable Materials, etc.

It shall be the contractor's responsibility to remove from the site of the works all surplus material, rubbish, debris and material unsuitable for inclusion in the works and dispose thereof at an approved location.

A1.15 Settlement

The contractor shall make good any settlement within the structure that may occur up to one year after completion of the works. Making good of settlement shall be with materials and in a manner approved by the client/designer.

A1.16 Preparation of Rock Surface for Crown Walls, etc.

Where the drawings indicate that the rock structure is to have a crown wall or other reinforced or mass concrete structure cast upon it, the interstices between the stones directly under the concrete structure are to be filled with selected core material to form an effective seal against leakage of concrete, leaving a natural key projection of not less than $0.25 \times D_{n50}$ between the irregular surface of the secondary armourstone and the concrete structure.

The thickness of concrete provided by the designer should, however, take account of the possibility of some minor grout loss locally from the underside of the slab. Where a shear key between the concrete structure and the rock structure beneath is not required for hydraulic stability purposes, the designer may eliminate the requirement for a significant natural key projection and, in addition, may specify a layer of polythene sheeting beneath the concrete to reduce concrete grout loss.

If settlement of the surface on which the concrete structure is to be cast has taken place, the use of a layer of small stones to bring it to the correct level will not be permitted. In such circumstances the thickness of concrete placed shall be increased to ensure that the correct surface levels of the concrete structure are attained.

Appendix 2: Standards for quarried rock materials applications in coastal and shoreline engineering

A2.1 Determination of Particle Size Distribution

Note: This standard is based on Draft NEN 5181 (March 1988).

A2.1.1 SUBJECT AND AREA OF APPLICABILITY

This standard presents the method for the particle size distribution of stone materials which are coarser than a nominal 31.5 mm size. The sieve sizes and requirements for standard gradings are given in Table A1.

A2.1.2 SAMPLE FOR ANALYSIS

The numerical value of the weight of the sample for analysis in kilograms shall be at least equal to the numerical value of the nominal upper limit of the particular grading if the nominal upper limit is smaller than or equal to 100 mm. The numerical value of the weight of the sample for analysis in kilograms shall be at least twice the numerical value of the nominal upper limit if this upper limit is greater than 100 mm.

A2.1.3 EQUIPMENT AND OTHER AIDS

A2.1.3.1 Sieves with square openings of 250×250 mm, 180×180 mm, 125×125 mm, 90×90 mm, 63×63 mm shall comprise steel rods of 12 mm diameter welded together at right angles, thus forming square openings with dimension according to the nominal sieve sizes with tolerances of ± 0.5 mm. Sieves from perforated metal plate shall be in accordance with ISO 565.

Of the sieves given above those which relate to the requirements of the particle size distribution of the particular gradation and, in addition, in all cases the 63×63 mm sieve shall be used.

A2.1.3.2 Buckets with a volume of at least $0.1 \, m^3$, on which the indicated sieves fit.
A2.1.3.3 Weighing equipment with a weighing limit of at least 150 kg, accurate to 0.1 kg.
A2.1.3.4 Shovel, brush.

A2.1.4 METHODS OF OPERATION
A2.1.4.1 Sieving and weighing

Place the steel rod sieves (i.e. 250 mm to 63 mm sizes) on the buckets. Pass the sample in successive parts over the sieves in order of increasing sieve size, starting with the 63 × 63 mm sieve.

Brush off, where necessary, any cohesive materials from the pieces of stone. Ensure that all stones which may pass the sieve in any orientation have so passed before the retained material is placed on the subsequent sieve.

Remove the fraction which passes the 63 × 63 mm sieve and determine its weight (m_1). If this is greater than 30 kg, split the fraction, taking and weighing a representative part of at least 15 kg (m_2). Execute the split by discharging the homogenised material over two adjoining buckets.

Pass the fraction which passes the 63 × 63 mm sieve, or the representative part thereof, in turn through the specified perforated metal plate sieves (i.e. 45, 31.5, 22.4 mm sizes) in reducing sequence of the sizes of the openings. Ensure that the stones are positioned in the most likely orientation for them to pass the sieves.

Weigh the material left on each sieve separately (m_n) and the fraction which passes the sieve with the smallest openings (m_3) accurately to 0.1 kg.

A2.1.4.2 Calculation

Determine, in the case that only a part of the fraction which passed the 63 × 63 mm sieve has been sieved further, the values of m_n and m_3 for the smaller aperture sieves by multiplying the individual fractional weights by

$$\frac{m_1}{m_2}$$

Calculate per sieve the cumulative retained weight (M_n) from the retained weights on the separate sieves (m_n).

Calculate per sieve the cumulative retained weight (C_n) in percentage (weight) and rounded up to 0.1% with the aid of the following formula:

$$C_n = \frac{M_n}{\Sigma m_n + m_3} \cdot 100$$

where

C_n = cumulative retained weight on the particular sieve in percentage weight,
m_1 = the weight of the fraction through a sieve 63 × 63 mm (kg),
m_2 = the weight of that part of m_1 which has been sieved further (kg),
m_3 = weight of material passing the smallest sieve (kg),
Σm_n = cumulative weight of material remaining on all separate sieves (kg),
M_n = weight of the cumulative remainder on the particular sieve (kg).

A2.1.5 REPORT

The report shall provide the following data:

1. The cumulative remainders on each sieve used in percentage weight;
2. References to this standard;
3. A description including weight of the sample;
4. The origin of the sample;
5. The data of inspection/test.

If it has been agreed beforehand, the particle size distribution shall be presented on a graph.

A2.2 Determination of Light and Heavy Standard Gradings

Part 1: Determination of the Weight Distribution of Quarried Stone

A2.2.1 SUBJECT AND AREA OF APPLICABILITY

This standard is based on the draft (March 1988) NEN 5182 Dutch standard and is used to verify a requirement for a sample of light or heavy gradings of a given class designation as defined in Appendix 1, Tables A1–3 and Figure A1, where the sample has been taken in accordance with Sections A1.8.2.3.2 and A1.8.2.4 of that appendix.

A2.2.2 EQUIPMENT AND OTHER AIDS

A2.2.2.1 Weighing equipment, accurate to 2% of the LCL.
A2.2.2.2 Lifting equipment and lifting aids for pieces that cannot be moved manually.

A2.2.3 METHOD OF WORKING

A2.2.3.1 Execution of inspection by weighing each stone

A2.2.3.1.1 Weighing

Weigh each stone heavier than the ELCL separately, W_i, and all pieces lighter than the ELCL (the stone fragments) together, W_s, accurate to 2% of the LCL. Record the total weights falling in each weight fraction together with the total number of stones, n, heavier than the ELCL.

A2.2.3.1.2 Calculation

Calculate the total weight, ΣW_i, of pieces equal to or heavier than the ELCL.

To obtain the cumulative curve where W_y is the weight for which the fraction y is lighter, calculate the successive points on the curve at weight intervals given in Table A5.

Table A5 Weight intervals for the cumulative weight plot

LCL of grading class (kg)	Weight interval (kg)
10–60	5
60–300	25
300–1000	50
1000–3000	200
3000 or greater	500

To obtain only those values of y for which a requirement has been set, i.e. the fractions corresponding to W_y at the ELCL, LCL, UCL and EUCL, calculate the total weight W_n, for W_y corresponding to each of the four class limits, with the formula:

$$y = \frac{W_n}{W_s + \Sigma W_i} \cdot 100$$

where

W_n = the total weight of stones lighter than W_y (kg),
ΣW_i = the total weight of all pieces heavier than the ELCL (kg),
W_s = the total weight of pieces lighter than the ELCL (kg).

Calculate the effective mean weight, W_{em}, to the nearest kilogram, using the formula:

$$W_{em} = \frac{\Sigma W_i}{n}$$

where:
W_{em} = the effective mean weight of the stone sample which equals the average weight of stones heavier than the ELCL,
n = the number of stones heavier than the ELCL.

A2.2.3.2 Inspection by bulk weighing

If it has been agreed beforehand, this method which is less precise than Section A2.2.3.1, but which requires fewer weighings, may be used. The cumulative weight distribution is deduced from bulk weighing of pieces in the LCL to UCL weight range as described in Sections A2.2.3.2.1 and A.2.2.3.2.2.

Explanation

Using this method, the complete or part sample could, for example, be weighed on a weighbridge, either in the bucket or a wheel loader or on a lorry. It forms an attractive system for the heavier gradings (of 60–300 kg and over). However, it demands careful working methods in counting stones and visual assessment of which stones fall within the class limit ranges ELCL to LCL, LCL to UCL, etc. The visual assessment may be controlled through check weighing. The partners involved in the inspection can demand inspection in accordance with Section A2.2.3.1, as that section is the decisive method. This is important if the weight distribution by method A2.2.3.2 just misses or just meets the requirement criteria.

A2.2.3.2.1 Weighing

Weigh the total sample, W_t, accurately to 1%, and all other weighings to an accuracy of 2% of the LCL:

1. Weigh separately, each stone heavier than the UCL, W_i;
2. Weigh separately, each stone with weight between the LCL and ELCL, W_i;
3. Determine the total weight of pieces lighter than the ELCL (stone fragments), W_i;
4. Determine the number of stones heavier than ELCL, n.

Calculations are then made with reference to Figure A7, which gives the convention for defining cumulative weight curves and class limits for the standard grading classes in Appendix 1.

A2.2.3.2.2 Calculations

From the W_i values, calculate W_2, W_4 and W_5. Calculate W_3 from the formula:

$$W_3 = W_t - (W_1 + W_2 + W_4 + W_5)$$

Calculate the cumulative percentage by weight of stones lighter, y, for each class limit as follows:

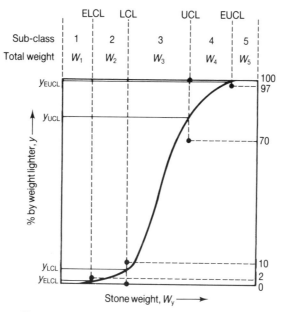

Figure A7 *Illustration and notation used in bulk weighing method*

$$y_{ELCL} = \frac{W_1}{W_t} \cdot 100$$

$$y_{LCL} = \frac{W_1 + W_2}{W_t} \cdot 100$$

$$y_{UCL} = \frac{W_1 + W_2 + W_3}{W_t} \cdot 100$$

$$y_{EUCL} = \frac{W_1 + W_2 + W_3 + W_4}{W_t} \cdot 100$$

Calculate the effective mean weight, W_{em}, to the nearest kilogram, defined in Section A2.2.3.1.2, as follows:

$$W_{em} = \frac{W_t - W_1}{n}$$

A2.2.4 REPORT

The following data shall be included in the report:

1. The measured weights and their associated weight percentages by weight lighter. For method A2.2.3.2, these are only given at the class limit weights: ELCL, LCL, UCL, EUCL;
2. The effective average weight of the stones;
3. A reference to this standard;
4. A description of the sample, including its weight;
5. The source of the sample;
6. The date of the inspection.

If it has been agreed beforehand, a graph showing the weight distribution shall be provided. For this purpose, the cumulative weights shall be plotted using the weight fraction intervals determined from Table A5.

If method A2.2.3.2 has been used, the precise form of the cumulative curve between the LCL and UCL is unknown and should be represented by a dotted straight-line segment.

Part 2: Determination of the Size Distribution and Average Weight of Light Gradings of Quarried Stone

A2.2.5 SUBJECT AND AREA OF APPLICABILITY

This standard is not intended for cover layer applications. It is used to verify a requirement for a sample of light gradings of a given class designated by size and average weight as defined in Appendix 1, Table A2, where the sample has been taken in accordance with Sections A1.8.2.3.4 and A1.8.2.4 of that appendix. The EU hole is defined as the square hole size through which between 100% and 97% by weight shall pass; the L and EL holes are similarly defined for 0–10% passing and 0–2% passing, respectively.

A2.2.6 EQUIPMENT AND OTHER AIDS

A2.2.6.1 Weighing equipment, accurate to 1% of the minimum average weight requirement.
A2.2.6.2 Lifting equipment (e.g. hydraulic) and lifting aids for pieces that cannot be moved manually.
A2.2.6.3 Hand-held hinged square guages with 5 mm thick sides and /or plates of 5 mm sheet metal with at least 20 square hole openings. The length of the sides of the square guages and/or square hole openings shall comply with the requirements of Appendix 1 for the EL, L and EU holes.

A2.2.7 METHOD OF WORKING

A2.2.7.1 Weighing and sorting

Weigh the complete sample $= W_t$

Spread the sample out evenly and thinly onto a tarpaulin or large, clean, flat working area. Physically separate the material passing the L hole using the plates or hand-held gauges. The latter indicates a passing piece when the borderline blocks are held with their long axes vertically and the hinged gauge can pass from the ground up to the top of the block with the guage hinge closed. Count the number, n_M, of pieces not passing the L hole and check that n_M is greater than 100 for a valid sample. Determine the total weight, W_L, of all pieces passing the L hole. Check that the largest pieces will pass the EU hole and determine the total weight, W_{EU}, of any that do not. Check the smallest pieces will not pass the EL hole and determine the total weight, W_{EL}, of all those that do.

A2.2.7.2 Calculations

Calculate the average weight of pieces not passing L hole, $\bar{W}_L$, where

$$\bar{W}_L = \frac{W_t - W_L}{n_M}$$

Calculate the percentage by weight passing the EL, L and EU holes as follows:

$$Z_{EL} = \frac{W_{EL}}{W_t} \cdot 100 \quad Z_L = \frac{W_L}{W_t} \cdot 100 \quad Z_{EU} = \frac{(W_t - W_{EU})}{W_t} \cdot 100$$

A2.2.8 REPORT

The following data shall be included in the report:

1. The measured cumulative percentage by weight passing the EL, L and EU holes;
2. The average weight of pieces not passing the L hole;
3. The rock density tested according to Section A2.6;
4. A reference to this standard;
5. A description of the sample, including its weight;
6. The source of the sample;
7. The date of the inspection.

A2.3 Determination of the Length-to-Thickness Ratio of Quarried Stone

A2.3.1 SUBJECT AND AREA OF APPLICABILITY

This standard is based on the draft (March 1988) NEN 5183 Dutch standard and is used to determine the content of stones with a length to thickness ratio greater than 3 and 2. It is used to verify the requirement given in Appendix 1, Section A1.5. For heavy gradings only, the weighing procedure is unnecessary, as only the number per cent of stones with length to thickness ratios of greater than 2 and 3 is required.

A2.3.2 SAMPLE FOR ANALYSIS

The sample shall contain at least 50 pieces taken at random from above the ELCL weight or EL size hole of the designated grading class and in accordance with Appendix 1, Section A1.8.2.3.3. The EL size hole for fine gradings is given by the 0–10% by weight passing size of Table A1.

A2.3.3 EQUIPMENT AND OTHER AIDS

A2.3.3.1 Measurement apparatus for the determination of length and thickness of stones with an accuracy of 3% or better.
A2.3.3.2 Weighing equipment, accurate to within 2% of the lightest piece to be weighed.

A2.3.4 METHOD OF WORKING

A2.3.4.1 Execution

Measure the length of each stone as the maximum distance between two points on the stone to within 3% accuracy. Measure the thickness of each stone defined as the minimum distance between two parallel straight lines through which the stone can just pass to within 3% accuracy. Weigh the total weight, W_3, of the stones with a length-to-thickness ratio of greater than 3 to within 2% accuracy. Determine the total weight, W_t, of the stones accurately to within 2%. Determine the number of stones with length-to-thickness ratio greater than 3, n_3, and the number of stones with length-to-thickness ratio greater than 2, n_2. Count the total number of stones, n.

A2.3.4.2 Calculations

Calculate the weight per cent rounded to the nearest 1% of stones with length-to-thickness ratio of greater than 3 using the formula:

$$C_{w3} = \frac{W_3}{W_t} \cdot 100$$

Calculate the number per cent of stones with length-to-thickness ratio greater than 3 and 2 using the formula:

$$C_{n_3} = \frac{n_3}{n} \cdot 100$$

$$C_{n_2} = \frac{n_2}{n} \cdot 100$$

A2.3.5 REPORT

The report must provide the following data:

1. The measured weight per cent of stones with length-to-thickness ratio greater than 3;
2. The measured number percent of stones with length-to-thickness ratio greater than 3, and greater than 2;
3. A reference to this standard;
4. A description of the sample, including the weight and the number of stones;
5. The source of the sample;
6. The date of the test.

Note: Box 35 in Section 3.6 of the main text gives practical guidance on taking length and thickness measurements.

A2.4 Determination of Resistance to Freeze/Thaw Cycles

Note: This standard is based on Draft NEN 5184 and B5812.

A2.4.1 SUBJECT AND AREA OF APPLICATION

This standard gives the method to determine the resistance against freeze/thaw cycles of a stone of a grading class with a nominal size greater than the 31.5 mm sieve size.

A2.4.2 SAMPLE FOR ANALYSIS

The stone must be taken at random from the largest fraction of stone material set by the requirements for gradings. If the stone is heavier than 20 kg, the test will have to be carried out on a representative part of at least 10 kg. The stone must have no cracks in it.

A2.4.3 EQUIPMENT AND OTHER AIDS

A2.4.3.1 Drying oven or other appropriate apparatus, capable of adjustable temperature of $(110 \pm 5)°C$;
A2.4.3.2 Weighing equipment, accurate up to 0.01% of the weight of the stone;
A2.4.3.3 Freezer-box with air circulation in which the stone can be exposed to the temperature described in Section A2.4.4.2;
A2.4.3.4 Vessel with a volume at least six times the volume of the stone;
A2.4.3.5 Saw for use in case the stone has a volume in excess of 150 ml;

A2.4.3.6 Plastic film, brush.

A2.4.4 METHOD OF OPERATION

A2.4.4.1 Water absorption at atmospheric pressure

Cut from the stone a representative piece, using the saw, if the stone has a volume in excess of 150 ml. The representative part of the stone should have a volume of at least 50 ml and, at most, 150 ml. Determine the water absorption, in accordance with Section A2.7, of the stone or part of the stone.

End the test if the water absorption does not exceed 0.5%, as in that case the stone is considered to be (satisfactorily) resistant to freeze/thaw cycles. Carry out freeze test in accordance with A2.4.4.2 below if the water absorption exceeds 0.5%

A2.4.4.2 Execution of the freeze test

Let the stone absorb water in accordance with Section A2.7. Wrap the stone in plastic film and place it in the freezer-box. Adjust the temperature control in such a way that the temperature in the stone reaches a level of $-15°C$ or lower in a time of about 5 hours. Maintain that temperature for at least 2 hours. Remove the plastic film and immerse the stone directly in the water in the vessel, which contains drinking water with at least five times the volume of the stone at a temperature of 15–20°C.

Leave the stone submerged for at least 2 hours. Repeat the freeze–thaw cycle 25 times. At the end of these tests, dry the stone in the oven at a temperature of $(110 \pm 5)°C$ until the stone reaches a stage when its weight remains constant.

Determine the weight loss of the stone and check to see if any cracks have developed.

A2.4.5 REPORT

1. Water absorption;
2. Weight loss in per cent and rounded to 0.1%;
3. The development of any cracks during the test;
4. Resistance against freeze/thaw cycles (weight loss less than 0.5% and no crack development);
5. Reference to this standard;
6. Description of the stone, including the weight loss;
7. Source of stone;
8. Duration of the tests.

A2.5 Determination of Dynamic Crushing Strength

Note: This standard is based on Draft NEN 5185.

A2.5.1 SUBJECT AND AREA OF APPLICATION

This standard provides the method for the determination of the dynamic crushing strength of natural stone and of other types of stone and stone-type materials. The dynamic crushing strength is determined as the average test result from a duplicated test.

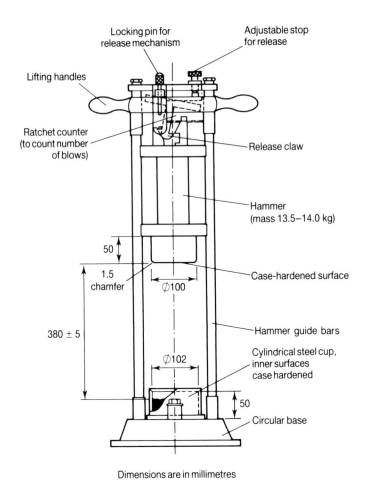

Figure A8 *Aggregate impact test machine*

A2.5.2 SAMPLE FOR ANALYSIS

The sample for analysis must contain sufficient material retained on an 8 mm square hole sieve for two determinations.

A2.5.3 APPARATUS

A2.5.3.1 An impact testing machine of the general form shown in Figure A8 and complying with the following:
A2.5.3.1.1 Total mass not more than 60 kg nor less than 45 kg. The machine shall have a circular metal base weighing between 22 and 30 kg, with a plane lower surface of minimum diameter 300 mm, and shall be supported on a level and plane concrete or stone block or floor at least 450 mm thick. The machine shall be prevented from rocking either by fixing it to the block or floor or by supporting it on a level and plane metal plate cast into the surface of the block or floor.
A2.5.3.1.2 A cylindrical steel cup with an internal diameter of 102 mm and an internal depth of 50 mm, walls not less than 6 mm thick and case-hardened inner surfaces. The cup shall be rigidly fastened at the centre of the base and be easily removed for emptying.
A2.5.3.1.3 A case-hardened metal hammer weighing 13.5–14.0 kg, with a cylindrical lower end diameter of 100 mm and length 50 mm, with a 1.5 mm chamber at the lower edge. The hammer shall slide freely between vertical guides so arranged that the lower (cylindrical) part of the hammer is above and concentric with the cup.

A2.5.3.1.4 Means for raising the hammer and allowing it to fall freely between the vertical guides from a height of 380 ± 5 mm on to the test sample in the cup, and means for adjusting the height of fall within 5 mm. (Note: Some means for automatically recording the number of blows is desirable.)

A2.5.3.1.5 Means for supporting the hammer while fastening or removing the cup.

A2.5.3.2 A straight metal tamping rod of circular cross-section, 10 mm diameter, 230 mm long, rounded at one end.

A2.5.3.3 A cylindrical metal measure of sufficient rigidity to retain its form under rough usage and with an internal diameter of 75 ± 1 mm and an internal depth of 50 ± 1 mm.

A2.5.3.4 Sieves as in NEN 2560 Nos C2, C8 and C11.2. The numbers refer to millimetres of the hole dimensions and C stands for square openings.

A2.5.3.5 Slot sieves S8 and S5.6, sheet metal sieves with slot-shaped apertures. Sheet metal thickness has to be 1.5 mm, and 1 mm, respectively, with slot widths 8 mm and 5.6 mm and slot lengths 40 mm and 30 mm.

A2.5.3.6 Drying oven or other appropriate apparatus capable of regulation to $(110\pm5)°C$.

A2.5.3.7 A balance of capacity not less than 500 g and accurate to 0.1 g.

A2.5.3.8 Jaw crusher.

A2.5.3.9 Scoop, rubber hammer, brush, bucket.

As there is uncertainty with respect to availability of sieves, the alternative standard sieve sizes may be substituted for those given above as follows: for NEN 2560 Nos C2, C8, C11.2, S5 and S8, substitute respectively 2.4 mm, 10 mm, 14 mm, 7.2 mm flakiness slot, 10.2 flakiness, slot, of the BS812 system throughout.

A2.5.4 METHOD OF WORKING

A2.5.4.1 Preparation of the test sample

Take from the sample for analysis a representative part, from which a sufficient quantity of material can be prepared for the strength determination. Reduce the stone sizes with a hammer if the pieces are too large for the available jaw breaker. Break the sample with the jaw breaker to a material size which can pass through sieve C11.2. Adjust the opening of the breaker to a gradually reducing size and, after sieving through C11.2, take the remainder again through the breaker. Continue the crushing process until all the required quantity can pass through sieve C11.2. Sieve the sample on sieve C8.

Sieve the fraction C11.2–C8 through the slot sieves S8 and S5.6. Take the fraction S8–S5.6 as the material for the test samples. Dry this material at a temperature of $(110\pm5)°C$ to a constant weight.

A constant weight is reached when during the drying process the loss of weight between two successive weighings (carried out at 24-hour intervals) is less than 0.5% of the total weight. Let the material cool off to room temperature.

Saturate the test samples in drinking water for 24 hours and dab the particle surfaces dry with a moist chamois leather until no shiny wet surfaces remain. The measure shall be filled about one-third full with the aggregate by means of a scoop, the aggregate being discharged from a height not exceeding 50 mm above the top of the container. The aggregate shall then be tamped with 25 blows of the rounded end of the tamping rod, each blow being given by allowing the tamping rod to fall freely from a height of about 50 mm above the surface of the aggregate, the blows being evenly distributed over the surface. A further similar quantity of aggregate shall be added in the same manner and a further tamping of 25 blows given. The measure shall finally be filled to overflowing, tamped 25 times and the surplus aggregate removed by rolling the tamping rod across, and in contact with, the top of the container, any aggregate which impedes its progress being removed by hand and aggregate being added to fill any obvious depressions. The net weight of aggregate in the measure shall be recorded (weight m_1) and the same weight used for the second test.

A2.5.4.2 Test procedure

Rest the impact machine, without wedging or packing, upon the level plate, block or floor, so that it is rigid and the hammer guide columns are vertical. Fix the cup firmly in position on the base of the machine and place the whole of the test sample in it and compact by a single tamping of 25 strokes of the tamping rod as above.

Adjust the height of the hammer so that its lower face is 380 ± 5 mm above the upper surface of the aggregate in the cup and then allow it to fall freely onto the aggregate. Subject the test sample to a total of 15 such blows, each being delivered at an interval of not less than 1 s. No adjustment for hammer height is required after the first blow.

Remove the crushed aggregate by holding the cup over a clean tray and hammering on the outside with the rubber mallet until the sample particles are sufficiently disturbed to enable the mass of the sample to fall freely onto the tray. Transfer fine particles adhering to the inside of the cup and the underside of the hammer to the tray by means of a stiff bristle brush. Dry the whole test sample in the oven at $110 \pm 5°C$ to constant weight as for initial preparation. Sieve the whole of the sample in the tray, for the standard test, on the 2 mm test sieve until no further significant amount passes in 1 min.

Weigh the fractions passing (m_2) and retained on (m_3) the 2 mm sieve with an accuracy of 0.1 g. If the sum of the weights m_2 and m_3 is less than the initial weight m_1 by more than 1 g, discard the result and make a fresh test. Repeat the whole procedure with the second test sample with the same weight as the first.

A2.5.4.3 Calculations

Calculate per test sample the ratio between the weight of the fraction through the 2 mm sieve and the weight of the whole test sample (m_1) in per cent and rounded to 0.1% with the aid of the formula:

$$C = \frac{m_2}{m_1} \times 100$$

where

C = the dynamic crushing value of a test sample (%),
m_1 = weight of the whole uncrushed test sample,
m_2 = weight of the fraction through the 2 mm sieve of the crushed test sample.

A2.5.5 REPORT

The report must contain the following data:

1. The mean dynamic crushing value of the two test results to the nearest 1%;
2. Reference to this standard;
3. A description of the sample, including the weight;
4. Source of the sample;
5. Date of the tests.

A2.6 Determination of Density of Rocks with a Volume of at Least 50 ml

Note: This standard is based on Draft NEN 5186.

A2.6.1 SUBJECT AND AREA OF APPLICATION

This standard gives the method for the determination of the density of a natural stone and of other types of stone and stone-type materials with a volume of at least 50 ml.

A2.6.2 SAMPLE FOR ANALYSIS

The stone shall have a volume of at least 50 ml. If the stones are very large, a representative part can be used, subject to the minimum volume required.

A2.6.3 EQUIPMENT AND OTHER AIDS

A2.6.3.1 Drying oven or other appropriate apparatus, adjustable to $(110\pm5)°C$.
A2.6.3.2 Weighing scales, accurate to 0.05% of the stone weight, suitable for weighing in air and under water.
A2.6.3.3 Water-bath, filled with tap water at room temperature and suitable for weighing stones under water.
A2.6.3.4 Thermometers, suitable for recording temperature in the water-bath, accurate to $1°C$.
A2.6.3.5 Moist chamois leather.

A2.6.4 METHOD OF OPERATION

A2.6.4.1 Execution of the investigation

Remove all loose parts and brush the stone clean with water. Measure the water temperature in the water-bath to $1°C$ accuracy. Keep the stone submerged in the tap water at room temperature for at least 5 min and then weigh it—submerged—(m_1) with an accuracy of 0.05% of the stone's weight.

Take the stone out of the bath, dry it with the moist chamois leather to the point that no shiny-wet surface remains and then weigh the stone (m_2) again with 0.05% accuracy.

Dry the stone in the oven to a constant (steady) weight, which is reached when two consecutive weighings with a 24-hour interval show less than 0.05% loss of total weight.

Weigh the stone again after cooling to room temperature (m_3) with 0.05% accuracy.

A2.6.4.2 Calculation

Calculate the density of the stone in kg/m^3 and rounded to $1 kg/m^3$ with the aid of the formula:

$$\rho_r = \frac{m_3 \times \rho}{m_2 - m_1}$$

where ρ_r = the density of the stone,
ρ = water density (g/ml) at the test temperature of the water-bath;
m_1 = apparent weight of the stone submerged (g),
m_2 = weight of the damp stone (g),
m_3 = weight of the dry stone (g).

A2.6.5 REPORT

The report must supply the following data:

1. The density of the stone;
2. Reference to this standard;
3. A description, including the weight of the stone and of the part of the stone that is used;
4. Source of the stone;
5. Date of testing.

A2.7 Determination of Water Absorption at Atmospheric Pressure

Note: This standard is based on Draft NEN 5187.

A2.7.1 SUBJECT AND AREA OF APPLICATION

This standard gives the method for the determination of the water absorption at atmospheric pressure of a natural stone or other stone material with a volume of at least 50 ml.

A2.7.2 SAMPLE FOR ANALYSIS

The stone must have a volume of at least 50 ml. If it has a volume in excess of 150 ml, break a part off to leave a volume under 150 ml.

A2.7.3 EQUIPMENT AND OTHER AIDS

A2.7.3.1 Drying oven or other appropriate apparatus, adjustable to $(110\pm5)°C$.
A2.7.3.2 Weighing scales, accurate to 0.05% of the weight of the stone.
A2.7.3.3 Water-bath filled with tap water at room temperature
A2.7.3.4 Moist chamois leather.

A.2.7.4 METHOD OF OPERATION

Remove loose parts and clean the stone by brushing with water. Place the stone submerged in the water-bath. Leave the stone submerged until the weight over a period of 24 hours does not increase more than 0.1%.

Take the stone from the bath, dry it with the most chamois leather until it leaves a dull surface and weigh it (m_1) to within an accuracy of 0.05%. Dry the stone in the oven to a constant weight (m_2), which is reached when the stone's weight over an interval of 24 hours does not reduce more than 0.05%.

A2.7.4.1 Calculation

Calculate the water absorption of the stone in percentage weight and rounded to 0.1% with the aid of the formula:

$$c = \frac{m_1 - m_2}{m_2} \times 100$$

where

c = water absorption at atmospheric pressure,
m_1 = weight of a moist stone after absorption (g),
m_2 = weight of a dry stone (g).

A.2.7.5 REPORT

The report must contain the following data:

1. The water absorption at atmospheric pressure;
2. Reference to this standard;
3. A description of the stone with its weight and, if used, of the part stone;
4. Source of the stone;
5. Date of testing.

A2.8 Determination of 'Sonnenbrand' in Basalt

Note: This standard is based on Draft NEN 5188.

A2.8.1 SUBJECT AND AREA OF APPLICATION

This standard presents a method for the determination of the presence of signs of 'Sonnenbrand' in a piece of basalt.

A2.8.2 SAMPLING

The stone must be sufficiently large to permit a sawn surface of at least $0.005 \, m^2$ to be cut.

A.2.8.3 EQUIPMENT AND OTHER AIDS

A2.8.3.1 Sawing machines.
A2.8.3.2 Heat source to boil water.
A2.8.3.3 Kettle, distilled water.

A2.8.4 METHOD OF OPERATION

Saw the stone into pieces so that one at least will have a sawn surface area equal to or greater than $0.005 \, m^2$. Boil that piece for a period of at least 36 hours in distilled water. Remove the stone from the water and leave it to dry until no more discoloration due to moisture in the sawn surface can be discerned. Lightly moisten the sawn surface by breathing on it or with a moist cloth. Check if, during subsequent drying of the surface, star-shaped spots with radiating hair cracks can be seen.

A2.8.5 REPORT

The report must give the following data:

1. If star-shaped spots have been observed;
2. Whether radiating hair cracks were observed;
3. Reference to this standard;
4. Source of the stone;
5. Date of the test.

A2.9 Determination of the Abrasion Resistance Index Using the Queen Mary and Westfield (QMW) Abrasion Mill

A2.9.1 GENERAL

The QMW mill abrasion test is designed to indicate the resistance of a rock to mutual attrition and grinding of water-saturated coarse aggregate size rock fragments. The mutual wear of rock fragments takes place as the pieces roll across the diametral free surface of the horizontally rotating cylindrical mill. The gentle rolling action is similar to that on a shingle beach, and does not result in whole-lump impact breakages typical of the Los Angeles Abrasion test (ASTM C131-76). The test procedure involves several increments of milling. The weight losses with time give a plot which is used to obtain the abrasion-resistance index. The test procedure has two important features which help towards maintaining a constant attrition environment inside the mill. These are (1) removal of fines and chippings in a suspension of flowing water and (2) addition of fresh material after each milling increment such that the weight of the fragments at the start of each milling increment equals the original weight at the start of the test. The advantage is that the abrasive environment conditions are practically linear because (i) no build-up of viscous slurry which would dampen the abrasive action can occur and (ii) the number of block diameters rolled per revolution is reasonably constant throughout the test. In addition, the results analysis method tends to filter out any influence of differing sample preparation methods and initial block shapes.

A2.9.2 TEST SAMPLE

The sample for testing shall consist of between 12 and 14 kg of crushed rock passing the 31.5 mm (ISO 565) standard square hole perforated plate sieve and retained on the 25.0 mm (ISO 565) standard sieve. The 26.5 mm (ISO 565) may replace the 25.0 mm sieve where the latter is unavailable. The percentage by weight of test sample material retained on the 19.7 mm flakiness slot (BS 812) is to be determined. This value shall be at least 50% for a valid test sample. The sample shall be prepared from one of the following options:

1. At least ten rock pieces of greater than 50 ml volume, taken or broken off from the rock representative of the source to be tested. These pieces are reduced to the required size with hammer and laboratory crusher by incrementally decreasing the crusher opening as necessary.
2. The production stockpile at the quarry if the client considers this to be representative or satisfactory. For example, 50 kg of 20–40 mm aggregate will generally contain sufficient material.

The sample shall contain between 300 and 400 pieces of the required sieve dimensions.

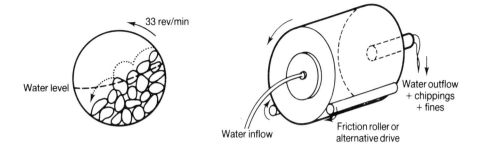

Mill specifications: Inside diameter 195 mm, inside length 230 mm
Outflow diameter 14 mm
Three plastic lifting ribs of 20 mm diameter, semicircular section, equally spaced and axially attached to inside

Figure A9 *The QMW abrasion mill (schematic)*

A2.9.3 ABRASION MILL APPARATUS

The QMW Abrasion Mill Cylinders shall conform to the design dimensions shown in the Figure A9 (see also Figures 60 and 61 of the main text). The materials in contact with the tumbling rock shall be of tough plastic polymers to prevent corrosion and reduce abrasive wear of the cylinder walls.

The gearing of the drive mechanism shall be such that the mill rotation rate is 33 ± 1 rev/min. The revolutions should be counted automatically and a means of presetting the increments of milling in revolutions or time shall be incorporated.

The water flow should be at a rate of at least 1 litre/min (of room-temperature tap water).

A2.9.4 OTHER EQUIPMENT AND AIDS

A2.9.4.1 Drying oven or other appropriate apparatus, adjustable to $(110 \pm 5)°C$.
A2.9.4.2 Weighing scales, accurate to within 1 g.
A2.9.4.3 Moist chamois leather.

A2.9.5 PROCEDURE

A2.9.5.1 Preparation

Obtain two identical test samples by filling one cylinder with the test aggregate, tap the full cylinder on a benchtop to allow moderate compaction and top up again. Empty this cylinder of material onto the bench and divide it equally into two representative samples, A and B, adjusting so that they have equal weights to within 50 g. The test will require about 50–60 additional lumps of test material, designated 'tracer' lumps, which shall be taken from the prepared sample for testing.

Dry the tracer lumps for 4 h in a well-ventilated oven at a temperature of $(110 \pm 5)°C$. When cool, immerse for 5 minutes in a solution of ultra-violet dye made up from ultra-violet powder dissolved in the appropriate solvent recommended by the dye manufacturer and allowed to dry (for certain solvents, a fume cupboard will be required).

Place the dried tracer lumps and samples A and B in three separate labelled plastic containers and immerse in tap water for 12 h.

A2.9.5.2 Determination of milling programme

All weights refer to a saturated surface-dried condition obtained using a consistent method from one increment to the next. This should be by dabbing dry with a chamois leather until no shiny-wet surface remains after spreading out all lumps on a towel-covered benchtop. Record the initial weight, W_0, of each sample. Place each sample in mills labelled A and B and set the first mill increment to 1000 revolutions. After milling, record the new weight, $W(t)$, of each sample. Use the table below to determine the subsequent milling programme MPI, MPII, or MPIII, making any adjustments to the increments indicated to ensure the test gives roughly equal increments of weight loss in the range $W(t)/W_0$ between 0.90 and 0.70. Ensure that the same programme is used for each sample.

Milling programmes (determined after first increment of $t = 1000$ revolutions)

	Programme increments in thousands of revolutions
MPI: for 'weak' rocks $W(t)/W_0 < 0.98$	1, 2, 3, 4, 6, 8, 10, 10, 10, 10 ...
MPII: for 'intermediate' rocks $W(t)/W_0$ between 0.98 and 0.99	1, 3, 6, 10, 15, 20, 20, 20, 20 ...
MPIII: for 'strong' rocks $Q(t)/W_0 > 0.99$	1, 5, 10, 20, 30, 40, 40, 40, 40 ...

A2.9.5.3 Testing

After the first milling increment, wash and weigh the stones and determine the weight loss from $\Delta W = W_0 - W(t)$. Add surface-dried tracer lumps of weight ΔW to the original material and return the sample to the mill for the second increment of milling, ensuring that the total weight of material inside the mill is within 10 g of the original weight W_0.

After the second milling increment, empty the material into a bowl and separate the tracer material using an ultra-violet lamp, knowing the expected number of tracer lumps. The weight of original matrerial, $W(t_2)$, is recorded together with the weight (and number) of the tracer lumps, W_{tracer}, and the weight of additional tracer lump(s) required, ΔW, is calculated from

$$\Delta W = W_0 - W(t_2) - W_{\text{tracer}}$$

Repeat the incremental milling procedure until $W(t)/W_0$ falls below 0.70 or the number of revolutions has exceeded 200 000. Between 20 and 30 tracer lumps per sample will be present at the end of the test. Tracer dye may need to be re-applied to enhance detection of tracer blocks during the test. Alternatively, old tracer blocks may be replaced by an equivalent weight of new tracer blocks if they are becoming hard to identify. Use careful work practices, including counting all blocks after each milling increment, to check that all stones are correctly sorted and accounted for. Ensure that the two samples are kept separate.

A2.9.6 CALCULATION

Weight (to 0.1 g) versus revolution results are tabulated and analysed according to the equation

$$W/W_0 = (1-b)\exp(-k_f t) + b\exp(-k_s t)$$

where t is in units of 1000 revolutions and k_s is the abrasion-resistance index, the material property to be reported. For completeness, it may be noted that k_f is the smoothing resistance index (a function of changing stone shape) and b is the fractional weight loss offset (a mixing term).

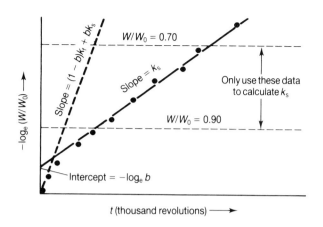

Figure A10 *Determination of abrasion resistance index, k_s*

The values of k_s and b can be determined using a plot of $-\log_e(W/W_0)$ versus t, as shown in Figure A10. The slope of the best straight line through all data points with W/W_0 between 0.9 and 0.7 is taken as the value of k_s and shall be calculated using a least-square regression technique. The test result to be reported as the abrasion-resistance index is the arithmetic mean of the two sample results.

A2.9.7 REPORT

The report must include the following:

1. The table of fractional weight loss and revolutions;
2. The logarithmic plot (Figure A10);
3. The direct plot of fractional weight loss versus t;
4. The test result for k_s to two significant figures;
5. Reference to this method;
6. Dates of testing;
7. Source and description of sample.

A2.10 Methylene Blue Absorption Test

A2.10.1 SCOPE OF THE TEST

This test is used to quantify the amount of clay mineral (smectite group) present in quarried rock and hence to indicate the soundness of the quarried rock.

A2.10.2 EQUIPMENT AND REAGENTS

25 ml burette mounted on stand
250 ml Erlenmeyer flasks
100 ml volumetric flask and stopper
250 ml beakers
Glass stirring rod
Magnetic stirrer
Small clock glasses
Sample containers
Spatula
Whatman No. 40 filter papers (12.5 cm diameter)
Distilled or de-ionised (demi) water
Methylene Blue
Chemical waste tank

Analytical balance; aggregate crusher; grinding apparatus; riffle boxes; scoop; metal trays; oven

A2.10.3 PREPARATION OF METHYLENE BLUE SOLUTION

The hygroscopic crystalline Methylene Blue (3.9-bis-dimethylamino-phenazothionium-chloride; $C_{16}H_{10}N_3ClS$) shall be oven dried for several hours at $110\pm5°C$ to expel water until it reaches constant weight. A 0.0094 N solution is prepared by dissolving 3.0 g of the oven-dried Methylene Blue crystals in 250 ml distilled water. The solution is cooled and diluted to 10% of the original concentration by adding further distilled water. The total volume of diluted solution is noted. The molecular weight of dry Methylene Blue is 319.9 g/mol.

A2.10.4 PREPARATION OF THE AGGREGATE SUSPENSION

Prepare from the rock or soil aggregate a very fine-grained powder by drying, crushing and sieving. The powder shall be representative of the aggregate, rock, or soil composition. The grain size shall be less than 100 mesh (0.15 mm). Make a suspension of about 2.0 g of this aggregate powder weighed to 0.01 g accuracy in the Erlenmeyer flask with 30 ml distilled or de-ionised water.

The crushed rock or soil may come from a small sample or hand specimen. If a representative aggregate sample is available, the following procedure shall be adopted to prepare the sample for testing.

Riffle the sample of aggregate down to a portion of about 1 kg and crush it in a jaw crusher to passing the 6.3 mm sieve. Then riffle the portion down to about 100 g and, using the jaw crusher, crush it as finely as possible (say, 0.425 mm). Follow this by grinding it in a mechanical pestle and mortar until nearly all of it passes the 0.075 mm sieve. The material retained on the 0.075 mm sieve after mechanical grinding shall be ground until it passes the 0.075 mm sieve using a hand agate pestle and mortar. The sample shall be air-dry and must not be exposed to temperatures higher than 35°C.

A2.10.5 TEST PROCEDURE

Add 0.5 ml of the Methylene Blue solution to the aggregate suspension by means of a 25 ml burette. Shaking during the addition is necessary, for which a magnetic stirrer may be used. To perform the titration, add successive volumes of 0.5 ml of the Methylene Blue solution to the Erlenmeyer flask. After each addition, agitate the flask for 1 min and remove a drop of the dispersion with the glass rod and dab it carefully on a sheet of filter paper. Initially, a circle of dust is formed which is coloured dark blue, has a distinct edge, and is surrounded by a ring of clear water.

When the edge of the dust circle appears fuzzy and/or is surrounded by a narrow light blue halo, agitate the flask for 1 min more and do another spot test. If the halo has disappeared, add more blue. If there is still a halo, agitate the flask for a further 2 min and carry out another spot test. Whatever the outcome of this test, add more blue, agitate for 2 min, carry out a spot test, then agitate for a further 2 min and perform another spot test. Repeat this sequence, with a total of 4 min of agitation, until there is a definite light-blue halo. Note the sample number and the amount of Methylene Blue added below each spot on the filter paper.

To determine the end-point of the titration, hold the filter paper up to the light while it is still damp, and compare the dust circles made after 4 min of agitation. It should then be possible to see where the halo first appears and thus where the end-point is. The corresponding volume of Methylene Blue solution added is noted. This procedure is called the 'spot' method.

After completing the test, pour the remaining Methylene Blue solution and the titrated suspension into a chemical waste container specially prepared for this purpose.

A2.10.6 CALCULATION OF MBA

The Methylene Blue Absorption (MBA) value is normally expressed in grams Methylene Blue absorbed by 100 g of sample material, given as g% or g/100 g:

$$\text{MBA} = \{(X/Y) \times p\}/(A/100)\,(g\%)$$

where

X = weight of dried Methylene Blue crystals (g),
Y = volume of diluted Methylene Blue solution (ml),
p = volume of Methylene Blue solution added (ml) to 0.5 ml accuracy,
A = weight of soil or rock powder (g).

At least two determinations of the MBA value shall be carried out on each sample and the average MBA value determined within an accuracy of 0.1 g/100 g.

A2.10.7 REPORT

The report must include the following:

1. The MBA value to 0.1 g/100 g;
2. Reference to this standard;
3. Source and description of sample;
4. Date of the test.

A2.11 Determination of the Drop Test Breakage Index

A2.11.1 SUBJECT AND AREA OF APPLICABILITY

This standard is used to determine the percentage of stone loss from heavy gradings of quarried stone in a standard drop test, this percentage being described as the Drop Test Breakage Index.

A2.11.2 SAMPLE FOR ANALYSIS

The sample shall contain at least 50 pieces taken at random from above the ELCL weight of the grading class in question.

A2.11.3 EQUIPMENT AND OTHER AIDS

A2.11.3.1 Suitable hydraulic grab (e.g. orange-peel type).
A2.11.3.2 Weighing equipment, accurate to within 2% of the lightest piece to be weighed.
A2.11.3.3 Bed of rocks of same grading as the sample to be tested.
A2.11.3.4 Sufficient volume of crushed rock aggregates to give a 0.5 m thick layer covering an area to support the bed of rocks.

A2.11.4 METHOD OF WORKING

A2.11.4.1 Execution

Determine the individual weights of the rock sample prior to the test in accordance with Section A2.2:Part 1 using the method described in Section A2.2.3.1.1. Prepare the bed of rocks by laying them out in a single compact layer on a 0.5 ± 0.05 m thick, layer of crushed rock aggregates. Subject each block in the test sample in turn to a drop of fall height $3\,\text{m} \pm 0.1\,\text{m}$ onto the bed of rocks. Record the result of each drop, such record to include the number and type of visible flaws in blocks and the number and type of blocks resulting.

Remove the block, or broken parts thereof, from the bed of rocks. Set aside all resulting pieces whose weight is greater than the ELCL weight, or whose weight is assessed to be close to the ELCL weight, for further weighings. Clear all rock fragments from the bed of rocks, leaving clean surfaces prior to dropping the next block in the test sample.

Individually weigh each stone piece in the test sample heavier than the ELCL on completion of drop testing accurate to within 2% of the LCL weight. Record the total weights in each weight fraction.

A2.11.4.2 Calculation

Calculate the cumulative weight distribution curves for the sample prior to drop testing and after drop testing for all pieces heavier than the ELCL and calculate the median sample weight before testing (W_{50i}) and after testing (W_{50f}), all in accordance with Section A2.2:Part 1, Section A2.2.3.1.2. Calculate the drop test breakage index, I_d, as

$$I_d = [(W_{50i} - W_{50f})/W_{50i}] * 100\%$$

A2.11.5 REPORT

The following data must be included in the report:

1. The Drop Test Breakage Index;
2. A reference to this standard;
3. A description of the sample, including its weight;
4. The source of the sample;
5. The date of the testing.

If agreed beforehand, the cumulative weight distributions before and after testing shall be provided and it is recommended that this be on a single graph.

Appendix 3: Measurement of quarried rock in coastal and shoreline engineering

A3.1 Introduction

There are many Standard Methods of Measurement. Each is designed to provide a suitable basis for the billing and admeasurement of an aspect of construction, i.e. building, roads and bridgeworks, civil engineering generally, etc. Rock slope protection, together with marine and tidal working which are often associated with it, are considered peripheral, and as a result guidance on billing is incomplete and/or unclear.

A3.2 Aim

The object of this appendix is *not* to produce yet another standard method of measurement but to provide a methodical approach to the billing of these works so that, regardless of the detailed format requirements of the standard method selected, the same basic points will be covered. This should make bills of quantities easier to produce, understand and measure for payment.

A3.3 Principles

Annexes A and B (see pages 534–539) provide the essence of the suggested approach to the measurement of rock structures. The principles presented therein are based on the 'top-down design' approach (an analytical approach starting with general considerations and working towards detail). This provides the flexibility required to both fit into standard methods (see Box A1) and cope with the very wide range of physical conditions which can be met within this type of work. As a bonus it also produces a good framework for computerised billing.

The approach is similar to that adopted by the British Civil Engineering Standard Method of Measurement (CESMM) produced by the Institution of Civil Engineers and the Federation of Civil Engineering Contractors. The CESMM was used as the reference method of measurement for preparing Annexes A and B. The bill compiler using this manual will, however, need to make some minor modifications to comply

---

Box A1 Example of use of principles in this appendix within the context of common UK methods of measurement

Requirement

To provide items for the mobilisation maintenance on-site and demobilisation of a dredger.

Solution

1. Under CESMM: no action required, as contractor can insert the items into the fixed and variable method related changes if required;
2. Under SMM (Building): check that general or specific items for
   (a) Bringing plant to and from site and
   (b) Maintenance on site
   are raised for the relevant section of the bill.
3. Under the Method of Measurement for Highway Works: separate items would need raising in Bill 1 for each of mobilisation, maintenance and demobilisation, plus the necessary definitions and item coverages in the preamble to the bill of quantities.

SCHEME:				Sheet:	E	
PART:	EARTHWORKS			Class:	4	
Item number	Item description	Unit	Qty	Rate	£	p
	PART 3 WORK OCCASIONALLY EXPOSED ABOVE WATER (MLWS-MLWN)					
	EXCAVATION FOR CUTTINGS					
	BEACH MATERIALS					
2.6.1	Maximum depth: not exceeding 0.25 m	m³	47			
2.6.2	Maximum depth: not exceeding 0.25–0.5 m	m³	57			
2.6.3	Maximum depth: not exceeding 0.5–1.0 m	m³	118			
2.6.4	Maximum depth: nor exceeding 1.0-2.0 m	m³	29			
	EXCAVATION ANCILLARIES					
	DISPOSAL OF EXCAVATED MATERIAL					
5.3.5	Beach materials	m³	251			
	FILLING					
	EMBANKMENTS					
	Imported Rock					
6.2.7.1	Bedstone (see specification)	m³	63			
6.2.7.2	Rock armouring (see specification)	t	69			
	PART 4 WORK PERMANENTLY BELOW WATER (TABLE)(BELOW MLWS)					
	EXCAVATION BY DREDGING					
	BEACH MATERIALS					
1.6.2	Maximum depth: 0.25–0.5 m	m³	14			
1.6.3	Maximum depth: 0.5–1.0 m	m³	82			
1.6.4	Maximum depth: 1.0–2.0 m	m³	48			
	Carried to			Summary £		

Figure A11 *Sample page of bill of quantities prepared according to CESMM (2nd edition) and Annexes A and B of this appendix*

with the format of the present (2nd edition) version of the CESMM, as illustrated in the example page of a bill of quantities given in Figure A11.

Although the CESMM was used as the basis for Annexes A and B, features have been adopted from several methods, and some are novel to this document. These should be used as a checklist by the compiler to ensure that due consideration has been given to possible variables. The details of exactly how and where these will be included in the bill of quantities will depend on the actual standard method adopted. Above all, the bill compiler should bear in mind that clarity and practicality of the end product are paramount requirements. Particularly on smaller contracts (under 250 000 t) the users of the compiled bill are not likely to be experienced in this type of work.

The preparation of bills of quantities using Annexes A and B is now discussed in more detail, considering first the main divisions and sub-divisions of the bill and then the need for preambles.

A3.3.1 MAIN DIVISIONS (EQUIVALENT OF CESMM CLASSES)

Two types of division should be considered, any particular project being potentially subject all or in part to any combination of the following divisions.

By bills and sub-bills

These can reflect the type of item, such as Method Related items in a preliminaries bill, or distinguish different structures or main areas of work.

By degree of influences by water

By their very nature, these works are frequently executed in areas subject to innundation. The latter can be either regular and predictable, such as tides; seasonal as in dam filling and drawdown; irregular as in flash flooding, surges and sieches and either natural or man-made. Main divisions should be:

Zone 1: Work unaffected by periodic flooding (in a tidal situation above highest astronomic tide (HAT));
Zone 2: Work affected by cyclical flooding (in a tidal situation between HAT and mean low-water neaps (NLWN));
Zone 3: Work occasionally exposed above water (in a tidal situation between MLWN and mean low-water springs (MLWS));
Zone 4: Work normally below water table (in a tidal situation below MLWS).

In respect of the above-suggested divisions in relation to *tidal working*, division boundaries relate to reference levels given in standard tide tables, and in situations of appreciable tidal range, and have been set for the following reasons:

Zone 1 is set as above HAT, as this limit allows for some run-up and surge effects for the great majority of tides.
Zone 2 has its lower limit set at MLWN, as this represents the lower limit of working during normal working hours. Over much of Northern Europe, spring and neap tides occur at weekends, with the lowest levels at night,
Zone 3 is occasionally exposed, and will require special planning if full advantage is to be taken of the times when it is exposed. This is particularly important on short-term contracts, as it is possible for there to be a period of months during which no tide falls as low as MLWS during normal working hours.

In zoning work, the bill compiler should consider which of the above factor(s) is likely to have the most effect on the working of the contract. For example, consideration should be given to the construction access to the work, and if access is via a lower level than the work itself then the whole should be billed at the lower level. When changes in level are small compared with the placed unit (e.g. 300 mm spring tide range to armour $D_{n50} = 1.0$ m), then only two zones would be required: above and below low water.

A3.3.2 SUB-DIVISIONS

The suggested sub-divisions worked out in Annexes A and B (see pages 534–539) are:

1. By work category (excavation and filling, piling, concrete work, etc.);
2. By different designs/sections;
3. By types within categories (dredging, bulk excavation, different fill types, etc.);
4. By operations within type (different dredge/excavation depths, preparing, variations within fill types, etc.).

Annex A. Method-related charges

First division	Second division	Third division
1. Fixed charges	1. Opening of dedicated quarry	1. Accesses 2. Proving and trials 3. Mobilisation of quarry plant
	2. Closing of dedicated quarry	1. Demobilisation of quarry plant 2. Reinstatement
	3. Mobilisation of floating plant	1. Dredgers 2. Tugs 3. Pontoons and barges 4. Specialist vessels 5. Ancillary equipment for floating plant
	4. Demobilisation of floating plant	
	5. Establishment of accesses	1. Temporary roads, bridges, weighbridges and wheelwashes 2. Causeways 3. Harbours, dolphins and anchorages 4. Haul roads within the finished works 5. Cofferdams
	6. Removal of accesses	
	7. Trials and testing	1. Testing of different rock and grading types nr 2. Trial areas nr
	8. Mobilisation of land-based plant	1. General earth-moving plant 2. Beach excavation plant 3. Specialist plant types (including rock handling)
	9. Demobilisation of land-based plant	
2. Time-related charges	1. Maintenance of dedicated quarry	
	2. Maintenance of floating plant	1. Dredgers 2. Tugs 3. Pontoons and barges 4. Specialist vessels 5. Ancillary equipment for floating plant
	3. Maintenance of accesses	1. Temporary roads, bridges, weighbridges and wheelwashes 2. Causeways 3. Harbours, dolphins and anchorages 4. Haul roads within the finished works
	4. Maintenance of land-based plant	1. General earth-moving plant 2. Specialist rock-handling plant 3. Beach excavation plant
	5. Engineer's quarry inspections	1. Within 50 km of the works nr 2. 50 to within 400 km of the works nr 3. Over 400 km from the works nr 4. Any distance overseas nr

Principles	Definition	Coverage	Additional description
P1 This is not intended as an exhaustive list of method-related items, but to give guidance as to types of item which it may be appropriate to include in this type of work			A1 *General* Individual types and operations shall be detailed where required by the nature of the works, i.e. rock dredgers and silt dredgers would warrant separate items if both were required for the execution of the works
P2 All units should be 'sum' unless otherwise noted or the plural is indicated by the contract		C1 1.1 will include all items detailed in 1.3, 1.5 and 1.8	
		C2 1.2 will include all items detailed in 1.4, 1.6 and 1.9	
		C3 1.3 and 1.4 will include inspections, surveys, insurance and movement to and from the site	
		C4 1.5.3 should include for provision of navigation, mooring and riding lights	
			A2 1.8 can include headings for other specialist plant required by the nature of the contract, i.e. piling equipment, casting yards, etc.
P3 All units should be 'week' except where noted 'number' or where other units would be more appropriate. Times and rates to be inserted by the contractors	D1 Maintenance items should be included for 1.1, 1.3, 1.5 and 1.8 where used		A3 2.2.1–5 See notes to 1.3.1–5, etc. A4 The bills should indicate a basic cost/day for working at various radii. The contractor should include for travel, accommodation, living expenses and any apparatus or facilities (to be listed in the specification) required by the engineer to make his assessments on which to base his approval

Annex B. Earthworks

First division	Second division		Third division	Fourth division
1. Work unaffected by periodic flooding (above HAT)	1. Excavation by dredging	m³	1. Topsoil 2. Stated beach materials	1. To different datum levels 2. To different datum grades
2. Work affected by periodic flooding (MLWN–HAT)	2. Bulk excavation 3. Excavation for structures	m³ m³	3. Stated rock type 4. Stated artificial hard material exposed at surface	3. In beaches including all necessary re-excavation to stated levels and/or grades
3. Work occasionally exposed above water (MLWS–MLWN)	4. Excavation in beaches 5. Excavations of constant cross-section	m³ m	5. Stated artificial hard material not exposed at surface	4. To maximum depths by increments normal to the method of measurement
4. Work permanently below water (table) (below MLWS)			6. Stated dredged material	
5. Different designs 6. Different sections			7. Material other than 1–6	

6. Excavation ancillaries	1. Trimming of excavated surfaces	m²	1. Topsoil 2. Stated beach materials	
	2. Preparation of excavated surfaces	m²	3. Stated rock type 4. Stated artificial hard material	
	3. Stated method of disposal	m³	5. Stated dredged material	
	4. Double handling of excavated material	m³	6. Material other than 1–5	
	5. Excavation below formation level and replacement with stated material	m³		
	6. Stated supports left in			

Principles	Definition	Coverage	Additional description
P1 Volumes, areas and lengths shall be calculated for bill preparation and measurement generally in accordance with the principles, definitions and coverages adopted in the standard method of measurement selected. The information listed below is to assist in the billing and measurement of additional items likely to be necessary in the environments frequently met with in this type of work			
P2 Each type and phase of dredging should be measured separately. Pre- and post-dredge levels should be established by soundings P3 Excavation of all works affected by water should be measured only once, except where there is the potential for a substantial time lapse between them. Measurement should be based on a survey of existing ground levels established prior to excavation P4 Excavations for constant cross-section should only be used where both the volume of excavation and its possible variation are very much smaller than the length. This would either be at least an order of magnitude smaller for excavation and two orders of magnitude for variation; or where it is to suit contractor's method of working (e.g. when driving sheet-type piles)	D1 Sections can mean either cross-sections or parts of the works D2 Dredging should be measured for all work permanently below water and may be measured for work occasionally exposed above water D3 A substantial time lapse will only occur when there is a planned break in construction (i.e. settlement period, season) or where two different structures (e.g. seawall and breakwater constructed under the same contract meet and the same area needs excavation for the construction of each) D4 Stated rock types may distinguish different degrees of weathering the same rock	C1 Dredging should include for all over-dredging and bottom levelling required to guarantee establishment of minimum dredge profile over the designated areas C2 Beach excavation should include for all re-excavation, cleaning and preparation of the works required by the effects of water or contractor's operations after the first occasion	A1 Where slopes are shown on the drawings vertical dredging should be permitted provided that this will not affect the design or utility of the finished works A2 Where re-excavation of heavy-graded quarry stone is required it may either be measured by number or tonne in stated weight bands A3 Where possible, material rendered unsuitable for use in a higher grade may be incorporated into the works where it conforms to the specification for a lower grade and will be measured as material of the lower grade

Annex B continued

First division	Second division	Third division		Fourth division
	7. Filling	1. Backfill to structures	m³	1. Excavated topsoil
		2. Bulk-placed embankments or structures	m³	2. Imported topsoil
				3. Unselected excavated material, excluding topsoil
		3. Heavy-graded quarry materials	t	4. Selected stated excavated material
		4. To stated depth or thickness excluding heavy graded quarry materials	m²	5. Imported stated grades of quarried stone
				6. Imported natural material
				7. Imported stated artificial materials
	8. Filling ancillaries	1. Trimming of filled surfaces	m²	1. Topsoil
		2. Preparation of filled surfaces	m²	2. Excavated rock
				3. Stated grades of quarried stone
				4. Natural materials
				5. Stated artificial materials
		3. Geotextiles	m²	1. Laid less than 15° to horizontal
				2. Laid at 15° or more to the horizontal
				3. Used in wrapping
	9. Landscaping	1. Turfing and dune grass planting	m²	1. Turfs of different quality
		2. Stated method and mix for grass seeding	m²	2. Different species of dune grass
		3. Stated plants, shrubs and trees —species-sizes	nr	3. Different planting patterns
				4. Work less than 15° to the horizontal
				5. Work at 15° or more to the horizontal
		4. Hedges—stated species, size, spacing	m	1. Single row
				2. Double row

Principles	Definition	Coverage	Additional description
P5 Measurement of heavy-graded materials shall be tonnes in stated weight grades against target tonnage per cubic metre	D5 The target tonnage shall be stated in the contract against a rock of given density	C3 Shall include for: (a) Any adjustments to the specification to counteract the effects of handling and transit damage between quarry and final position. (b) Any adjustments to the quantity arising from a difference in density between the tender target density and that of the material supplied. (c) Disposal of any material rendered unsuitable between initial approval and final placing	

A3.3.3 PREAMBLES

The compiler will need to consider introducing preambles to the bills of quantity to cover a number of matters which are discussed in this section.

A3.3.3.1 Beach excavation

This should only be measured once for a particular construction operation based on a survey of beach levels taken immediately prior to the start of excavation (generally not more than 1 week in advance of the point of excavation). All re-excavation to remove material deposited by tides/inundations and cleaning off of the works is deemed to be included. The contractor's attention should be drawn to suitable method-related items which may be used to cover the difference between the actual excavation cost and the overall cost of meeting this requirement.

A3.3.3.2 Measurement of armourstone (heavy-graded quarry materials) by the tonne (see Box A2).

When measuring by weight

There will still normally be a requirement in this situation for a contractor to build to the lines and levels, including finished surface profile, shown on the drawing unless the armour is specified in the preambles on a weight per unit area (t/m^2) basis. Where a finished surface profile is specified, in order to measure by weight and also to provide control on cost, the preambles to the bills of quantities should specify a target tonnage of stone required per cubic or square metre. This tonnage should be quoted for each grade of rock and be consistent with the porosity of the structure and the packing arrangement required by the designer. A tolerance on this target tonnage should be stated of, say, $\pm 0.1\,t/m^3$ along with the assumed saturated surface dry density of the rock. As a guidance, using a rock of saturated surface dry density of $2.68\,t/m^3$, the tonnage of rock required per cubic metre of two-layer armour across a large number of prototypes ranged from $1.5\,t/m^3$ to $1.7\,t/m^3$. For this purpose:

1. The contractor must keep accurate records of tonnages of armourstone delivered and the location of the deliveries in the works.
2. The contractor's placing procedures must also be agreed in detail with the designer.

In this connection it is recommended that allowance be made for the construction of trial sections, which will normally become part of the finished works (see main text, Box 39), as layer thickness and porosity coefficients used by the designers to

estimate finished profiles are notoriously unreliable. The trial section allows the actual weight per cubic metre to be determined and the target tonnage per cubic metre adjusted accordingly, in fairness to all parties to the contract. Adjustments to the target tonnage outside the initial tolerance $\pm 0.1\,m^3$ should be at the client's risk (cost saving or cost increase). It also allows the designer to ensure that the packing arrangement and porosity required is being achieved.

Once the target tonnage is established fairly as described above, the contractor then must comply with this and will not be paid for excess stone.

When measuring by area

As for measurement by weight, the target tonnage should be stated and adjusted using a trial section of the works, with adjustments to the target tonnage being reflected in a pro-rata adjustment to the per cubic metre rate for the as-placed rock quoted by the contractor. Such adjustments should, however, only be made to reflect differences in packing and not in the intrinsic saturated surface dry density, adjustments for which, if permitted, should be made separately (see below).

> **Box A2** Measurement of armourstone by weight rather than volume
>
> It has been traditional to measure large rock armour structures by volume, and many people have tried to transfer this method of measurement to the smaller structures used in sea and coast defences—with variable results!
>
> In large structures, armour normally accounts for about 10% of the total volume, whereas in smaller ones it is normally greater than 80%. Hence changes in armour quantities have a far greater bearing on the financial success or otherwise of the contract. Furthermore, there are often overriding requirements of aesthetics, environmental compatibility and public safety which will lead to variation from optimum hydraulically controlled packing on all or part of the structure. Such variations are extremely difficult to specify clearly in a contract document, although they can be demonstrated on the ground, even though it may need some trial and error to achieve the exact result.
>
> Experience of a number of coastal authorities, built up since the early 1980s, suggests that, especially when coupled with inspection and approval of armour at the quarry (which is also strongly advocated in this manual), measurement of armour by the tonne has the following advantages.
>
> 1. *For the client*
>
>   (a) Better relations with the contractor;
>   (b) Availability of more contractors to carry out the work, as less specialist knowledge is required.
>
> These have been shown to lead to keener pricing and less cost overrun.
>
> 2. *For the designer*
>
>   (a) Greater flexibility in meeting the client's brief;
>   (b) Allows more sophistication and complexity within the design of the various parts and components of the structure;
>   (c) Permits greater on-site flexibility to adjust the structure to meet unforeseen physical conditions or other conditions without leading to extensive claims.
>
> 3. *For the contractor and supplier*
>
>   (a) Reduces risk;
>   (b) Improves the smooth running of the contract;
>   (c) Reduces the need for claims; and hence
>   (d) Improves cash flow;
>   (e) Gives opportunities to enter this market to those who would not otherwise be able to do so.
>
> These advantages can also be carried over to larger structures which, with the constant development of more sophisticated design tools, will increasingly benefit from the use of these methods. Furthermore, where quality assurance (see Section 2.1.8 of the main text) is required in the contract, the procedure set out in this appendix should form a viable basis for the quality plan.

A3.3.3.3 Rock density adjustments

Where armour stone is not from nominated or dedicated quarries, the contractor should normally be permitted to submit a tender based on a rock density higher than that given with the target tonneage without offering an alternative design. Bids will then require adjustment to keep them on a strictly comparable basis, which may be achieved through the following procedures:

1. Calculate an adjustment factor by dividing the density of the offered rock by that given with the target tonneage.
2. Multiply the target tonneage by the adjustment factor to arrive at a revised target tonneage.
3. Where measurement of armourstone has been selected to be by weight, multiply the relevant bill item quantities by the adjustment factor to arrive at revised quantities for use in the tender total.

Provision should be made in the preamble and the relevant bills for the contractor to insert the adjustment factor and revised target tonneage, density and bill item quantities. If the density of the offered rock differs substantially from that used in the original design, the designer should be consulted to determine if the works can be redesigned to take advantage of this property. For example, with increased density, the reduced D_{n50} of the armourstone may permit reductions in armour layer thickness (see, for example, Box 82 in Section 6.2.4.1 of the main text) unless total volume of voids is important for energy dissipation.

These procedures will ensure fair control of the project and that the client knows what the scheme is going to cost, and is not faced with an 'open-ended' bill for the supply of rock armour where dense material is offered.

A3.3.3.4 Inspection and approval

Quality control should be centred at the producing quarry(s). For this purpose, the designer should indicate in the specification what facilities he requires for carrying out his assessment of the materials. Except in the case of a nominated source of materials, he should also indicate the quantity per hour of each type of material he would expect to assess and a day rate for his services while so doing.

Inspection on-site should be limited to ensuring that the materials delivered are those approved at the quarry(s) and that they have not suffered damage in transit which would render them unsuitable for inclusion in the works. Where such inspection or previous experience with a particular source of supply indicates that the rock is of a friable or brittle nature it may be necessary to adjust the criteria of assessment at the quarry(s) to ensure that the materials delivered are as specified on arrival at site (see main text, Section 3.6.2.3). Such adjustments should be deemed to be allowed for in the rate for supply.

Items should be provided in the bills of quantity for inspection of quarry(s) at different radii. In pricing these, the contractor shall allow for all the costs of providing the specified facilities, transport and the designer's time in carrying out the assessments, including time spent travelling, rejecting sub-standard material and waiting through lack of material to inspect in the quantities indicated. Where necessary, the items should also allow for overnight accommodation and living expenses. The actual time spent by the designer in carrying out his inspection as detailed above should be charged to the contract at the rates given.

Appendix 4: Hydraulic data measurement and instrumentation

A4.1 Bathymetric Surveys

The following devices are commonly used for bathymetric surveys.

Pole

The use of a graduated pole lowered vertically onto the bottom is the simplest method, limited in applicability only by large water depths, strong currents and vertical movements of the vessel. A bottom plate can be adjusted to ballast the pole and to prevent it from being pushed into the bottom.

Sounding line

Provided with a lead weight and markers, a sounding line can be used except for conditions with vertical movements of the vessel. To a certain degree, deviations of the line from the vertical, due to a given current, can be accounted for using correction factors.

Echo sounding

The principle of echo sounding is that a transmitted signal is reflected by a pronounced density interface (e.g. the seabottom). The reflected signal is received and the measured two-way travelling time of the signal is an indication of the water depth (see Figure A12).Calibration should be carried out regularly, particularly when variations in the water density and corresponding changes of signal propagation velocity can be expected.

The principle of echo sounding may be unreliable (or even unapplicable) when 'false' echos are generated. This should be accounted for in cases with strong transitions of density, due to a bottom layer with suspended sediment or layers of different temperature or salinity.

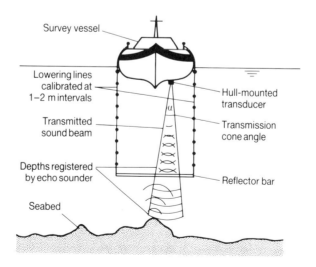

Figure A12 *Principle of echo sounding*

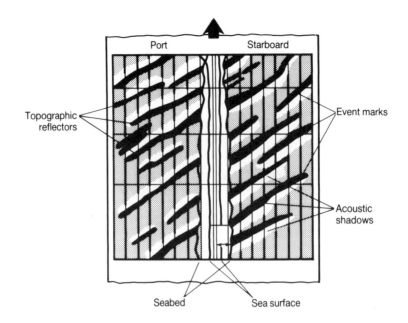

Figure A13 *The principle (a) and typical product (b) of side-scan sonar*

Remote-controlled profiling

When the co-ordinate of a vehicle, driving remote controlled on the seafloor, is received on a ship the bottom profile can be obtained along the track of the vehicle. Corrections on the horizontal scale of the cross-section obtained should be made for variations in the speed of the vehicle.

Side-scan sonar

Side-scan sonar provides a means of detailed surveying of seabottom features. A strip with a width of 500 m on either side of the track of the vessel can surveyed by sonar signals transmitted obliquely to the seafloor from a submerged 'fish' towed by the vessel (Figure A13). The received signals can be plotted, taking into account their respective direction of transmission, to obtain a sonograph of the seabed. This is a seafloor map, showing bottom features or objects on the seabed by means of acoustic shadows. The shadow lengths are a measure for the height of the bottom features (e.g. sand ripples, wrecks, boulders).

Remote sensing (satellites)

The transmission and reception of signal by a satellite, positioned at a given height above the area of interest, is basically similar to echo sounding. Calibration should include the velocities of propagation of the signal in both air and seawater and the height of the satellite. The method is particularly advantageous for large and distant areas, where mobilization of alternative equipment is very expensive.

With the Global Positioning System (GPS) the position of any point can be determined in three dimensions, provided that a receiver can be placed on the seafloor. The system is based on the known instantaneous positions of selected satellites. The position of the receiver is determined by a built-in computer from the propagation delays of the signals received.

A4.2 Water Level Measurement

Water levels can be determined easily by visual observation using a fixed tide board or by use of any one of the measuring devices discussed below. For repetitive measurements, especially those extending over a long period of time, the reference or datum level of the gauge should be checked regularly. Levels may be obtained

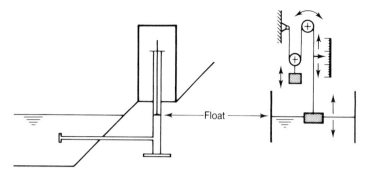

Figure A14 *Principle of a float water level measuring system*

by reference to local benchmarks on rigid and stable exposed rock or on structures (e.g. churches, bridge piers, lighthouses, etc.). These references levels are related in turn to the national datum level and values are updated regularly by the National Survey organisation to take account of any geological movements.

Water-level measuring devices available include floats, radar, hydrostatic pressure meters or through acoustic, laser, resistive or capacitive measurements. Float measurements are generally most reliable. When continuous recording of the instantaneous water level is needed an automatic system for measurements and data logging should be installed. The most common concepts are discussed below.

Float

Automatic measurement systems are usually based on the principle of measuring the vertical position of a float in a communicating water column (see Figure A14). In order to provide filtering of high-frequency disturbances (exceeding 0.1–0.01 Hz) damping should be guaranteed by installing sufficient hydraulic resistance.

When connected with systems for digital data transfer and storage, the float is a most reliable concept of a automatic tidal station. The use of a copper float is recommended to prevent growth of organic material, which might gradually affect the measurements. The maximum inaccuracies of the present digital level meters is approximately 0.01 m, mainly due to temperature and density effects.

Radar

Radar systems use the return time of a transmitted signal reflected by the water surface as a measure for the distance from the transmitter to the water level.

Pressure meter

At a fixed and known position under water the instantaneous water pressure is measured. As far as no significant vertical accelerations of the water occur, the pressure fluctuations registered are linearly dependent on the water-level fluctuations. Absence of vertical accelerations (e.g. due to the vertical component of orbital movement) can be assumed close to the seabed, so the instrument is usually fixed upon or near the seabed. The accuracy may be limited by sensitivity to density and temperature variations.

Resistance and capacitance meters

Although more suited to use in the laboratory, resistance and capacitance concepts can be applied. The first concept involves measuring the vertically integrated conductivity of the water column between the submerged part of two electrodes. The second concept involves induction of electric capacity to a capacitor by submergence of an electrode. Both methods are sensitive to the instantaneous ionic composition of the water and hence good calibration is required.

Remote sensing (satellites)

The use of satellites can be combined with other measuring instruments to form an effective measurement system. Such systems are particularly advantageous for large and distance areas, where mobilisation of alternative equipment is very expensive. With the Global Positioning System (GPS) the position of any point can be determined in three dimensions. Receivers should therefore follow the sea surface (for instance, by fixing these to floats or buoys). A built-in computer determines the instantaneous position relative to the known position of a number of selected satellites. The calculations are based upon propagation delays of the signals received. The data thus obtained can be stored by the instrument or transferred to a central station.

A4.3 Current Measurements

The following devices are commonly used for current measurements.

Float and tracers

Floats and tracers only give information about the current velocities at the surface and are highly sensitive to wind. Except for these limitations and so long as a reliable means to determine the instantaneous position of the floats or tracers is available, these methods have the advantage of simplicity.

Propeller current meter

Propellers (e.g. the Ott propeller) are driven by the pressure exerted on the blades. Because this pressure is related to the flow velocity, the rotation speed of the propeller is directly related to the flow velocity. A table or graph relating current velocity to rotation speed is provided by the manufacturer of the propeller. Calibration of the instrument should be done regularly in a known current (for example, in a flume with a known discharge) and in water of a similar temperature to the seawater. Depending on the specific type, the functioning of propellers may be sensitive to suspended materials such as silt or sand. In general, however, a properly maintained propeller meter is a reliable instrument for use in uniform and quasi-stationary flows, inaccuracies generally being in the range of a few per cent. Use of propeller meters is limited to quasi-stationary (e.g. tidal) currents, but by mounting the propeller horizontally with a tail fin and providing free rotation around a vertical axis, the instrument can be applied in directionally varying currents.

Pressure tube (Pitot tube)

Pitot or pressure tubes are measuring devices based on the relation between flow velocity and the stagnation pressure exerted by the flow upon the head of a tube that is directed opposite to the flow direction. The actual relation for a specific instrument is obtained from a table or graph provided by the manufacturer of the instrument. The instruments are very sensitive to the direction of the flow relative to the orientation of the instrument and to suspended material.

Laser-doppler

Laser-doppler systems can be used if little or no material is suspended in the water. The method is more suitable for the laboratory but can be used in prototype. They are particularly useful where there is a need for detailed records of the instantaneous velocity or where turbulence characteristics are required. With three-beam laser-doppler systems velocities can be measured along three orthogonal axes. The principle of the system is the (Doppler) shift in phase between two (optical) laser beams, which is a deterministic and known function of the flow velocity. The system consists of a transmitter and a receiver located at a distance from the transmitter, generally of the order of a metre, governed by the strength of the transmitted signal. The received signal can either be converted into an analogue plot or recorded digitally.

Acoustic doppler meter

Acoustic doppler meters use acoustic signals to measure current velocities. The Doppler shift between the frequencies of the transmitted and received signals is proportional to the instantaneous flow velocity in the control volume of water. Simultaneous measurement of the transport of suspended sediments by measurement of the intensity of the received signal is also possible, the device then being called an Acoustic Sediment Transport Meter.

Electromagnetic flow meter

The concept of electromagnetic flow meters is based upon the interference between the current velocity and an electromagnetic field. The system must be positioned parallel to the dominant flow, and this can be achieved by building it inside the head of a 'fish' fixed to a line or by placing the systems on the bed. The received signal can either be converted into an analogue plot or recorded digitally. Unless the fish can be fixed effectively to the bed, its sensitivity to changing current directions may be a disadvantage.

A4.4 Wave Measurements

Wave recordings usually imply sampling of the instantaneous surface elevation, the standard unit-recording period being 20 min. The resolution of the wave spectrum to be estimated (σ_f, also indicated as the bandwidth) is limited by the record length (T_{rec}) according to:

$$\sigma_f = 1/T_{rec} \qquad (A4.1)$$

The highest frequency (f_{max}), to which the spectrum can be estimated is related to the sampling frequency (f_{sampl}) or sampling interval ($T_{sampl} = 1/f_{sampl}$):

$$f_{sampl} \geq 2 f_{max} \qquad (A4.2)$$

This frequency is also known as the Nyquist frequency (f_N) or folding frequency. The latter refers to aliasing, the 'folding' (with respect to f_{max}) of energy contained in frequencies $f > f_{max}$ into the spectrum leading to energy being added to the lower frequencies of the spectrum and the spectral shape being distorted. A sampling interval of 1/10 to 1/20 of T_s (T_s being the anticipated significant wave period) is therefore required to obtain a reliable spectrum from the sample record.

The following devices may be available for making wave measurements.

Wave guide (visual)

By fixing a graduated pole in or upon the seabed, one can make a reasonable estimate of height of the (one-third) largest waves in a wave field and the corresponding period. The method has the advantage of simplicity. Limitations are for use in shallow water only and the disregarding of all but the higher waves. Besides, information obtained on wave periods is very poor, except for conditions with prevailing swell. Additional video or film recordings, however, can provide extra means to obtain better data from this method.

Step gauge

A number of electrodes are attached with equal distances (0.05 m) to a vertical pile. The pile can be connected to any structure that is fixed in the seabed. Although rather insensitive to mechanical impact, the instrument may be protected by placing it within a steel frame (see Figure A15). The number of submerged electrodes corresponds to the instantaneous water level. Submergence is detected through the electrical response (a step function) to submergence. The signals of the electrodes are further transferred in a digital form, giving the height of the instantaneous water level. By combining standard sections, the step gauge can be applied for a

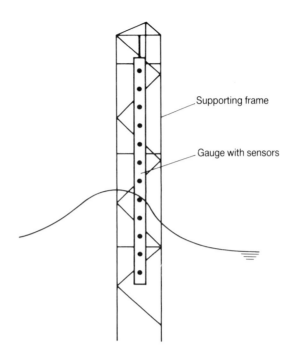

Figure A15 *Wave measurement using a step gauge*

wide range of water levels and wave heights. With the attached software a wave spectrum is obtained while wave height precisions of approximately 0.01 m can be realized.

Resistive and capacitative surface elevation meters

Using electromagnetic principles, these methods have proved to be suitable for wave measurements. Application has been limited to laboratory conditions in the past, but recent experience with resistive devices has shown (Chadwick, 1989) that so long as the supporting wave pole is sufficiently robust for the local wave conditions, good-quality digital records of similar precision to those from step-gauges can be obtained for subsequent spectral analysis and determination of wave parameters.

Both step gauges and resistive meters have the advantage of being able to operate in shallow water, if required, without the disadvantages of filtering out of short-period waves which arise in the case of a seabed pressure meter discussed below.

Wave rider

A wave rider is a buoy that can follow the instantaneous movements of the water surface. Positioning is realised with a flexible connection with the seabottom that allows for the largest expected wave.

By a built-in device, instantaneous accelerations of the buoy are measured and this signal can be transmitted or recorded internally. Double integration of the digitized signal gives the instantaneous velocity and position, which corresponds to the water surface movement. A major advantage of the wave rider buoy is its capability to take measurements over several months independently and without maintenance. Maximum resolution is 0.5–1 Hz (corresponding to a smallest wave period of 2 to 1 s). Wave rider buoys are probably the most useful type of deep-water wave-measuring device available. However, they cannot sensibly be used in water depths less than a low-tide minimum of 5 m.

Pressure meter

The method to measure wave heights through wave-induced deviations from hydrostatic pressures has been discussed under water-level measurements (Section

4.1.4.2). Although the concept can be used for wave measurements, application is limited to shallow-water waves. This is because with a transducer placed on the seabed, pressure fluctuations with higher frequencies decrease rapidly with increasing water depth. This effect also means that many short-period waves are filtered out in the measure process, leading to the measured average wave period, T_m, being higher than the true value, a noticeable effect whern T_m is less than about 8 s. Nevertheless, these devices have proved relatively robust and inexpensive and measure wave heights within an accuracy of 10%. They are therefore worthwhile for use in long-term monitoring of wave conditions at coastal sites.

Radar

Radar techniques can be used to survey the sea state over a wide area, but need to be calibrated using recordings from another reference device such as a wave rider buoy.

A4.4.1 DIRECTIONAL WAVE MEASUREMENTS

Where the design is sensitive to wave direction (for example, where phenomena such as wave refraction and diffraction and wave-induced longshore sediment transport are involved), a directional wave climate is required which may need the acquisition and analysis of directional wave data. Such an analysis may range from a relatively simple determination of the main wave direction to a full two-dimensional spectral analysis of the spectrum $E_{\eta\eta}(f, \theta)$. Whereas a frequency analysis needs only a record of water surface elevations from one single point, directional analysis requires additional data. These can be obtained either by using a tilting buoy or by mesurements from several (usually three) locations. A summary of available methods has been provided by Goda (1985), while an extended overview is given by Panicker (1974).

The principal methods available at present can be sub-divided as follows.

Wave gauge array

The simultaneous use of several wave gauges (or buoys) enables determination of the cross-correlation (covariance analysis) of the surface elevations. Thus the two-dimensional spectrum, $S_{\eta\eta}(f, \theta)$, is obtained. The gauges should be placed in positions that are carefully chosen for optimal resolution. This means that:

1. No two pairs of gauges may have the same relative orientation or distance between them.
2. The differences between the orientations of pairs of gauges should be maximised.
3. No pair of gauges should have a distance between them of more than $L_{min}/2$, where L_{min} is the wave length of the highest wave frequency (shortest wave period) for which the directions are of interest for the design.

Directional wave rider

Tilting buoys have also been developed, recordings of which permit a directional analysis. By additionally measuring the angle of the buoy relative to a plane horizontal reference these devices enable the gradients of the buoy relative to north and east to be determined. By combining these gradients, the directional spread can then be calculated.

Two-dimensional current meter

The directional wave spectrum can be estimated from analysis of two-dimensional measurements of flow velocity at one fixed level (outlined in Section 4.1.5.2).

Stereophotogrammetry

A large series of consecutive photographs taken from above the sea surface can be used to gather data on the instantaneous water surface within a certain spatial range. The spatial differences in the surface elevations must be determined by composition of photographs. Covariance analysis between the water surface elevations in a number of discrete points reveal the directional wave spectrum. Photographs may be taken from aircraft or satellite, but the required resolution may impose limits to the height of the camera above the water surface. An important limitation of the method is that wave components, propagating in opposing directions (difference, 180°) cannot be distinguished. The method can be very costly due to the large number of photographs needed to cover a certain range of time.

Microwave (radar) technique

Directional wave climates can also be assessed from measurements of the water surface elevation high-frequency radar techniques. Radar transmission is possible using aircraft satellites or land-based coastal stations.

Appendix 5: Instrumentation for geotechnical data collection

A5.1 Site Investigation Methods

In general, four classes of site-investigation methods are available:

1. Geophysical measurements from the soil surface;
2. Penetration test, such as the standard penetration test and cone penetration testing;
3. Borings, including sampling and installation of piezometers;
4. Specific geotechnical measurements, such as plate loading tests, nuclear density measurements, pressuremeter tests and vane shear measurements.

Further discussion on these methods (such as those described in BS 5930: 1980 and ASTM (1987) can be found in such useful references as Hawkins *et al.* (1985) and Power and Paisley (1986).

A5.1.1 GEOPHYSICAL METHODS

The following geophysical methods are most commonly used:
1. Seismic—refraction
   —reflection
2. Electric resistivity
3. Electromagnetic

A5.1.1.1 Seismic methods

If a seismic impulse is generated at the surface by artificial means (e.g. explosives or hammer blows), energy radiates into the ground and is reflected (following similar laws to optics) back to the surface from any distinct geological boundaries that may be below the ground. Each material allows the seismic wave to pass through it at a certain velocity. Generally, the stronger the material, the greater the velocity (e.g. velocity through granite is greater than that through clay). If the contrast in velocity between two materials is good there will be good refraction or reflection from the boundary between the two materials. The principle of the seismic methods is in Figures A16 and A17. Seismic methods can also be carried out from boreholes (as cross-hole testing) and from penetration tests (for example, the seismic cone).

A5.1.1.2 Electrical resistivity techniques

In theory, electrical resistivity is more versatile than other shallow geophysical exploration methods in respect of the range of geological situations to which it can be applied. There are, however, practical and interpretive limitations which tend to restrict its application to comparatively simple cases on land. There are many variations in technique, but all are based on passing a current through a set of electrodes and along the ground, measuring the potential set up in the ground using another set of electrodes. Based on the current and measured potential, the

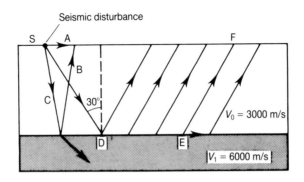

Figure A16 *Principle of seismic reflection (SCB) and refraction (SDEF)*

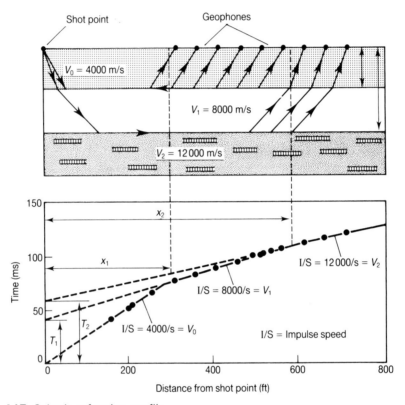

Figure A17 *Seismic refraction profile*

'apparent resistivity' can be calculated. Changes in apparent resistivity should reflect changes in soil or rock type, moisture content and groundwater quality (saline or fresh water). The principal of the most frequently used electrode array (known as the Wenner configuration) comprising four equally spaced electrodes in line is illustrated in Figure A18.

A5.1.1.3 Electromagnetic techniques

Electromagnetic methods are based on the mreasurement of terrain conductivity using an inductive electromagnetic technique. Very small currents are induced in the earth from a magnetic dipole transmitter. These produce a weak secondary magnetic field which the instrument then measures. The ratio of the natural primary field and the secondary field value is read from a voltmeter which is calibrated in ms/m. The measurement are performed with a portable instrument as shown in Figure A19, which is able to investigate to a depth of approximately 6 m.

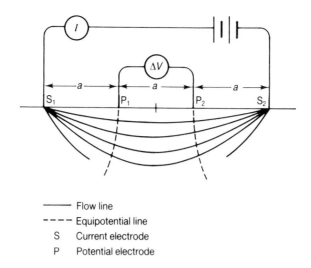

Flow line
---- Equipotential line
S   Current electrode
P   Potential electrode

Figure A18 *Principle of electrical resistivity surveying*

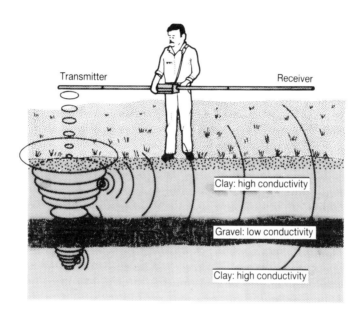

Figure A19 *Principle of electromagnetic surveying*

The main advantage of electromagnetic methods is the speed with which a survey can be performed. This speed, depending on accessibility of terrain and the intensity of measurements, achieved approxiamtely 4 to 6 km/h, when measuring along a line. When a square of side 20 m is measured, about 4–6 ha (8 to 12 acres) per hour can be investigated. The absolute value of terrain conductivity (or resistivity) is seldom diagnostic; great value is placed on lateral variation of resistivity for survey interpretation. A faster survey technique produces, for the same expenditure, a more detailed survey or a survey over a larger area, compared with other methods of measuring terrain conductivity. In Figure A20 a comparison is presented of electrical resistivity and electromagnetic measurements.

A5.1.2 PENETRATION TESTS

Penetration tests are increasingly used because of their economy compared to borings, their reliability and increased sophistication. The most commonly used tests are:

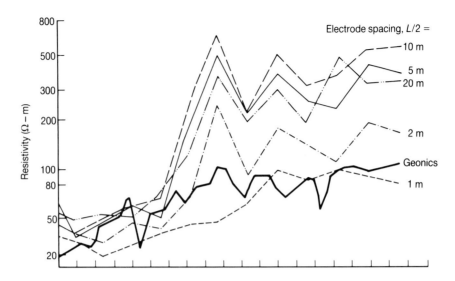

Figure A20 *Comparison of electrical resistivity and electromagnetic measurements*

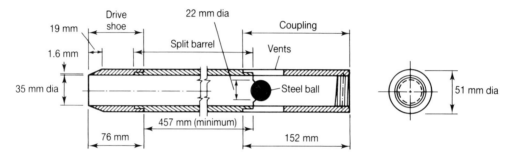

Figure A21 *Cross-section of SPT sampler*

- Standard penetration tests (SPT)
- Static cone penetrometer tests (CPT)
- Field vane shear tests (VST)

Although less common, the application of the following tests is growing:

- Pressuremeter test (PMT)
- Dilatometer test (DMT)

A review is given of the main penetration tests and their perceived applicability in Table 30 of the main text. A short description test of the three first mentioned type of test will be presented. In addition, some information on the last-mentioned type of tests is given.

A5.1.2.1 Standard penetration test (SPT)

This is a dynamic penetration test, carried out in a borehole using a standard procedure and equipment. The test was initially introduced by the Raymond Pile Company in the United States to provide an indication of the *in-situ* relative density of sand. It is widely used internationally. The equipment consists of a 50 mm diameter split spoon (barrel) sample of standard dimensions, weighing 66.7 N (see Figure A21).

This split spoon is driven into the soil at the bottom of a borehole by an automatic triphammer, which allows a 622.7 N weight to drop through a height of 762 mm. The number of blows is counted to give a penetration of 300 mm after the

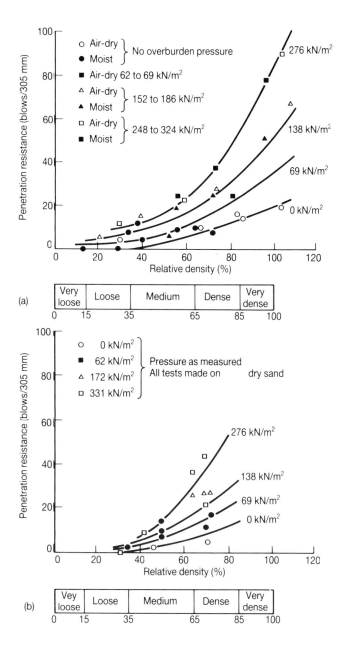

Figure A22 SPT results in (a) coarse sand, and (b) fine sand

tool has been driven through an initial depth of 150 mm of soil, which is assumed to have been disturbed by the boring tools.

The test, which is empirical, has been widely practised and there is much published information linking the results of the tests with other soil parameters and the performance of structures. The main purpose of the test is to obtain an indication of the relative density of sands and gravels. It has also been used to assess the consistency of other soils such as silts and clays and of the strength of weak rocks. Other correlations have been produced relating the number of blows N to deformation moduli. One such correlation is illustrated in Figure A22.

The principal advantage of the test and the main reason for its widespread use is its simplicity and cheapness. There is also a great body of information, built up by experience, on the use of SPT results in relation to foundation design. The test is undoubtedly primitive and the values obtained are somewhat approximate and unreliable, but many engineering projects have been successfully designed and constructed using SPT results.

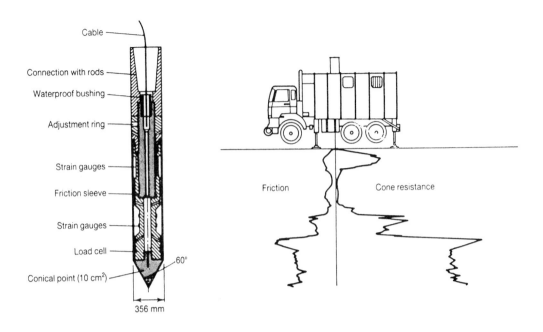

Figure A23 *Principle of the electrical CPT on land*

A5.1.2.2 The static cone penetration test (CPT)

All static cone penetration tests (CPT) consist of pushing a rod into the ground and assessing the physical properties of the material through which the rod is pushed by measuring the resistance at the cone. A very wide variety of static cone penetrometers are available throughout the world but is most widely used in the Netherlands, where the mechanical and electrical cones were developed (see Figure A23).

Mechanical and electrical cones are now standardised. The test was designed to enable a cheap and rapid assessment to be made of the length and diameter of piles necessary to carry a given load. In the original mechanical version the apparatus consisted of a rod with a cone end and a tube that surrounds the rod, so arranged that both cone and tube may be advanced together or separately. They are both inserted into the ground by means of a static load. The cone is then advanced ahead of the tube to measure end resistance alone, and if the friction jacket type of cone is used the local friction and cone resistance are also measured. The results are expressed graphically in depth of penetration versus cone resistance plus local friction charts as shown in Figure A24.

This test does not provide samples to allow direct assessment of the type of soil being penetrated but it may be possible to estimate this from the resistance values and the local friction values. By comparing boring logs and CPTs it was found that the ratio of local friction and cone resistance provides an identification of the lithology (see Figure A25).

An important development in penetration cone design was one in which the original purely mechanical arrangements for measuring cone and tube resistance were replaced by load cells operating electrically (electric cone). The introduction of electically operated load cells at the tip of the sounding rods has opened the way for other improvements, such as the electrical conductivity cone, the piezometer cone, the seismic cone and the density penetrometer (see e.g. Figure A26). With the piezometer cone the water pressure during penetration can be registered. This type of information provides a very accurate stratigraphy of the soil (see Figure A27). The density probe provides information about the *in-situ* density or porosity, which

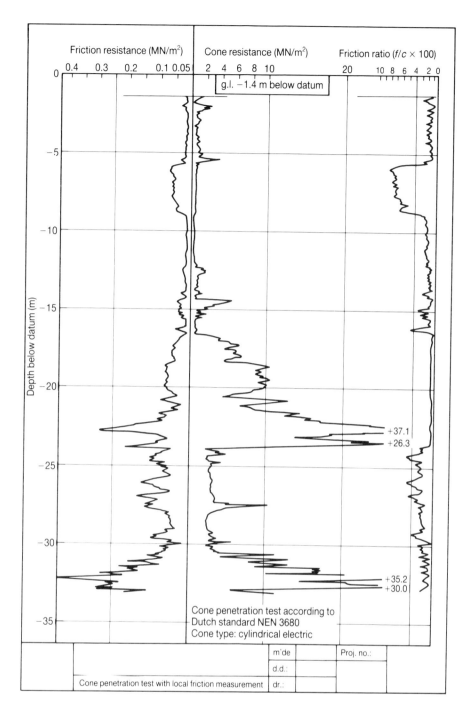

Figure A24 *Presentation of CPT results of electrical cone with local friction measurement*

cannot always be provided by normal boring techniques. By means of the density probe, correct measurement can be made and reliable values presented. The electric cone has also opened the way to *in-situ* testing in shallow and deep water. Developments in that direction have led to remotely controlled cone penetration systems (see Section A5.2).

Application of results of cone penetration testing

Applications of the results of core penetration testing include the assessment of stratification and soil identification mentioned above. In addition, the measured cone resistance, g_c, and local friction, f, also provides a basis for the assessment of a number of parametric values for geotechnical computational models:

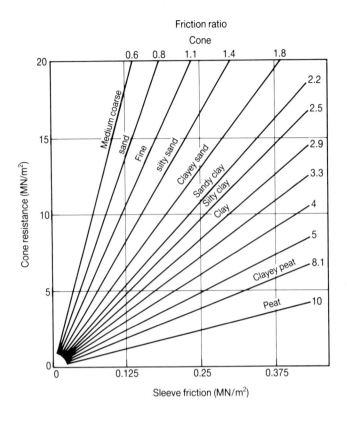

Figure A25 *Soil identification by cylindrical electrical cone resistance versus local friction*

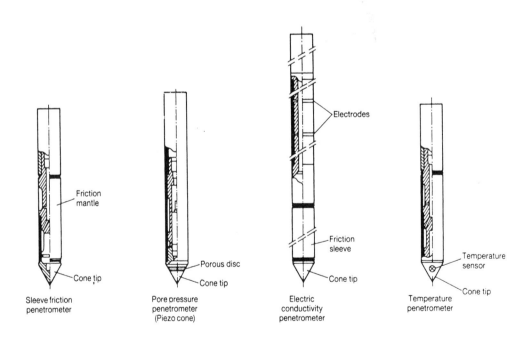

Figure A26 *New developments in cone penetrometers by Fugro McClelland*

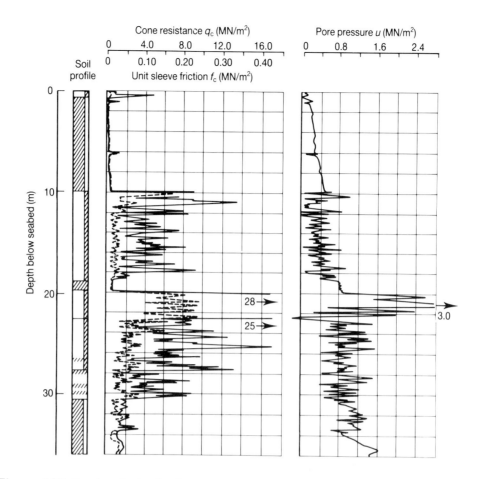

Figure A27 *Results obtained using the piezocone penetrometer*

1. Parameters for the bearing capacity of a pile at the point, which is directly related to q_c and the friction along the pile, which is also directly related to f, in the case of a straight pile;
2. Parameters for the prediction of settlement. The compression modulus, C, is related to the q_c by $A \cdot q_c$, in which A depends on the type of soil as follows:

A	Soil
10–5	Coarse sand
6.6–3.3	Fine sand
5–2.5	Sand clay, loam
2.5–1.25	Clay
1.25–0.6	Peat

3. In addition, other parameters can be roughly correlated with q_c from empirical relationships, such as:

   (a) The angle of internal friction, ϕ, in granular soils;
   (b) Young's Modulus of elasticity, $E (E \approx 1.5 q_c$ to $2q_c)$;
   (c) The apparent (undrained) cohesion, s_u, in clay ($q_c \approx 14 s_u$);
   (d) Density and porosity (n).

Advantages and limitations of cone penetration testing

A great advantage of the CPT is that measurements are independent of the operator. Others are that many test can be performed in a short time, making them economic compared to boring. Also very important is that the test gives little disturbance of the soil, thus providing accurate *in-situ* measurements. A limitation, like all penetration tests, is that hard, consolidated formations cannot be tested and to obtain soil samples, boreholes will also be required.

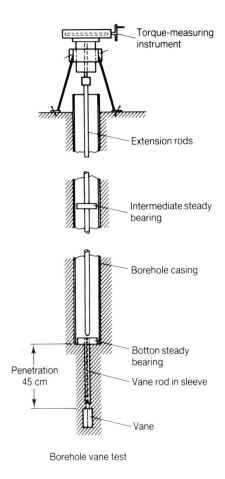

Figure A28 *Field vane test performed in a borehole on land*

A5.1.2.3 The field vane shear test (VST)

A cruciform vane on the end of a solid drill rod is forced into the soil below the bottom of the borehole and then rotated. The torque required to rotate the vane can be related to the shear strength of the soil. The test is normally restricted to fairly uniform cohesive fully saturated soil, and is used mainly for clays having shear strengths of up to about $100 \, kN/m2$. The results are of doubtful value in stronger or fissured clays. The field vane used at the bottom of a borehole is shown in Figure A28.

Advantages and limitations of the field vane shear test

The main advantage of the field vane shear test is that the test itself causes little disturbance to the soil and is carried out below the bottom of the borehole in virtually undisturbed ground. This is particularly apparent in sensitive clays, where the vane tends to give higher shear strengths than those derived from laboratory tests on samples obtained with a tube sampling device. Under these conditions, the vane test results are generally considered to be more realistic. Vane equipment is also available for use in the laboratory and hand vane equipment may be used in the field to determine the strength of soils exposed in trial pits. Recently, remote *in-situ* vane testing equipment has been developed by Fugro–McClelland for use on marine soil investigations (see Figure A29).

A5.1.2.4 The pressuremeter (PMT) and dilatometer (DMT) test

In these tests a measuring device consisting of a probe (measuring cell—PMT) or a metal blade provided with a vertical flexible membrane (DMT) is inserted into the

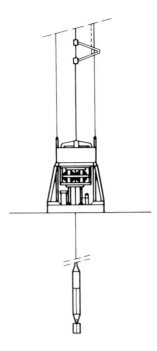

Figure A29 *Field vane test in seabed mode as developed by Fugro McClelland*

soil. At the required depth the cell or membrane is extended horizontally by pneumatic tubing. The horizontal displacement of the cell or membrane is measured as a function of the exerted pressure. From the measured load-displacement relationship the horizontal modulus of elasticity, E_h, can be drived.

The pressuremeter cell can be placed within a pilot-hole (Menard PMT, see Figure A30) or can be pushed into the soil from the ground level by means of a penetration system (push-in type pressuremeter). The dilatometer blade (see Figure A31) is directly pushed into the soil to the required depth in a way similar to the CPT system.

A5.1.3 BORINGS

Most site investigation include boreholes, which are drilled to gather information about the nature and distribution of ground materials and the discontinuities that ramify the ground mass. They may also be used as a means for *in-situ* tests, to allow inspection of the strata exposed in the walls of the borehole, or to install *in-situ* instrumentation which records changes in the condition of the ground. Descriptions of the geological nature of the materials penetrated by the boreholes can sometimes be achieved by taking 'disturbed' samples, that is, samples whose condition in no way represents the condition of the material *in situ* as the result of disturbance due to the sampling procedure.

Samples required for description in terms of both geology and geotechnical properties are taken by special sampling devices and are then termed 'undisturbed'. No samples are truly undisturbed, for extraction from the borehole almost always implies a change in the conditions of stress under which the sample existed *in situ*.

Ideally, sampling should be continuous and provide a complete sample of all the strata penetrated by the borehole. This is possible to limited depths in soft soils by means of special long samplers, pushed into the soil, such as the Delft continuous sampler or the longer piston samplers, and in rocks by rotary core drilling. Unfortunately, most soils are too strong for deep continuous sampling and too weak or bouldery for successful coring, while the soil-like layer of weathered rock overlying fresh bedrock is also difficult to core. In such circumstances, recourse

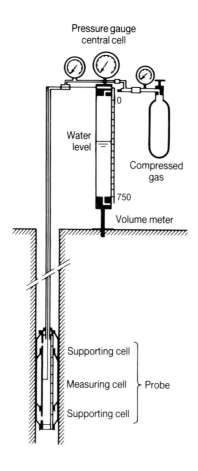

Figure A30 *Schematic diagram of a pressuremeter test set-up*

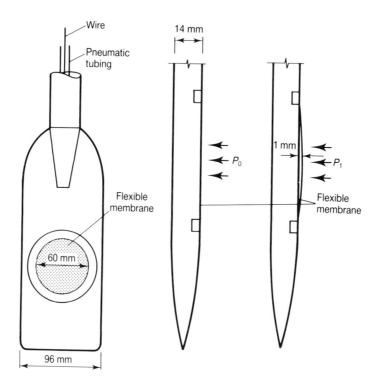

Figure A31 *Dilatometer (Marchatti)*

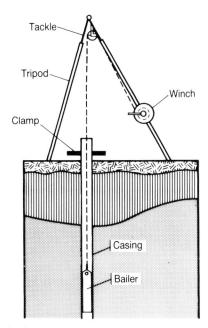

Figure A32 *Percussion boring*

must be decided before drilling the borehole, or is left to the discretion of the driller acting within a framework of instructions given by the investigator.

A wide range of drilling systems is available that can be used for soils varying from soft to rock:

Boring system	Soil type
Rotary core drilling	Rock
Non-core rock drilling	Rock
Percussion boring	Soft soil
Continuous flight auger	Soft soil
Hydraulic boring	Soft soil
Soil Delft continuous sampler	Soft soil

More information of two of these boring systems is provided in Figures A32 and A33.

Borehole records

Most materials are described from samples taken from boreholes. Their descriptors are given in borehole records which should contain a complete account of:

- The method and progress of boring;
- The samples taken;
- *In-situ* tests made in the borehole;
- The nature of the strata encountered.

An example of a good-quality borehole record is given in Figure A34.

A5.2 Site Investigations on Water

The methods for soil exploration as described above are, in principle, applicable to both land and marine conditions. On land, drilling rigs (detached or mounted on trolleys or trucks) and penetration trucks can be used. For low accessible terrains special vehicles with low wheel pressure have been developed. In addition, detached penetration and drilling equipment can be used. For costal areas special high-wheel vehicles are available which are suitable for penetration testing during low tides.

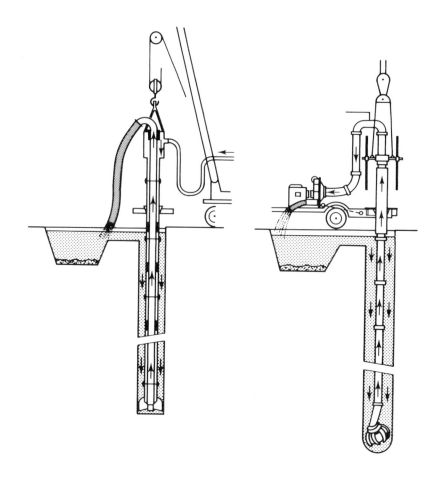

Figure A33 Hydraulic rotary drilling in unconsolidated soil

If the test equipment cannot be used from the (temporary) dry ground level, special operational facilities should be applied. The principal facilities that may be used are:

- *Temporary fixed platforms*, supplied with detached drilling rigs and penetration systems with truck-mounted systems (incuding the truck itself);
- *Floating facilities*, such as pontoons, self-navigating ships and special-purpose ships. Operation take place from the deck, while the reaction force for penetrating and boring can be obtained from the floating body, the self-weight and friction between the drill string and the soil;
- *Remotely controlled systems*, managed from ships. The reaction force is exerted at seafloor level mainly from the self-weight of the submersible system.

In the case of fixed platforms or floating facilites it is usually a matter of moving the workfloor from the ground surface to a floating surface or the fixed deck. As long as the water is not deep or moving, the condition for site investigation can be dealt with. If water depths become great and strong currents or high waves and swell occur, soil exploration becomes more difficult and expensive.

In most cases of investigations on water a pontoon may be used. When exploration has to be executed offshore a seaworthy ship becomes necessary. The major handicap with both type of floating facilities is that the deck moves in respect to the seabed. This means that all activities in the borehole must be done independently from the movement of the deck, and can be achieved by putting a casing provided with a heavy weight on the seabottom. This casing extends above the deck and stands freely from the pontoon. When the water depth exceeds about 30 m, investigations with such an arrangement becomes almost impossible.

The rotary drill system is instead the appropriate solution. This method, derived

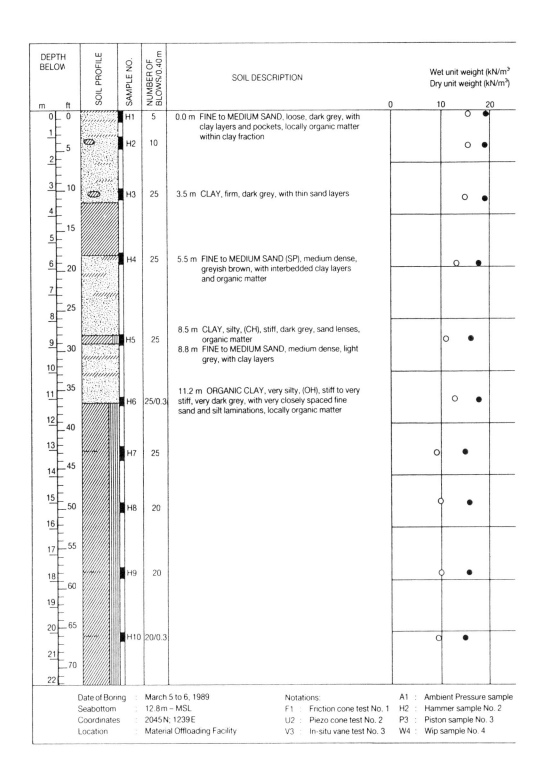

Figure A34 *Good-quality borehole record*

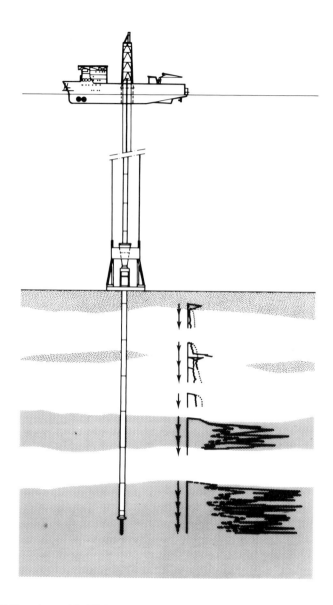

Figure A35 *Drill string with Weissen cone penetrometer*

from the oil-exploration industry, makes use of a drill string which is lowered from the deck of the ship and penetrates the soil. The string has enough flexibility to allow for some ship movement (see Figure A35). At intervals, drilling can be stopped and rive samplers lowered to the bottom of the borehole through the hollow drill string to take samples.

Instead of a sampler, a wire line penetrometer can be lowered. With such a penetrometer a CPT test of 1.5 m in length can be performed at the bottom of the hole. The reaction froce is obtained from the weight of the drill string and the mobilised shear resistance of the drill string and the adjacent soil layers.

Remotely controlled systems for soil exploration have been developed for the oil production in the North Sea and the Gulf of Mexico. These are based on the idea that testing should be done from the seabed, independently of the ship's movement. Several operational boring and penetration systems are available (see Figure A36). The reaction forces for penetration are derived from the self-weight of the submersible system on the seafloor.

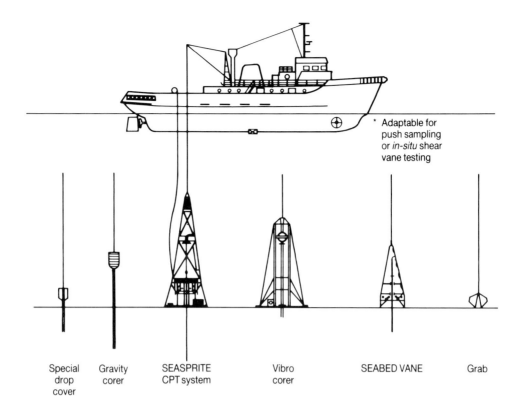

Figure A36 *Remote control boring and penetration systems for seabed investigation*

A5.3 Laboratory Tests

Several types of tests may be carried out in order to predict the behaviour of the soil in future conditions. Often these tests are not all required, especially when the general geological condition is known. A distinction can be made between:

1. Mineralogical testing by means of microscopy, X-ray and chemical analysis;
2. Mechanical testing, through which the properties of the soil samples are determined;
3. Model testing on a small scale either under normal gravity condition (1 G) or under increased gravity conditions, i.e. in a centrifuge (more than 100 G). This type of test is mostly applied when the configuration of soil and structure is complicated and a good prediction of the interaction between soil and structure is wanted.

It is beyond the scope of this manual to describe all the kinds of tests which are possible, and reference should be made to publications such as Head (1980). Some routine tests, mostly of category 2, will, however, be mentioned briefly. These types of tests are classified as (a) composition and classification tests (Section A5.3.1) and (b) stress–strain tests (Section A5.3.2).

A5.3.1 COMPOSITION AND CLASSIFICATION TESTS

A5.3.1.1 Granular analyses

The purpose of the test is to define the grain-distribution diagram. The test is performed, *inter alia*, to investigate the soil compaction possibilities and to examine to what extent the soild can be used for hydraulic fill, and can be applied to both sand and clay. The grain-distribution diagram is defined on the basis of results from sieve tests and/or hydrometer tests. The practical significance of the sieve curve is that, from the curvature, data can be derived on the grain size and the distribution of the material, and a general insight can be gained on the permeability.

A5.3.1.2 Chemical analysis

To supplement the granular analysis, the humus and lime contents can be defined by a chemical analysis.

A5.3.1.3 Density

It is possible to determine:

1. The specific, weight, volume and water content of clay samples; simultaneously, the consistency limits are also defined. This is done in order to assess the workability of clay and to investigate its water-retaining capacity. The natural moisture content is measured and its value relative to the liquid and plastic limits will determine its suitability as a water-retaining material.
Optimum density. In order to obtain information on the degree of compaction of a sand, densities can be determined for a given compactive effort and varying moisture contents. These can also be determined for different voids ratios or porosity percentages.

A5.3.1.4 Permeability

The permeability of the soil is also important for determining the settlement rate of the soil under the influence of an applied load. Permeability is also important when designing de-watering wells. It is strongly dependent on the composition of the soil and can therefore differ from one area to another. In cohesive soils, the permeability tests are performed on undisturbed samples obtained from borings specially taken for such tests. For non-cohesive soils, it is practically impossible to take undisturbed samples. Therefore the permeability is determined from samples recompacted at varying density.

The horizontal and vertical permeability of soils normally differs; rarely are they equal. Permeability tests in cohesive soils are therefore generally performed both horizontally and vertically. Permeability can also be determined indirectly from the results of the consolidation tests, but scale effects mean that the values obtained tend to be significantly lower than those measured directly or in the field.

A5.3.3 STRESS–STRAIN LABORATORY TESTS

A5.3.2.1 Consolidation

The soil-consolidation properties are investigated by oedometer tests, in which a 2.0 cm high sample is subjected to various vertical loads. If the sample is water saturated, the settlement resulting from the load will be retarded, depending on the permeability of the soil. During the test, the progress of settlement for each top load is constantly monitored. Apart from the settlement constants C and C_s, according to Terzaghi's formula, the consolidation coefficient, c_v, and the hydrodynamic period can also be determined. The consolidation constant, c_v, depends on the load and is therefore, in fact, not a constant. The settlement calculations will gain in accuracy if this factor is taken into account.

A5.3.2.2 Friction properties

The aim of measuring soil friction properties is to enable the assessment of the stability of potential or pre-existing shear planes in order to determine, for example,

- The stability of earthfill slopes;
- The soil pressure on earth-retaining structures;
- The bearing capacity of soil.

In order to define the friction properties, the following tests can be peformed in which the vertical load is varied and the relevant shear resistance is measured.

Direct consolidated shear test

For each top load, the sample is consolidated prior to the performance of the shear test. The shear plane is horizontally oriented. This need not necessarily be the most unfavourable direction. Consequently, the values from this test may yield too favourable a result. For peat, however, which usually is found in horizontal layers, the test can give a good indication of the shear resistance.

Direct not (completely) consolidated shear test

If no consolidation has yet taken place, then $\phi_u = 0$ occurs. This situation compares with the phase immediately after a load has been applied. In this situation the undrained shear strength, s_u, is also determined. In order to obtain more accurate test results on the shear resistance of the soil, cylindrical samples are taken. The advantage of this method as compared to that of the direct shear test are:

- That most unfavourable shear plane can develop freely. Its direction is not, as in the direct shear test, determined by the test procedure.
- It is possible to obtain an image of the total stress situation of the sample.

The cylindrical samples are vertically loaded, making it possible to exert horizontal confining pressure. The purpose of this test is to investigate the horizontal/vertical pressure combination at which the sample collapses. Tests performed on cylindrical samples are (1) the unconfined compression test and (2) the triaxial test:

1. The *unconfined compression test* is performed on samples without any confining pressures. The sample is vertically loaded until it collapses. When carrying out an unconfined compression test on sand the consolidated ϕ' is measured. When tests are performed on clay or peat the completely unconsolidated situation is measured: $\phi_u = 0$. The values for the friction properties obtained from this test can be used for preliminary calculations. The accuracy is limited, as the pressure of the samples is generally not in accordance with the pressure situation of the terrain. An advantage of the unconfined compression test is that it can be performed rapidly and easily in the field.
2. In a *triaxial test* horizontal confining pressures are exerted by the water pressure in the cell. The drainage of pore water can be blocked completely in order to measure the water pressures in the sample.

There are three different types of triaxial tests:

1. *Consolidated drained (CD) test*. Here the pore water can be discharged. The test is performed after the sample has been completely adjusted to the applied vertical load. In cohesive soil samples this test takes a relatively long time.
2. *Unconsolidated undrained (UU) test*. In this test, the drainage of the pore water is blocked completely. As a result, excess water pressures will arise in the sample.
3. *Consolidated undrained (CU)*. The sample is consolidated according to the load. However, when the load is increased, the outflow of water is impeded, as a result of which excess water pressures occur.

Figures A37 to A39 show the set-up of the triaxial test and its results in the form of a stress–strain diagram as well as the determination of ϕ' and c' when two tests have been performed.

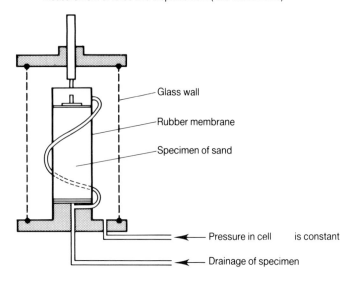

Figure A37 *Principle of triaxial test apparatus*

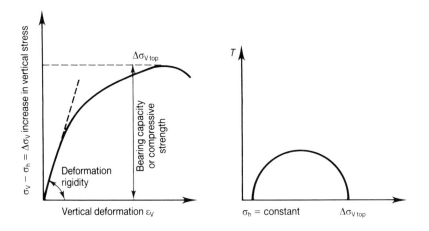

Figure A38 *Stress/strain diagram and critical Mohr circle obtained in triaxial tests*

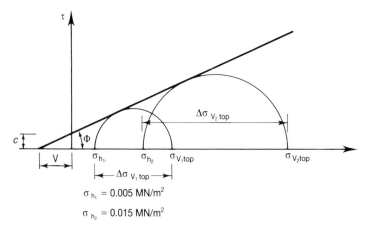

$\sigma_{h_1} = 0.005$ MN/m^2

$\sigma_{h_2} = 0.015$ MN/m^2

Figure A39 *Determination of internal angle of friction, φ, and cohesion, c, from triaxial tests*

Appendix 6: Structure-monitoring techniques

A6.1 Introduction

A6.1.1 MONITORING OF RUBBLE STRUCTURES

Rubble structures respond to the destructive mechanisms of wave and tidal action by changes in the profile of the structure and to the size and shape of its component parts. Typically, rock armour may be displaced, abraded and/or fractured. Any quantiative description of the state of the structure must be able to identify these different responses.

Major failures by storm action are easily identified. Gradual desgradation caused by settlement, solution, abrasion or fracture of the armour may be markedly more difficult to identify. Such damage often remains unquantified until major rehabilitation is needed or a significant failure has occurred.

Methods are needed to quantify the state of the structure so that the owner can:

1. Assess damage caused by particular events;
2. Predict the future working life of the structure;
3. Plan maintenance or rehabilitation expenditure.

All owners and designers need the data derived from regular monitoring to check the efficacy of the design methods in use; to assess the safety of existing structures; and hence to improve future design and construction standards. The performance of a structure is assessed by comparing measures of its state at a number of points in time. Ideally, a monitoring programme would be designed at the time of the structure design. The first few surveys would be conducted shortly after construction, and then at set intervals through the structure life. The techniques used must be repeatable and tolerant of slight operator or procedural variations. The interpretation of the measurements should also following a clearly defined specification, as it will inevitably be done by staff unfamiliar with many of the original design assumptions, particularly in the later stages of the structure life.

All survey methods must be designed to yield repeatable results that can be interpreted and compared unambiguously with those of previous surveys. The measurement and working methods must be well considered and simple to operate. Rubble structures are potentially hazardous areas on which to operate. All concerned should be familiar with intertidal working, and well informed on the wave and tidal conditions anticiapted during the survey period. Periods of spring tides will allow access to lower levels, but can lead to high flow velocities and rapid water-level changes. Access to the armour face is a common problem. The intertidal zone is frequently covered in weed and/or algal growth, and movement in this area requires great care. Large, often smooth, armour units will exacerbate this problem, offering large voids into which the surveyor may fall.

Monitoring programmes should be considered at the design stage so that references points can be incorporated into the construction. If the structure includes a crown wall it will often be convenient to build reference points into the wall or roadway. It should be noted that such elements may be displaced, or even destroyed in extreme cases, so it will again be important to relate the positions of these reference points back to land-based datum points. On structures without a crown wall it will be necessary to establish other reference points. Navigation or warning beacons or lights will often be required on such structures and will, if strongly founded, provide suitable bases.

The time interval between surveys requires careful consideration, and some flexibility. A rubble structure will settle and adjust most during the first few years. Thereafter most changes will arise during the major storm periods. However, many sources of deteroration to the armour are gradual. Armour can suffer deterioration due to chemical attack, abrasion, weathering, freeze/thaw and related effects. It will therefore be important that monitoring intervals be appropriate to the rate of degradation and damage arising. At particularly sensitive structures some monitoring may be on a daily basis, and it is seldom possible or worthwhile to make detailed observations at such intervals. It is likely, however, that annual surveys are justified to ensure that survey staff are familiair with the structure and to ensure continuity of data. Additional inspections should also be made after all major storms, perhaps whenever the storm wave heights have exceeded 75% of the design value. The threshold value should be set in relation to the design wave climate and associated return periods and the damage response characteristics of the structure.

Not all surveys will be at the same level of detail, and its is custormary for owners to make frequency inspections of a rapid and relatively cursory nature, with less frequency surveys at a more detailed level.

A6.2 Land-based Survey Methods

Land-based survey methods of interest here include conventional engineering surveying techniques. We also include visual assessment made by the surveyor, particularly of the state of the armour layuer. All these methods require access onto and over the structure.

A6.2.1 CONVENTIONAL ENGINEERING SURVEYS

The measurement of the location of the structure in relation to an external reference system will require conventional engineering survey equipment such as levels, theodolites or their electronic equivalents. The techniques used will be essentially as on other surveys, and will not be discussed in detail here. Points specific to rubble coastal structures include the use and interpretation of measurements; resolution levels; control and reference points; setting out and relocation of profile lines.

Conventional survey methods will generally be used to define the position of each of the reference points; the main structural elements; control points used for photogrammetric measurements (see Chapter 3); and each point along the profile lines selected. The results from levelling surveys can be used to determine settlement or other displacement of the crown wall and/or armouring. Profile surveys should also locate and quantify areas of armour displacement, local settlement and toe erosion or accretion (Figure A40). Downslope settlement of armour may not be well described unless supported by additional data from photographs or by the identification of the position of individual armour units.

All such surveys require the identidication of survey points. At Santa Cruz, California, brass discs were set into the breakwater crown wall and steel pins were cast into 44 armour units to allow precise identification of their position and level. At Manasquan Inlet and Crescent City anti-fouling and epoxy paints were used to form two or three targets per unit. Painted targets wear away, but repainting at each survey may suffice for frequent surveys.

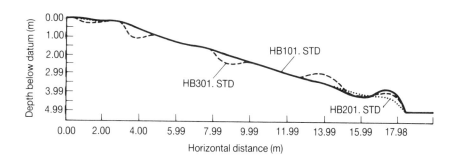

Figure A40 *Comparison of armour slope profiles*

The primary survey line is usually along the crest of the structure, and will be set out from the land-based reference points. The survey setting-out lines will be defined by fixed targets. On the breakwater or seawall targets will generally be painted at each profile line, often with numerals large enough to be read from the air, from the structure toe, or from a boat. Targets on both vertical and horizontal faces of a crown wall may be needed. Profile lines will generally be set out perpendicular to the crown wall or setting-out line using simple sighting aids. Along the trunk the interval will be set by the resolution needed; the structure complexity; and the resources allowed for the monitoring work. In the exercise on the Wash trial bank described by Young *et al.* (1980), profile lines were spaced 0.75 m apart, up to three stone diameters. At Manasquan Inlet, nine survey lines along the structure trunk were set at around 30 m intervals. At Cleveland, ten survey lines in 1340 m were spaced at intervals between 91 and 274 m. In the construction of a new rubble-mound breakwater in Iceland, Read (1988) reports profile lines spaced at 15 m and 7.5 m intervals

It is often difficult to locate a fixed reference point for each profile at the base of the armour slope, in which event the survey line will be controlled by precise survey methods on each survey. At the Wash trial bank, pipe frames were concreted into the structure at toe and crest. At each survey a wire was then tensioned from frames at toe and crest. Measurements were then made at fixed intervals along the wire. The interval up the profle varied for the different armour surveyed, and was generally 0.5–1.0 stone diameters. At this site a hemispherical foot was used to prevent the survey staff entering interstices between the stones. The foot diamater was 0.5 stone diameter.

On large structures, and where access is difficult, the survey staff may be replaced by a heavy plumbing rode or tube marked off in height intervals. This can be suspended from a crane. At the foot of this 'staff' a spherical cage replaces the hemispherical foot. Provided that crane reach is adequate, this 'staff' can be used down much of the structure face. The plan position of the survey point, as well as its level, is then confirmed by theodolite or similar measurements.

Generally, profile measurements should be made at intervals of less than 1.0 stone diameter, when the results can be compared with those derived from hydraulic model tests. It is not often practical to enforce a fixed sampling interval along the profile line, although this may be desirable for interpretation. It may therefore be a reasonable compromise to ask the surveyor to take a level, and position, on the centre of each stone along the profile line. The comparison of profile lines from different surveys will then require interpolation between survey points. A profile analysis method using a cubic spline has been used by Bradbury *et al.* (1988) in hydraulic model tests.

The results of profile surveys may be used to calculate areas of erosion, and hence damage levels, after the methods used in hydraulic model tests. Profile surveys may also be used to generate contoured plans of the structure. Areas where levels fall below (or above) design values may then be identified. The data may also be used

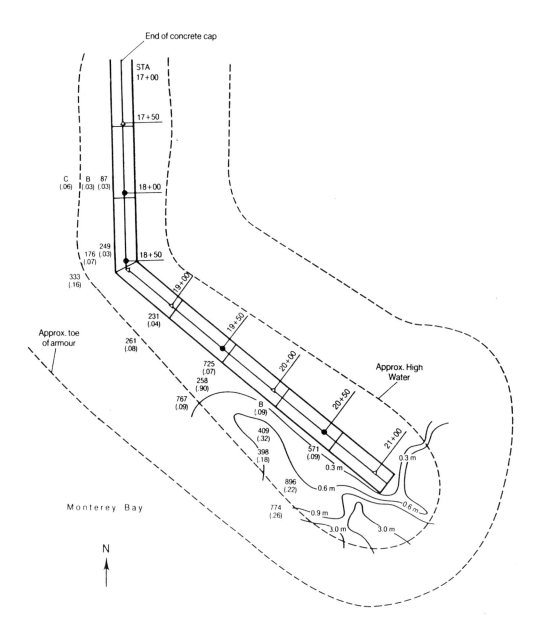

Figure A41 *Contours of equal settlement on a breakwater*

to prepare plans showing areas of settlement. An example drawn from work by Weymouth and Magoon (1968) is shown in Figure A41.

A6.2.2 ARMOUR DEGRADATION

Monitoring techniques considered so far have concentrated on the position of elements forming the structure. For rubble structures the condition of the armour layers, and hence their performance, will depend upon the size, shape and surface texture of the armour units. Rock armour deterioration may arise from abrasion or attrition, spalling and fracture. In some extreme cases chemical attack may also lead to solution of elements of the rock. Research at Queen Mary College supported by HR Wallingford Ltd has addressed this, and has led to methods to identify rocks particularly susceptible to degradation (see Allsop *et al.*, 1985; Poole *et al.*, 1983). During that research a number of assessment methods were developed and used. The techniques initially relied upon the judgement of skilled researchers, but have been developed to provide more generally applicable methods. The original appraisal methods, described by Allsop *et al.* (1985), were used on sites in the UK, around the Arabian Gulf and on the east coast of Australia. A simplified approach was taken by Fookes and Thomas (1986) in a very brief study on an island off the coast of Iran.

The monitoring procedures is intended to identify armour layer damage as given by:

1. Cavities;
2. Fractured armour;
3. Sub-size armour;
4. Unstable armour.

In each location an area containing at least 100 armourstones is marked out. Each survey area should run from crest to as low a level as practical. Typically, the survey area should be at least 5 m wide. It must be possible to relocate the survey area in subsequent years. The area could be set out either side of a profile line.

The first state is to count the total number of armourstones falling within the survey area. Generally, it is convenient only to count those in the uppermost layer. Then each example falling into categories (1)–(4) above are recorded. A cavity is recorded where a void could be filled by an armourstone of design size. Fractured armour includes all examples where the stone has broken in place. Sub-size armour is all armour smaller than the specified lower limit.

The state of interlock of the armour layer may be assessed by two methods. Unstable armour may be defined where the restraint given by adjacent units is reduced and it is estimated that the stone could be moved easily by storm waves. Such armour is often charcterized by rounded or abraded edges resulting from rocking movements, and if frequently observed on newly constructed structures where initial placement may allow the armour to move freely. Such a condition is usually temporary as unstable armour tends to stabilise or is removed completely from the armour layers by wave action. An alternative parameter is the co-ordination number. This is defined as the average number of stones in contact with each stone in the sample. Its assessment is somewhat laborious, and may be subject to variation between surveyors. The co-ordination number will be a function of the grading width, and it is probably only practical for narrow-graded armour rock rather than rip-rap.

For the main damage categories identified above the number of armour units may be expressed as a proportion or percentage of the total counted. These assessment methods have been used at three UK sites and the results are summarised in Table A6.

A6.3 Photographic Methods

A variety of photographic techniques have been used to complement surveys of coastal structures. Photographs provide a cheap, permanent record of the condition of a structure at a given instant, and are therefore used frequently when monitoring structures. The degree of sophistication of photographic survey methods varies quite considerably. At their most sophisticated, techniques may include complex photogrammetric methods, and at their simplest, record photographs taken on general inspections.

Table A6 Summary of survey data at three UK sites

Location	Survey date	No. units	Fractures (%)	Cavities (%)	Sub-Size	Unstable (%)	Total damage	Co-ordination number
Stornoway	10/85	693	1.4	1.7	2.2	3.0	8.3	4.6
Stornoway Area 1	11/89	100	4.5	3.5	14.0	2.0	24	3.8
Stornoway Area 2	11/89	106	1.9	15.1	15.1	3.8	22	4.0
Stornoway Area 3	11/89	96	2.9	4.4	16.7	1.8	26	3.6
Herne Bay+	10/86	828	0.2	2.2	0.0	1.7	(4.1)	—
Herne Bay+	2/87	860	0.6	2.4	0.0	0.7	(3.7)	—
Port Talbot	7/83	2712	1.1	9.4	1.5	0.8	12.8	4.2
Total								
Area 1 supratidal		492	0.4	4.1	0.2	0.4	5.1	4.4
intertidal		368	0.8	10.6	0.3	5.4	17.1	4.1
Area 2 supratidal		468	0.9	7.1	0.2	1.3	9.4	4.8
intertidal		408	1.2	14.0	2.2	0.7	18.1	4.2
Area 3 supratidal		223	3.6	9.9	0.4	0.9	14.8	3.7
intertidal		177	4.0	33.3	4.0	2.8	44.1	4.0

A6.3.1 COMPARATIVE PHOTOGRAPHY

The most basic photographic survey technique is comparative photography. Photographs of the same view are repeated on each survey and the images compared to detect differences. The value and accuracy of such methods is dependent upon the location from which the picture is taken, the field of view and the precision with which they are matched on each subsequent survey. For optimum results, photographs should be taken from the same location, and the same object at the same angle. Records must therefore be made of the location of the point from which the photograph is taken, distance and orientation of object from that point, and focal length of the lens used.

It is relatively easy to obtain photographs of a structure which is fully surface emergent at some stage of the tide. The most useful angle of view of a structure would, however, be normal to the slope. This would necessitate the taking of photographs from an elevated platform, which is not usually possible. This procedure was, however, adopted with some success at Table Bay harbour breakwater (see Kluger, 1988). The photographer was suspended from a crane hook in a cradle and the attitude of the cradle was controled by light lines led diagonally back to the crown wall. The predetermined camera positions were then controlled from two theodolite positions on the breakwater. The camera used was a Hasselblad 500 EL with a 40 mm focal length lens. The configuration used to generate stereo photographs may be summarised as follows:

- Camera elevation above MSL   21.5 m
- Horizontal distance of camera from wall   7.5 m
- Camera distance perpendicular to slope   18 m
- Separation in stereo photographs   6.0 m
- View angle of camera (both planes)   70°
- Overlap of stereopairs   70%
- Stereo cover area (each pair)   24 m × 17 m

A more complex procedure is required for photographs from the water. A procedure suggested by Kluger (1982) involves photographing the breakwater in sections from a boat at a distance such that individual armour units can be identified easily. The whole of the surface-emergent section of the structure can be covered in a series of photographs. The accuracy of this method depends largely on the accuracy of the position fixing, of both the section of structure photographed and the position from which the photograph is taken. The method used by Kluger does not incorporate sophisticated positioning equipment but relies on a relatively simple alignment system.

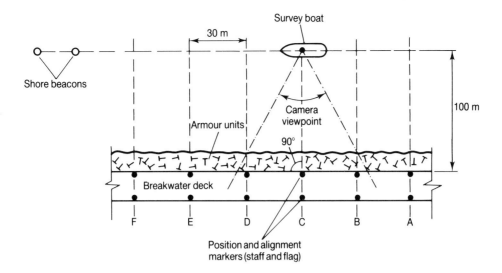

Figure A42 *Taking comparative photographs from a boat*

Two beacons are erected onshore, providing a fixed reference line parallel to the line of the breakwater. The principle of 'leading lights' is followed by simply keeping the two beacons aligned. Each breakwater section to be photographed is identified with markings on the crest of the structure at equidistant spacings. Two marks, making a line perpendicular to the breakwater, are located at the crest of the structure at each marked section. Markers are located above each point at the time of the survey. These provide the other line of sight for the boat, fixing its position by the intersection of two lines. This is illustrated in Figure A42.

It is recommended that survey sections are marked so that photographs overlap. This provides for the posssibility of three-dimensional viewing under a stereoscope. The horizontal aspect photographs taken at sea level can be complemented by aerial photographs enlarged to the same scale. Surveys should be carried out on low-water spring tides where possible to ensure maximum coverage of the emergent section of the structure. This is particularly important, as the inter-tidal zone is the area which is most susceptible to damage under wave conditions.

As with all other survey techniques, it is important that the survey is well documented so that the survey may be repeated precisely. For initial comparison of consecutive surveys, photographs of corresponding breakwater sections are enlarged to the same scale. Photographs are then examined to detect any major changes to the armouring and may be compared by overlaying photographs printed on transparent paper. Changes in location of armouring can be identified using this technique. Alternatively, outlines of armour units may be traced onto paper, and the two outlines overlaid for comparison. If finer detail or closer examination is required, photographs can be enlarged. Photographs printed at a scale of 1:250 are usually suitable for initial comparison.

One specific application of comparative photography is the measurement of the degree of roundness of armourstones. Visual comparisons may be made between the differing degrees of rounding in the intertidal and supratidal zones of a structure, giving some idea of the durability of the armourstone. A more detailed assessment may be made with photographs of the rock armour. Photographs allow statistical data to be obtained showing shape, size, rounding and damage to rock armour. Fookes and Poole (1981) and Dibb *et al.* (1983) discuss methods of assessing block shape. Applications of these methods are further discussed by Poole *et al.* (1983) and by Allsop *et al.* (1985). Rates of rounding of rock and changes in breakwater stability due to subsequent weight loss may be estimated using these techniques.

Photographs of the structure are required, taken perpendicular to the face, and must be sharp, defining the edges of armourstones. It is useful to sample a series of locations along the breakwater. Photographs should be enlarged to A4 or more. Automated methods such as video imaging may be used to calculate the average roundness of the armour. The results of the assessment may be compared with laboratory-derived data in order to estimate the rate of rounding of the rock and thus quantify the likely life of the rock in the marine environment.

A6.3.2 PHOTOGRAMMETRIC METHODS

Aerial photographs have been used to provide valuable information about the condition of a breakwater. The quality of results is, like any other survey methods, dependent on the scaling control and on the method of interpolation of the photographs. For photgrammetric analysis it is necessary to establish an accurate ground-control system. Ground controls should be clearly defined by markings which can be located in photographic images. The distance between controls must be carefully measured, and care should be taken to level the controls periodically from nearby geodetic aznd vertical control benchmarks, since the crest capping may be subject to settlement. These primary controls are used to define horizontal and vertical data from which all measurements are scaled. At Cleveland, large circles divided into alternate black and white quadrants were painted at intervals along the crest of the structure. The controls were spaced between 90 m and 280 m apart. A typical layout and style of ground control markings for aerial photography is shown in Figure A43.

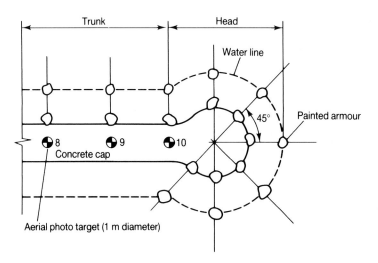

Figure A43 *Armour survey scheme*

The aerial survey flight must also be carefully controlled. The main problem encountered when taking low-level aerial photographs is reaching a compromise between the altitude and the speed of the plane. Low-level flights are desirable for production of maximum photograph definition. However, the lower the altitude of the fight, the slower the aircraft speed must be. Low-level flights have been carried out at a speed of 80 mph at an altitude of 180 ,, using a twin-engine Cessna 320, for the surveys at Mansquan inlet. At Crescent City, a Cessna TU-206 was used with a 12-inch focal length camera at a height of 370 m. This was later revised to improve vertical definition and a Wild 6-inch camera was used at 180 m. Photographs were taken at intervals of about 110 m.

Selection of appropriate photographic equipment is important, since a high level of detail is usually required. An accurately calibrated cartographic camera is necessary to achieve good results. The aerial surveys at Cleveland and Manasquan Inlet both used a Zeiss RMK A 15/23 cartographic camera. In both cases photographs were taken at a contact sale of 1:1200, as was also used at Crescent City.

Lighting and weather conditions are also important factors and lighting conditions for aerial photography are best usually between 10 am and 2 pm. The timing of photographs should also coincide with low-water spring tides so that exposure of the structure is maximised. Low visibility caused by cloud or rain makes conditions unsuitable for detailed aerial photography. Consideration should also be given to the time of year, since shadow length will vary according to the position of the sun.

Photogrammetric techniques may be used to quantify movement of individual armour units and/or to describe the outer surface of the structure. To date, these detailed methods have been applied principally to breakwaters armoured with concrete units. The same techniques can be used to rock-armoured structures. First, selected features must be combined from stereoscopic pairs of photographs to produce photogrammatic maps. These may include location and elevation of individual armour units and other elements of the breakwater, such as crown-wall sections. These features should be superimposed on to a grid which defines location and orientation of the features in the horizontal plane. Vertical data are recorded as spot-height elevations at selected points on the same features. The survey at Manasquan inlet used an enlargement factor of 20 times the contact scale to produce maps at a scale of 1:60. For work at Crescent City, prints were made at a scale of 1:120. A Kern PG Z-AT stereo restitution instrument was used to compile the features, interfaced with a digitising graphic enhancer. Contours were not compiled, but special attention was paid to recording levels close to static water level.

The stereo compilation of the photogrammetric maps showing complex shapes such as armour units requires a highly competent operator to produce the plots. The final plots were produced on transparent paper showing the location, orientation and elevation of armour units. Comparison between surveys were made visually by registering plots of different surveys to determine both horizontal and vertical armour movements.

Some estimates of the accuracy of photogrammetry for breakwater analysis have been given by Gerbert and Clausener (1984) and Nale (1983). Conventional levelling techniques have been used in direct comparison with photogrammetric results in order to assess accuracy. In both instances, measurements have been recorded to the nearest 3 mm. A levelling check on a sample of 160 armour units suggests that photogrammetry is capable of resolving to better than 90 mm. The structures at Manasquan Inlet jetties have been monitored on two occasions using both photogrammetry and conventional levelling. The results on the first survey indicated that discrepencies of the order of ±90 mm occurred. This survey was not, however, carefully controlled, as levelling was carried out over a period of two or three months and levelling positions on dolos were not clearly defined. The second survey, which was undertaken within eight storm-free days of the aerial photography, demonstrated that much greater accuracy was possible. Care was taken to ensure that levels were taken on controlled markings on the dolos. Comparison of the two data sets indicated that 84% of the levels recorded were within ±30 mm of each other and 98% were within ±60 mm.

Photogrammetric analysis has a number of advantages over conventional levelling techniques. The whole of the visible area of the structure can be recorded in one flight. Horizontal movements of armouring are easily defined. The area which is close to static water level can be monitored without risk. No time limit on data translation is needed, and the final product presents easily interpreted graphical data. Precise locations of the magnitude of movements and directions of movements may all be interpreted with relative ease.

Compilation of the photogrammetric plots requires specialist skill. Although it has been suggested that photogrammetric techniques might be applicable to rock armouring, no specific assessment has yet been made for this purpose. While compilation costs may be quite high for a detailed survey showing all visible armour, these may be reduced if only selected areas of the structure are analysed in depth.

The initial costs of a photogrammetric survey of a structure may be high relative to a conventional survey. Periodic photogrammetric mapping will, however, permit detection of incipient or progressive failure along any visible portion of a structure before this is detected by other means. Correlations of photogrammetric surveys with standard levelling surveys indicate that photogrammetry may be used as a substitute for levelling surveys where armour movement is to be monitored.

Photographs provide a permanent record of the state of the structure. The level of later analysis may be tailored to the owner's needs, but only if the photographs have been taken.

A6.4 Underwater Techniques

A6.4.1 GENERAL

Considerable damage may occur beneath the water level, particularly immediately below it, where wave impacts and damage are often greatest. The toe of the structure is also often susceptible to damage, and is an area from which damage can spread rapidly. Additional techniques are therefore required to analyse the performance of continuously immersed sections. These techniques should be capable of identifying both large- and small-scale irregularities (or damage) to the structure.

A6.4.2 BATHYMETRIC SURVEYS

Acoustic sounding equipment such as echo sounders have been relied upon for some time to determine the shape of submerged structures and the adjacent seabed. These surveys provide information about variations in depth to the seabed or structures on it. On their own they do not provide enough information about the structure to allow damage to be assessed. However, by conducting bathymetric surveys in conjunction with other remote-sensing techniques it is possible to build up a more complete description of the condition of the submerged portion of the structure.

Bathymetric charts are compiled with the aid of three components of data: location in the horizontal plane, depth of soundings and the water level at the time of sounding. Position fixing may be carried out using a variety of techniques, including optical methods, electromagnetic fixing systems and laser systems. Soundings may also be made with range echo sounders or side-scan sonar. All depth-measuring systems operate on the principles of transmitting and receiving acoustic signals of known frequency between transducer and target, measuring the time lapse between transmitted and received signals. Normally, soundings are made along parallel courses across the survey area. Sounding lines are usually run in a direction as nearly as possible at right angles to the depth contours. The interval between soundings and sounding lines varies according to the accuracy required, but generally soundings should never be further apart than 25 mm on the scale of the intended survey. Lines spacings should not normally exceed 10 mm. If a detailed profile of a breakwater or the seabed is required, soundings should be more frequent. The transmission frequency of the transducer is also important. High-resolution narrow-beam signals provide the most accurate measurements and are most suitable for measuring breakwater profiles. Relatively high-frequency signals (in excess of 120 kHz) are normally used in bathymetric surveys of this type.

Precision depth recorders may also be employed in bathymetric surveys. These are similar to normal echo sounders but use a more sophisticated signal pulse-generation system with a continuously oscillating source. The precision depth recorded is of particular use for surveying the slopes of the structure and the seabed slopes close to it, where most detail is required. Detailed features can be resolved by precision depth recorders, including scour holes and small rock outcrops on the seabed.

Wave action reduces the quality of records from acoustic sounding equipment and records collected on calm days show more detail. Such action can cause large offsets to the bottom traces, and this may therefore prevent surveying over relatively steep sections of the structure. Waves may also restrict positioning of the vessel. It is possible to filter wave noise from the traces, although line detail will be lost.

A6.4.3 SIDE-SCAN SONAR

Side-can sonar systems operate in a similar manner to bottom-surveying echo sounders except that the signals transmitted by the transducers are directed laterally by two side-surveying beams. A pair of transducers in an underwater housing known as a fish are linled to a recorder (or computer) by a conductive cable. The recorder initiates the signal which is reflected back and appears as a darkened area on the chart recorder. The more reflective the object illuminated by the signal, the darker the record. Shadow areas are shown according to the angle of the signal and the size of the object. Various shades of grey indicate changes in texture and relief. Dingler and Anima (1984) Patterson and Pope (1983) investigated the use of side-scan sonar techniques as a method of underwater inspection of breakwaters.

The degree of resolution of a side-scan survey is dependent upon the equipment and survey techniques. High-frequency transducers (500 Hz) are capable of giving better resolution and finer detail than 100 Hz equipment. While pattern-placed or regular armour can be easily located, randomly orientated units are more difficult to assess. Individual armour units are not easily identified. Irregularities in the profile may be seen on the sonograph, thus allowing areas requiring detailed inspection to be located. If the armour is laid in a regular manner, gaps, irregularities and placement trends may be identified on the sonar traces. Variations such a slope changes, depressons, other irregularities and the line of the toe can usually be detected.

The image projected onto the sonograph is not a true representation of the slope scanned and must be corrected for distortions. Wave conditions and the boat speed can affect the quality of results. Boat speeds of less than 1 m/s are required to identify features of about 1 m size. Similarly, the frequency of the transducer signal is most important. Transducers with a frequency of 500 Hz may be capable of resolving variations in size of armour units, although they cannot identify the precise location of individual units. The resolution of a sonograph also depends to a large extent upon the beam width and method of towing the fish. If the fish is towed low in the water, offshore from the structure, the line of the toe may be well defined on the sonograph. If it is towed close to water level, viewing down the slope, shadowing effects are accentuated, allowing high areas and depressions to be identified. Similarly, steep zones on a structure may reduce the definition of other parts of the structure. Smearing of the sonograph may occur if the fish is allowed to yaw while being towed. The effect of different types or armour placement on side-scan signals and shadow effects are illustrated in Figure A44.

A6.4.4 SUB-BOTTOM SURVEYS

Sub-surface surveys are used to identify problems in connection with foundations that would otherwise go undetected. For example, subsidence due to sub-surface faulting may be detected by shallow seismic survey work. A surface-towed sub-bottom profiler may be employed to identify the structure of the sub-surface. This instrument is of use in determining the depth of bedrock and type of surface cover, and normally operates at a frequency of about 3.5 Hz. The seismic profiler will indicate whether the shallow sub-surface is sandy or dense rock, and can also identify the location of faults. Shallow seismic surveys at Humboldt Bay identified faults 20 m beneath unfaulted sediments. Indication of such features may be of importance if extensions or repairs to a structure have been proposed.

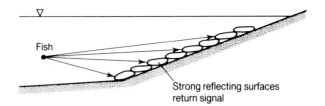

(a) Profile of side-scan sonar view of idealised smooth placement along slope

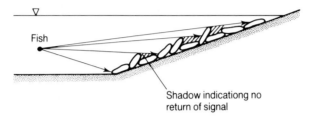

(b) Effects of shadows on side-scan sonar records due to random armour placement

Figure A44 *Effect of armour placement on side-scan sonar return signal*

A6.4.5 DIVING INSPECTIONS

Diving inspections are frequently used to survey submerged sections of breakwaters. General inspections of breakwaters by divers can provide useful detailed information about the state of the structure beneath the water surface. Such an inspection may include observations of broken armour and cavities caused by removal of armour. The stability of the toe (and the toe trench, if applicable) may be examined and small-scale movements in this area identified. A diving survey is particularly useful in providing confirmatory evidence or details of damage identified using other survey techniques such as bathymetric and side-scan sonar surveys. Diving surveys were carried out at Humboldt and Crescent City to confirm and further investigate irregularities at the toe of structures, originally identified on sonographs of the structures.

Divers may also be used to identify locations of damaged or displaced armouring. A detailed underwater survey is, however, labour-intensive and may therefore be costly.

A typical survey procedure for the examination of a breakwater would be as follows. Three engineer divers form the diving team—a diver, a stand-by diver and a supervisor. This number of divers is required to meet safety standards. Additionally, a further two surveyors are required on land for position fixing. The diver often works from a boat, attached to it by a safety line/communication rope, and is also linked to a floating marker buoy with a separate line. Inspection of the armour slope is carried out by ascending and descending on a compass bearing at fixed chainage points marked at predetermined positions along the crest of the structure. The spacing of these points would typically be at 5 m centres, assuming that visibility is sufficient to see 2.5 m or more. On observation of a fault or point of interest on the structure (i.e. location of voids, broken armour or exposure of core, the diver maintains position and communicates to the shore party, who record the buoy position, held by the diver directly over the point of interest. Position fixing of the buoy may use two intersecting theodolite transits, theodolite transit and chainage, sextant angles or electronic position-fixing equipment. The level of the subject of interest can be measured by a helium pressure-depth gauge on the diver's wrist which can be measured to an accuracy of between 0.1–0.3 m. Alternatively, the line to the floating market buoy can be calibrated for direct depth reading. A tide graph and times of measurements are required to relate the depth to the datum used. This method should identify a position within 1 or 2 m radius in plan position and ±0.3 m in level. This should be sufficient to pinpoint an individual armour unit, allowing relocation at a later date.

Safety consideration are very important in diving inspections and all diving operations should be carried out in accordance with the appropriate local and national safety standards. Diving inspection methods have limitations, not least the range angle of underwater visibility. Such visibility is generally poor and reduces with depth and concentration of material in suspension. The angle of view often allows no more than a 2–3 m width of the structure to be viewed at any time, therefore large irregularities in the profile of the structure are not easily identified by divers. Observations may also be limited by weed on or around the armouring. Weathering conditions also impose restrictions on diving. In areas of high tidal ranges, diving surveys are best carried out at neap tides. These allow safer investigation of the intertidal zone, which is the potentially most vulnerable part of the structure.

A6.5 Acknowledgement

This appendix was adapted from a draft report by Bradbury and Allsop (1989).

Appendix A7:
European, British and Dutch legislation/authorities/designated sites relevant to environmental assessment of projects involving the use of rock in coastal and shoreline engineering

A7.1 Legislation/Guidance Documents

A7.1.1 EUROPEAN COMMUNITY (EC) DIRECTIVES

86/280 on limit values and quality objectives for discharges of certain dangerous substances included in list I of the Annex to Directive 76/464
85/337 on the assessment of certain public and private projects on the environment
84/491 on limit values and quality objectives for discharges of hexachlorocyclohexane
84/156 on limit values and quality objectives for mercury discharged by sectors other than the chlor-alkali electrolysis industry
83/513 on limit values and quality objectives for cadmium discharges
80/68 on the protection of groundwater against pollution caused by certain dangerous substances
79/923 on the quality required of shellfish waters
79/409 on the conservation of wild birds
78/176 on waste from the titanium dioxide industry
76/464 on pollution caused by certain dangerous substances discharged into the aquatic environment
76/160 concerning the quality of bathing water

A7.1.2 UNITED KINGDOM

In the UK and EC directive 85/337 is integrated into existing planning and related legislation by various statutory instruments. Exceptions to this general rule are as follows.

Environmental assessment for works authorised by private Act of Parliament

It has been recommended that the requirements of EC directive 85/337 be applied to developments which are the subject of Private Acts of Parliament and are therefore exempt from the requirements of the directive.

Environmental assessment for works authorised by the Crown

UK Crown development is not subject to the provisions of EC directive 85/337, although the Crown may choose to follow a 'shadow procedure' for planning its development—including environmental assessment where this would be required for non-Crown development of a similar type.

A7.1.2.1 Acts of Parliament

Food and Environment Protection Act 1985 (Part II)
Coast Protection Act 1949
Wildlife and Countryside Act 1981
Control of Pollution Act 1974
Water Act 1989
Town and Country Planning Acts 1971–1984
Harbours Act 1964

A7.1.2.2 Orders

The Town and Country Planning General Development Order 1977

A7.1.2.3 Statutory Instruments

Planning: 1988/1199 (England and Wales)
            1988/1221 (Scotland)
Other: 1988/1217
            1988/1336
Control of Substances Hazardous to Health Regulations 1988
            1988/1657

A7.1.2.4 Guidance documents

Environmental Assessment, a guide to the procedures, HMSO, 1989
Ramsar Sites (Wetlands of International Importance), HMSO Command Paper 6465, Treaty Series number 34 (1976)

UK Department of the Environment Circulars:
15/88 *Environmental Assessment*
27/87 *Nature Conservation*

Note: Comparable documents to the above are available for Wales and Scotland.

A7.1.3 THE NETHERLANDS

de Rivierenwet (River Act) Wet van 9 november 1908, Stb. 339, laatselijk gewijzigd bij Wet van 23 janurari 1984, Stb. 19.
de Wet op de Ruimtelijke Orderning (Planning Act) Wet van 5 juli 1962, Stb. 286, laatselijk gewijzigd bij Wet van 21 november 1985, Stb. 623, 624 en 625.
de Wet Algemene Bepalingen Milieuhygiene (General Environmental Terms Act) Wet van 13 juni 1979, Stb. 442, laatselijk gewijzigd bij Wet van 30 juni 1982, Stb. 456.
de Wet Bodembescherming (Benthos Protection Act) Wet van 3 juli 1986, Stb. 374.
de Wet Verontreiniging Oppervlaktewateren (Surface Water Pollution Act) Wet van 13 november 1969, Stb. 536, laatselijk gewijzigd bij Wet van 12 december 1985, Stb. 683.
de Wet Verontreiniging Zeewater (Sea Water Pollution Act) Wet van 5 juni 1975, Stb. 352, laatselijk gewijzigd bij Wet van 14 december 1983, Stb. 683.
de Afvalstoffenwet (Waste Products Act) Wet van 23 juni 1977, Stb. 455, laatselijk gewijzigd bij Wet van 3 juli 1986, Stb. 374.
de Wet Chemische Afvalstoffen (Chemical Waste Products Act) Wet van 11 februari 1976, Stb. 214, laatselijk gewijzigd bij Wet van 3 juli 1986, Stb. 374.

de Hinderwet (Nuisances Act) Wet van 15 mei 1952, Stb. 274, laatselijk gewijzigd bij Wet van 29 augustus 1985, Stb. 494.
de Wet Milieugevaarlijke Stoffen (Environmental Hazards Chemical Act) Wet van 5 december 1985, Stb. 639.
de Wet Gevaarlijke Stoffen (Hazardous Chemical Act) Wet van 20 juni 1963, Stb. 313, laatselijk gewijzigd bij Wet van 21 maart 1979, Stb. 202 en 328.
de Natuurbeschermingswet (Nature Conservation Act) Wet van 15 november 1967, Stb. 572.
de Ontgrondingenwet (Mining [excavation] and Dredging Act) Wet van oktober 1965, Stb. 509, laatselijk gewijzigd bit Wet van 23 juli 1976, Stb. 377.

A7.2 Statutory Consultees (UK)

United Kingdom statutory consultees comprise:

1. The Nature Conservancy Council (NCC) (projects affecting SSSIs require negotiation with the NCC under the terms of the Wildlife and Countryside Act 1981);
2. The Countryside Commission

and, if the proposed works are likely to impinge on their areas interest:

3. The Health and Safety Executive (HSE);
4. HM Inspectorate of Pollution (HMIP);
5. The Ministry of Agriculture, Fisheries and Food (MAFF) (DAFS in Scotland);
6. The affected Coast Protection Authority;
7. The National Rivers Authority (NRA) (River Purification Boards in Scotland)

Shoreline works may require a licence for placing material on the seabed or approval for navigation. In such cases in the UK, it would be obligatory to consult with the following:

1. Crown Estate Commissioners (CEC)—responsible for the administration of the seabed; and/or
2. MAFF.

These bodies may also require an environmental assessment as part of the application for their consent, licence or approval.

A7.3 Sensitive Locations (UK)

Projects which impinge on the following particular sensitive locations will require special attention:

1. Wetlands of International Importance:
   (a) Special Protection Area (SPA) (Conservation of Wild Birds EC/79/409);
   (b) Ramsar Site (sites designated under the terms of the Convention on Wetlands of International Importance, 1971);
2. National Park/Nature Reserve;
3. Site of major archaeological/historical interest;
4. Site of Special Scientific Interest (SSSI). Geological Sites of Special Scientific Interest on the coast are often noted for the fresh exposures caused by continuing erosion. In such cases the NCC will have an interest in preserving some rate of coastal erosion, contrary to the engineering imperative of stopping it;
5. Area of Outstanding Natural Beauty (AONB).

de Hinderwet (Nuisances Act) Wet van 15 mei 1952, Stb. 274, laatselijk gewijzigd bij Wet van 29 augustus 1985, Stb. 494.

de Wet Milieugevaarlijke Stoffen (Environmental Hazards Chemical Act) Wet van 5 december 1985, Stb. 639.

de Wet Gevaarlijke Stoffen (Hazardous Chemical Act) Wet van 20 juni 1963, Stb. 313, laatselijk gewijzigd bij Wet van 21 maart 1979, Stb. 202 en 328.

de Natuurbeschermingswet (Nature Conservation Act) Wet van 15 november 1967, Stb. 572.

de Ontgrondingenwet (Mining [excavation] and Dredging Act) Wet van oktober 1965, Stb. 509, laatselijk gewijzigd bit Wet van 23 juli 1976, Stb. 377.

A7.2 Statutory Consultees (UK)

United Kingdom statutory consultees comprise:

1. The Nature Conservancy Council (NCC) (projects affecting SSSIs require negotiation with the NCC under the terms of the Wildlife and Countryside Act 1981);
2. The Countryside Commission

and, if the proposed works are likely to impinge on their areas interest:

3. The Health and Safety Executive (HSE);
4. HM Inspectorate of Pollution (HMIP);
5. The Ministry of Agriculture, Fisheries and Food (MAFF) (DAFS in Scotland);
6. The affected Coast Protection Authority;
7. The National Rivers Authority (NRA) (River Purification Boards in Scotland)

Shoreline works may require a licence for placing material on the seabed or approval for navigation. In such cases in the UK, it would be obligatory to consult with the following:

1. Crown Estate Commissioners (CEC)—responsible for the administration of the seabed; and/or
2. MAFF.

These bodies may also require an environmental assessment as part of the application for their consent, licence or approval.

A7.3 Sensitive Locations (UK)

Projects which impinge on the following particular sensitive locations will require special attention:

1. Wetlands of International Importance:
   (a) Special Protection Area (SPA) (Conservation of Wild Birds EC/79/409);
   (b) Ramsar Site (sites designated under the terms of the Convention on Wetlands of International Importance, 1971);
2. National Park/Nature Reserve;
3. Site of major archaeological/historical interest;
4. Site of Special Scientific Interest (SSSI). Geological Sites of Special Scientific Interest on the coast are often noted for the fresh exposures caused by continuing erosion. In such cases the NCC will have an interest in preserving some rate of coastal erosion, contrary to the engineering imperative of stopping it;
5. Area of Outstanding Natural Beauty (AONB).

References

ABBOTT, M.B. (1979)
Computational Hydraulics: elements of the theory of free surface flows
Pitman, London

ABRAHAM, G., KARELSE, M. and VAN OS, A.G. (1979)
The magnitude of interfacial shear of sub-critical stratified flows in relation to interfacial stability

AHRENS, J.P. (1981)
Irregular wave run-up on smooth slopes. Technical Paper No. 81-17
US Army Corps of Engineers, Coastal Engng Res. Center, Fort Belvoir

AHRENS, J.P. (1987)
Characteristics of reef breakwaters. Technical Report CERC-87-17
US Army Corps of Engineers, Coastal Engng Res. Center, Vicksburg

ALLISON, D.M. and SAVAGE, R.P. (1976)
Tests of low density marine limestone for use in breakwaters. Tech. Paper No. 76-4
US Army Corps of Engineers, Coastal Engng Res. Center

ALLSOP, N.W.H. (1990)
Rock armouring for coastal and shoreline structures: hydraulic model studies on the effects of armour grading. Report EX 1989
Hydraulics Research Ltd, Wallingford

ALLSOP, N.W.H., BRADBURY, A.P., POOLE, A.B., DIBB, T.E. and HUGHES, D.W. (1985a)
Rock durability in a marine environment. Report No. SR11
Hydraulics Research Ltd, Wallingford

ALLSOP, N.W.H. and CHANNELL, A.R. (1989)
Wave reflections in harbours: reflection performance of rock armoured slopes in random waves. Report OD 102
Hydraulics Research Ltd, Wallingford

ALLSOP, N.W.H., HAWKES, P.J., JACKSON, F.A. and FRANCO, L. (1985b)
Waves run-up on steep slopes—model tests under random waves. Report SR 2
Hydraulics Research Ltd, Wallingford

ALLSOP, N.W.H., SMALLMAN, J.V. and STEPHENS, R.V. (1988)
Development and application of a mathematical model of wave action on steep slopes
Proc. 21st Int. Coastal Engng Conf.
Am. Soc. Civ. Engrs

AMERICAN SOCIETY FOR TESTING MATERIALS (1959)
Specific gravity and absorption of fine aggregate
ASTM Test Designation C127

AMERICAN SOCIETY FOR TESTING MATERIALS (1963)
Abrasion of graded coarse aggregate by use of the Deval machine
ASTM Test Designation C289

AMERICAN SOCIETY FOR TESTING MATERIALS (1963)
Soundness of aggregates by use of sodium sulphate or magnesium sulphate
ASTM Test Designation C88

AMERICAN SOCIETY FOR TESTING MATERIALS (1965a)
Resistance to abrasion of large size coarse aggregate by use of the Los Angeles machine
ASTM Test Designation C535

AMERICAN SOCIETY FOR TESTING MATERIALS (1965b)
Petrographic examination of aggregates for concrete
ASTM Test Designation C295

AMERICAN SOCIETY FOR TESTING MATERIALS (1971)
Unconfirmed compressive strength of rock core specimens
ASTM Test Designation D2938

AMERICAN SOCIETY FOR TESTING MATERIALS (1981)
Abrasion resistance of concrete by sandblasting
ASTM Test Designation C418

AMERICAN SOCIETY FOR TESTING MATERIALS (1984)
Resistance of concrete to rapid freezing and thawing
ASTM Test Designation C666

AMINTHI, P. and FRANCO, L. (1988)
Wave overtopping on rubble mound breakwaters
Proc. 21st Int. Coastal Engng Conf.
Am. Soc. Civ. Engrs

ANGEREMOND, K.D. (1970)
Use of asphalt in breakwater construction
Coastal Engineering Conference, Washington DC

ASCE (1984)
Proc. Conf. Dredging and Dredged Materials Disposal
Am. Soc. Civ. Engrs, Florida

ASTM (1987)
Annual Book of ASTM Standards, Vol. 04.08, *Soil and Rock; Building stones; Geotextiles*
Philadelphia
ATKINSON, B.K. (Ed.) (1987)
Fracture Mechanics of Rock
Academic Press, London
BAIRD, W.F. and WOODROW, K.W. (1988)
The development of a design for a breakwater at Keflavik, Iceland
In *Design of Breakwaters*
Thomas Telford, London
BARBER, P.C. and DAVIES, P.C. (1985)
Offshore breakwaters—Leasowe Bay
Proc. Instn Civ. Engrs, **77,** February 85–109
BARENDS, F.B.J. et al. (1983)
Westbreakwater—sines—dynamic—geotechnical stability of breakwaters
Proc. Conf. Coastal Structures
Arlington
BARENDS, F.B.J. (1991)
A general review: Interaction between ocean waves and sea bed
PHR1, Geo-Coast '91, Yokohama
BARENDS, F.B.J. and CALLE, E.O.F. (1985)
A method to evaluate the geotechnical stability of offshore structures founded on a loosely packed seabed sand under a wave loading environment
BOSS 85, Delft
BARRETT, P.J. (1980)
The shape of rock particles, a critical review
Sedimentology, **27,** 291–303
BARTON, N. and KJAERNSLI, N. (1981)
Shear strength of rockfill
Proc. ASCE, Geot. Eng. Divn, **107,** No. GT7
BATTJES, J.A. (1974) *Computation of set-up, longshore currents, run-up and overtopping due to wind-generated waves*
Comm. on Hydraulics, Report 74-2
Dept of Civil Engineering, Delft University of Technology
BATTJES, J.A. and JANSSEN, J.P.F.M. (1978)
Energy loss and set-up due to breaking of random waves
Proc. 16th Int. Coastal Engng Conf.
Am. Soc. Civ. Engrs
BEIL, N.J. and SORENSEN, R.M. (1988)
Perched beach profile response to wave action
Proc. 21st Int. Coastal Engng Conf., 1482–92,
Am. Soc. Civ. Engrs
BELL, F.G. (1983)
Engineering Properties of Soils and Rocks,
2nd edition
Butterworths, London
BENJAMIN, J.R. and CORNELL, C.A. (1969) *Probability Statistics and Decision for Civil Engineers*
McGraw-Hall, New York
BERENGUER, J.M. and ENRIQUEZ, J. (1988)
Design of pocket beaches: the Spanish case
Proc. 21st Int. Coastal Engng Conf., 1843–57
Am. Soc. Civ. Engrs
BIJKER, E.W. (1967)
Some considerations about scales for coastal models with movable beds. Publication 50
Delft Hydraulics Laboratory
BISS, M.A. and CRAIG, C.W. (1987)
Clacton sea defences
Ministry of Agriculture Fisheries and Food Conference of River and Coastal Engineers, Loughborough
BLACKMAN, D.L. (1985)
New estimates of annual sea level maxima in the Bristol Channel
Estuarine, Coastal and Shelf Science, **20,** 229–32
BOER, S. and HULSBERGEN, C.H. (1989)
Thermal aspects of trenching, burial and covering of hot submarine pipelines
Offshore Mechanics and Arctic Engng Conf., The Hague
BOER, S., HULSBERGEN, C.H., RICHARDS, D.M., KLOK, A. and BIAGGI, J.P. (1986)
Buckling consideration in the design of the gravel cover for a high-temperature oil line
Offshore Technology Conf., Houston, Paper no. OTC 5294
BOND, F.C. (1959)
Confirmation of the third theory
Am. Inst. Mining Metall. Engineers, San Francisco Annual Meeting
BOND, F.C. and WHITNEY, D.D. (1959)
The work index in blasting
3rd Symp. on Rock Mechanics, Quarterly of Colorado School of Mines, **54,** No. 3, July
BOODT, C. (1981)
Bescherming van zinker tegen ankers, Onderzoek en beproeving van constructies
Polytechnisch Tijdschrift, Bouwkunde, Wegen-en waterbouw, nr 5

BOUWMEESTER, J., KAA, E.J. van der, NUHOFF, H.A. and ORDEN, R.G.J. van (1977)
Recent studies on push-towing as a base for dimensioning waterways. Publication 194
Delft Hydraulics Laboratory, November
BOWEN, A.J. (1969)
The generation of longshore currents on a plane beach
J. of Marine Research, **27**
BRADBURY, A.P., ALLSOP, N.W.H. and STEPHENS, R.V. (1988)
Hydraulic performance of breakwater crown wall. Report SR 146
Hydraulics Research Ltd, Wallingford
BRADBURY, A.P., ALLSOP, N.W.H., LATHAM, J.-P., MANNION, M.B. and POOLE, A.B. (1988)
Rock armour for rubble mound breakwaters, sea walls and revetments: recent progress. Report SR 150
Hydraulics Research Ltd, Wallingford (published in collaboration with Queen Mary College, London)
BRADBURY, A.P. and ALLSOP, N.W.H. (1989)
Monitoring techniques for armoured coastal structures. Report IT 343
Hydraulics Research Ltd, Wallingford
BRADBURY, A.P., LATHAM, J.-P. and ALLSOP, N.W.H. (1990)
Rock armour stability formulae—influence of stone shape and layer thickness
Proc. 22nd Int. Coastal Engng Conf.
Am. Soc. Civ. Engrs
BREAKWATERS (1983) *Design and construction*
Thomas Telford, London
BREAKWATERS (1985) *Development in breakwaters*
Thomas Telford, London
BREAKWATERS (1988) *Design of breakwaters*
Thomas Telford, London
BREBNER, A. and DONNELLY, P. (1962)
Laboratory study of rubble foundations for vertical breakwater. Engineer Report No. 23
Queen's University, Kingston, Ontario, Canada
BREBNER, B. and DONNELLY, P. (1962)
Laboratory study of rubble foundations for vertical breakwaters
Proc. 8th Int. Coastal Engng Conf.
Am. Soc. Civ. Engrs
BRETSCHNEIDER, C.L. (1954)
Generation of wind waves over a shallow bottom. Technical Memorandum No. 51
Beach Erosion Board, Office of the Chief of Engineers
BRETSCHNEIDER, C.L. and REID, R.O. (1954)
Changes in wave height due to bottom friction, percolation and refraction. Technical Memorandum No. 45
Beach Erosion Board, US Army Corps of Engineers
BREUSERS, H.N.C. and SCHUKKING, W.H.P. (1971)
Incipient motion of bed material (begin van beweging van bodemmateriaal). Speurwerkverslag S 159-1
Delft Hydraulics Laboratory
BRITISH STANDARDS INSTITUTION (1981)
Code of Practice for site investigations
BS 5930
BRITISH STANDARDS INSTITUTION (1951, 1975)
Sampling and testing of mineral aggregates, sands and fillers
BS 812
BRITISH STANDARDS INSTITUTION (1981)
Code of Practice for site investigations
BS 5930 (formerly CP 2001)
BRITISH STANDARDS INSTITUTION (1989)
Testing aggregates: Part 121, *Method for determination of soundness*
BS 812
BRITISH STANDARDS INSTITUTION (1984)
Code of Practice for maritime structures: Part 1 *General criteria*
BS 6349
BROCARD, D.N. and HARLEMAN, D.R.F. (1980)
Two layer model for shallow horizontal convective circulation
J. of Fluid Mechanics, **100**, Part 1
BROOK, N. (1985)
The equivalent core diameter size and shape correction in point load testing
Int. J. Rock Mech. Min. Sci. and Geotech Abst., **22**, No. 2, 61–70
BROWN, E.T. (1981)
Rock Characterization Testing and Monitoring—ISRM Suggested Methods
Pergamon Press, Oxford
BRUUN, P. (1985)
Design and Construction of Mounds for Breakwaters and Coastal Protection
Elsevier, Amsterdam
BURCHARTH, H.F. and FRIGAARD, P. (1987)
On the stability of berm breakwater roundheads and trunk erosion in oblique waves
Seminar on Unconventional Rubble-mound Breakwaters, Ottawa
BURCHARTH, H.F. and RIETVELD, C.F.W. (1987)
Construction, maintenance and repair as elements in rubble mound breakwater design
2nd Int. Conf. on Coastal and Port Engng in Developing Countries, Beijing

BURCHARTH, H.F. and THOMPSON, A.G. (1983)
Stability of armour units in oscillating flow
Proc. Coastal Structures '83 Conf.,
Am. Soc. Civ. Engrs
CARLYLE, W.J. (1987)
Wave damage to upstream slope protection of reservoir in the UK
British National Committee on Large Dams Conference
CARMICHAEL, R.S. (1982)
Handbook of Physical Properties of Rocks, Vols I–III
CRC Press, New York
CELIK, I. and RODI, W. (1985)
Calculation of wave induced turbulent flows in estuaries
Ocean Engineering, **12**
CHADWICK, A.J. (1989)
Field measurements and numerical model verification of coastal shingle transport
In M.H. Palmer (Ed.), *Advances in Water Modelling and Measurement*
BHRA
CIAD (1985)
Computer aided evaluation of the reliability of a breakwater design. Report of the CIAD project group on breakwaters
CIAD Zoetermeer, Netherlands
CIRIA (1986)
Seawalls, survey of performance and design practice. Technical Note 128
CIRIA, London
CLARK, A.R. (1988)
The use of Portland stone rock armour in coastal protection and sea defence works
Q.J. Engng Geol., London, **21**, 13–136
COLE, W.F. and SANDY, M.J. (1980)
A proposed secondary mineral rating for basalt road aggregate durability
Aust. Road Rest., **10**, No. 3, 27–37
COLLIS, L. and FOX, R.A. (Eds) (1985)
Aggregates: sand, gravel and crushed rock aggregates for construction purposes. Engng Geol. Sp. Pubn. No. 1
Geological Society, London
COST, R. (1985)
Fracture toughness and rock durability
Memoirs of the Centre for Engineering Geology in The Netherlands, **30**
CUNNINGHAM, C. (1963)
The Kuz-Ram model for prediction of fragmentation from blasting
1st Symp. on Rock Fragmentation by Blasting, August, 439–53, Lulea, Sweden
CUNNINGHAM, C.V.B. (1987)
Fragmentation estimations and the Kuz-Ram model—four years on
2nd Int. Symp. on Rock Fragmentation by Blasting, 475–87, Colorado
CUR (1987)
Manual on Artifical Beach Nourishment. Publication No. 130
Gouda, The Netherlands
DAEMRICH, K.F. and KAHLE, W. (1985)
Shutzwirkung von Unterwasserwellen brechern unter dem einfluss unregelmassiger Seegangswellen
Eigenverlag des Franzius-Instituts für Wasserbau und Küsteningenieurswesen, Heft 61
DA GAMA, C.D. (1983)
Use of comminution theory to predict fragmentation of jointed rock masses subjected to blasting
1st Int. Symp. on Rock Fragmentation by Blasting, 565–79, Lulea, Sweden
DEAN, R.G. (1986)
Coastal armouring: effects, principles and mitigation
Proc. 20th Int. Coastal Engng Conf., 1843–57, Am. Soc. Civ. Engrs
DEAN, R.G. (1988)
Evaluation of shore protection structures (including beach nourishment)
In Short Course on Planning and Designing Maritime Structures, Malaga, Spain, June
DEARMAN, W.F., TURK, N., IRFAN, Y. and ROWSHANEI, H. (1987)
Detection or rock material variation by sonic velocity zoning
Bulletin of the International Association of Engineering Geology, **35**, 1–8
DELFT HYDRAULICS-M2006 (1986)
Taluds van losgestorte materialen. Stabiliteit van stortsteen-bermen en teenkonstrukties. Verslag literatuurstudie en modelonderzoek. (Slopes of loose materials. Stability of rubble mound berm and toe structures. Report on literature and model investigation.) (in Dutch)
DELFT HYDRAULICS-M1983 (1987)
Taluds van losgestorte materialen. Statische stabiliteit van stortsteen taluds onder golfaanval. Ontwerp formules. Verslag modelonderzoek, deel I. (Slopes of loose materials. Static stability of rubble mound slopes undes wave attack. Design formula. Report on model investigation, Part I.) (In Dutch)
DELFT HYDRAULICS-M1983 (1987)
Taluds van losgestorte materialen. Dynamische stabiliteit van grind- en stortsteen taluds onder golfaanval. Model voor profielvorming. Verslag modelonderzoek, deel II. (Slopes of loose materials. Dynamic stability of gravel beaches and rubble mound slopes under wave attack. Model for profile formation. Report on model investigation, Part II.) (in Dutch)

DELFT HYDRAULICS-M1983 (1988)
Rock slopes, static stability of rock slopes under wave attack, Part I
DELFT HYDRAULICS-M1983 (1989)
Taluds van losgestorte materialen. Golfoploop of statisch stabiele stortsteen taluds onder golfaanval. Verslag modelonderzoek, deel III. (Slopes of loose materials. Wave run-up on statically stable rock slopes under wave attack. Report on model investigation, Part III.) (in Dutch)
DELFT UNIVERSITY PRESS (1987)
The Closure of Tidal Basins
The Netherlands
DEPARTMENT OF MAIN ROADS (1974)
Test method T214—Washington Degradation Test
Manual of Testing Procedures, New South Wales
DEPARTMENT OF THE ENVIRONMENT/WELSH OFFICE (1988)
Guidelines for aggregates provision in England and Wales
HMSO, London
DIBB, T.E., HUGHES, D.W. and POOLE, A.B. (1983)
Controls of size and shape of natural armourstone
Q. Jnl Engng Geol.
DIBB, T.E., HUGHES, D.W. and POOLE, A.B. (1983)
The identification of critical factors affecting rock durability in marine environments
Q. Jnl Engng Geol., **16**
DINGLER, J.R. and ANIMA, R.J. (1984)
Surveys of coastal structures using geophysical techniques
Proc. 19th Int. Coastal Engng. Conf.
Am. Soc. Civ. Engrs
DITLEVSEN, O. (1979)
Narrow reliability bounds for structural systems
Jnl Struct. Mechanics, **7**, No. 4
DE QUELERIJ, L. and VAN HIJUM, E. (1990)
Maintenance and monitoring of water retaining structures
In Pilarczyk, K.W. (ed.), *Coastal Protection*, 369–401
Balkema, Rotterdam
DRAPER, L. (1963)
Derivation of a 'design wave' from instrumental records of sea waves
Proc. Instn Civ. Engrs, **26**, 201–303
DUNCAN, N. (1969)
Engineering Geology and Rock Mechanics. Vol. 1
International Textbook Co., London
EHRLICH, R. and WEINBERG, B. (1970)
An exact method for characterising grain shape
Jnl of Sedimentary Petrology, **40**, 205–12
FIESSLER, B. (1979)
The program FORM on computation of failure probabilities of system components. Research Report 43
(in German)
Technical University, Munich
FISCHHOFF, B., LICHTENSTEIN, S., SLOVIC, P., DARBY, S.L. and KEENEY, R.L. (1983)
Acceptable Risk
Cambridge University Press, Cambridge
FLEMING, C.A. (1990a)
Guide on the use of groynes in coastal engineering. CIRIA Report 119
London
FLEMING, C.A. (1990b)
Principles and effectiveness of groynes
In K.W. Pilarczyk (Ed.), *Coastal Protection*, 121–56
Balkema, Rotterdam
FOOKES, P.G., DEARMAN, W.R. and FRANKLIN, J.A. (1971)
Some engineering aspects of rock weathering with field examples from Dartmoor and elsewhere
Q. Jnl Engng Geol., **4**, 139–85
FOOKES, P.G. (1980)
An introduction to the influence of natural aggregates on the performance and durability of concrete
Q. Jnl Engng Geol., **13**, 207–29
FOOKES, P.G. and POOLE, A.B. (1981)
Some preliminary consideration on the selection and durability of rock and concrete materials for breakwaters and coastal protection works
Q. Jnl Engng Geol., **14**, 97–128
FOOKES, P.G.and THOMAS, R.S. (1986)
Rapid site appraisal of potential breakwater rock at Qeshm, Iran
Proc. Instn Civ. Engrs, **80**, October, 1297–1325
FOOKES, P.G., GOURLEY, C.S. and OHIKERE, C. (1988)
Rock weathering in engineering time
Q. Jnl Engng Geol., **21**, 33–57
FRANKLIN, J.A. (1970)
Observations and tests for engineering description and mapping of rocks
Proc. 2nd Congress Int. Soc. for Rock Mech, Belgrade **1**, Paper 1–3

GEISE, J.M. and KOLK, H.J. (1983)
The use of submersibles for geotechnical investigations
Proc. Subtech '83, London, Paper 7.2, 13
GRADY, D.E. and KIPP, M.E. (1987)
Dynamic rock fragmentation
In B.K. Atkinson (Ed.), *Fracture Mechanics of Rock*
Academic Press, London
GRAFF, J. (1981)
An investigation of the frequency of distributions of annual sea level maxima at ports around Great Britain
Estuarine, Coastal and Shelf Science, **12**, 389–449
GEOLOGICAL SOCIETY OF LONDON (1977)
The description of rock masses for engineering purposes. Working Party Report
Q. Jnl Engng Geol., **10**, 355–88
GEOTEXTILE (1986)
IIIrd Int. Conf. on Geotextiles, Vienna
GERBERT, J.A. and CLAUSENER, J. (1984)
Photogrammetric monitoring of dolos stability, Manasquan inlet N.J.
Proc. 19th Int. Coastal Engng Conf., Am. Soc. Civ. Engrs
GIVLER, L.D. and SORENSEN, R.M. (1986)
An investigation of the stability of submerged homogeneous rubble-mound structures under wave attack. Report IHL-110-86
Lehigh University, H.R. IMBT Hydraulics
GODA, Y. (1974)
A new method of wave pressure calculation for the design of composite breakwaters
Proc. 14th Int. Coastal Engng Conf.
Am. Soc. Civ. Engrs
GODA, Y. (1976)
On Wave Groups
Proc. BOSS'76, Trondheim
GODA, Y. (1985)
Random Seas and Design of Maritime Structures
Tokyo
GRAVESEN, H. and SORENSEN, T. (1977)
Stability of rubble mound breakwaters
Proc. 24th Int. Navigation Congress
GROEN, P. and DORRESTEIN, R. (1976)
Sea Waves (in Dutch)
The Hague
GUNSALLUS, K.L. and KULHAWY, F.H. (1984)
A comparative evaluation off rock strength measures
Int. J. Rock Mech. Min. Sci. & Geomech. Absts, **21**, No. 5, 233–48
HALES, L.Z. (1985)
Erosion control of scour during construction, summary report. Technical Report HL-80-3, 8
Waterways Experiment Station, Vicksburg
HAMMOND, T.M. and COLLINS, M.B. (1979)
On the threshold of transport of sand-sized sediment under the combined influence of unidirectional and oscillatory flow
Sedimentology, **26**, 795–812
HANNOURA, A. and BARENDS, F.B.J. (1981)
Non-Darcy flow; a state-of-the-art
Euromech:143, Flow and Transport in Porous Media, Balkema, Rotterdam
HASSELMANN, K. et al. (1973)
Measurements of wind wave growth and swell decay during the joint North Sea wave project
Deutsche Hydr. Zeit, **12**
HAWKINS, A.B. et al. (Eds)
Site investigation practice
Proc. 20th Reg. Mtg. Engng Group Geol. Soc.
HEAD, K. (1980)
Manual of Laboratory Testing (3 vols)
HEMPHILL, R.W. and BRAMLEY, M.E. (1989)
Protection of River and Canal Banks
CIRIA/Butterworths, London
van HERPEN, J.A. (1990)
Asphalt mixtures for revetments of water defences and embankments In K.W. Pilarczyk (Ed.), *Coastal Protection*, 327–67
Balkema, Rotterdam
HERBICH, J. et al. (1984) *Seafloor Scour*
Marcel Dekker, New York
HERBICH, J.B. and KO, S.C. (1986)
A scour of sand beaches in front of seawalls
Proc. 11th Int. Coastal Eng. Conf.
Am. Soc. Civ. Engrs
HEUZÉ, F.E. (1990)
An overview of projectile penetration into geological materials, with emphasis on rocks
Int. Jnl of Rock Mechanics Min. Sci. & Geomech. Abstr., **27**, No. 1, 1–14

HIGGS, N.B. (1986)
Preliminary studies of Methylene Blue absorption as a method of evaluating degradable smectite-bearing concrete aggregate sands
Cement and Concrete Research, **16,** 525–534

HM STATIONERY OFFICE (1975)
The Reservoir Act
London

HOBBS, R.E. (1984)
In-service buckling of heated pipelines
Jnl of Transportation Engineering, **110,** No. 2, 175–89

HOGBEN, N., DACUNHA, N.M.C. and OLLIVER, G.F. (1986)
Global Wave Statistics
British Maritime Technology, Teddington

HOSKING, J.R. and TUBEY, L.W. (1969)
Research on low grade and unsound aggregates. Road Research Laboratory Report LR 293
Crowthorne

HUDSON, J.A. and PRIEST, S.D. (1979)
Discontinuities and rock mass geometry
Int. Jnl Rock Mech. Min. Sci. & Geomech. Abstr., **16,** 339–62

HUGHES, S.A. (1984)
The T.M.A. shallow water spectrum descriptions and applications. Techn. Report 84-7
US Army Corps of Engineers

HUIS in't VELD, J. et al. (Eds) (1984)
The Closure of Tidal Basins
Delft University Press, The Netherlands

IAHR/PIANC (1986)
List of sea state parameters
Supplement to PIANC Bulletin No. 52, January

INTERIM COUNCIL OF THE EXPLORATION OF SEA (ICES) (1980)
Interaction between the fishing industry and the offshore gas/oil industries. Co-operative Research Report, No. 94
Copenhagen, Denmark

INSTITUTION OF CIVIL ENGINEERS (1978)
Floods and Reservoir Safety

IPPEN, A.T. (1986)
Estuary and Coastline Hydrodynamics
McGraw-Hill, New York

IRFAN, T.Y. and DEARMAN, W.R. (1978)
The engineering petrography of a weathered granite in Cornwall, England
Q. Jnl Engng Geol., **11,** 233–44

ISRM (1985)
Commission on Testing Methods. Suggested method for determining Point Load Strength (revised version)
Int. Jnl Rock Mech. Min. Sci. & Geomech. Abstr., **22,** 51–60

ISRM (1988)
Commission on Testing Methods. Suggested method for determining the fracture toughness of rock
Int. Jnl Rock Mech. Min. Sci. & Geotech. Abstr., **25,** No. 2, 71–96

JAMES, W.R. (1975)
Techniques in evaluating suitability of borrow material for beach nourishment. Report TM60
US Army Corps of Engineers, Coastal Engng Res. Center, Vicksburg

JENNINGS, J.N. (1985)
Karst Geomorphology
Blackwell, Oxford

JENSEN, O.J. (1984)
A monograph on rubble mound breakwaters
Danish Hydraulic Institute

JONSSON, I.G. (1966)
Wave boundary layers and friction factors
Proc. 10th Int. Coastal Engng Conf., 127–48
Am. Soc. Civ. Engrs

KAMPHUIS, J.W. (1990)
Sediment transport rate
Proc. 22nd Int. Coastal Engng Conf.
Am. Soc. Civ. Engrs

KAO, J.S. and HALL, K.R. (1990)
Trends in stability of dynamically stable (berm) breakwaters
Proc. 22nd Int. Coastal Engng Conf.
Am. Soc. Civ. Engrs

KENNEY, T.C. et al. (1985)
Internal stability of granular filters
Canadian Geotech. Jnl, **22,** 215–25

KEVELING-BUISMAN, A.S. (1941)
Soil Mechanics (in Dutch)
Klopper PC

KIMURA, A. (1980)
Statistical Properties of Random Wave Groups
Proc. 17th Int. Conf. on Coastal Eng., Sydney

KIRKGÖZ, M.S. (1986)
Particle velocity prediction at the transformation point of plunging breakers
Coastal Engng, **10**, 139–47
KLUGER, J.W. (1982)
Monitoring of rubble-mound breakwater stability using a photographic survey method
Proc. 18th Int. Coastal Engng Conf.
Am. Soc. Civ. Engrs
KLUGER, J.W. (1988)
Monitoring of 9t dolos test section on Table Bay harbour breakwater, November 1986 to July 1987.
CSIR Report EMA-T 8810
EMA, Stellenbosch
KOBAYASHI, M. *et al.* (1987)
Bearing capacity of a rubble mound supporting gravity structure. Report 26(5)
Port and Harbour Research Institute, Ministry of Transport, Nagase, Yokosuka
KOBAYASHI, N., OTTA, A.K. and ROY, I. (1987)
Wave reflection and run-up on rough slopes
Jnl Waterway, Port, Coastal and Ocean Engineering, Proc. Am. Soc. Civ. Engrs, **113**, No. WW3, May
KOEMER, R.M. (1986)
Designing with Geosynthetics
Prentice-Hall, Englewood Cliffs, NJ
KOMAR, P.D. (1969)
The Longshore Transport of Sand on Beaches
University of California, San Diego
KOMAR, P.D. and MILLER, M.C. (1974)
Sediment transport threshold under oscillatory waves
Proc. 14th Int. Coastal Conf., 756–75
Am. Soc. Civ. Engrs
KOSTER, J. (1974)
Digging in of anchors into the bottom of the North Sea, Publication No. 129
Delft Hydraulics Laboratory, The Netherlands
KRAUS, N.C. and PICKLEY, O.H. (Eds) (1988)
The effects of seawalls on the beach
Jnl Coastal Res., Special Issue No. 4
LAAN, G.J. (1981)
The relation between shape and weight of pieces of rock. Report MAW-R-81079 (in Dutch)
Rijkswaterstaat, Delft, The Netherlands
LAMBE, T.W. and WHITMAN, R.V. (Eds) (1964)
Soil Mechanics
Massachusetts Institute of Technology/John Wiley, New York
LATHAM, J.-P. and POOLE, A.B. (1987)
Pilot study of an aggregate abrasion test for breakwater armourstone
Q. Jnl Engng Geol., **20**, 311–16
LATHAM, J.-P. and POOLE, A.B. (1987)
The application of shape descriptor analysis to the study of aggregate wear
Q. Jnl Engng Geol., **20**, 297–310
LATHAM, J.-P., MANNION, M.B., POOLE, A.B., BRADBURY, A.P. and ALLSOP, N.W.H. (1988)
The influence of armourstone shape and rounding on the stability of breakwater armour layers. Report No. 1
Coastal Engineering Research Group, Queen Mary College, University of London, and Hydraulics Research Ltd, Wallingford
LATHAM, J.-P. and POOLE, A.B. (1988)
Abrasion testing and armourstone degradation
Coastal Engng, **12**, 233–55
LATHAM, J.-P., POOLE, A.B., LAAN, G.J. and VERHOEFF, P.N.W. (1980)
Geological constraints on quarried rock for use in coastal structures
Proc. 6th Congr. Int. Assoc. Engng Geol., Amsterdam
LATHAM, J.-P. (in press 1991)
Degradation model for rock armour in coastal engineering
Proc. Conf. Geol. Materials in Construction, Geol. Soc. of London, November 1989: *Q. Jnl Engng Geol.*, **24**, 91–99
LCPC (Laboratoire Central des Ponts et Chaussées) (1989)
Les Enrochments. Rapport de la Ministère de l'equipment, du logement, des transports et de la mer
LCPC (in French)
LEE, S.G. and DE FREITAS, M.H. (1989)
A revision of the description and classification of weathered granite and its application to granites in Korea
Q. Jnl Engng Geol., **22**, 31–48
LIENHART, D.A. and STRANSKY, T.E. (1981)
Evaluation of potential sources of riprap and armourstone—methods and considerations
Bulletin Assoc. Engng Geol., **18**, No. 3, 323–32
MADSEN, O.S. and GRANT, W.D. (1975)
The threshold of sedimentary movement under oscillatory water waves: a discussion
Jnl Sed. Petrology, **45**, No. 1

MADSEN, O.S. and GRANT, W.D. (1976)
Sediment threshold in the coastal environment. Report 209
Department of Civil Engineering, Massachusetts Institute of Technology, Cambridge, Massachusetts
McCAMY, R.C. and FUCHS, R.A. (1954)
Wave forces on piles: a diffraction theory. Tech. Memo No. 69
US Army Corps of Engng, Beach Erosion Board
MEI, C.C. (1983)
The Applied Dynamics of Ocean Surface Waves
Wiley-Interscience, New York
MEREDITH, P.G. (1988)
Comparative fracture toughness testing of rock
Proc. Int. Workshop on Fracture Toughness and Fracture Energy, Tohoku University, Sendai, Japan, 211–23
MORFETT, J.C. (1988)
Modelling shingle beach evolution
Proc. Symp. Mathematical Modelling of Sediment Transport Coastal Zone
IAHR, Copenhagen
NALE, D.K. (1983)
Photogrammetic mapping and monitoring of the Manasquan inlet and dolosse
Proc. ACSM-ASP Fall Convention, Salt Lake City
NIELSEN, N.J.R., PEDERSEN, P.T., GRUNDY, A.K. and LYNBERG, B.S. (1988)
New design criteria for upheaval creep of buried sub-sea pipelines
Offshore Mech. and Arctic Engng Conf., Houston, Paper No. OMAE-88-861
NIESE, M.S.T., VAN EIJK, F.C.A.A., LAAN, G.J. and VERHOEF, P.N.W. (in press)
Quality assessment of large armourstone using an acoustic velocity analysis method
NORDENSTROM, N. (1971)
Methods for predicting long term distributions of wave loads and possibility failure of ships. DNV Report 71-2-5, Part I
Oslo
NORME, N.F.P. (1980)
Determination of the continuity index. Editée par l'Association Française Normalisation, 18-556 AFNOR
OWEN, M.W. (1988)
Wave prediction in reservoirs; comparison of available methods. Report No. EX 1809
Hydraulics Research Ltd, Wallingford
OWEN, M.W. (1980)
Design of seawalls allowing for wave overtopping. Report No. EX 924
Hydraulics Research Ltd, Wallingford
OWEN, M.W. and ALLSOP, N.W.H. (1983)
Hydraulic modelling of rubble mound breakwaters
Proc. Conf. Breakwaters '83—Design and Construction
Thomas Telford, London
OWEN, M.W. and BRIGGS, M.G. (1985)
Limitations of modelling
Proc. Conf. Breakwaters '85—Developments in Breakwaters
Thomas Telford, London
PAINTAL, A.S. (1971)
Concept of critical shear stress in loose boundary open channels
Jnl Hydr. Res., **9**, No. 1, 91–113
PANICKER, N.N. and BORGMAN, L.E. (1974)
Enhancement of directional wave spectrum estimates
Proc. 14th Int., Coastal Engng Conf.
Am. Soc. Civ. Engrs
PAPOULIS, A. (1965)
Probability, Random Variables and Stochastic Processes
McGraw-Hill, Kogakusha
PARTENSCKY, H.W. (1987)
New investigations on vertical and rubble mound breakwaters
Proc. 2nd Chinese German Symp. on Hydrology and Coastal Engng
University of Hannover
PATTERSON, D.R. and POPE, J. (1983)
Coastal applications of side scan sonar
Proc. Conf. Coastal Structures '83
Am. Soc. Civ. Engrs
PEDERSEN, P.T. and JEBSENS, J.J. (1988) Upheaval creep of buried heated pipelines with initial imperfections
Marine Structures, Design, Construction and Safety, **1**, 11–22
PEDERSEN, P.T. and MICHELSEN, J. (1988)
Large deflection upheaval buckling of marine pipelines
BOSS Conf., Stavanger, 965–80
PELNARD-CONSIDERE, R. (1956) *Essai de theorie de l'évolution des formes de rivage en plages de sable et de galets* (Essay on the theory of the evolution of the form of beaches of sand and shingle).
4th Journées de l'Hydraulique, Les Energies de la Mer, Question II, Rapport No. 1
PERLIN, M. and DEAN, R.G. (1985)
3-D model of bathymetric response to structures
Jnl Am. Soc. Civ. Engrs, Waterway, Port, Coastal and Ocean Engng Divn, **111**, 2

PIANC (1987a)
Guidelines for the design and construction of flexible revetments incorp. geotextiles for inland waterways
Supplement to PIANC Bulletin No. 57, Brussels

PIANC (1987b)
Risk consideration when determining bank protection requirements
Supplement to PIANC Bulletin No. 57, Brussels

PIANC PTC II (1990)
Analysis of rubble mound breakwaters
Working Group 12, reports in preparation

PILARCZYK, K.W. (1984)
Filters
In Huis in't Veld (Ed.), *The Closure of Tidal Basins*
Delft University Press

PILARCZYK, K.W. (1989)
Design of coastal protection structures. Short Course
Asian Institute of Technology, Bangkok

PILARCZYK, K.W. (1990)
Design of seawalls and dikes—including overview of revetments
In K.W. Pilarczyk (Ed.), *Coastal Protection*, 197–288
Balkema, Rotterdam

PIERSON, W.J. and MOSKOVITZ, L. (1964)
A proposed spectral form for fully developed wind seas based on the similarity law of S.A. Kitaigorodskii
Jnl Geophys. Res., **69**, No. 24

POOLE, A.B., FOOKES, P.G., DIBB, T.E. and HUGHES, D.W. (1983)
Durability of rock in breakwaters
Prof. Conf. Breakwaters '83—Design and Construction, London, 31–42

POSTMA, G.M. (1989)
Wave reflection from rock slopes under random wave attack
MSc thesis, Delft University of Technology

POWELL, K.A. (1987)
Toe scour at seawalls subject to wave action—literature review. Report SR 119
Hydraulics Research Ltd, Wallingford

POWELL, K.A. (1989)
The scouring of coarse sediments at the toe of seawalls
Proc. Seminar—Seawall Design and SWALLOW
Hydraulics Research Ltd, Wallingford

POWELL, K.A. and ALLSOP, N.W.H. (1985)
Low-crest breakwater, hydraulic performance and stability, Report SR 57
Hydraulics Research Ltd, Wallingford

POWER, P.T. and PAISLY, J.M. (1986)
The collection, interpretation and presentation of geotechnical data for marine pipeline projects, *Oceanology*, 301–4
Graham and Trotman, London

PRAIRIE FARM REHABILITATION ADMINISTRATION (1972)
Upstream Slope Protection for Earth Dams in the Prairie Provinces
Soil Mechanics and Materials Division

PRIESTLEY, M.B. (1981)
Spectral Analysis and Time Series
Academic Press, London

PUGH, D.T. (1990)
Is there a sea-level problem?
Proc. Instn Civ. Engrs, Part 1, **88**, 347–66.

RANCE, P.J. and WARREN N.F. (1968)
The threshold of movement of coarse material in oscillatory flow
Proc. 11th Conf. on Coastal Engng
Am. Soc. Civ. Engrs

READ, J. (1988)
The control of rubble-mound construction with particular reference to Helguvik breakwater in Iceland
Proc. Conf. Breakwaters '88—Design of Breakwaters
Thomas Telford, London

REDEKER, F. (1985)
Probabilistic approach of cover layer material
Delft University of Technology

RENGERS, N., BERGHNIS, R. and ROSINGH, J.W. (1988)
Recent developments in the preparation of maps and plans of steep rockslopes with help of stereo-photogrammetry
Delft Progress Report, **13**, No. 1–2, 297–306

RICHARDS, A.F. and ZUIDBERG, H.M. (1985)
In-situ determination of the strength of marine soils
ASTM 11–40

RICHARDS, D.M. and ANDRONICOU, A. (1986)
Seabed irregularity effects on the buckling of heated submarine pipelines
Holland Offshore 86, Advance in Offshore Technology
Amsterdam

RIJKSWATERSTAAT (1985)
The use of asphalt in hydraulic engineering
The Hague
ROBERTSON, P.K. and CAMPANELLA, R.G. (1983)
Interpretation of cone penetration tests; parts I and II
Canadian Geotechnical Journal, **20,** No. 4, 718–45
RUITER, J. De (1975)
The use of *in-situ* testing for North Sea soil studies
Proc. Offshore Europe Conf. Aberdeen, 219.1–10
RUITER, J. DE (Ed.) (1988)
Penetration testing
Proc. ISOPTI, Orlando, US
Balkema, Rotterdam
ROULEAU, A. and GALE, J.E. (1985)
Statistical characterisation of the fracture system in the Stripa Granite, Sweden
Int. Jnl Rock Mech. Min. Sci. & Geomech. Abstr., **22,** No. 6, 353–67
SAVILLE, T. et al. (1962)
Freeboard allowance for waves in inland reservoirs
Proc. Am. Soc. Civ. Engrs, **18,** No. WW2, May
SCHÄLE, E. (1962)
Ankerverschuche zur Ermittlung der notwendigen Schuzshicht uber der Dichtung von Schiffahrttskanälen, 48
Mitteilung der Versuchsanstalt fur Binnenschiffbau eV, Duisburg
SCHIJF, J.B. and JANSEN, P.P. (1953)
PIANC Communications 18th Int. Congress, Rome
SCHUURMANS, S.TH., BOER, S. and LINDENBERG, J. (1989)
Hot pipelines, a hot issue; Pipeline–pipecover interaction related to upheaval buckling
Offshore Pipeline Technology Seminar (OPT), Amsterdam
SEED, H.B. and RAHMAN, M.S. (1978)
Wave induced pore pressure in relation to ocean floor stability of cohesive soils
Jnl Marine Geotech., **3,** No. 2
SEELIG, W.N. (1980)
Two-dimensional tests of wave transmission and reflection characteristics of laboratory breakwaters.
CERC Technical Report No. 80–1
Vicksburg
SEELIG, W.N. (1983)
Wave reflection from coastal structures
Proc. Conf. Coastal Structures '83
Am. Soc. Civ. Engrs
SEYMOUR, R.J. (1977)
Estimating wave generation on restricted fetches
Proc. Am. Soc. Civ. Engrs, **103,** No. WW2, May
SEYMOUR, E.V., CRAZE, D.J. and RUINEN, W. (1984)
Design and intallation of the North Rankin trunkline and slugcatcher
Proc. 5th Offshore South East Asia Conf., Singapore
SHIELDS, A. (1936)
Use of similarity mechanics of turbulence to shear movement (in German)
Inst. of Hydraulics and Ship Building, Berlin
SHIELDS, A. (1936)
Anwendung der Ahnlichkeitsmechanik und der Turbulenzforschung aud die Geschiebebewegung, Mitt. der Preuss
Versuchanst. für Wasserbau und Schiffbau, Heft 26, Berlin
SIMM, J.D. and FOOKES, P.G. (1989)
Improving reinforced concrete durability in the Middle East during the period 1960–1985: an analytical review
Proc. Instn Civ. Engrs, **86,** 333–58
SLEATH, J.F.A. (1978)
Measurements of bed load in oscillatory flow, modified shields
Proc. Am. Soc. Civ. Engrs, J. Waterw. Port Coastal and Ocean Div., **104,** No. WW3
SLEATH, J.F.A. (1984)
Applied Mechanics
John Wiley, New York
SMALLMAN, J.V., TOZER, N.P. and JONES, D.K. (1989)
Mathematical modelling of wave climate near offshore breakwaters. Hydraulics Research Report SR 211 (July)
SMITH, A.W. and CHAPMAN, D.M. (1982)
The behaviour of prototype boulder walls
Proc. 18th Int. Coastal Engng. Conf., 1914–28
Am. Soc. Civ. Engrs
SOMMERFIELD, A. (1986)
Mathematical theory on diffraction (in German)
SONG, W.O. and SCHILLER, R.E. (1973)
Experimental studies on beach scour due to wave action. Report No. 166, TAMU-SG-73-211
Texas A&M University
SOULSBY, R.L. (1987)
Calculating bottom orbital velocity beneath waves
Coastal Engng, **11,** 371–80

SPM (1978)
Shore Protection Manual, 3rd edition
US Army Corps of Engrs, Coastal Engng Res, Center, US Govt Printing Office, Washington, DC
SPM (1984)
Shore Protection Manual, 4th edition
US Army Corps of Engrs, Coastal Engng Res, Center, US Govt Printing Office, Washington, DC
STAPEL, E.E. and VERHOEF, P.N.W. (1989)
The use of Methylene Blue absorption test in assessing the quality of basaltic tuff rock aggregate
Engng Geol., **26**, 223–46
SWART, D.H. (1974)
Offshore sediment transport and equilibrium beach profiles. Publication 131
Delft Hydraulics Laboratory
TANIMOTO, K. et al. (1982)
Irregular wave tests for composite breakwater foundation
Proc. 18th Int. Coastal Engng Conf.
Am. Soc. Civ. Engrs
TAUTENHAIN, E., KOHLHASE, S. and PARTENSCKY, H.W. (1982)
Wave run-up at sea dikes under oblique wave approach
Proc. 18th Int. Coastal Engng Conf.
Am. Soc. Civ. Engrs
TAWN, J.A. and VASSIE, J.M. (1989)
Extreme sea levels: the joint probabilities method revisited and revised
Proc. Instn Civ. Engrs, Part 2, **87**, September, 429–42
TAW/Technical Advisory Committee on Waterdefences (1985)
The use of asphalt in hydraulic engineering. Rijkswaterstaat Communications No. 37
TAW/Technical Advisory Committee on Protection against Inundation (1974)
Wave run-up and overtopping
Government Publishing Office, The Hague
The Closure of Tidal Basins (1987)
Delft University Press, The Netherlands
THOMAS, R.S. and HALL, B. (forthcoming 1991)
Seawall Design. CIRIA Water Engineering Report
CIRIA/Butterworth–Heinemann, London
THOMPSON, A.C. (1988)
Numerical model of breakwater wave flows
Proc. 21st Int. Coastal Engng Conf.
Am. Soc. Civ. Engrs
THOMPSON, D.M. (1973)
Hydraulic design of sea dikes. Report DE 6
Hydraulics Research Ltd, Wallingford
THOMPSON, D.M. and SHUTTLER, R.M. (1975)
Riprap design for wind-wave attack—a laboratory study in random waves. Report EX 707
Hydraulics Research, Wallingford
THOMPSON, D.M. and SHUTTLER, R.M. (1976)
Design of riprap slope protection against wind waves. Report 61
CIRIA, London
TOTI, M., CUCCIOLETTA, P. and FERRANTE, A. (1990)
Beach nourishment at Lido di Ostia (Rome)
Proc. 27th Int. Nav. Congr. Osaka, Brussels, S.II, 23–8
TRAN NGOG LAN (1983)
Deux nouveaux essais d'identification des sols argileux
Laboratoire des Ponts et Chaussées, Paris, France
(Tropical '85 Congress)
TYHURST, M.F. (1986)
Rubble groynes at Christchurch, Dorset—four case histories
Paper presented to Standing Conference on Problems Associated with the Coastline (SCOPAC)
US ARMY CORPS OF ENGINEERS (1985)
Engineer Manual EM 1110-2-1614, 1985
VAN DER MEER, J.W. (1987)
Stability of breakwater armour layers—design formulae
Coastal Engng, **11**, 219–39
VAN DER MEER, J.W. (1988a)
Rock slopes and gravel beaches under wave attack. Doctoral thesis
Delft University of Technology; also Delft Hydraulics Communication No. 396
VAN DER MEER, J.W. (1988b)
Deterministic and probabilistic design of breakwater armour layers
Proc. Am. Soc. Civ. Engrs Jnl Waterways Ports, Coastal and Ocean Engng Divn, **114**, No. 1
VAN DER MEER, J.W. (1990a)
Low-crested and reef breakwaters. Report H198/Q638
Delft Hydraulics
VAN DER MEER, J.W. (1990b)
Data on wave transmission due to overtopping. Report H 986
Delft Hydraulics
VAN DER MEER, J.W. (1990c)
Extreme shallow water wave conditions. Report H 198
Delft Hydraulics

VAN DER MEER, J.W. (1990d)
Taluds van losgestorte materialen. Stabiliteit van lage dammen en overgangskonstrukties bij stortsteen onder golfaanval. (Slopes of loose materials. Stability of low-crested and composite structures of rock under wave attack, in Dutch). Report M 1983 Part V
Delft Hydraulics
VAN DER MEER, J.W. (1990e)
ENDEC verification on slope 1:10. Report H 986
Delft Hydraulics
VAN DER MEER, J.W. and KOSTER, M.J. (1988)
Application of computational model on dynamic stability
Proc. Conf. Breakwaters '88—Design of Breakwaters
Thomas Telford, London
VAN DER MEER, J.W. and PILARCZYK, K.W. (1987)
Stability of breakwater armourlayers—Deterministic and probabilistic design. Communication No. 378
Delft Hydraulics
VAN HIJUM, E. and PILARCZYK, K.W. (1982)
Equilibrium profile and longshore transport of coarse material under regular and irregular wave attack. Publication No. 274
Delft Hydraulics
VAN OORSCHOT, J.H. and D'ANGREMOND, K. (1968)
The effect of wave energy spectra on wave run-up
Proc. 11th Inst. Coastal Engng Conf.
Am. Soc. Civ. Engrs
VAN REE et al. (1991)
A stochastic approach to soil environmental problems
Probabilistic safety assessment and management conference. Los Angeles
VAN RIJN, L.C. (1982) Equivalent roughness of alluvial bed
Proc. Am. Soc. Civ. Engrs Jnl Hydr. Divn, **108,** No. HY10
VELDEN, E.T. VAN DER (1989)
Coastal Engineering
Delft University of Technology, The Netherlands
VELDHUIZEN VAN ZANTEN, R. (1986)
Geotextiles and Geomembranes in Civil Engineering
Balkema, Rotterdam/Boston
VELLINGA, P. (1986)
Beach and dune erosion during storm surges. Doctoral thesis
Delft University of Technology
VERHEY, H.J. and BOUGAERTS, M.P. (1989)
Ship waves and the stability of armour layers protecting slopes. Publication No. 428
Delft Hydraulics
VERHOEF, P.N.W. (1987)
Sandblast testing of rock
Int. Jnl Rock Mech. Min. Sci. & Geomech. Abstr., **24,** No. 3, 185–92
VERRUYT, A., et al. (Eds) (1982) Penetration testing
Proc. ESOPT 11
Balkema, Rotterdam
VISSCHER, J.T. (1980)
Burial depth of ships' and work anchors (in Dutch). MaTS PL-2
Part V, Netherlands Industrial Council for Oceanology
VRIJLING, J.K. (1990)
Probabilistic design of flood defences
Proc. 22nd Int. Coastal Engng Conf.
Am. Soc. Civ. Engrs
VRIJLING, J.K. and NOOY VAN DER KOLFF, A.H. (1990)
Quarry yield and breakwater demand
Proc. 6th Int. Cong. Int. Assoc. Engng Geol. 2927–34
Balkema Rotterdam
WAKELING, H.L. (1977)
The design of rubble mound breakwaters
Proc. Symp. Design of Rubble Breakwaters
British Hovercraft Corp, 1–18
WANG, H., LATHAM, J.-P. and POOLE, A.B. (1990)
In-situ block size assessment from discontinuity spacing data
Proc. 6th Cong. Int. Assoc. of Engng. Geol.
Amsterdam
WANG, H., LATHAM, J.-P. and POOLE, A.B. (1991)
Predictions of block size distribution for quarrying
Proc. Conf. on Geol. Materials in Construction, Geol. Soc. of London, November 1989. To be published in *Q.J. Engng Geol.,* **24,** 91–99. London
WEYMOUTH, O.F. and MAGOON, O.T. (1968)
Stability of quadripod coverlayers
Proc. 11th Int. Coastal Engng. Conf.
Am. Soc. Civ. Engrs
WILSON, B.W. (1963).
Generation and dispersion characteristics of tsunamis
NESCO, October

WILSON, B.W., WEBB, L.M. and HENDRICKSON, J.A. (1962)
The nature of tsunamis: their generation and dispersion in water of finite depth. Technical report No. SN57-2.
NESCO
WINKLER, E.M. (1973)
Stone: Properties and durability in man's environment
Springer-Verlag, New York
WU, G. and JENSEN, O.J. (1983)
Stability of rubble foundations for composite breakwaters
Proc. Conf. Coastal and Port Engng in Developing Countries, Colombo, Sri Lanka
YOO, D., O'CONNER, B.A. and McDOWELL, D.M. (1989)
Mathematical models of wave climate for port design
Proc. Instn Civ. Engrs, Part II, **86,** 513–30
YOUNG, R.M., PITT, J.D., ACKERS, P. and THOMPSON, D.M. (1980)
Rip-rap design for wind-wave attack: long term observations on the offshore bank in the Wash. Technical note 101
CIRIA, London

INDEX

(Numerals marked in **bold type** indicate a chapter/section devoted to the subject entry.)

A

Abrasion test, Los Angeles (ASTM), 524, 551
   Mill apparatus, 524–525
      Test, Queen Mary and Westfield (QMW), 524–527
   Resistance Index, 524–527
   rock resistance to, 496
Access channel, provision of protection for, 355–356
Accidents, fatal, 11–12
Acoustic doppler meter, 547
Aerial surveys, 576–578
Aggregate suspension, preparation of the, 528
Air quality, 63–64
Amenity considerations, 411
Anchors and fishing gear, dragging, 468–470
Archaeological considerations, 63
Area of Outstanding Natural Beauty (AONB), 444, 587
Armour, crest and rear, 272–276
   degradation, 574–575
   layers, 264–271, 503–505
      selection and design of, 418–420
      toe protection to, 277–278
Armourstone breakage, 83–86
   degradation rate, durability and, **106–113**
   grading, 175–177
   handling and placing of, 440–443
   measurement of, 540–541
   shape, 175–177
   stocks, 170
ASTM, 524, 551
Armouring, concerns relating to coastal, 410

B

Bastion groyne, 396, 399–403, 433–434
Bathymetric surveys, **543–544, 580–583**
Bathymetry, 186
Beach(es), amenity value of, 411
   classification, 400–403
   design of pocket, 405
   downdrift, 411
   excavation, 540
   gravel, **453–463**
   -nourishment losses, 462
   profiles, parametric model for shingle, 286–287
   sill and perched, 440
Benefit cost analysis (BCA), 20
Berth protection, 354–355
Bishops' method, 466
Bituminous systems, 292
Blasting, 137–139
Block (*see also* Rock)
   aspect ratio data, 89
   integrity, 86–87, 133–134, 170, 496
   porosity, 103–104
   shape, 87, 92–93
   weight and size, 94–103
Borings, **561–563**
Boundary conditions and data collection, **277–233**, 358
Breakwater, Berm or S-slope 367–368
   Caisson-type, 353–354, 369–373
   construction logistic simulation, 387
   cost of, 384–385, 388–395
   cross-section, design considerations for, 358
   fishtailed, 397, 406–407, 436
   head, 281–285

Breakwater (cont.)
   L- and T-shaped, 397, 407, 438
   layout, 354, 357
   low-crested and submerged, 368
   offshore, 396–397, 403–406, 435
   reef-type, 353, 368
   rubble-mound, **351–395**
   sill or submerged, 397, **407**, 438
British Civil Engineering Standard Method of Measurement (CESMM), 531–532

C

Cavities, 318
Cement mortar, colloidal, 159–160
CESMM, 531–532
Channel siltation, littoral transport and, 356
CIAD (1985) report, 279, 349–350
Clacton Sea Defence Scheme, UK, 408
Coastal armouring, 410
   shoreline defence structures, 32, **397–407**
      engineering (*see also* Rock structures and Beaches), **479–487**
         environmental assessment of, **585–587**
         hydraulic interactions and, **237–308**
         materials availability for, 415–416
         measurement of quarried rock in, **531–536**
         model specification for quarried rock in, **489–507**
         models of, **304–306**
         parameters, **239–308**
         plan layout and concept selection for, **397–443**
         response, assessment of, 235
         standards for quarried rock materials in, **509–530**
         structures, exposure zones for, 9
Cohesion c, 309
Colcrete, 160
Composite systems, **168–170**
Composition and classifications tests, soil, 567–568
Concrete, 155–156
Cone penetration tests, the static (CPT), 556–559
Construction considerations, rock, 17
Core construction, 373–374, 378–380, 388–395
   material, placement of, 503
Cranes, comparison of land-based and floating, 395
Critical shear concept, 295–297
Crown Estates (UK), 136, 585, 587
   walls, 276–277, 507
Crushing strength of stone, dynamic, 509–511
Cultural aspects, rock structures and, 63

D

Dam: construction aspects, 451–452
   design considerations, 445–451
   face protection, 32, **443–453**
   maintenance aspects, 452–453
   measurement and cost aspects, 452
   physical site conditions and data collection, 445–449
Data collection, **183–233**
   collection, equipment for, 185
      instrumentation for geotechnical, **551–570**
      options for, 185
   for feasibility study and design, 183

Data collection (cont.)
  for maintenance, 185
  to be supplied, 490
Deformation, rock and soil, 316
Degradation of rock materials, 105–107, 109–113
Delft Hydraulics, 280, 285
Design approach, data specification and use, 183
  approach, rock structure, 45–46
  concepts, alternative, 14
  considerations, general, 445–451
    and detailing, final, 21
    with joint waves and water levels, 223–226
  methods, structure-specific, 16
  process, 2–7
  tools, physical processes and, **235–350**
Designer's Quality plan, 28
Deterministic approach, 47–48
Dike Height, Dutch practice for determination of, 431
Dilatancy, 313
Dilatometer (DMT) test, 544, 560–561
DIN (Germany), 115, 118, 122
Discontinuities, 80–83
Diving survey inspections, 582–583
Dredging costs, reducing maintenance, 356
Drop Test Breakage Index, 529–530
Dynamic effects, 317–318

E
Earthquake effects, 335, 451
  magnitude, 192–193
Echo sounding, 543
Ecology (*see also* Environment), 61–62
Electrical resistivity techniques, 551–552
Electromagnetic flow meter, 547
  techniques, 552–553
ENDEC (ENergy DECay) model, 210, 216, 227
Engineering experience, practical formulae and, **327–350**
Environment: Acts of Parliament, 586
  Guidance Documents, 585
  Sensitive Locations (UK), 587
  Statutory Consultees (UK), 587
  Instruments, 586
Environmental assessment (EA), 14, 56–59, 444, **585–587**
  assessment regulation, EC, 58
  conditions or loadings, measures of, 82
  considerations, 411
  impact, 3, 10, 15
  Impact Assessment (EIA), 14–15
  legislation, European, British and Dutch, 585–587
  monitoring, 60
  Statement (ES), 66
Equation, Bond's third theory, 140
  Rosin–Rammler, 141
Equipment, drying, 516, 522
  handling, **141–155**
  lifting and weighing, 511, 514–516, 521–522, 525, 529
  and reagents, 527
Erosion/fatigue, internal, 317
European Community (EC), 58
  Community (EC) Directives, 585

F
Failure mechanisms, initial, 42–43
  modes, principal, 31, 33–37
Fatigue, internal erosion, 317
Fault tree analysis, 38–44
  tree systems and mechanisms, 41, 427

F (cont.)
Federation of Civil Engineering Contractors, 531
Filter rules, 343
Fishing gear, 469–470
Float, 545
  and tracers, 546
Flood waves, 191
Fracture toughness, 122–124
Freespans, 470
Freeze/Thaw cycles, 516–517
Friction angle, 311–312
Functional analysis, 12–13

G
Gabions, 168–169
  and mattresses, use of, 422
Geomechanic principles, 314
Geometric and materials requirements, **490–502**
Geomorphology, 61
Geophysical methods, 551–563
Geotechnical aspects, scope of, **313–350**
  boundary conditions and data collection, **227–233**
  data collection, instrumentation for, **551–570**
  interactions, 307
  models, parameters used in, 308
    of rock structure behaviour, 324–325
  processes, 16
  stability of offshore rock structure, 466
Geotextiles, 160–161
  design aspects of, 346
  non-woven, 164–166
  woven, 161–163
Global Positioning System (GPS), 544, 546
Gradations, size and weight, 115–117
Grading(s), armourstone, 175–182
  block, 94–103
  class limit system of standard, 97–98
  control, 179–180
  non-standard, 101–103
  plan: 3–6 tonne, 178
  quarried stone, **490–493, 511–530**
  size and weight, 100
Gravel beaches, **453–463**
  beaches, coastal processes, 455–456
    construction and cost, 457–460
    cost optimisation of, 461
    design process, 454–455
    monitoring and renourishment, 460–461
    profile, 456–457
  production, 145
Grout(s), 157
  Asphalt, 157–159
Groynes, comparison of timber and rock for, 415

H
HADEER code, the, 327
Haul roads and tracks, temporary, 505
Historical aspects, 63
Human sensory considerations, 63
Hydraulic boundary conditions, **186–227**
  data measurement and instrumentation, **543–550**
  and geotechnical processes, 16
  interactions, **237–308**
  parameters, 308
Hydrocrete mortar, 160

I
Impact resistance, 123
Impacts from choice of materials and construction, 64–65
Impurities, quarried rock, 497
Industrial by-products, 156–157
  waste, 445

I (cont.)
Inspection by bulk weighing, 512–514
  diving, 580–583
  on site, 503, 542
  rock, 18, **497–502**
Institution of Civil Engineers, 531
Interlocking and armour layer porosity, 103–104
ISO (International Standards Organisation), 2
ISRM International Society for Rock Mechanics), 86, 115, 118, 122, 127–128

L
Landslides, 451
Laser-doppler systems, 546
Layer thickness, porosity and, 104–105
  topography, roughness of, 104
Legislation, Acts of Parliament, 586
  environmental assessment, **585–587**
  European Community (EC), 585
  Netherlands, 586–587
  Statutory Instruments, 586
  United Kingdom, 585
Liquefaction, 38, 317–318
  potential of seabed sand, 339
Loading and boundary conditions, 314
Losses, categories of, 11

M
Maintenance considerations, 18, **479–487**
  programme flow chart, 480
Marine sources, 135
Maritime structures, British Standard Code for, 115
  structures, maintenance of, **479–487**
    monitoring programme for, **471–484**
Materials, **67–182**
  availability and properties, 15–16
  composite, 168–170
  and construction, impacts from choice of, 64–65
  sources of, 64
Mattresses, 168–169
Measurement(s), current, 546–547
  and instrumentation, hydraulic data, **543–550**
  of quarried rock, **531–536**
  standard methods of, 531
  water level, 544–547
  wave, 547–550
  UK methods of, 531
Methylene Blue Absorption Test, 527–529
Meters, acoustic doppler, 547
  electromagnetic flow, 547
  pressure, 545, 548
  propeller current, 546
  resistance and capacitance, 545
  resistive and capacitative surface elevation, 548
Microwave (radar) technique, 550
Migration of sublayer, 36–37
Milling programme, determination of, 526
Mineral(s), breakdown due to clay, 132–133
  fabric breakage, resistance to, 496
Model(s), mathematical background of available, 321–324
  methods and, 324–325
  specification for quarried rock in coastal engineering, **489–507**
  types, computer, 325
Modelling and simulation, **319–350**
Monitoring, 2, 18
  environmental, 60–61
  frequency of, 481–484
  of rubble structures, **571–577**
  type of, 480–481
Morphology, 186

M (cont.)
Multi-Criteria Analysis (MCA), 18–19

N
National Physical Laboratory, 160
NEN (Netherlands), 2, 115, 118, 128, 130, 509–530
Netherlands, environmental legislation in the, 586–587

O
Offshore construction methods, 474–478
  dumping stone, 474–478
  engineering, rockfill in, **463–478**
  fields, upheaval and buckling in, 470–474
  platforms, falling objects and, 466–468
  structure, cost aspects of, 478
    geotechnical stability of, 466
  survey, 476–478
Oscillatory flow, 297–300

P
Palaeontological aspects, 63
Parameters, 308–313
Paylines and surface profiles, **505–506**
Penetration tests, **553–559**
  test, standard (SPT), 554–555
    static cone (CPT), 556–559
Permeability k, 309–310
Petrographic examination, 118
Photogrammetric survey methods, **578–580**
Photographic survey methods, **575–580**
Physical processes and design tools, **235–350**
  placement of core materials, bulk, 182
    of armour, 182
Pipeline(s), accidents and, 466–468
  and dragging anchors and fishing gear, 468–470
  rock cover protection/stabilisation, 464
Placed materials, disturbance to, 505
  materials, protection of, 505
Placement of armour, 182
  /construction requirements, **503–505**
  of core materials, 182, 503
Placing equipment, armourstone, 440–443
  tolerances, 383
Planning authorities, 10
  and designing, 7
  permission, 10
  policy, 10
Porosity n, 309
Porous flow, 315
Pressure meter (PMT) test, 545, 554, 560–561
  tube (Pitot tube), 546
Probabilistic analysis, 222–226, 348–350
  approach, 48–53
Problem identification, 7
Production and handling, **134–155**
Propeller current meter, 546
Protection of placed materials, 505

Q
Quality, air, water and soil, 63–64
  assurance, 26–27
  control systems, 29–31, 169, 536
  plan, 28
  requirements, rock, **494–497**
Quantities, bill of, 532–536
Quarry faces and production, 170
  quality control in the, 169–171
  selection processes in the, 140
  types of, 136
Queen Mary and Westfield (QMW) Abrasion Mill Test, 524–527

R
Radar systems, 545
Ramsar Convention, the, 444

R (cont.)
Ras Lanuf, accessibility evaluation at, 394–395
Remote-controlled profiling, 544
    sensing (satellites), 544–545
Repair costs, calculation of, 24
Resistance and capacitance meters, 545
Resistive and capacitative surface elevation meters, 548
Resource appraisal, 135
Revetment, conventional, 396, **421–443**
    crest detailing, 425
    land-reclamation, 423
    selection and design of, 418–420
    toe protection, 432–433
Risk, acceptable, 11
    analysis, 348
    -level assessment, 44–45
    personal, 12
Rock(s) abrasion resistance, 126–127
    applications, model specification for quarried, **489–507**
    -based revetment systems, failure of, 419
    batch, homogeneity of the, 498
    breakage, 83–85
    breakwater, functions of component parts of, 13
    coastal structures, cross-section design for, **413–443**
    crushing resistance, 125–126, 517–520
    definitions and requirements for, 489
    degradation, 105–117, 109–113
    density, 78, 106, 118, 494, 521, 534–536
    determination of 'Sonnenbrand' in Basalt, 523
    dumping, simulation of, 381–382
    durability, 117
    and armourstone degradation rates, 105–106, 109
    fabric strength, 82
    -fill, construction aspects, 373–395
        local stability, 338
        in offshore engineering, **463–478**
        settlement of, 338
        structure, design of offshore, **465–474**
    fracture toughness test, 124–125
    geotechnical stability of, 466
    grouted, 292
    hardness, 127–128
    hydraulic stability of, 466
    igneous, 68–70, 72, 75
    impurities, 497
    inspection, **497–502**
    length-to-thickness ratio of, 515–516
    materials, alluvial, glacial and marine sources of, 75–79
        particle size distribution, **509–530**
        samples, size and composition of, 498–499
        sampling, **498–502,** 509, 515–516, 518, 521–524, 529
        specifications and bills of quantities, 23
        standards for quarried, **509–530**
    measurement of quarried, **531–536**
    metamorphic, 68, 71–72, 75
    pore pressure build-up in, 330–331
    porosity, water absorption and, 79–80, 118
    production and handling, **134–155**
    protection systems, stability comparison for, 291
    quality requirements, 494–497
        tests, 174–180
    resistance to abrasion, 496–497
        to Freeze/Thaw Cycles, 516–517
        impact and mineral breakage, 495
        weathering, 494–495
    rubble structures, monitoring of, **571–577**
    rules for inspection, 489
    sedimentary, 68, 70, 72, 75

Rock(s) (cont.)
    shape, 117–118, 493
    size and weight distribution of, 514–515
    and weight gradations, 115–117
    slope, stability of, 327
    strength, 122–125
    structures, **351–478**
        amenity value of, 411
        construction aspects, 440–443
        ecological value of, 412–413
        low-crested, 272–276
        maintenance of, 479–487
        monitoring of, **471–484**
            techniques, **571–577**
        performance, appraisal of, 484–485
        repair/replacement construction methods, 486–487
        visual impact of, 412
    testing and evaluation, 114–115, 175–176
    types, 68–72
        evaluation of, 68–69
    and water absorption, 495
        at atmospheric pressure, 522
    weathering, 69, 73–75, 81, 84, 128
Rubble-mound breakwaters, **351–395**
    mound breakwaters construction costs, 25
    conventional, 361
    with monolithic crown wall, 365–367
    structures, monitoring of, **571–577**

S
Sampling, quarried stone, **497–502**
Satellites, (remote-sensing), 544–545
SATURN code, the, 323, 325
Schijf's chart, 223
Schmidt impact hammer test, 128–129
Scour, 301–304
    protection, 396, 399, 433
Sea bed, dredging from the, 64
    bed ecology, 64
    bottom orbital velocities, 220
    currents, 220–221
    development and storm and wave forecasting, **202–227**
    -walls, crown, 276–279
        management, safety control panel for, 485
        rehabilitation, sizing armour stone for, 425
        and scour mats, 409
        and shoreline protection structures, 395–413
Seaford gravel beach-nourishment project (1987), 460
Seiches, 191, 449
Seismic methods, 551
Sensory aspects, human, 63
Settlement, 34, 507
Shields criterion, 295–299
Ship-manoeuvring simulation, 356–357
Shoaling, 210–214
Shoreline erosion, 2
    protection structures, 2, **395–413**
Side-scan sonar, 544, 581
Sieving and weighing, 509–510
Site conditions, **183–233**
    investigations on water, 563–567
    of Special Scientific Interest (SSSI), 444
SLS (serviceability limit states), 21, 34
Sliding of structure, 37–38
Slip failure, 316
Slope angle, upstream, 451
Slopes, stepped and composite, 285–290
Social and socio-economic considerations, 62–63
Soil consolidated shear test, 563
    consolidation properties, 568
    data collection, **551–570**
    density, 568

Soil (cont.)
   friction, 568
   investigation programme, planning a, 228–233
   permeability, 568
   quality, 63
   settlement of compressible, 336–337
   stress-strain laboratory tests, 568–570
   triaxial tests, 569
   -water-structure interaction, 323
'Sonnenbrand' in Basalt, 523
   effect, 132–133
Sound velocity, 84, 86, 128
Sounding, acoustic, 580
   echo, 543, 580
   line, 543, 580
SPM (1984), the, 450
Squeezing of soft layers, 340–342
Stability analysis, Bishops', 328
   coverlayer, 333
   slope, 332
Standard(s), national and international, 26, 48, 115, 118, 122, 126, 130
   Penetration Tests (SPT), 231–233
   for quarried rock materials, **509–530**
Static cone penetrometer tests (CPT), 554, 556–557
Stereophotogrammetry, 549–550
Storm surges, 189
Sulphate soundness test, 130
Surplus materials, disposal of, 507
Survey(s), bathymetric, **543–544, 580–583**
   comparative photography, 576–578
   conventional engineering, 572–574
   inspections, diving, 582–583
   methods, land-based, **572–580**
   offshore, 476–478
   photogrammetric, 578–580
   photographic, **575–580**
   side-scan sonar, 581
   sub-bottom, 581–583
   technique, 506
   underwater, 580–583

T
Terrestrial sources, 135
Test, aggregate abrasion test (AAV), 127
   abrasion mill test, 126
   ASTM C88, 130
   BS 6349, 130
   coverlayer stability, 40
   Breakage Index, Drop, 529–530
   Freeze/Thaw, 132
   Los Angeles (LAV), 126
   methods, *in-situ*, 232
   Methylene Blue Absorption (MBA), 133, 527–529
   procedure, 520
   PMT, 544, 560–561
   QMW mill abrasion test, 127
   requirements, 130
   rock quality, 175–176
   sample, preparation of, 519–522
   sandblast, 127
   soil composition and classification, 567–568
   the 'Sonnenbrand' effect, 132–133
   sulphate soundness, 130
   triaxial, 569
   Washington degradation, 133
   weathering resistance, 128
   wet Deval, 127
   wetting-drying, 133
Tides, 188–189
Tolerances, 505–506
Transport, and handling, quarried rock, 503
   and identification of quarried stone samples, 502

Transport (cont.)
   longshore, 293–295
Transportation, road, rail and sea, **148–155**
   cost implications of, 153–155
   quality control and, 169
Tsunami, 191

U
UCL (upper class limit), 98, 101, 176, 511–513
ULS (ultimate limit scale), 21, 34
Underlayer and core materials, 179
Underwater survey techniques, 580–583
United Kingdom, environmental legislation and the, 585–586
   Kingdom, sensitive locations in the, 587
   Statutory Consultees, 587

V
Vane shear test, field (VST), 560–561
Veritec Worldwide Offshore Accident Databank, 467

W
Waste disposal, 507
   material in shoreline works, use of, 64–65
Water absorption, 122, 494, 517, 522
   current measurement, 546–547
   levels, 186–188, 193
      and currents, numerical and physical modelling of, 226
      measurement, 544–546
   quality, 63–64
   site investigations on, 563–567
Wave(s) attack, 264–271, 295–301
   breaking and diffraction, 207, 214–216, 450–451
   conditions, deep-water, 194–195
   -current interaction, 218
   forecasting, 204–207
   gauge array, 549
   height and period, 197, 242
      in shallow water, extreme, 217
   measurements, directional, 547–550
   modelling, numerical and physical, 226–227, 304–306
   observations, visual, 199
   overtopping, 253, 259, 272, 449
   penetration, 328–331
   reflection, 217–219, 239, 259–263, 449
   refraction, 208–210, 449
   rider, 548–549
   run-up and run-down, 246–253
   set-up, 190–191
   in shallow water, transformation of, 207
   ship-induced, 221–222
   spectral description of, 200–202
   statistical properties of, 196–197
   structural parameters related to, 241–246
   transmission, 239, 256–259
   and water levels, design with joint, 223–226
Weathering resistance, quarried rock's, 495–496
   tests, 128
Weighing equipment, 181, 511–529
Weight distribution of quarried stone, **511–530**
   median and effective, 99
Wind set-up, 189–190
   statistics, 199

Y
Young's Modulus, 137, 160

Z
Zeebrugge, 394–395
   accessibility evaluations at, 394
   breakwaters, construction costs for, 389